AF352873

Advances in Sustainable Concrete System

Advances in Sustainable Concrete System

Editors

Yifeng Ling
Chuanqing Fu
Peng Zhang
Peter Taylor

MDPI • Basel • Beijing • Wuhan • Barcelona • Belgrade • Manchester • Tokyo • Cluj • Tianjin

Editors
Yifeng Ling
School of Qilu Transportation
Shandong University
Jinan
China

Chuanqing Fu
Civil Engineering
Zhejiang University of
Technology
Hangzhou
China

Peng Zhang
School of Water Conservancy
Engineering
Zhengzhou University
Zhengzhou
China

Peter Taylor
National Concrete Pavement
Technology Center
Iowa State University
Ames
United States

Editorial Office
MDPI
St. Alban-Anlage 66
4052 Basel, Switzerland

This is a reprint of articles from the Special Issue published online in the open access journal *Crystals* (ISSN 2073-4352) (available at: www.mdpi.com/journal/crystals/special_issues/sustainable_concrete_system).

For citation purposes, cite each article independently as indicated on the article page online and as indicated below:

LastName, A.A.; LastName, B.B.; LastName, C.C. Article Title. *Journal Name* **Year**, *Volume Number*, Page Range.

ISBN 978-3-0365-4444-1 (Hbk)
ISBN 978-3-0365-4443-4 (PDF)

Contents

About the Editors ix

Preface to "Advances in Sustainable Concrete System" xi

Yifeng Ling, Chuanqing Fu, Peng Zhang and Peter Taylor
Advances in Sustainable Concrete System
Reprinted from: *Crystals* **2022**, *12*, 698, doi:10.3390/cryst12050698 1

Yunyun Tong, Abdel-Okash Seibou, Mengya Li, Abdelhak Kaci and Jinjian Ye
Bamboo Sawdust as a Partial Replacement of Cement for the Production of Sustainable
Cementitious Materials
Reprinted from: *Crystals* **2021**, *11*, 1593, doi:10.3390/cryst11121593 5

Qihui Chai, Fang Wan, Lingfeng Xiao and Feng Wu
The Influence of Fly Ash Content on the Compressive Strength of Cemented Sand and Gravel
Material
Reprinted from: *Crystals* **2021**, *11*, 1426, doi:10.3390/cryst11111426 19

Zongxian Huang, Yuqi Zhou and Yong Cui
Effect of Different NaOH Solution Concentrations on Mechanical Properties and Microstructure
of Alkali-Activated Blast Furnace Ferronickel Slag
Reprinted from: *Crystals* **2021**, *11*, 1301, doi:10.3390/cryst11111301 29

Kuisheng Liu and Yong Cui
Effects of Different Content of Phosphorus Slag Composite Concrete: Heat Evolution,
Sulphate-Corrosion Resistance and Volume Deformation
Reprinted from: *Crystals* **2021**, *11*, 1293, doi:10.3390/cryst11111293 41

Junfeng Guan, Meng Lu, Xianhua Yao, Qing Wang, Decai Wang and Biao Yang et al.
An Experimental Study of the Road Performance of Cement Stabilized Coal Gangue
Reprinted from: *Crystals* **2021**, *11*, 993, doi:10.3390/cryst11080993 51

Hossein Mohammadhosseini, Rayed Alyousef, Shek Poi Ngian and Mahmood Md. Tahir
Performance Evaluation of Sustainable Concrete Comprising Waste Polypropylene Food Tray
Fibers and Palm Oil Fuel Ash Exposed to Sulfate and Acid Attacks
Reprinted from: *Crystals* **2021**, *11*, 966, doi:10.3390/cryst11080966 69

Shengli Zhang, Yuqi Zhou, Jianwei Sun and Fanghui Han
Effect of Ultrafine Metakaolin on the Properties of Mortar and Concrete
Reprinted from: *Crystals* **2021**, *11*, 665, doi:10.3390/cryst11060665 91

Junfeng Guan, Yulong Zhang, Xianhua Yao, Lielie Li, Lei Zhang and Jinhua Yi
Experimental Study on the Effect of Compound Activator on the Mechanical Properties of Steel
Slag Cement Mortar
Reprinted from: *Crystals* **2021**, *11*, 658, doi:10.3390/cryst11060658 103

**Hamad Hassan Awan, Muhammad Faisal Javed, Adnan Yousaf, Fahid Aslam, Hisham
Alabduljabbar and Amir Mosavi**
Experimental Evaluation of Untreated and Pretreated Crumb Rubber Used in Concrete
Reprinted from: *Crystals* **2021**, *11*, 558, doi:10.3390/cryst11050558 121

Muhammad Faisal Javed, Afaq Ahmad Durrani, Sardar Kashif Ur Rehman, Fahid Aslam, Hisham Alabduljabbar and Amir Mosavi
Effect of Recycled Coarse Aggregate and Bagasse Ash on Two-Stage Concrete
Reprinted from: *Crystals* **2021**, *11*, 556, doi:10.3390/cryst11050556 135

Ammar Iqtidar, Niaz Bahadur Khan, Sardar Kashif-ur-Rehman, Muhmmad Faisal Javed, Fahid Aslam and Rayed Alyousef et al.
Prediction of Compressive Strength of Rice Husk Ash Concrete through Different Machine Learning Processes
Reprinted from: *Crystals* **2021**, *11*, 352, doi:10.3390/cryst11040352 145

Yuqi Zhou, Jianwei Sun and Zengqi Zhang
Effects of High-Volume Ground Slag Powder on the Properties of High-Strength Concrete under Different Curing Conditions
Reprinted from: *Crystals* **2021**, *11*, 348, doi:10.3390/cryst11040348 169

Yuanxie Shen, Junhao Sun and Shixue Liang
Interpretable Machine Learning Models for Punching Shear Strength Estimation of FRP Reinforced Concrete Slabs
Reprinted from: *Crystals* **2022**, *12*, 259, doi:10.3390/cryst12020259 185

Afnan Nafees, Muhammad Faisal Javed, Muhammad Ali Musarat, Mujahid Ali, Fahid Aslam and Nikolai Ivanovich Vatin
FE Modelling and Analysis of Beam Column Joint Using Reactive Powder Concrete
Reprinted from: *Crystals* **2021**, *11*, 1372, doi:10.3390/cryst11111372 207

Ruihe Zhou, Longhui Guo and Rongbao Hong
Study on Energy Evolution and Damage Constitutive Model of Siltstone
Reprinted from: *Crystals* **2021**, *11*, 1271, doi:10.3390/cryst11111271 229

Weifeng Bai, Xiaofeng Lu, Junfeng Guan, Shuang Huang, Chenyang Yuan and Cundong Xu
Stress–Strain Behavior of FRC in Uniaxial Tension Based on Mesoscopic Damage Model
Reprinted from: *Crystals* **2021**, *11*, 689, doi:10.3390/cryst11060689 245

Xuefeng Zhang, Huiming Li, Shixue Liang and Hao Zhang
Experimental and Numerical Study of Lattice Girder Composite Slabs with Monolithic Joint
Reprinted from: *Crystals* **2021**, *11*, 219, doi:10.3390/cryst11020219 269

Ye Chen, Shuang Zhu, Shenghua Ye, Yifeng Ling, Dan Wu and Geqiang Zhang et al.
Acoustic Emission Study on the Damage Evolution of a Corroded Reinforced Concrete Column under Axial Loads
Reprinted from: *Crystals* **2021**, *11*, 67, doi:10.3390/cryst11010067 285

Ruihe Zhou, Hua Cheng, Mingjing Li, Liangliang Zhang and Rongbao Hong
Energy Evolution Analysis and Brittleness Evaluation of High-Strength Concrete Considering the Whole Failure Process
Reprinted from: *Crystals* **2020**, *10*, 1099, doi:10.3390/cryst10121099 301

Qiannan Wang, Guoshuai Zhang, Yunyun Tong and Chunping Gu
A Numerical Study on Chloride Diffusion in Cracked Concrete
Reprinted from: *Crystals* **2021**, *11*, 742, doi:10.3390/cryst11070742 317

Qingzhang Zhang, Zihan Kang, Yifeng Ling, Hui Chen and Kangzong Li
Influence of Temperature on the Moisture Transport in Concrete
Reprinted from: *Crystals* **2020**, *11*, 8, doi:10.3390/cryst11010008 329

Jiangfeng Wang, Mengting Chen, Xiaohui Li, Xuexuan Yin and Wenlong Zheng
Effect of Synthetic Quadripolymer on Rheological and Filtration Properties of Bentonite-Free
Drilling Fluid at High Temperature
Reprinted from: *Crystals* **2022**, *12*, 257, doi:10.3390/cryst12020257 . . . . **347**

Haichao Liu, Yisa Fan and Han Peng
Effect of Full Temperature Field Environment on Bonding Strength of Aluminum Alloy
Reprinted from: *Crystals* **2021**, *11*, 657, doi:10.3390/cryst11060657 . . . . **361**

Lang Pang, Zhenguo Liu, Dengquan Wang and Mingzhe An
Review on the Application of Supplementary Cementitious Materials in Self-Compacting
Concrete
Reprinted from: *Crystals* **2022**, *12*, 180, doi:10.3390/cryst12020180 . . . . **381**

About the Editors

Yifeng Ling

Yifeng Ling is a Professor at the School of Qilu Transportation at Shandong University. He achieved his PhD at Iowa State University in 2018, following a post-doctoral fellowship at the National Concrete Pavement Technology Center. He is a member of ACI Committee 309. His research interests include concrete materials, material characterization, mechanical and durability testing, and concrete consolidation. He has published over 30 papers in peer-reviewed journals.

Chuanqing Fu

Chuanqing Fu achieved his Ph.D. at Zhejiang University in 2012. He has been working at Zhejiang University of Technology since 2012, and was promoted to Professor in 2019. His current research interests focus on the corrosion and deterioration of reinforcement concrete structures. He has made some creative achievements. He has published more than 90 scientific research papers, and almost 30 patents have been authorized.

Peng Zhang

Dr. Peng Zhang is currently a Professor and doctoral supervisor at the School of Water Conservancy Engineering in Zhengzhou University. He has been acknowledged as an Academic and Technical Leader of Education Department of Henan Province in China, a Science & Technology Innovation Talent in Universities of Henan Province, China, and Outstanding Young Backbone Teacher of Henan Province, China. His research interests include fiber- and nano-particle-reinforced high-performance concrete and cementitious composites. He has published more than 170 academic papers and 5 books. As the principal investigator, he has hosted more than 20 research projects, which have been sponsored by the National Natural Sciences Foundation of China, Chinese Postdoctoral Science Special Foundation, Chinese Postdoctoral Science Foundation, etc. He has won four prizes of Provincial Science & Technology Awards and 15 other Science & Technology Awards.

Peter Taylor

Peter Taylor, FACI, is the Director of the National Concrete Pavement Technology Center of the Institute for Transportation at Iowa State University. He achieved his PhD at the University of Cape Town, Cape Town, Western Cape, South Africa. He is Chair of ACI Subcommittee 325-F, Concrete Pavement Overlays. His research interests include concrete characterization, concrete durability, concrete pavement construction, and concrete consolidation.

Preface to "Advances in Sustainable Concrete System"

Concrete is the most widely used construction material in the world, and is typically produced using Portland cement (PC) as the binder. The mass of PC used in concrete construction represents a critical environmental issue, due to the high emissions of carbon dioxide gas during its manufacture from the calcination of limestone and the combustion of fossil fuel. On the other hand, the rising demands to reduce the cost of binders in concrete necessitate the search for an alternative source of PC. This book aims to select current advanced concrete studies on environmentally friendly, cost-effective concretes and waste recycling. The relevant research focuses on, but is not limited to, the properties, evaluation, novel manufacturing/experimental techniques, analytical methods, microstructure, modeling, design, production, and practical applications of new binders/aggregates in concrete, and their behaviors in the concrete structures of in situ performance, renovation, maintenance, recycling, durability, and sustainability.

This reprint, edited by Yifeng Ling, Chuanqing Fu, Peng Zhang, and Peter Taylor, will be of interest to scholars, engineers, and government agencies in the relevant fields. The Guest Editors would like to express their acknowledgement to all authors of the papers published in the Special Issue entitled "Advances in Sustainable Concrete System".

Yifeng Ling, Chuanqing Fu, Peng Zhang, and Peter Taylor
Editors

Editorial

Advances in Sustainable Concrete System

Yifeng Ling [1,2,*], Chuanqing Fu [3,*], Peng Zhang [4,*] and Peter Taylor [5,*]

[1] School of Qilu Transportation, Shandong University, Jinan 250002, China
[2] Key Laboratory of Concrete and Pre-stressed Concrete Structures of Ministry of Education, Southeast University, Nanjing 211189, China
[3] College of Civil Engineering and Architecture, Zhejiang University of Technology, Hangzhou 310023, China
[4] School of Water Conservancy Engineering, Zhengzhou University, Zhengzhou 450001, China
[5] National Concrete Pavement Technology Center, Iowa State University, Ames, IA 50011, USA
* Correspondence: yfling@sdu.edu.cn (Y.L.); chuanqingfu@126.com (C.F.); zhangpeng@zzu.edu.cn (P.Z.); ptaylor@iastate.edu (P.T.)

Citation: Ling, Y.; Fu, C.; Zhang, P.; Taylor, P. Advances in Sustainable Concrete System. *Crystals* 2022, 12, 698. https://doi.org/10.3390/cryst12050698

Received: 11 May 2022
Accepted: 12 May 2022
Published: 14 May 2022

Publisher's Note: MDPI stays neutral with regard to jurisdictional claims in published maps and institutional affiliations.

In recent years, the implementation of a sustainable concrete system has been a great topic of interest in the field of construction engineering worldwide as a result of the large and rapid increase in carbon emissions and environmental problems from traditional concrete production and industry. For example, the uses of supplementary cementitious materials, geopolymer binder, recycled aggregate and industrial/agricultural wastes in concrete are all approaches to building a sustainable concrete system. However, such materials have inherent flaws due to their variety of sources, and exhibit very different properties compared to traditional concrete. Therefore, they require specific modifications in preprocessing, design and evaluation before use in concrete. This Special Issue, entitled "Advances in Sustainable Concrete System", covers a broad range of advanced concrete research in environmentally friendly concretes, cost-effective admixtures and waste recycling, specifically including the design methods, mechanical properties, durability, microstructure, various models, hydration mechanism and practical applications of solid wastes in the concrete system.

A vast amount of research is concentrated on how to effectively use solid wastes. The utilization of recycled moso bamboo sawdust (BS) as a substitute in a new bio-based cementitious material was investigated by Tong et al. [1]. They first focused on the effect of pretreatment methods (cold water, hot water and alkaline solution) on the mechanical properties and microstructure of BS. Since the alkali-treated BS had a more favorable bonding interface in the cementitious matrix, both compressive and flexural strength were higher than those of the other two treatments. However, an increased proportion of BS (1–7%) led to a reduction in workability and strength with a more porous structure, but which still met the minimum strength requirements of masonry construction. The influence of fly ash content on the compressive strength of cemented sand and gravel (CSG) materials was evaluated by Chai et al. [2]. They found that 90 d or 180 d strength should be used as the design strength in the design of CSG dam material. There is an optimal content (50%) of fly ash in CSG materials. The effects of NaOH concentrations on the mechanical properties and microstructure of alkali-activated blast furnace ferronickel slag (BFFS) were assessed by Huang et al. [3]. Less C-A-S-H gel at a low concentration resulted in low compressive strength. A high concentration produced more hydrotalcite than C-A-S-H as intensive reaction at the early stage hindered the growth of C-A-S-H in the later stage, and decreased the Al/Si ratio and polymerization of C-A-S-H, which led to a low strength. Phosphorus slag and limestone composite (PLC) can greatly reduce the adiabatic temperature rise, chloride permeability and drying shrinkage of concrete, but do not affect the long-term strength of concrete [4]. Guan et al. [5] identified the possible problems of coal gangue as a road base via an unconfined compressive strength test, a splitting test, a freeze–thaw test and a drying shrinkage test of cement-stabilized gangue with varying cement amounts. The maximum dry density and optimum moisture content of the optimized cement-stabilized gangue and cement-stabilized macadam increased with the increase in cement content. A cement content of 4% is optimal for cement-stabilized coal gangue, which can be used to produce light traffic bases and heavy traffic subbases of class II below highways. The performance of concrete composites comprising waste plastic food trays (WPFTs) as low-cost fibers and palm oil fuel ash (POFA) exposed to acid and sulfate solutions was evaluated in an immersion period of 12 months by Mohammadhosseini et al. [6]. It was discovered that adding WPFT fibers and POFA to the concrete reduced its workability, but increased its long-term compressive strength. As a result of the positive interaction between POFA and WPFT fibers, both the crack formation and spalling of concrete samples exposed to acid and sulfate solutions were reduced. WPFT fibers have a significant protective effect on concrete against chemical attacks. Zhang et al. [7] investigated the influence of ultrafine metakaolin, replacing cement with a cementitious material, on the properties of concrete and mortar. Adding ultrafine metakaolin or silica fume can effectively increase the compressive strength, splitting tensile strength, resistance to chloride ion penetration and freeze–thaw properties of concrete due to the improved pore structure. The sulphate attack resistance of mortar can be improved more obviously by simultaneously adding ultrafine metakaolin and prolonging the initial moisture curing time. Guan et al. [8] investigated the influence of activator, steel slag powder, metakaolin and silica fume on the resulting strength of steel slag cement mortar through orthogonal experiments. The optimal dosages of activator, steel slag powder, metakaolin and silica fume were suggested. Awan et al. [9] evaluated the mechanical performance of untreated and treated (NaOH, lime and common detergent) crumb rubber concrete (CRC). These treatments were performed to enhance the mechanical properties of concrete that are affected by adding CR. The mechanical properties were improved after the treatment of CR. Lime treatment was found to be the best treatment. In two-stage concrete (TSC), Javed et al. [10] placed coarse aggregates in formwork, and then injected grout at a high pressure to fill up the voids between the coarse aggregates. Ten percent and twenty percent bagasse ash were used as a fractional substitution of cement, along with recycled coarse aggregate (RCA). The compressive and tensile strength of the TSC that had RCA was improved by the addition of bagasse ash. Machine learning was used to successfully predict the properties of concrete containing rice husk ash by Iqtidar et al. [11]. Input parameters include age, amount of cement, rice husk ash, super plasticizer, water and aggregates. Artificial neural networks (ANNs), the adaptive neuro-fuzzy inference system (ANFIS), multiple nonlinear regression (NLR) and linear regression were employed to evaluate the properties of concrete containing rice husk ash. The ANN and ANFIS outperformed other methods. Zhou et al. [12] carried out an experimental study of the effects of ground slag powder (GSP) on the hydration, pore structure, compressive strength and chloride ion penetrability resistance of high-strength concrete. Adding 25% GSP increases the adiabatic temperature rise of high-strength concrete, whereas adding 45% GSP decreases the initial temperature rise. Incorporating GSP refines the pore structure to the greatest extent and improves the compressive strength and chloride ion penetrability resistance of high-strength concrete. Increasing curing temperature has a more obvious impact on the pozzolanic reaction of GSP than cement hydration.

Some studies have examined mechanical properties using machine learning, numerical and constitutive models. Shen et al. [13] established machine learning-based models to accurately predict the punching shear strength of fiber-reinforced polymer (FRP)-reinforced concrete slabs. Artificial neural network, support vector machine, decision tree and adaptive boosting were selected to build models. SHapley Additive exPlanation (SHAP) was adopted to provide global and individual interpretations, and feature dependency analysis for each input variable. Nafees et al. [14] provided a new modeling approach using reactive powder concrete (RPC) beam-column joint to predict the behavior and response of structures and to improve the shear strength deformation against different structural loading. RPC in the joint region increased the overall strength by more than 10%, as well as the ductility of the structures. Zhou et al. [15] studied the energy evolution characteristics and damage constitutive relationship of siltstone. The damage constitutive equation of siltstone was developed based on the damage mechanics theory through the principle of minimum energy consumption and by considering the residual strength of rock. Bai et al. [16] investigated the mesoscopic damage mechanism of fiber-reinforced concrete (FRC) under uniaxial tension. The damage constitutive model for FRC under uniaxial tension was established to reflect the potential bearing capacity of materials. The influence of fiber content on the initiation and propagation of micro-cracks and the damage evolution of concrete was evaluated. Zhang et al. [17] examined the behavior of lattice girder composite slabs with monolithic joint under bending using a finite element model and found that lattice girder significantly increased the stiffness of the slab. Additionally, the damage of a reinforced concrete (RC) column with various levels of reinforcement corrosion under axial loads was characterized using the acoustic emission (AE) technique of Chen et al. [18]. Zhou et al. [19] performed a brittleness evaluation of high-strength concrete through a triaxial compression test of C60 and C70 high-strength concrete. With the increase in the confining pressure, the proportion of elastic energy in the whole process of high-strength concrete failure gradually decreased.

Finally, several investigations of durability concrete have been conducted. Wang et al. [20] revealed the effect of crack geometry on chloride diffusion in cracked concrete. The crack depth showed a more significant influence on the chloride penetration depth in cracked concrete than crack geometry did. Compared with rectangular and V-shaped cracks, the chloride diffusion process in cracked concrete with a tortuous crack was slower at the early immersion age, while the crack geometry had a marginal influence on the chloride penetration depth in cracked concrete during long-term immersion. Zhang et al. [21] built a moisture saturation equilibrium relationship of concrete under different temperatures and relative humidity conditions to develop moisture absorption and desorption curves. They concluded that the moisture absorption rate was lower at higher temperatures and largely dependent on the saturation gradient, while the desorption was increased at higher temperatures and mostly affected by the saturation gradient.

The present Special Issue on "Advances in Sustainable Concrete System" collects the current research progress in construction material reforming into a green and sustainable system.

Funding: This research was funded by the Natural Science Foundation of Shandong Province (grant number: ZR2021QE174), the Open Project Fund of Key Laboratory of Concrete and Pre-stressed Concrete Structures of Ministry of Education in Southeast University (grant number: CPCSME2022-03), the Natural Science Foundation of Henan Province of China (grant number: 212300410018), the Program for Innovative Research Team (in Science and Technology) in University of Henan Province of China (grant number: 20IRTSTHN009), the Natural Science Foundation of Zhejiang Province of China (grant number: LR21E080002), and the National Natural Science Foundation of China (grant number: 51978620).

Conflicts of Interest: The authors declare no conflict of interest.

References

1. Tong, Y.; Seibou, A.-O.; Li, M.; Kaci, A.; Ye, J. Bamboo Sawdust as a Partial Replacement of Cement for the Production of Sustainable Cementitious Materials. *Crystals* **2021**, *11*, 1593. [CrossRef]
2. Chai, Q.; Wan, F.; Xiao, L.; Wu, F. The Influence of Fly Ash Content on the Compressive Strength of Cemented Sand and Gravel Material. *Crystals* **2021**, *11*, 1426. [CrossRef]
3. Huang, Z.; Zhou, Y.; Cui, Y. Effect of Different NaOH Solution Concentrations on Mechanical Properties and Microstructure of Alkali-Activated Blast Furnace Ferronickel Slag. *Crystals* **2021**, *11*, 1301. [CrossRef]
4. Liu, K.; Cui, Y. Effects of Different Content of Phosphorus Slag Composite Concrete: Heat Evolution, Sulphate-Corrosion Resistance and Volume Deformation. *Crystals* **2021**, *11*, 1293. [CrossRef]
5. Guan, J.; Lu, M.; Yao, X.; Wang, Q.; Wang, D.; Yang, B.; Liu, H. An Experimental Study of the Road Performance of Cement Stabilized Coal Gangue. *Crystals* **2021**, *11*, 993. [CrossRef]
6. Mohammadhosseini, H.; Alyousef, R.; Poi Ngian, S.; Tahir, M.M. Performance Evaluation of Sustainable Concrete Comprising Waste Polypropylene Food Tray Fibers and Palm Oil Fuel Ash Exposed to Sulfate and Acid Attacks. *Crystals* **2021**, *11*, 966. [CrossRef]

7.   Zhang, S.; Zhou, Y.; Sun, J.; Han, F. Effect of Ultrafine Metakaolin on the Properties of Mortar and Concrete. *Crystals* **2021**, *11*, 665. [CrossRef]
8.   Guan, J.; Zhang, Y.; Yao, X.; Li, L.; Zhang, L.; Yi, J. Experimental Study on the Effect of Compound Activator on the Mechanical Properties of Steel Slag Cement Mortar. *Crystals* **2021**, *11*, 658. [CrossRef]
9.   Awan, H.H.; Javed, M.F.; Yousaf, A.; Aslam, F.; Alabduljabbar, H.; Mosavi, A. Experimental Evaluation of Untreated and Pretreated Crumb Rubber Used in Concrete. *Crystals* **2021**, *11*, 558. [CrossRef]
10.   Javed, M.F.; Durrani, A.A.; Kashif Ur Rehman, S.; Aslam, F.; Alabduljabbar, H.; Mosavi, A. Effect of Recycled Coarse Aggregate and Bagasse Ash on Two-Stage Concrete. *Crystals* **2021**, *11*, 556. [CrossRef]
11.   Iqtidar, A.; Bahadur Khan, N.; Kashif-ur-Rehman, S.; Faisal Javed, M.; Aslam, F.; Alyousef, R.; Alabduljabbar, H.; Mosavi, A. Prediction of Compressive Strength of Rice Husk Ash Concrete through Different Machine Learning Processes. *Crystals* **2021**, *11*, 352. [CrossRef]
12.   Zhou, Y.; Sun, J.; Zhang, Z. Effects of High-Volume Ground Slag Powder on the Properties of High-Strength Concrete under Different Curing Conditions. *Crystals* **2021**, *11*, 348. [CrossRef]
13.   Shen, Y.; Sun, J.; Liang, S. Interpretable Machine Learning Models for Punching Shear Strength Estimation of FRP Reinforced Concrete Slabs. *Crystals* **2022**, *12*, 259. [CrossRef]
14.   Nafees, A.; Javed, M.F.; Musarat, M.A.; Ali, M.; Aslam, F.; Vatin, N.I. FE Modelling and Analysis of Beam Column Joint Using Reactive Powder Concrete. *Crystals* **2021**, *11*, 1372. [CrossRef]
15.   Zhou, R.; Guo, L.; Hong, R. Study on Energy Evolution and Damage Constitutive Model of Siltstone. *Crystals* **2021**, *11*, 1271. [CrossRef]
16.   Bai, W.; Lu, X.; Guan, J.; Huang, S.; Yuan, C.; Xu, C. Stress–Strain Behavior of FRC in Uniaxial Tension Based on Mesoscopic Damage Model. *Crystals* **2021**, *11*, 689. [CrossRef]
17.   Zhang, X.; Li, H.; Liang, S.; Zhang, H. Experimental and Numerical Study of Lattice Girder Composite Slabs with Monolithic Joint. *Crystals* **2021**, *11*, 219. [CrossRef]
18.   Chen, Y.; Zhu, S.; Ye, S.; Ling, Y.; Wu, D.; Zhang, G.; Du, N.; Jin, X.; Fu, C. Acoustic Emission Study on the Damage Evolution of a Corroded Reinforced Concrete Column under Axial Loads. *Crystals* **2021**, *11*, 67. [CrossRef]
19.   Zhou, R.; Cheng, H.; Li, M.; Zhang, L.; Hong, R. Energy Evolution Analysis and Brittleness Evaluation of High-Strength Concrete Considering the Whole Failure Process. *Crystals* **2020**, *10*, 1099. [CrossRef]
20.   Wang, Q.; Zhang, G.; Tong, Y.; Gu, C. A Numerical Study on Chloride Diffusion in Cracked Concrete. *Crystals* **2021**, *11*, 742. [CrossRef]
21.   Zhang, Q.; Kang, Z.; Ling, Y.; Chen, H.; Li, K. Influence of Temperature on the Moisture Transport in Concrete. *Crystals* **2021**, *11*, 8. [CrossRef]

Article

Bamboo Sawdust as a Partial Replacement of Cement for the Production of Sustainable Cementitious Materials

Yunyun Tong [1], Abdel-Okash Seibou [1], Mengya Li [2,*], Abdelhak Kaci [2] and Jinjian Ye [1]

1 School of Civil Engineering and Architecture, Zhejiang University of Science & Technology, Hangzhou 310023, China; 112013@zust.edu.cn (Y.T.); 911902814005@zust.edu.cn (A.-O.S.); 211902814006@zust.edu.cn (J.Y.)
2 Laboratory of Mechanics and Materials of Civil Engineering (L2MGC), CY Cergy Paris Université, F-95000 Cergy, France; abdelhak.kaci@cyu.fr
* Correspondence: mengya.li1@cyu.fr; Tel.: +33-1-34-25-69-07

Abstract: This paper reports on the utilization of recycled moso bamboo sawdust (BS) as a substitute in a new bio-based cementitious material. In order to improve the incompatibility between biomass and cement matrix, the study firstly investigated the effect of pretreatment methods on the BS. Cold water, hot water, and alkaline solution were used. The SEM images and mechanical results showed that alkali-treated BS presented a more favorable bonding interface in the cementitious matrix, while both compressive and flexural strength were higher than for the other two treatments. Hence, the alkaline treatment method was adopted for additional studies on the effect of BS content on the microstructural, physical, rheological, and mechanical properties of composite mortar. Cement was replaced by alkali-treated BS at 1%, 3%, 5%, and 7% by mass in the mortar mixture. An increased proportion of BS led to a delayed cement setting and a reduction in workability, but a lighter and more porous structure compared to the conventional mortar. Meanwhile, the mechanical performance of composite decreased with BS content, while the compressive and flexural strength ranged between 14.1 and 37.8 MPa and 2.4 and 4.5 MPa, respectively, but still met the minimum strength requirements of masonry construction. The cement matrix incorporated 3% and 5% BS can be classified as load-bearing lightweight concrete. This result confirms that recycled BS can be a sustainable component to produce a lightweight and structural bio-based cementitious material.

Keywords: bamboo; sawdust; pretreatment; bio-based material; mechanical property

Citation: Tong, Y.; Seibou, A.-O.; Li, M.; Kaci, A.; Ye, J. Bamboo Sawdust as a Partial Replacement of Cement for the Production of Sustainable Cementitious Materials. *Crystals* **2021**, *11*, 1593. https://doi.org/10.3390/cryst11121593

Academic Editors: Chongchong Qi, Yifeng Ling, Chuanqing Fu, Peng Zhang and Peter Taylor

Received: 25 November 2021
Accepted: 17 December 2021
Published: 20 December 2021

Publisher's Note: MDPI stays neutral with regard to jurisdictional claims in published maps and institutional affiliations.

1. Introduction

The Paris 2024 Olympics have committed to reducing their carbon footprint, with the Olympic and Paralympic Villages being built using 100% bio-based materials. This example illustrates that the trend in the construction industry in the twenty-first century is towards sustainable and environmentally friendly building materials. To date, concrete has been the most widely used building material in the world, with ordinary Portland cement being the key ingredient in concrete. Cement production exceeds 4 billion tons per year and leads to ~8% of the world's carbon dioxide emissions [1], which is the main driver of the greenhouse effect. According to the Paris Agreement on climate change, emissions from the cement sector should decrease by at least 16% by 2030. These factors explain our passion for investigating a natural and sustainable substitute for cementitious materials.

China contains 5.4 million hectares of bamboo planting area where the environment and climate are suitable for bamboo growing [2]. Bamboo belongs to the grass family, which consists of more than 1600 species. As the fastest-growing plant in the world, bamboo can grow up to 90 cm per day and reach maturity in only 2–3 years. Mature bamboo has almost the same strength and hardness as hardwood, which take more than 50 years to mature [3]. Its rapid maturity and sustainability have made bamboo the main rival for wood in the construction industry over the past 20 years. Applications of bamboo in various fields are increasing, and include framework, cladding, posts, and furniture. The main environmental problem in the booming bamboo industry is solid waste disposal. According to a report from the Cultural and Tourism Office of Anji, China, in 2015, the city produced more than 0.16 million tons of bamboo waste per year. Most bamboo waste is generated as clean biomass energy, but limited amounts of bamboo are reused. The efficient recycling of these wastes and their reuse as an ingredient in bio-based materials was the initial objective of this research.

Many studies have been carried out on mortar or concrete that incorporates lignocellulose aggregates, including hemp [4–6], wood [7], sisal [8], jute [9], coconut [10], palm oil [11], and sugarcane bagasse [12]. Cellulose, hemicellulose, and lignin are the primary biochemical compositions of these natural fibers. Other components as impurities, such as pectin and wax substances, also exist on the fiber surface. Cellulose is the main composition of cell wall, which provides the strength of fiber. It is insoluble in water, organic solvents and alkaline solution [13]. Hemicellulose comprises the sugar component, which bridges cellulose fibers with hydrogen bonds. Lignin is a cementing material. Bamboo fibers contain 40–55% cellulose, 18–20 hemicellulose, and 15–32% lignin [14]. The high proportion of cellulose makes bamboo a potential component for building materials. However, limited literature has focused on the valorization of bamboo in cementitious materials. Fias et al. [15] studied the replacement of 10–20 wt.% cement by Brazilian bamboo leaf ash (BLA) calcined at 600 °C. The same compressive strength resulted for the control mortar and BLA blended mortar, which confirmed the use of BLA as a substitute for cement. Xie et al. [16] observed that the flexural strength and impact resistance of mortar could be improved by reinforcement with 4–16 wt.% bamboo fibers (BF). Shrinkage could be inhibited by 12% because of the BF reinforcement [17]. However, efficient incorporation of vegetable products in the cementitious matrix requires the overcoming of problems of workability, compatibility, and durability. The workability problem relates to the high water absorption of vegetable fibers [18]; the compatibility problem relates to a weak bonding between natural fibers and the cement matrix [19]; and the durability problem relates to biodeterioration and a resistance to freezing and thawing [20]. To overcome these problems, the fibers are pretreated before fabricating the composite. The pretreatment methods, such as cold water, hot water, and alkali solution, have been investigated in the literature in order to modify the structure and morphology of natural fibers and improve the matrix interface, as well as increasing the durability of fibers in the alkaline environment of the cement matrix [21–24].

Sawdust is a by-product generated during the process of manufacturing. The recycling of such waste provides the benefits of reducing the need to extract new raw materials and limiting the air pollution due to incineration. Ahmed et al. investigated the potential of wood sawdust as a replacement of fine aggregates in concrete [25], and confirmed the utilization of this material for structure application. Besides, lightweight concrete incorporated sawdust presented a thermal conductivity 23% lower than conventional concrete [26].

To the best of our knowledge, few studies have reported incorporating bamboo sawdust (BS) in a cementitious matrix. This study contributes to the design of a new cementitious material containing local BS. The effect of different pretreatment methods for BS on the composite were firstly assessed through their morphological, physical, and mechanical behaviors, which aimed to choose a more efficient treatment. Furthermore, the characteristics of composites incorporating different proportions of alkali-treated BS were analyzed.

2. Materials and Methods

2.1. Raw Materials

The bamboo particles that were used in this study were recycled sawdust from a local bamboo furniture manufacturing industry (Zhejiang, China). The species of bamboo was moso (Phyllostachys edulis), which is most common in China.

The particle size distribution of the recycled BS is presented in Figure 1. The average particle size (D50) of the BS was 0.09 mm. A BS particle size of between 0.037 and 0.16 mm was used. The basic characteristics of the moso BS are summarized in Table 1. The particle density was determined by the Archimedes method, where the apparent mass of BS was measured in air and after immersion in distilled water with a pycnometer. The bulk density was calculated from the BS mass divided by the volume occupied.

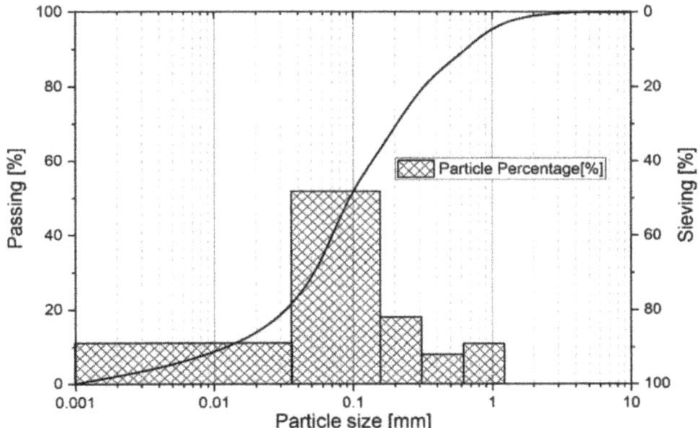

Figure 1. Recycled bamboo particle size distribution.

Table 1. Characteristics of moso BS.

Particle density (kg/m^3)	1563
Bulk density (kg/m^3)	390
Initial moisture content	6.089%
Water absorption capacity (1 min)	277.5%
Water absorption capacity (1 h)	320%

Sawdust treatment modifies the rheological, physical, and mechanical properties of mortar that is reinforced with vegetable particles. Three types of treatment were used to remove excessive lignin and hemicellulose, and to improve the wettability [22]:

- Aqueous treatment: BS was soaked in cold water for 24 h.
- Hot aqueous treatment: BS was soaked in hot water at 100 °C for 16 h.
- Alkali treatment: BS was soaked in 10% sodium hydroxide solution for 30 min at room temperature and rinsed with distilled water until neutral pH.

After treatment, the BS was dried in an oven at 60 °C for 48 h, and then stored in a desiccator.

2.2. Mixture Proportion

Ordinary Portland cement CEMI 42.5 and standard sand with a proportion of 1:3 were used to prepare the control mortar with no added BS in accordance with EN 196 [27]. The relative specific gravity of cement and sand was 3.12 and 2.64, respectively. The water to binder (W/C) ratio was fixed at 0.5 for mortar and 0.25 for cement paste. Similar to most cellulosic fibers, the BS adsorbs large amounts of water. The BS was pre-wetted to prevent the inner structure from absorbing water from the mixture. The quantity of pre-wetting water (PW) was calculated from:

$$\text{Mass of pre-wetting water} = \frac{\text{mass at saturation state} - \text{dry mass}}{\text{dry mass}} \times \text{BS dosage} \qquad (1)$$

The pre-wetting time was determined by the saturation state of the bamboo particles, as shown in Figure 2. The water absorption increased to 86.7% saturation in 1 min, the growth rate slowed in the following 4 min, and then the growth was stable until complete saturation. The study of Monreal et al. [28] showed that the pre-wetting water should be below complete saturation, otherwise excessive water may result in the mixture. From a perspective of mixing time and energy saving, the pre-wetting time was set to 3 min, at 95% particle saturation, where the pre-wetting water quantity was three times the BS. With the same cement and sand ratio, 1 wt.%, 3 wt.%, 5 wt.%, and 7 wt.% cement was substituted by BS, and the new biomaterials were designated as bamboo sawdust cement mortar BSC1, BSC3, BSC5, and BSC7, respectively. The limit of substitution was set at 7 wt.% due to the very low workability of composite at fresh state (nearly 0), which is not suitable for construction. The results of the slump test are discussed in Section 3.2.1. Details of mixture proportions for formulations are recapitulated in Table 2.

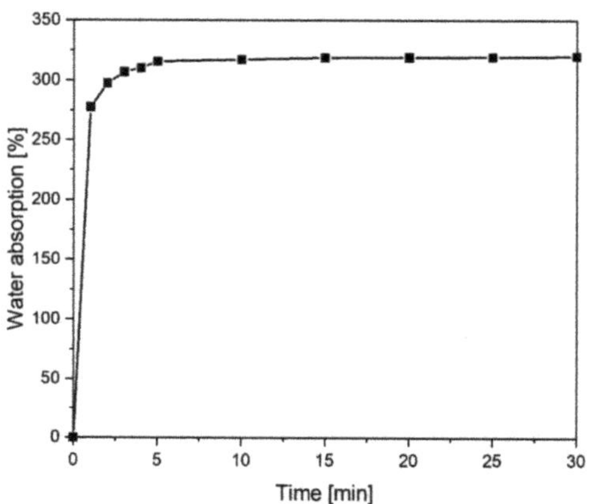

Figure 2. BS particle water-absorption rate.

Table 2. Compositions of different formulations.

Designation	Cement [kg/m³]	Fine Aggregate [kg/m³]	Bamboo Sawdust [kg/m³]	Pre-Wetting Water [l/m³]	Water [l/m³]
Control	511.018	1533.055	0.000	0.000	255.509
BSC 1	509. 336	1528.009	5.145	16.463	254.668
BSC 3	505. 903	1517. 709	15.646	50.069	252.951
BSC 5	502. 374	1507. 121	26.441	84.610	251.187
BSC 7	498. 745	1496. 235	37.540	120.128	249.372

2.3. Mixture Preparation

The BSC composite was prepared by using a mortar mixer. The bamboo particles were pre-wetted for 3 min, and then mixed at a low speed of 140 ± 5 r/min^{-1} for 3 min. Cement and sand were added and mixed for 2 min at a low speed. Mixing continued for 1 min 30 s, after which water was added. After 1 min 30 s, the mixer was stopped as it was scraping the bowl. The last step involved restarting the mixer and running at 285 ± 10 r/min^{-1} for 1 min. The mixing procedure is summarized in Table 3.

Table 3. Mixture procedure.

Operation	Introduction of Pre-Wetting Water and BS	Addition of Cement and Sand	Addition of Water	Scrape the Bowl	Mixture
Duration	3 min	2 min	1 min 30 s	30 s	1 min
State of mixer	Stop	Slow speed		Stop	High speed

All molds that contained specimens were kept in a humid atmosphere (50 ± 5 RH%) for 24 h at 20 ± 2 °C before being demolded, and the demolded specimens were kept under water in a temperature-controlled wet preservation cabinet until the tests. A general view of the BSC specimens is presented in Figure 3.

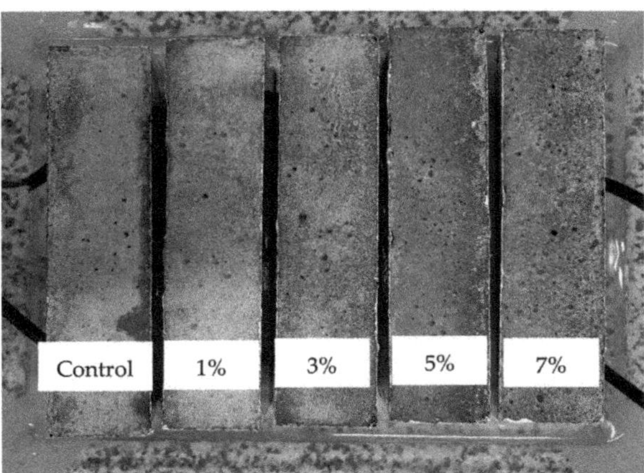

Figure 3. BSC mortar specimens.

2.4. Morphology

The BS microstructure with and without treatment, and the composite mortar at 28 days were investigated by scanning electron microscopy (SEM, HITACHI Model TM3000). The BS samples were placed directly on aluminum stub with a diameter of 25 mm. The BSC samples were sputter coated with Au alloy in order to reduce white regions on the image, which were caused by electron charging. The observation condition mode 15kV was applied.

2.5. Rheological Properties

The cement paste and mortar that contained BS were evaluated according to their setting time and workability in the fresh state, respectively. The setting time of the cement paste was measured by using a Vicat apparatus in accordance with NF P15-431 [29]. Slump tests were performed by using a mini slump cone with diameters of 50 mm and 100 mm at the top and base, respectively, and a height of 150 mm, in accordance with MBE (method to design mortar concrete containing admixture) [30] to evaluate the workability of the composite mortar.

2.6. Physical and Mechanical Properties

The density, porosity, and compressive and flexural strength in the hardened state were determined for all specimens. The specimen density was calculated from the measured dry mass at 28 days and the volume. The porosity was characterized according to the AFPC-AFREM testing protocol [31]. The specimens that were cured at 28 days were placed in a desiccator, where a maximum internal constant pressure of 25 mbar was maintained by a vacuum pump. After 4 h, water was introduced to immerse the specimens, and the same pressure was maintained in the desiccator for 68 h. The water-accessible porosity was calculated from:

$$\text{Porosity} = \frac{M_{air} - M_{dry}}{M_{air} - M_{sat}} \times 100\% \tag{2}$$

where M_{air} and M_{sat} are the mass of vacuum-saturated specimens measured in air and water, respectively, and M_{dry} is the mass of specimen that was oven-dried at 105 °C for 24 h.

The compressive and flexural strengths were measured according to EN 196 [27] at 3, 7, 14, and 28 d curing. Cubic specimens (40 × 40 × 40 mm) were prepared for a study of their compressive behavior. Tests were carried out using a compressive machine model STYE-1000, with a force-controlled rate of 2500 N/s. The flexural behavior was evaluated using an electric three-point bending testing machine, model DKZ-6000. The set-up was force controlled with a rate of 50 N/s. Specimens of 40 × 40 × 160 mm were characterized for flexural strength.

3. Results and Discussion

3.1. Effect of BS Treatment on BSC Composite

3.1.1. Mass Loss of BS

A remarkable mass loss was observed during the BS treatment. The percentage mass loss because of the treatment was calculated from:

$$\text{Percentage weight loss} = \frac{W_1 - W_2}{W_1} \times 100\% \tag{3}$$

where W_1 (g) and W_2 (g) represent the BS dry mass before and after treatment, respectively.

Figure 4 shows the BS mass variation between the different treatments. The mass loss was rapid in 1 min, and then slowed to a plateau. The BS lost 13.7% of its mass after aqueous treatment. The mass stabilized after 8 h. However, the hot aqueous accelerated the component removal, and improved the efficiency with 17.5% of the components eliminated in 4 h. The mass variation of alkali-treated particles was most important, with an earlier plateau and a mass loss of 28.2% higher than in the aqueous treatment and 23.6% higher than in the hot aqueous treatment. Das et al. [22] reported the similar significant mass loss of bamboo fibers occurred by alkali attack. It can be noted that the fiber cell wall contains an amount of hydroxyl groups (-OH), which can react with alkaline solution [32]:

$$\text{Cell-OH} + \text{NaOH} \rightarrow \text{Cell-O-NA}^+ + \text{H}_2\text{O} + \text{Surface impurities} \tag{4}$$

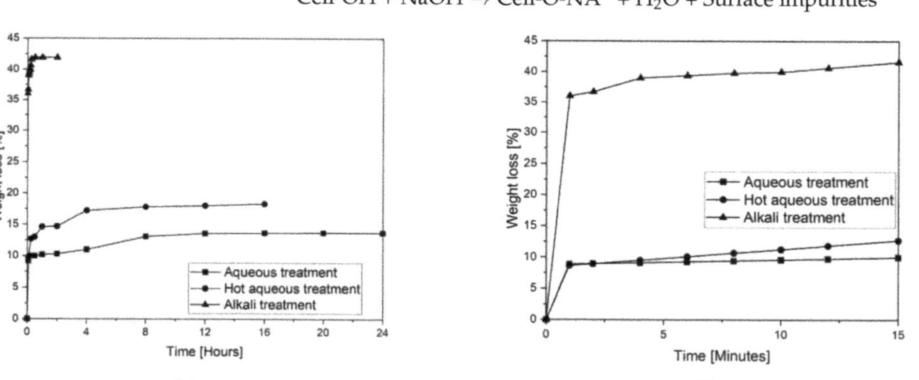

Figure 4. BS mass variation during different treatments (**a**) in 24 h (**b**) in the first 15 min.

The removal of alkali-sensitive components (hemicellulose and lignin) results in the mass loss of fiber. The good side is to provide the mechanical and thermal stability of fiber in the composite matrix. The alkali concentration and immersion time both affect mass loss [33].

3.1.2. Morphological Properties

The morphological changes to the bamboo particles from the different treatments are presented in Figure 5. Compared with (hot) aqueous treated particles (Figure 5b,c), cell walls of the untreated particles were wrapped in wax, pectin, and impurities (Figure 5a).

Figure 5b shows that cold water removed the impurities, where the particle surface was neater and smoother compared with the untreated particles. Cellulosic defibrillation is shown in Figure 5c,d, where rougher and larger surfaces are visible. Alkali treatment was more violent and intensive than hot water. The surface area increased and the network porosity decreased because of the disruption in hydrogen bonding in the network structure by alkali attack, so the surface roughness increased [34]. This effect may improve the particle compatibility with cement by creating a more effective bonding area. Some authors reported that alkali treatment could break the microfibril bundles and produce individual fibers, so that the mechanical interaction between the particles and the matrix can be improved [35,36].

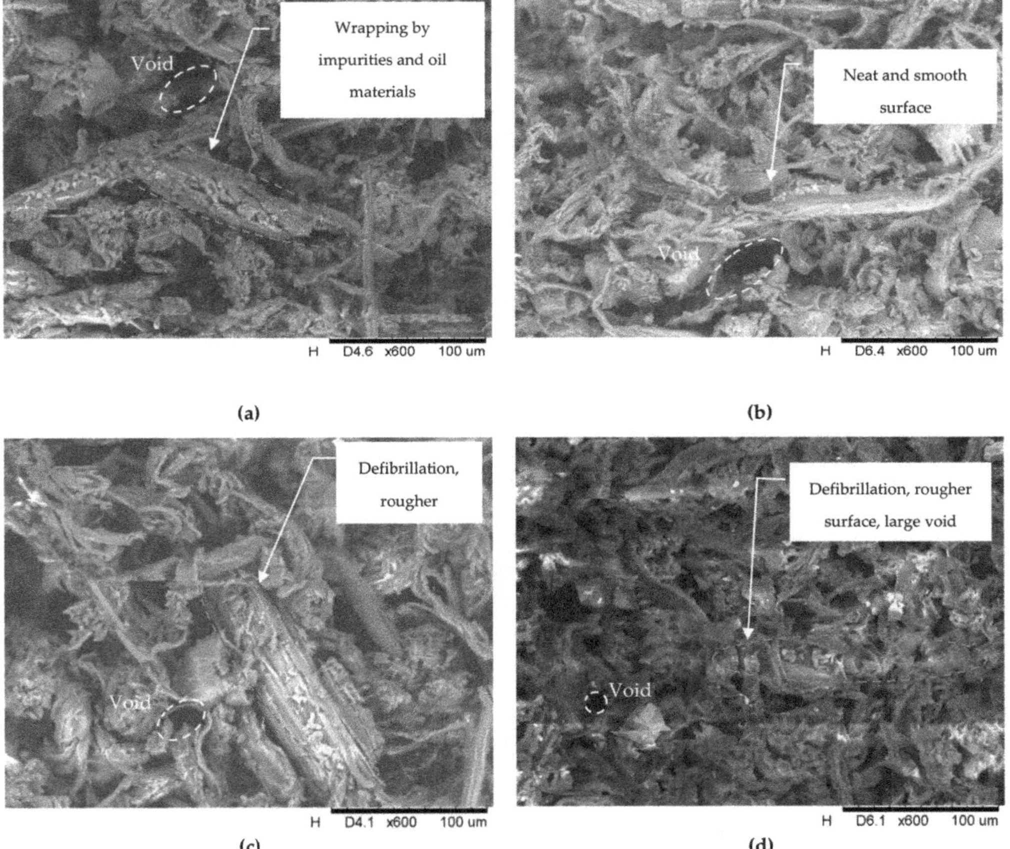

Figure 5. SEM analysis of BS (**a**) without treatment, (**b**) with aqueous treatment, (**c**) with hot aqueous treatment, (**d**) 10% concentration alkali treatment.

3.1.3. Setting Time and Workability

Plant particle addition delays the setting and inhibits strength development due to the presence of hemicellulose and lignin [37]. The initial and final setting times of cement paste with 3% BS by mass with different treatments were measured to meet construction material requirements. Table 4 shows the initial and final setting time of BSC paste using different treatments BS. Meanwhile, the final setting of aqueous-treated BSC paste occurred more than 2 h after the control. Owing to the removal of lignin, which is soluble in water, the final setting time of hot aqueous-treated was 40 min faster than the aqueous-treated. The setting was accelerated by approximately 1 h for the alkali-treated particle compared with the control paste due to the dissolution of hemicellulose in alkaline solution. These results confirmed that adding vegetable particles may inhibit the cement hydration and delay the setting time. Cold and hot water treatment can remove some extractives. However, plant particles require a more violent alkaline extraction to remove hemicellulose.

Table 4. Evolution of setting time with different treatments.

	Control	Aqueous Treatment	Hot Aqueous Treatment	Alkali Treatment
Initial setting time (min)	110	230 (+109%) *	190 (+73%) *	160 (+45%) *
Final setting time (min)	140	310 (+121%) *	270 (+93%) *	200 (+43%) *

* Value compared with control cement paste.

Even though the same amount of pre-wetting water was used in each treatment, slump was measured at 41 mm, 40.7 mm, 34.5 mm, and 9.8 mm for the control mortar, cold aqueous, hot aqueous, and alkali-treated BSC3, respectively. Hot aqueous and alkali-treated BSC3 showed a weak slump. A possible explanation for these results is cellulosic defibrillation, whereas the expanded rougher fiber surface probably absorbed more water. The same results have been reported previously in [35]. The water absorption was more significant for the sodium-hydroxide treatment because of the removal of the lignin layer, which is an impermeable layer of vegetable fiber.

3.1.4. Physical and Mechanical Properties

Despite the various treatments, no significant difference resulted for the density and porosity, which were ~1994 $\pm$ 10 kg/m^3 and 16.4 $\pm$ 0.04%, respectively. The same results were reported in [38] that only 0.7% difference was found between mortar incorporated alkali-treated and non-treated jute fiber.

The experimental results for the mechanical behavior are shown in Figure 6; Figure 7. Figure 6 shows that the compressive strength of BSC3 was lower than that of the control mortar. The compressive performance between the cold and hot water treatments were close, with only a 3% difference observed. The alkali-treated BSC3 exhibited a higher compressive strength of 16.1% and 13.9% than the cold and hot aqueous treatments, respectively. The alkali-treated BSC3 strength was 73% of the control mortar strength.

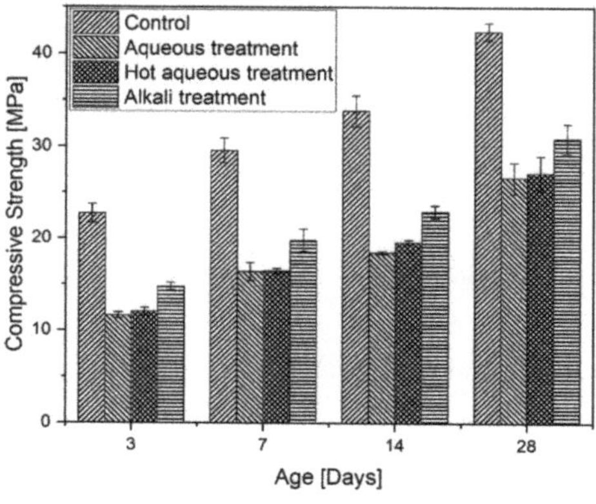

Figure 6. Compressive strength of BSC3 with different treatments.

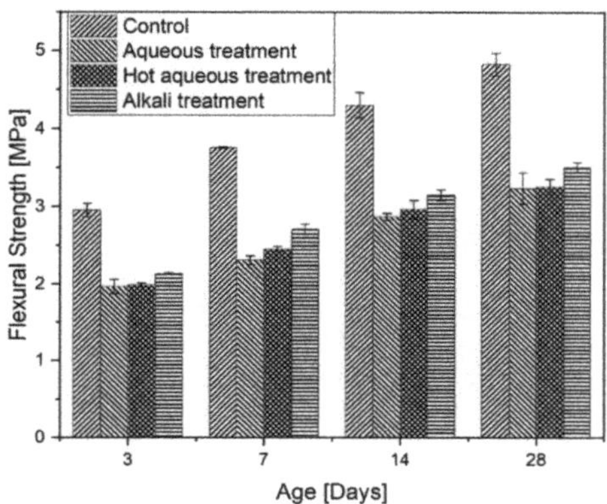

Figure 7. Flexural strength of BSC3 with different treatments.

The flexural strength f_b was obtained from:

$$f_b = \frac{3F \times l}{2b \times h^2} \tag{5}$$

where F is the applied load and l, b, and h are the specimen length, width, and height, respectively.

Data from the three-point bending tests indicate that the flexural strength of the cement matrix was attenuated because of the addition of BS (Figure 7). The degraded mechanical behavior of the BSC is related to the lower modulus of the bamboo particles compared with the cement. No significant difference in bending performance resulted between the three treatments using the ANOVA analysis. The alkali-treated BSC3 had an 8.2% and 7.6% higher flexural strength than the cold and hot aqueous treatments, respectively. The flexural strength of the alkali-treated BSC3 decreased by 37% compared with the control mortar.

Both the compressive and flexural strength value of alkali-treated BSC were higher than for the two other treatments, which confirmed the efficiency of alkali treatment. Besides, a more obvious advantage of alkali-treated BSC can be noticed on compressive behavior. It can be explained that after alkaline attack, the particles bundle into microfibrils (Figure 5d), which simplified their dispersion in the matrix, making the composite more homogenous.

3.2. Effect of BS Content on Composite

Results in Section 3.1 show that the BSC with alkaline treatment had a quicker setting and better mechanical behavior than the other treatments, which indicates that a better compatibility exists between BS and the cementitious matrix. Hence, alkaline treatment was chosen in the following studies to evaluate the effect of particle proportion on the physical and mechanical properties.

3.2.1. Setting Time and Workability

Table 5 provides the setting time of the cement paste with BS and indicates that cement setting was delayed. BSC7 showed the most significant delay, at 64% later than the control. The slowest final setting occurred for BSC7, which was postponed by 230 min (~3.8 h). As mentioned in [39], all cement specimens with an initial setting time of less than 45 min or a final setting time of longer than 6.5 h should be considered as unqualified and sub-quality products. Hence, even though the final setting time was delayed, BSC pastes still achieved a satisfactory setting time.

Table 5. Evolution of setting time with different treatments.

	Control	BSC1	BSC3	BSC5	BSC7
Initial setting time (min)	110	140 (+27%) *	160 (+45%) *	160 (+45%) *	170 (+55%) *
Final setting time (min)	140	170 (+21%) *	200 (+43%) *	220 (+57%) *	230 (+64%) *

* Value compared with control cement paste.

The workability is an important parameter in concrete mixture design. Therefore, slump tests were performed for all BSC specimens, and the results are shown in Figure 8. A marked loss in workability resulted in an increase in the number of particles in the matrix. A higher replacement of BP with cement led to a weaker slump. Nearly zero slump was observed for BSC5 and BSC7 because the increasing content of BS may absorb additional water from the mixture, which requires further investigation.

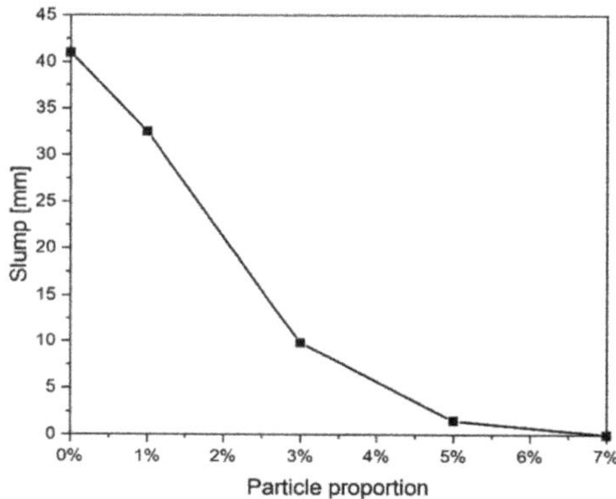

Figure 8. Slump evolution of alkali-treated BSC.

3.2.2. Morphology and Physical Properties

An analysis of BSC at 28 days by SEM in Figure 9 showed a favorable interface between the BS and matrix. The particles were spherical or cylindrical, distributed nearly uniformly in the matrix, and no large pores were observed. Some microcracks were visible, which may be because of the dehydration of cement.

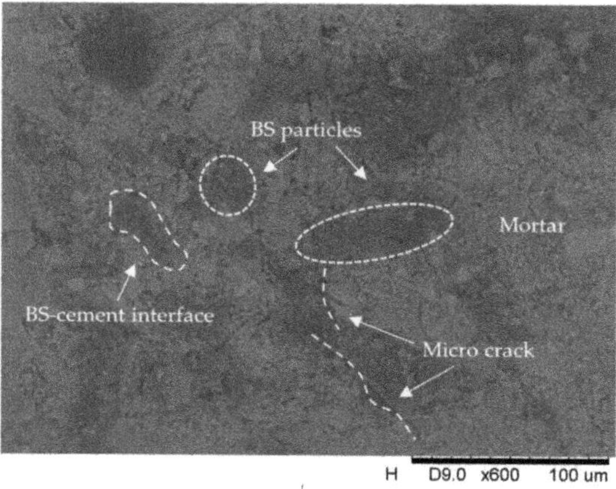

Figure 9. SEM analysis of alkali-treated BSC.

Figure 10 compares the density and porosity for different BSC contents. A correlation between the density and porosity showed that the specimens became lighter and more porous with an increase in BS proportion. The same observation is possible in Figure 3. BS addition to the mixture resulted in an increased void exposure at the specimen surface. The substitution of 7 wt.% cement by BS yielded a matrix that was 14.1% lighter and 6.1% more porous than the control mortar, which agrees with previous research [40].

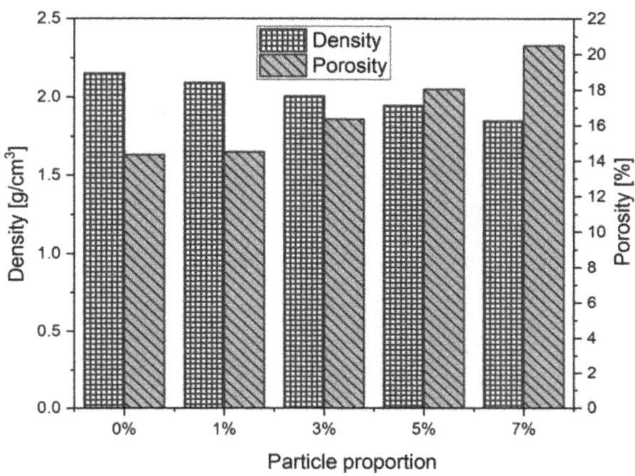

Figure 10. Density and porosity of alkali-treated BSC.

3.2.3. Mechanical Properties

The compressive strength and flexural strength evolution of BSC with different BS contents at three, seven, 14, and 28 days are presented in Figure 11; Figure 12. An increasing substitution of alkali-treated BS content from 1% to 7% decreased the composite compressive and flexural strengths in comparison to the control mortar. The reason for this is that BS addition increased the porosity of matrix, while the strength of porous matrix was consequently weakened. The composite with 1% BS behaved like the control mortar. Only a 10.9% reduction in compressive strength and 7.3% bending strength resulted after 28 days. However, the composite with 7% BS showed a decrease of 66.7% and 50.5% in compression and bending, respectively. The BSC was more resistant to bending than to compression. The difference in strength reduction between the bending and compression was significant with an increase in particle proportion. Ren et al. [41] reported similar results for cement-blended bamboo charcoal with a particle size between 23 and 359 µm, and the mechanical strength decreased after particle addition. All designed specimens met the minimum strength required in Chinese specifications for masonry mortar [42], whereas the compressive strength exceeded 5 MPa. Besides, according to RILEM [43], concrete with a unit weight range of 1600–2000 kg/m^3 and a compressive strength minimum at 15 MPa can be classified as lightweight concrete and can be applied as a load bearing wall. In our case, BSC7 has a compressive strength lower than 15 MPa, which cannot apply to structural components, while both BSC3 and BSC5 are in the light weight concrete class. The density of BSC1 was superior than 2000 kg/m^3, classified as normal concrete.

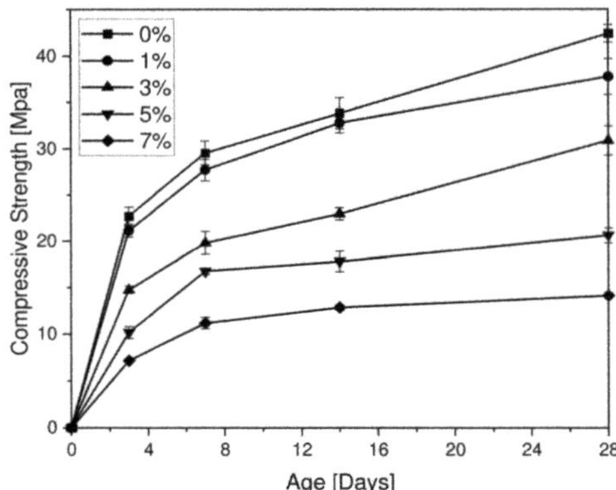

Figure 11. Compressive strength of alkali-treated BSC.

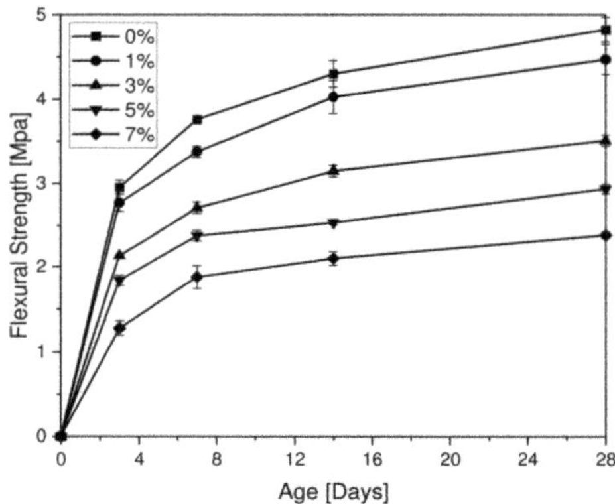

Figure 12. Flexural strength of alkali-treated BSC.

4. Conclusions

This work evaluated the potential of BS as a sustainable substitute for cementitious materials. Bamboo incorporation into mortar as a recycled byproduct of the bamboo industry has not been described previously in the literature. In this work, a new bio-based material that contained recycled BS (BSC) was prepared. The effect of different treatments and particle proportions on the physical, rheological, and mechanical properties of the BSC was evaluated. On the basis of the experimental results, the following conclusions were drawn:

1. Different BS pretreatment (cold, hot aqueous, and alkali) methods have no influence on the composite density and porosity. Unlike the physical properties, the addition of BS in the composite results in increasing the setting time and decreasing the workability and compressive and flexural strength compared with the control mortar.

2. Alkali-treated particles exhibited superior compatibility compared with other pretreatment methods. The SEM results showed a better microstructural interface between the alkali-treated BS and cement matrix. The mechanical behavior of the alkali-treated BSC was higher than that for the other two treatments. Hence, alkaline treatment is recommended in future work.

3. The BS content in the matrix affects the physical, rheological, and mechanical behavior of the composite. The replacement of BS with cement in the mortar makes the composite lighter and more porous. However, a greater particle content addition increases the setting time, reduces the slump, and the composite compressive and flexural strength

decrease. All BSC composite mortars satisfied the strength requirements for masonry mortar. BSC3 and BSC5 can be classified as load-bearing lightweight concrete.

Compared with traditional building materials, the cementitious composite with BS valorizes local waste, reduces cement consumption, and decreases the carbon emissions. The replacement of recycled BS with cement in the mortar yields a new lightweight and structural material. Further studies should focus on humidity control performance and building comfort regulations of this bio-based composite with recycled bamboo wastes.

Author Contributions: Conceptualization, Y.T. and M.L.; methodology, Y.T. and A.-O.S.; validation, Y.T. and A.K.; Investigation, Y.T and A.-O.S.; resources, J.Y.; data curation, A.-O.S.; writing—original draft preparation, Y.T.; writing—review and editing, Y.T., A.-O.S., M.L., A.K., and J.Y. All authors have read and agreed to the published version of the manuscript.

Funding: This research was funded by Natural Science Foundation of Zhejiang Province, grant number LY19E080002.

Institutional Review Board Statement: Not applicable.

Informed Consent Statement: Not applicable.

Data Availability Statement: Data is contained within the article.

Acknowledgments: The authors wish to thank CY Cergy Paris Université, L2MGC, for their technical support.

Conflicts of Interest: The authors declare no conflict of interest.

References

1. Gagg, C.R. Cement and concrete as an engineering material: An historic appraisal and case study analysis. *Eng. Fail. Anal.* **2014**, *40*, 114–140. [CrossRef]
2. Piazza, M.; Lobovikov, M.; Paudel, S.; Ren, H.; Wu, J. *World bamboo Resources—A Thematic study Prepared in the Framework of the Global Forest Resources Assessment 2005*; Food and Agriculture Organization of the United Nations: Rome, Italy, 2007.
3. Yadav, M.; Mathur, A. Bamboo as a sustainable material in the construction industry: An overview. *Mater. Today Proc.* **2021**, *43*, 2872–2876. [CrossRef]
4. Collet, F.; Chamoin, J.; Pretot, S.; Lanos, C. Comparison of the hygric behaviour of three hemp concretes. *Energy Build.* **2013**, *62*, 294–303. [CrossRef]
5. Dhakal, U.; Berardi, U.; Gorgolewski, M.; Richman, R. Hygrothermal performance of hempcrete for Ontario (Canada) buildings. *J. Clean. Prod.* **2017**, *142*, 3655–3664. [CrossRef]
6. Arnaud, L. Mechanical and thermal properties of hemp mortars and wools: Experimental and theoretical approaches. In Proceedings of the 3rd International Symposium on Bioresource hemp, Wolfburg, Germany, 13–16 September 2000.
7. Li, M.; Khelifa, M.; El Ganaoui, M. Mechanical characterization of concrete containing wood shavings as aggregates. *Int. J. Sustain. Built Environ.* **2017**, *6*, 587–596. [CrossRef]
8. Savastano, H.; Warden, P.G.; Coutts, R.S.P. Brazilian waste fibres as reinforcement for cement-based composites. *Cem. Concr. Compos.* **2000**, *22*, 379–384. [CrossRef]
9. Ramakrishna, G.; Sundararajan, T. Studies on the durability of natural fibres and the effect of corroded fibres on the strength of mortar. *Cem. Concr. Compos.* **2005**, *27*, 575–582. [CrossRef]
10. Galicia-Aldama, E.; Mayorga, M.; Arteaga-Arcos, J.C.; Romero-Salazar, L. Rheological behaviour of cement paste added with natural fibres. *Constr. Build. Mater.* **2019**, *198*, 148–157. [CrossRef]
11. Thomas, B.S.; Kumar, S.; Arel, H.S. Sustainable concrete containing palm oil fuel ash as a supplementary cementitious material—A review. *Renew. Sustain. Energy Rev.* **2017**, *80*, 550–561. [CrossRef]
12. Rodier, L.; Bilba, K.; Onésippe, C.; Arsène, M.-A. Utilization of bio-chars from sugarcane bagasse pyrolysis in cement-based composites. *Ind. Crop. Prod.* **2019**, *141*, 111731. [CrossRef]
13. Vaickelionis, G.; Vaickelioniene, R. Cement hydration in the presence of wood extractive and pozzolan mineral additives. *Ceram. Silik.* **2006**, *50*, 115.
14. Gowthaman, S.; Nakashima, K.; Kawasaki, S. A State-of-the-Art Review on Soil Reinforcement Technology Using Natural Plant Fiber Materials: Past Findings, Present Trends and Future Directions. *Materials* **2018**, *11*, 553. [CrossRef] [PubMed]
15. Frías, M.; Savastano, H.; Villar, E.; Sánchez de Rojas, M.I.; Santos, S. Characterization and properties of blended cement matrices containing activated bamboo leaf wastes. *Cem. Concr. Compos.* **2012**, *34*, 1019–1023. [CrossRef]
16. Xie, X.; Zhou, Z.; Yan, Y. Flexural properties and impact behaviour analysis of bamboo cellulosic fibers filled cement based composites. *Constr. Build. Mater.* **2019**, *220*, 403–414. [CrossRef]

17. Ramaswamy, H.S.; Ahuja, B.M.; Krishnamoorthy, S. Behaviour of concrete reinforced with jute, coir and bamboo fibres. *Int. J. Cem. Compos. Lightweight Concr.* **1983**, *5*, 3–13. [CrossRef]
18. Coutts, R.S.P. A review of Australian research into natural fibre cement composites. *Cem. Concr. Compos.* **2005**, *27*, 518–526. [CrossRef]
19. Fan, M.; Ndikontar, M.K.; Zhou, X.; Ngamveng, J.N. Cement-bonded composites made from tropical woods: Compatibility of wood and cement. *Constr. Build. Mater.* **2012**, *36*, 135–140. [CrossRef]
20. Walker, R.; Pavia, S.; Mitchell, R. Mechanical properties and durability of hemp-lime concretes. *Constr. Build. Mater.* **2014**, *61*, 340–348. [CrossRef]
21. Akinyemi, A.B.; Omoniyi, E.T.; Onuzulike, G. Effect of microwave assisted alkali pretreatment and other pretreatment methods on some properties of bamboo fibre reinforced cement composites. *Constr. Build. Mater.* **2020**, *245*, 118405. [CrossRef]
22. Das, M.; Chakraborty, D. Evaluation of improvement of physical and mechanical properties of bamboo fibers due to alkali treatment. *J. Appl. Polym. Sci.* **2008**, *107*, 522–527. [CrossRef]
23. Li, M.; Zhou, S.; Guo, X. Effects of alkali-treated bamboo fibers on the morphology and mechanical properties of oil well cement. *Constr. Build. Mater.* **2017**, *150*, 619–625. [CrossRef]
24. Agarwal, A.; Nanda, B.; Maity, D. Experimental investigation on chemically treated bamboo reinforced concrete beams and columns. *Constr. Build. Mater.* **2014**, *71*, 610–617. [CrossRef]
25. Ahmed, W.; Khushnood, R.A.; Memon, S.A.; Ahmad, S.; Baloch, W.L.; Usman, M. Effective use of sawdust for the production of eco-friendly and thermal-energy efficient normal weight and lightweight concretes with tailored fracture properties. *J. Clean. Prod.* **2018**, *184*, 1016–1027. [CrossRef]
26. Sales, A.; de Souza, F.R.; dos Santos, W.N.; Zimer, A.M.; do Couto Rosa Almeida, F. Lightweight composite concrete produced with water treatment sludge and sawdust: Thermal properties and potential application. *Constr. Build. Mater.* **2010**, *24*, 2446–2453. [CrossRef]
27. NF EN 196-1. *Méthodes D'essais des Ciments*; Association Francaise de Normalisation: Paris, France, 1990.
28. Monreal, P.; Mboumba-Mamboundou, L.B.; Dheilly, R.M.; Quéneudec, M. Effects of aggregate coating on the hygral properties of lignocellulosic composites. *Cem. Concr. Compos.* **2011**, *33*, 301–308. [CrossRef]
29. NF P15-431. *Liant Hydraulique-Technique des Essais-Détermination du Temps de Prise sur Mortier Normal*; AFNOR Editions: Saint-Denis, France, 1994.
30. Schwartzentruber, A.; Catherine, C. La méthode du mortier de béton équivalent (MBE)—Un nouvel outil d'aide à la formulation des bétons adjuvantés. *Mat. Struct.* **2000**, *33*, 475–482. [CrossRef]
31. Laboratoire Central des Ponts et Chaussée. *Maîtrise de la Durabilité des Ouvrages D'art en Béton: Application de L'approche Performantielle*; Techniques et Méthodes; LCPC: Paris, France, 2010.
32. Mwaikambo, L.Y.; Ansell, M.P. Chemical modification of hemp, sisal, jute, and kapok fibers by alkalization. *J. Appl. Polym. Sci.* **2002**, *84*, 2222–2234. [CrossRef]
33. Ray, D.; Sarkar, B.K. Characterization of alkali-treated jute fibers for physical and mechanical properties. *J. Appl. Polym. Sci.* **2001**, *80*, 1013–1020. [CrossRef]
34. Li, X.; Tabil, L.G.; Panigrahi, S. Chemical Treatments of Natural Fiber for Use in Natural Fiber-Reinforced Composites: A Review. *J. Polym. Environ.* **2007**, *15*, 25–33. [CrossRef]
35. Braga Costa Santos, E.; Moreno, C.; Barros, J.; Moura, D.; Fim, F.; Ries, A.; Wellen, R.; Silva, L. Effect of alkaline and hot water treatments on the structure and morphology of piassava fibers. *Mater. Res.* **2018**, *21*. [CrossRef]
36. Hamidon, M.H.; Sultan, M.T.H.; Ariffin, A.H.; Shah, A.U.M. Effects of fibre treatment on mechanical properties of kenaf fibre reinforced composites: A review. *J. Mater. Res. Technol.* **2019**, *8*, 3327–3337. [CrossRef]
37. Vo, L.T.T.; Navard, P. Treatments of plant biomass for cementitious building materials—A review. *Constr. Build. Mater.* **2016**, *121*, 161–176. [CrossRef]
38. Jo, B.W.; Chakraborty, S.; Kim, H. Efficacy of alkali-treated jute as fibre reinforcement in enhancing the mechanical properties of cement mortar. *Mater. Struct.* **2016**, *49*, 1093–1104. [CrossRef]
39. Zhang, H. Cement. In *Building Materials in Civil Engineering*; Zhang, H., Ed.; Woodhead Publishing Series in Civil and Structural Engineering: Southston, UK, 2011; pp. 46–423. ISBN 978-1-84569-955-0.
40. Xie, X.; Zhou, Z.; Jiang, M.; Xu, X.; Wang, Z.; Hui, D. Cellulosic fibers from rice straw and bamboo used as reinforcement of cement-based composites for remarkably improving mechanical properties. *Compos. Part B Eng.* **2015**, *78*, 153–161. [CrossRef]
41. Ren, Q.; Zeng, Z.; Jiang, Z.; Chen, Q. Incorporation of bamboo charcoal for cement-based humidity adsorption material. *Constr. Build. Mater.* **2019**, *215*, 244–251. [CrossRef]
42. JGJT 98-2010. *Specification for Mix Proportion Design of Mansonary Mortar*; Ministry of Housing and Urban-Rural Development of the People: Beijing, China, 2011.
43. RILEM. *Recommandations: Functional Classification of Lightweight Concretes*; RILEM Publications SARL: Bagneux, France, 1978.

crystals

Article

The Influence of Fly Ash Content on the Compressive Strength of Cemented Sand and Gravel Material

Qihui Chai [1,2], Fang Wan [1,2,*], Lingfeng Xiao [1] and Feng Wu [1]

[1] School of Water Resources, North China University of Water Resources and Electric Power, Zhengzhou 450046, China; chaiqihui@ncwu.edu.cn (Q.C.); xlf19980616@163.com (L.X.); wufeng@ncwu.edu.cn (F.W.)
[2] Henan Key Laboratory of Water Resources Conservation and Intensive Utilization in the Yellow River Basin, Zhengzhou 450046, China
* Correspondence: wanxf1023@163.com

Abstract: Cemented sand and gravel (CSG) material is a new type of dam material developed on the basis of roller compacted concrete, hardfill, and ultra-poor cementing materials. Its main feature is a wide range of sources of aggregate (aggregate is not screened but by simply removing the large particles it can be fully graded on the dam filling) and low amounts of cementitious materials per unit volume. This dam construction material is not only economical and practical, but also green and environmentally friendly. There are many factors affecting the mechanical properties of CSG materials, such as aggregate gradation, sand ratio, water content, water–binder ratio, fly ash content, admixture content, etc. Based on the existing research results of the team, this paper focuses on the influence of fly ash content on the compressive strength of CSG materials. Through a large number of laboratory measured data, we found: (1) The compressive strength law of materials at different ages; the compressive strength of CSG material at age 90 d is generally 10%~30% higher than that at 28 d, and it is proposed that 90 d or 180 d strength should be used as the design strength in the design of CSG material dam; (2) There is an optimal value of fly ash content in CSG materials: when the fly ash content is 50% of the total amount of cementitious materials (cement + fly ash), the fly ash content is defined as the optimal content, and the test data are verified by regression analysis. The discovery of an 'optimal dosage' of fly ash provides an important reference for the design and construction of CSG dams.

Keywords: cementitious gravel; fly ash; age; optimal dosage

Citation: Chai, Q.; Wan, F.; Xiao, L.; Wu, F. The Influence of Fly Ash Content on the Compressive Strength of Cemented Sand and Gravel Material. *Crystals* **2021**, *11*, 1426. https://doi.org/10.3390/cryst11111426

Academic Editors: Peter Taylor, Yifeng Ling, Chuanqing Fu, Peng Zhang and Tomasz Sadowski

Received: 23 October 2021
Accepted: 18 November 2021
Published: 21 November 2021

Publisher's Note: MDPI stays neutral with regard to jurisdictional claims in published maps and institutional affiliations.

1. Introduction

Cemented sand and gravel (CSG) is a new material for dam construction works, produced by adding cementitious materials and water to easily accessible rock based material, including sand and gravel at the river bed or excavation muck near the dam site, and mixing them with simple equipment and process. The cemented sand gravel dam (CSGD) built on CSG material has the advantages of both a roller compacted concrete dam (RCC dam) and a rockfill dam. Compared with an RCC dam, it demands less cement, with simplified aggregate preparation and mixing facilities. Moreover, temperature control measures are not necessary in the construction. As a result, it can effectively speed up the construction and lower the project cost. Compared with rockfill dams, a CSG dam requires a significantly lower amount of engineering works, but has batter capacity to withstand seepage deformation and scouring. Through the utilization of waste materials, CSG technology lowers the demand for artificial materials and high-quality aggregate, and thus promotes the efficient use of resources, less destruction of land vegetation, and a lower impact on the natural environment. Therefore, CSGD is a new kind of easily constructed dam that is economic, safe, low carbon, and eco-friendly.

In recent years, the relationship between dams and the natural environment has attracted increasingly public attention. It has become a trend of dam technology development to strike a balance between the low-cost and efficient construction of modern dams as well as a lower impact on the natural environment. As the basis and carrier of water resources and hydropower development, reservoir dams play a major role in the comprehensive utilization of water and hydropower resources. Their position will be further strengthened for the sustainable social and economic development of China, to which the CSGD development is the key technology.

According to statistics, dozens of CSGDs have been built around the world since the 1980s. Japan, Greece, Dominica, the Philippines, Pakistan, and Turkey are among countries that have already carried out engineering explorations and applied the technology [1–7]. China began the study of CSGD at the end of last century. By the beginning of this century, many colleges and research institutes have made extensive studies and discussions on the property, constitutive model, calculation and analysis methods, design principles, and standards of CSG material. The representative researchers include Jia Jinsheng, Zheng Cuiying et al. of China Institute of Water Resources and Hydropower Research [8–11], Sun Mingquan et al. of North China University of Water Resources and Electric Power [12–16], He Yunlong et al. of Wuhan University [17–20], and Cai et al. [21,22] of Hohai University. The previous studies believe that as CSG material only uses a small amount of cement, fly ash can be added in large quantities to improve its strength. However, the specific increased value of material strength and the optimal content of fly ash are not fully discussed. Therefore, the primary purpose of this study is to examine the influence of fly ash content on the compressive strength of CSG material. This quantitative examination provides various crucial insights for building dams of this type in the future.

2. Test Design

2.1. Test Materials

CSG material is composed of water, sand (fine aggregate), stone (coarse aggregate), and cementitious materials (cement and fly ash). The proportions of them can vary. In this test, (1) water: Zhengzhou tap water; (2) Sand: Natural river sand in Ruzhou County section of Ruhe River was adopted, the fineness modulus FM = 2.57, in accordance with the requirements of the Technical Guidelines for Damming with Cemented Granular Materials (SL 678-2014) that " ... in natural materials, the fineness modulus of sand should range from 2.0 to 3.3". (3) Stone: to study the effect of aggregate gradation on the strength of CSG material, natural graded aggregates in Ruzhou County section of Ruhe River were used. After artificial screening, aggregates were divided into 5–20 mm, 20–40 mm, 40–80 mm, 80–150 mm and distributed in silos, as illustrated in Table 1. (4) Cementitious materials: the 425# ordinary Portland cement (P.O.42.5) was produced by Henan Duoyangda Cement Co., Ltd., Zhengzhou, China, which met the requirements for cement in the CSG material in the Technical Guidelines for Damming with Cemented Granular Materials (SL 678-2014) that "all Portland cement series conforming to GB175 and GB200 can be used to build dams with cemented granular materials. When mineral admixtures such as fly ash are added into cementitious materials, Portland cement, ordinary Portland cement, medium or low heat Portland cement should be preferentially selected." The dry discharged F Class II fly ash from Zhengzhou Thermal Power Plant was used in the test. The performance index is shown in Table 2.

Table 1. Aggregate gradation.

Gradation	Coarse Aggregate/%					Fine Aggregate/%
	5–20 mm	20–40 mm	40–80 mm	80–150 mm	>150 mm	<5 mm
Proportion	22.91	36.52	23.09	5.75	8.92	2.81

Table 2. Performance index of fly ash.

Apparent Density/(g/cm^3)	45 μm Sieving Residue (%)	Water Demand (%)	Chemical Composition (%)				
			SiO_2	Fe_2O_3	Al_2O_3	CaO	Loss of Burning
2.11	17	102	59.61	7.41	21.33	4.24	1.78

2.2. Mix Proportion in the Test

In this test, "weight method" was used in a mixed proportion design. According to the Technical Guidelines for Damming with Cemented Granular Materials (SL 678-2014) and previous research [23–30], the apparent density of CSG material was selected as 2350 kg/m^3 (the apparent density was checked after the specimen test. The samples were all about 2350 kg/m^3, and the maximum fluctuation was within $\pm 2\%$). According to the research results, factors such as water–binder ratio, sand ratio, and aggregate gradation have great influence on the material strength, and all factors have optimal values. This paper mainly focused on the influence of fly ash content on the strength of CSG material, so the proportion of other materials followed the reference of previous studies with the water–binder ratio at 1.0, sand ratio at 20%, coarse aggregate ranging from 20–40 mm, and 5–20 mm, mass ratio at 6:4, and cement content of 50 kg/m^3 and 60 kg/m^3. The fly ash content was designed to be 0 kg/m^3, 20 kg/m^3, 30 kg/m^3, 40 kg/m^3, 50 kg/m^3, 60 kg/m^3, 70 kg/m^3, 80 kg/m^3, 90 kg/m^3, and 100 kg/m^3, respectively as the admixture for cementitious materials. The designed mix proportion is shown in Table 3.

Table 3. Mix proportion of CSG material.

Sample Code	Volume of Material Per Unit Volume/(kg/m^3)					
	C	F	W	S	NA1 (20–40 mm)	NA2 (5–20 mm)
C50F0		0	50	450	1080	720
C50F20		20	70	442	1061	707
C50F30		30	80	438	1051	701
C50F40		40	90	434	1042	694
C50F50	50	50	100	430	1032	688
C50F60		60	110	426	1022	682
C50F70		70	120	422	1013	675
C50F80		80	130	418	1003	669
C50F90		90	140	414	994	662
C50F100		100	150	410	984	656
C60F0		0	60	446	1070	714
C60F20		20	80	438	1051	701
C60F30		30	90	434	1042	694
C60F40	60	40	100	430	1032	688
C60F50		50	110	426	1022	682
C60F60		60	120	422	1013	675
C60F80		80	140	414	994	662
C60F100		100	160	406	974	650

Note: Volume of material per unit volume/(kg/m^3); C, cement; F, fly ash; W, water demand; S, sand; NA1, coarse aggregate (20–40 mm); and NA2, coarse aggregate (5–20 mm).

2.3. Specimen Preparation

Considering that no standard CSG material test exists, and the composition of CSG is similar to concrete, this test followed the Test Code for Hydraulic Concrete (SL352-2006). As the construction and dam filling of CSGDs home and abroad have all adopted roller compaction, the CSG specimens in this paper were also formed through rolling compacted concrete, with the upper part of the specimens compacted and the lower part vibrated. After the vibration, the surface of specimens was smoothed and covered with a film (to prevent water evaporation). Before demolding, they were placed in a room with the temperature of 20 °C $\pm$ 5 °C for 48 h [31–34]. After demolding, the specimens were sent to the standard maintenance room for maintenance lasting 28 d and 90 d. Eighteen groups were formed, with eight cubic specimens of 150 mm $\times$ 150 mm $\times$ 150 mm in each group. The preparation process is shown in Figure 1.

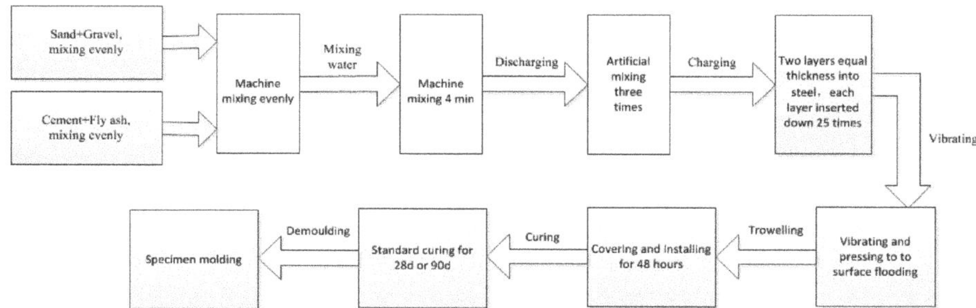

Figure 1. Flow chart of CSG sample preparation.

Preparation process of cemented sand and gravel material: Premixed stone and sand weighed according to test mix ratio, adding cement and fly ash, mixing evenly, and then adding water to mix, and finally mixing evenly and charging and vibrating. After the vibration is completed, the surface of the specimen is smoothed, and the film is covered (to prevent water evaporation). After standing at 20 °C $\pm$ 5 °C for 48 h, the demolded specimen is sent to the standard maintenance room for maintenance until the test age.

3. Test Results and Analysis

3.1. Determination of Compressive Strength

The cubic compressive strength of CSG material was followed the Test Rules for Hydraulic Concrete (SL352-2006).

In this test, four specimens were prepared as one group. After testing, the largest discrete data were removed and the other three groups were averaged. The test results were in accordance with the Rules.

3.2. Analysis of the Test Data

3.2.1. Effect of Fly Ash Content on the Compressive Strength of CSG Material

Fly ash is a type of artificial pozzolanic material made of silica or silica-alumina. It possesses minimal or no cementing value. However, in the presence of water, fly ash powder will react with $Ca(OH)_2$ at room temperature to form a compound with cementing properties. The pozzolanic activity of fly ash is reflected by its cementing.

The test results are shown in Figure 2 for the analysis of the relationship between age, fly ash content, and strength.

As detailed in Figure 2, the rules can be found as follows: (1) the compressive strength of CSG increases with age. The compressive strength of CSG at 90 d is generally 10–30% higher than that at 28 d. (2) When the cement content is 50 kg/m^3 and 60 kg/m^3, respectively, the compressive strength of CSG increases with the addition of fly ash content. When the cement content is 50 kg/m^3, the increase of compressive strength at 28 d is greater than that at 90 d. When the cement content is 60 kg/m^3, the increase of compressive strength at 90 d is greater than that at 28 d [35].

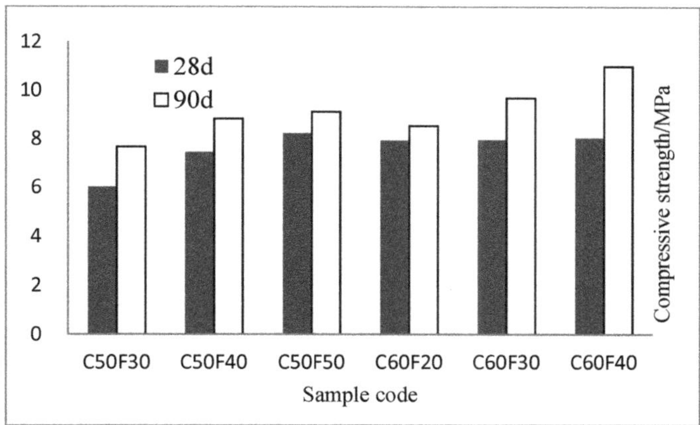

Figure 2. The relationship between the compressive strength of CSG material and the age/the content of fly ash.

The test results can be explained with two-phase hydration reaction of the fly ash-cement system. First, the induction stage, during which the soluble ions on the surface of fly ash particles dissolve, which affects the nucleation of $Ca(OH)_2$ and C-S-H hydrates and retards the initial level of C_3A in cement. Meanwhile, the hydration of cement is delayed, as the fine size of fly ash particles leads to their easy adhesion to the surface of cement particles under physical action. In addition, $Ca(OH)_2$ and C-S-H produced by cement hydration wrap the surface of fly ash and prevent the hydration of fly ash particles. Therefore, the fly ash with early age has low self-activity. Second, the acceleration stage, during which $Ca(OH)_2$ and C-S-H begin to nucleate and grow. Fly ash particles provide additional accommodation for the precipitation of C-S phase hydrates in cement through many new surfaces. In this way, fewer C-S phase hydrates will precipitate on the surface of cement particles, and the dissolution of C_3S will accelerate. As a result, the hydration of cement is promoted. Therefore, during the second phase, the presence of fly ash facilitates the hydration of cement. Moreover, due to the nucleation of cement hydration products, the permeability of fly ash coating increases and leads to faster hydration of the fly ash. As fly as hydration consumes more $Ca(OH)_2$, the hydration of cement is promoted further. During this stage, cement and fly ash promote and accelerate each other's hydration at the same time. This phenomenon become obvious with the development of age, which is highlighted by the significant increase of strength of CSG material at the later stage. Therefore, considering the gradual development of material strength, it is appropriate to choose 90 d or 180 d strength as the design strength of CSG dam.

3.2.2. Optimal Fly Ash Content of CSG Material

Most hydraulic projects will add fly ash to cementitious materials, and the RCC dam has the highest proportion of fly ash addition (70%) (Jiangya Hydropower Station). Technical Guidelines for Damming with Cemented Granular Materials (SL678-2014) also proposes including fly ash into cement gravel and sand material for dam construction, but without mentioning the most optimal and economical fly ash content.

During the test, prepared the cement content of 50 kg/m^3 and 60 kg/m^3, added fly ash of 20 kg/m^3, 30 kg/m^3, 40 kg/m^3, 50 kg/m^3, 60 kg/m^3, 80 kg/m^3, or 100 kg/m^3, respectively, into the cement and waited for 90 d maintenance to make CSG materials to see the difference of their compressive strengths, with results drawn in the figure below. According to the figure, as the fly ash content increases, the compressive strength reaches a peak before falling down to a level. The CSG material test proves that an optimal content of fly ash could be found. The compressive strength of materials will reach its maximum when both cement and fly ash contents stand at 50 kg/m^3 or 60 kg/m^3, as highlighted in Figure 3.

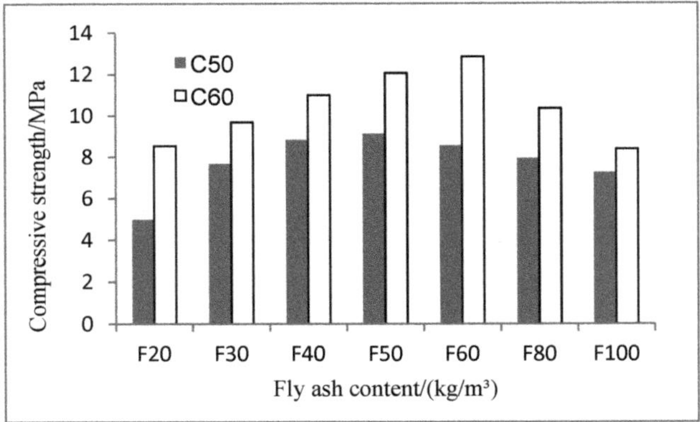

Figure 3. Optimal content of fly ash of CSG material.

According to our analysis, as the design of mixed proportions in this test is based on previous research results, the total water consumption of the mixture is fixed by selecting a water–binder ratio of 1.0 and sand ratio of 20%. [35,36] At the initial stage, the strength of CSG material increases with the increase of fly ash content and the water amount is enough to support the hydration of cement and fly ash. The increase of fly ash facilitates C-S-H cementing; when all the water is consumed during the hydration of cement and fly ash, the fly ash added will only serve as a filler, which fill in the concrete voids or attach to the surface of cement particles. At this stage, more fly ash will affect the formation of Ca(OH)$_2$, C-S-H, which leads to the declining strength of CSG materials. Therefore, there is an optimal fly ash content in CSG material, and it is defined to be 50% of the contentious material (cement + fly ash).

4. Numerical Regression Analysis

Regression analysis is an important branch of mathematical statistics and an important statistical tool to study the correlation between variables. It is well applied in many areas, such as seeking empirical formulas or establishing mathematical models.

According to the test, fly ash content has significant influence on the compressive strength of CSG material. Based on the theory of linear regression model, Formula (1) is built to reflect the statistical relationship between 90 d compressive strength and fly ash content in CSG material:

$$X_F = a_f x_f^2 + b_f x_f + c_f \tag{1}$$

where a_f, b_f, and c_f are statistical constants.

After calculation, the regression equation parameters of fly ash content and compressive strength are obtained and shown in Table 4.

As illustrated in Table 4, the fitting correlation coefficients are greater than 0.90 and the standard errors are less than 0.58 in all cases. The fitting strength of each regression equation and measured values have demonstrated excellent correlation with regard to 90 d compressive strength of CSG material, with small deviation of regression values and measured values, and high fitting accuracy. The comprehensive analysis shows that the test results suit the mathematical model of linear regression.

Through numerical analysis, the maximum value of compressive strength under different fly ash content can be obtained, where x represents the optimal fly ash content and y represents the maximum compressive strength of CSG material, as detailed in Table 4.

The numerical analysis shows that when (1) 50 kg/m^3 of cement content is combined with 51 kg/m^3 of fly ash content, or (2) the cement and fly ash contents are 60 kg/m^3 and 59 kg/m^3, respectively, the compressive strength of CSG material is the highest in 90 d. Therefore, when fly ash content: cement content $\approx$ 1:1 or when the fly ash content is 50% of cementitious materials (cement + fly ash), the "optimal content" is achieved. The "optimal fly ash content" of CSG material was verified again through numerical analysis.

Table 4. Regression analysis parameter table.

Cement/ (kg/m^3)	Fly Ash/(kg/m^3)	Compressive Strength/(MPa)	Fitting Strength/ (MPa)	Absolute Error	Correlation Coefficient	Standard Error	a_f	b_f	c_f	Optimal Fly Ash Content/ (kg/m^3)	Corresponding Strength/ (MPa)
50	30	7.94	8.236	−0.3	0.9	0.39	−0.00082	0.09	6.273	51	8.22
	40	8.82	8.563	0.26							
	50	9.11	8.726	0.38							
	60	8.53	8.726	−0.2							
	80	7.93	8.233	−0.3							
	100	7.24	7.086	0.15							
60	20	8.53	8.35	0.18	0.96	0.58	−0.0024	0.288	3.567	59	12.11
	30	9.68	10.015	−0.34							
	40	10.99	11.198	−0.21							
	50	12.05	11.896	0.15							
	60	12.83	12.111	0.72							
	80	10.34	11.089	−0.75							
	100	8.37	8.132	0.24							

5. Conclusions

As a new type of eco-friendly material for dam construction, CSG is a low-cost material as it requires a lower amount of cement content and higher fly ash content than traditional ones. This paper briefly introduced the steps of CSG material test. The influence of fly ash content on the strength of CSG material was studied from two perspectives—test work and numerical analysis. The conclusions are as follows:

(1) The compressive strength of CSG increases with age. The compressive strength of CSG at 90 d is generally higher by 10–30% compared with that at 28 d. After taking into account the gradual development of material strength, 90 d or 180 d strength is recommended as the design strength for the CSG dam.

(2) Test work and numerical analysis have verified that an optimal fly ash content for CSG material exists, and is defined as 50% of the total cementitious material (cement + fly ash). The results are of material significance in the solution of problems during the design and construction of CSG dam.

(3) In the new material of CSG, the gravel material of natural river is selected to study the tensile, compressive, and shear tests under different particle gradation, sand rate, water–cement ratio, and cementitious material dosage, and the tensile, compressive, and shear strength indexes are obtained. This paper mainly takes the content of fly ash as an example to study the compressive strength characteristics of materials. In the follow-up study, the mechanical properties of materials such as compressive strength, tensile strength, and shear strength under multiple factors will be closely analyzed.

Author Contributions: All authors contributed to the study conception and design. Q.C. performed all experiments and analysis along with data collection and discussion of the results. F.W. (Fang Wan) wrote the manuscript. L.X. edited the manuscript. F.W. (Feng Wu) reviewed the manuscript. All authors have read and agreed to the published version of the manuscript.

Funding: The research was supported by Central Plains Science and Technology Innovation Leading Talent Funding Project (194200510008); Major Science and Technology Special Projects in Henan Province (201300311400).

Data Availability Statement: The data used to support the findings of this study are included within the article.

Conflicts of Interest: We declare that we have no conflict of interest or the authors do not have any possible conflicts of interest.

References

1. Hirose, T.; Fujisawa, T.; Kawasaki, H.; Kondo, M.; Hirayama, D.; Sasaki, T. Design Criteria for Trapezoid-Shaped CSG Dams. In Proceedings of the ICOLD 69th Annual Meeting, Dresden, Germany, 13 September 2001.
2. Hirose, T.; Fujisawa, T.; Yoshida, H. Concept of CSG and Its Material Properties. In Proceedings of the 4th international symposium on RCCD, Madrid, Spain, 17–19 November 2003; pp. 465–473.
3. Fujisawa, T.; Nakamura, A.; Kawasaki, H.; Hirayama, D.; Yamaguchi, Y.; Sasaki, T. Material Properties of CSG for the Seismic Design of Trapezoid-Shaped CSG Dam. In Proceedings of the 13th World Conference on Earthquake Engineering, Vancouver, BC, Canada, 1–6 August 2004; pp. 391–394.
4. Tokmechi, Z. Structural safety studies of Klahir dam in Iran. *Middle East J. Sci. Res.* **2010**, *6*, 500–504.
5. Kongsukprasert, L.; Tatsuoka, F.; Tateyama, M. Several factors affecting the strength and deformation characteristics of cement-mixed gravel. *Soils Found.* **2005**, *45*, 107–124. [CrossRef]
6. Kongsukprasert, L.; Sano, Y.; Tatsuoka, F. Compaction-Induced Anisotropy in the Strength and Deformation Characteristics of Cement-Mixed Gravelly Soils. In *Soil Stress-Strain Behavior, Measurement, Modeling and Analysis—Geotechnical Symposium in Roma*; Springer: Berlin/Heidelberg, Germany, 2007; pp. 479–490.
7. Lohani, T.N.; Kongsukprasert, L.; Watanabe, K.; Tatsuoka, F. Strength and Deformation Properties of Compacted Cement-Mixed Gravel Evaluated by Triaxial Compression Tests. *Soils Found.* **2004**, *44*, 95–108. [CrossRef]
8. Jia, J.S.; Ma, F.L.; Li, X.Y.; Chen, Z.P. Study on material characteristics of cement-sand-gravel dam and engineering application. *J. Hydraul. Eng.* **2006**, *37*, 578–582. [CrossRef]
9. Jia, J.S.; Chen, Z.P.; Ma, F.L. *Study on Material Characteristics of Cemented Sand and Gravel Dam and Its Effect on Face-Slab Seepage-Proof Body*; China Water Conservancy Hydropower Science Academe: Beijing, China, 2004.
10. Jia, J.S.; Liu, N.; Zheng, C.Y.; Ma, F.L.; Du, Z.K.; Wang, Y. Research on progress and engineering application of cemented granular material dam. *J. Hydraul. Eng.* **2016**, *47*, 315–323.
11. Feng, W.; Jia, J.S.; Ma, F.L. Study on durability of cemented gravel dam materials and development of new protective materials. *J. Hydraul. Eng.* **2013**, *4*, 500–504.
12. Li, Y.L.; Hou, J.K.; Sun, M.Q.; Xing, Z.X.; Chen, Y.; Li, M.Z. Test and Study on Mechanical Property of Super-Short Cement Concrete. *Yellow River* **2007**, *29*, 59–60.
13. Sun, M.Q.; Yang, S.F.; Zhang, J.J. Constitutive model of super-short cement concrete. *Technol. Water Resour.* **2007**, *27*, 35–37.
14. Sun, M.Q.; Meng, X.M.; Xiao, X.C. Study on profile form of super-short cement concrete. *Technol. Water Resour.* **2007**, *27*, 40–41, 45.
15. Sun, M.Q.; Peng, C.S.; Chen, J.H. Nonlinear analysis of super-short cement concrete. *Technol. Water Resour.* **2007**, *27*, 42–45.
16. Sun, M.Q.; Peng, C.S.; Li, Y.L. Triaxial test of super-short cement concrete. *Technol. Water Resour.* **2007**, *27*, 46–49.
17. Peng, Y.F.; He, Y.L.; Wan, B. Hardfill Dam: A new concept of roller compacted concrete dam. *Water Power* **2008**, *34*, 61–63.
18. He, Y.L.; Liu, J.L.; Li, J.C. Study on stress-strain characteristics and constitutive model of hardfill dam. *J. Sichuan Univ. (Eng. Sci. Ed.)* **2011**, *6*, 40–47.
19. He, Y.L.; Zhang, S.H.; Shi, X.R. Seismic characteristics of cemented gravel dam and calculation method of its seismic action. *J. Hydraul. Eng.* **2016**, *47*, 589–598.
20. Zhang, S.H.; He, Y.L.; Sun, W. Seismic behavior of Shoukoupu cemented gravel dam. *Eng. J. Wuhan Univ.* **2016**, *2*, 193–200.
21. Cai, X.; Wu, Y.; Guo, X.; Ming, Y. Research review of the cement sand and gravel (CSG) dam. *Front. Struct. Civ. Eng.* **2012**, *6*, 19–24. [CrossRef]
22. Cai, X.; Yang, J.; Guo, X.W. Summary of research on cemented gravel dam. *J. Hehai Univ. (Nat. Sci.)* **2015**, *43*, 431–441.
23. Yang, H.C. *Structural Design and Engineering Application of Cemented Sand and Gravel Dam*; China Water Conservancy Hydropower Science Academe: Beijing, China, 2013.
24. Sun, M.Q.; Yang, S.F.; Chai, Q.H. Study on foundation theory of cemented sand and gravel dam. *J. N. China Univ. Water Resour. Electr. Power (Nat. Sci. Ed.)* **2014**, *35*, 43–46.
25. Chen, C.L. Discussion on construction concrete from the mix proportion design angle of hydraulic concrete. *Concrete* **2015**, *11*, 91–94.
26. Fu, H.; Chen, S.S.; Han, H.Q.; Ling, H.; Cai, X. Static and dynamic triaxial shear tests of cemented sand and gravel. *Chin. J. Geotech. Eng.* **2015**, *2*, 357–362. [CrossRef]
27. Li, Z.L.; Deng, C.L.; Zhang, G.H. Prediction model of effective elastic modulus of concrete considering aggregate gradation. *J. Hydraul. Eng.* **2016**, *47*, 575–581. [CrossRef]
28. Zhu, X.L.; Ding, J.T.; Cai, Y.B.; Bai, Y. Experimental study on strength and elastic modulus of cemented sand and gravel. *Yellow River* **2016**, *28*, 126–128.
29. Guo, L.; Zhang, Y.; Zhong, L.; Wang, M.; Zhu, X. Study on Macroscopic and Mesoscopic Mechanical Behavior of CSG based on Inversion of Mesoscopic Material Parameters. *Sci. Eng. Compos. Mater.* **2020**, *27*, 280. [CrossRef]
30. Guo, L.; Zhang, Y.; Zhong, L.; Zhang, F.; Wang, M.; Li, S. CSG Elastic Modulus Model Prediction Considering Meso-components and its Effect. *Sci. Eng. Compos. Mater.* **2020**, *27*, 264–271. [CrossRef]

31. Cheng, C.; Liu, Z.; Guo, Z.; Debnath, N. Waste-to-energy policy in China: A national strategy for management of domestic energy reserves. *Energy Sources Part B Econ. Plan. Policy* **2017**, *12*, 925–929. [CrossRef]

32. Liu, Z.; Baghban, A. Application of LSSVM for biodiesel production using supercritical ethanol solvent. *Energy Sources Part A Recover. Util. Environ. Eff.* **2017**, *39*, 1869–1874. [CrossRef]

33. Peng, W.; Wang, L.; Mirzaee, M.; Ahmadi, H.; Esfahani, M.; Fremaux, S. Hydrogen and syngas production by catalytic biomass gasification. *Energy Convers. Manag.* **2017**, *135*, 270–273. [CrossRef]

34. Peng, W.; Ge, S.; Ebadi, A.G.; Hisoriev, H.; Esfahani, M.J. Syngas production by catalytic co-gasification of coal-biomass blends in a circulating fluidized bed gasifier. *J. Clean. Prod.* **2017**, *168*, 1513–1517. [CrossRef]

35. Chai, Q.H.; Yang, S.F.; Sun, M.Q. Study on the Influence Factors of Compressive Strength of CSG Material. *Yellow River* **2016**, *7*, 86–88.

36. Sun, M.Q.; Yang, S.F.; Tian, Q.Q. Summary of research on mechanical properties, durability and dam type of cemented sand and gravel material. *Yellow River* **2016**, *7*, 83–85.

crystals

Article

Effect of Different NaOH Solution Concentrations on Mechanical Properties and Microstructure of Alkali-Activated Blast Furnace Ferronickel Slag

Zongxian Huang [1,2], Yuqi Zhou [1] and Yong Cui [2,*]

[1] China Construction First Group Construction and Development Co., Ltd., Beijing 100102, China; huang-zx18@mails.tsinghua.edu.cn (Z.H.); zhouyuqi@chinaonebuild.com (Y.Z.)
[2] Department of Civil Engineering, Tsinghua University, Beijing 100084, China
* Correspondence: cuiyong20@mails.tsinghua.edu.cn; Tel.: +86-186-1082-8446

Abstract: Blast furnace ferronickel slag (BFFS) is a kind of industrial solid waste that has not been effectively utilized in construction industry. The effects of different NaOH concentrations on the mechanical properties and microstructure of alkali-activated blast furnace ferronickel slag were investigated in this study. The results show that an optimal concentration for compressive strength is found, both higher and lower concentrations cause strength degradation. The pore structure, phase composition and hydration heat revealed that less C-A-S-H gel is produced at low concentration and result in the low compressive strength The phase composition and hydration heat revealed that more hydrotalcite is produced than C-A-S-H at high concentration due to more violent reaction at the early age hinders the growth of C-A-S-H in the later stage. FT-IR also shows that high concentration decreased the Al/Si ratio and polymerization of C-A-S-H which also leads to the low strength at high concentration.

Keywords: blast furnace ferronickel slag; alkali-activated material; compressive strength; dosage of activator

Citation: Huang, Z.; Zhou, Y.; Cui, Y. Effect of Different NaOH Solution Concentrations on Mechanical Properties and Microstructure of Alkali-Activated Blast Furnace Ferronickel Slag. *Crystals* **2021**, *11*, 1301. https://doi.org/10.3390/cryst11111301

Academic Editors: George Z. Voyiadjis, Yifeng Ling, Chuanqing Fu, Peng Zhang and Peter Taylor

Received: 15 September 2021
Accepted: 23 October 2021
Published: 26 October 2021

Publisher's Note: MDPI stays neutral with regard to jurisdictional claims in published maps and institutional affiliations.

1. Introduction

Ferronickel slag is a byproduct of ferronickel alloy production. Approximately 0.64 million tons of nickel–iron alloys are produced globally each year [1]. China accounts for 48% of all ferronickel alloy production. According to different ferronickel alloy production methods, ferronickel slag can be classified as electric furnace ferronickel slag (EEFS) or blast furnace ferronickel slag (BFFS). The electric furnace method is the main method for producing ferronickel alloys and is used worldwide [2]. With the lack of nickel-rich minerals and the demand for ferronickel alloys in China, the blast furnace method is still the main production method [3]. The chemical and mineral composition of BFFS and EFFS are different with different production methods. EFFS is mainly composed of SiO_2, Fe_2O_3 and MgO; BFFS is mainly composed of CaO, Al_2O_3 and SiO_2. The main minerals in EFFS are crystalline substances such as forsterite and diopside; BFFS is mainly composed of an amorphous phase [4]. The resource utilization of EFFS has been extensively studied by many researchers. EFFS can be used as a supplementary cementitious material [5–8], as an aggregate [9–11] in concrete, and as a precursor in geopolymer synthesis [12,13]. Compared with EFFS, there are relatively few studies on BFFS.

Alkali-activated materials have recently become popular as green building materials in cement. Alkali-activated materials use solid waste as a precursor in the preparation of cementitious paste, effectively transforming solid waste into a construction resource. Researchers have studied different industrial solid wastes including blast furnace slag [14–16], red mud [17–19], ground coal bottom ash [20], silico-manganese (Si–Mn) slag [21], electrolytic manganese residue [22], Cu–Ni slag [23] and lead slag [24]. Alkali-activated materials have greater strength and durability than Portland cement [25]. However, alkali-activated materials also have workability problems such as large shrinkage and a fast setting time [26].

Using BFFS as a precursor to produce alkali-activated materials is a potential way to utilize BBFS. The amorphous aluminosilicate phase in solid waste is the active source of the precursor to produce alkali-activated material [27]. The calcium content in the amorphous phase has a great impact on the solid waste activity; the presence of calcium lowers the polymerization degree of the aluminosilicate framework in the amorphous phase and increase the activity for alkali-activated reaction. The most commonly used precursors are blast furnace slag (BFS), which has a high calcium content. BFFS is mainly composed of CaO, SiO_2, Al_2O_3 and MgO, which has a chemical composition similar to BFS. However, BFFS has a higher CaO content and a lower Al_2O_3 content than BFS. Thus, BFFS can be classified as a medium-calcium slag for alkali-activated materials and has potential activity for alkali-activation. Few investigations were conducted on BFFS due to most BFFS is adopted only in some parts of China [28]. Wang [28] compared the reaction mechanism and mechanical properties of the alkali-activated BFFS and alkali-activated BFS. The early compressive strength of alkali-activated BFFS is lower than that of alkali-activated BFS, while the strength development of alkali-activated BFFS is better.

This study investigates the effect of NaOH concentration on the mechanical properties and microstructure of alkali-activated BFFS. The mechanical property is measured by the compressive strength. The pore structure is tested to understand the trend of strength on the microstructure scale. The phase composition and hydration heat are used to analyze the effect of reaction product on the pore structure. The polymerization degree and chemical bonding of C-A-S-H is tested by FT-IR.

2. Raw Materials and Test Methods

2.1. Raw Materials

BFFS was used as the precursor of the alkali-activated material. The specific surface area of the FFS was 439 m^2/kg. The chemical composition of the FFS was determined by X-ray fluorescence (XRF), as shown in Table 1. FFS consists mainly of CaO, SiO_2, Al_2O_3 and MgO.

Table 1. Chemical composition of raw materials (wt.%).

Compound	CaO	SiO$_2$	Al$_2$O$_3$	MgO	MnO	Cr$_2$O$_3$	Fe$_2$O$_3$	SO$_3$	LOI
BFFS	32.72	27.31	21.82	8.64	1.99	1.95	1.57	1.58	2.41

The X-ray diffraction (XRD) patterns of the FFS are shown in Figure 1. The slag consists mainly of amorphous phases, indicated by the broad peaks from 17° to 37° in the XRD patterns. The main crystalline phases in BFFS are calcite ($CaCO_3$), spinel ($MgO \cdot Al_2O_3$) and gypsum ($CaSO_4 \cdot 2H_2O$). It is widely accepted that the reactive CaO, SiO_2 and Al_2O_3 contents in amorphous phases determine the precursor reactivity. The CaO content in BFFS is lower, and the Al_2O_3 content is higher than that of BFS. The MgO content in BFFS is mainly in the form of spinel, which is stable and does not easily react.

The alkali hydroxide solutions of five NaOH concentrations (2 M, 5 M, 8 M, 10 M, 12 M) were prepared by dissolving sodium hydroxide pellets (99.9% purity) in water. The water-to-slag ratio was 0.5 for the mortar used in this study. Hydroxide was added to the water to obtain the required concentration. The slag-to-sand ratio for the mortar was 1:3. The paste and mortar specimens were cured in standard curing conditions of 20 $\pm$ 2 °C and 95% RH.

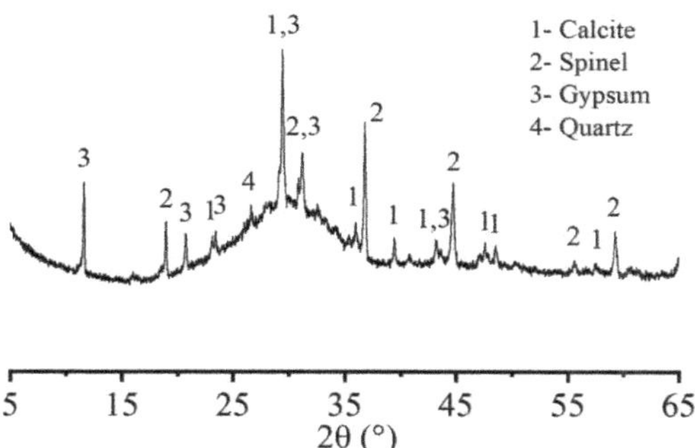

Figure 1. XRD pattern for BFFS.

2.2. Test Method

2.2.1. Compressive Strength

Mortar samples were prepared to test the compressive strength by mixing the raw materials with a planet mixer. The pastes were mixed for 30 s at low speed after the solution was added. The sands were added for another 4 min. The fresh mortar was cast in 40 mm × 40 mm × 160 mm molds and cured in standard curing conditions of 20 ± 2 °C and 95% RH until testing. The specimens were tested at 1 d, 3 d, 7 d and 28 d.

2.2.2. Isothermal Calorimetry

The hydration heat of alkali-activated BBFS with a water/binder ratio of 0.5 was measured by isothermal conduction calorimetry (TA instrument, TAM Air) at 25 °C. Two slags were weighed and placed in plastic bowls. The bowls were set in the calorimeter until thermal balance was reached. After 10 h of thermal equilibration, the activator was mixed with the precursors for 3 min. The heat evolution was recorded. Compared with the reaction heat, the heat disturbance caused by mixing can be neglected.

2.2.3. X-ray Diffraction (XRD)

The alkali-activated pastes were mixed with a planet mixer for 2 min. The water-to-slag ratio for all paste in this study is 0.5 after mixing, the suspensions were cast into 50-mL sealed centrifuge tubes and cured at 25 °C. After a specified duration, hydration of the sample was terminated using the isopropanol replacement method. The detailed method is described in previous studies [29,30]. The sample powder was ground to <100 μm for XRD. XRD analysis was performed with a Bruker D8 advance using CuKα radiation (scanning range from 5° to 65° for 40 min).

2.2.4. Mercury Intrusion Porosimeter

A mercury intrusion porosimeter (MIP) was used to characterize the pore structure of the paste specimens. The paste was obtained from the same sample analyzed by XRD. The surface tension and contact angle were 135° and 0.485 N/m, respectively. The paste samples for the MIP were cut into small cubes with a maximum size of 0.5 mm.

2.2.5. ATR–FTIR Spectroscopy

Attenuated total internal reflectance Fourier transform infrared spectroscopy (ATR–FTIR) was used to determine the chemical microstructure of the paste. The measurements were performed in the mid-infrared spectral region of 550–4000 cm^{-1} with a spectral resolution of 2 cm^{-1} using a BRUKER Tensor 27 spectrometer equipped with a diamond crystal as the ATR element (PIKE Miracle™). The paste was obtained from the sample analyzed by XRD. Same powder sample analyzed by XRD was used for ATR-FTIR.

3. Results and Discussion

3.1. Compressive Strength Development

The compressive strength results are presented in Figure 2. NaOH-activated BFFS demonstrated a good activating effect. The lowest 28-d compressive strength was observed for 12 M activated BFFS. The highest compressive strength was observed for 8 M activated BFFS. As the concentration increased, the strength of the mortar first increased and then decreased, indicating that when the concentration of the activator is too high, the mechanical properties of the alkali-activated BFFS are decreased. This is consistent with the results of previous research [31,32]. A high concentration of NaOH increases the alkalinity of the activator and the pore solution, which is critical for BFFS dissolution. However, a higher pH in the pore solution also promotes the formation of crystal products in the paste such as portlandite and hydrotalcite, which decrease the strength. According to Bondar [33] higher concentration of NaOH results in a higher viscosity of the solution. The pastes require more time for excess water to evaporate; this is the key step in the development of a network of aluminosilicates in which strength is developed.

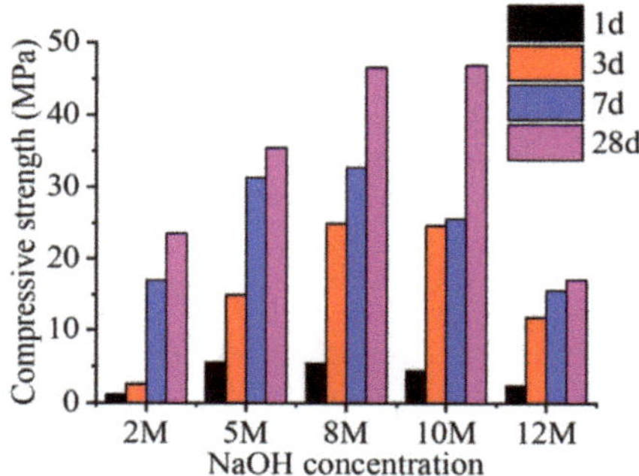

Figure 2. Compressive strength of AA–BFFS mortar at different concentrations.

3.2. Pore Structure of Hardened Pastes

Pore structure is analyzed by MIP in order to understand the trend of compressive strength on the microstructure scale. Figure 3 shows the cumulative pore volumes of the alkali-activated BFFS at 3 d and 7 d. The pore structure results are consistent with the compressive strength. The 2 M activated paste had the weakest pore structure of all concentration groups. Figure 3a shows that the 2 M activated sample has many pores from 100 to 1000 nm in size, which are considered to be harmful in porous materials [34], and are mainly due to the relatively low concentration of 2 M NaOH, which slows down the early dissolution of the BFFS and reduces the production of C-A-S-H gel. The pore structure is fragile, and the compressive strength is low. The 12 M activated paste has a pore structure similar to that of the 8 M activated paste but does not match the compressive strength. Figure 3a shows that the pore structure curves are similar for 8 M and 12 M; the total pore volume for 8 M (0.249 mg/L) is slightly smaller than that for 12 M (0.265 mg/L). However, the compressive strength of the 8 M paste (24.9 MPa) is more than twice that of the 12 M paste (11.9 MPa) at 3 d. According to the section below, more crystal phases are formed at high pore concentrations. The possible reason for the optimal concentration is that although the 12 M NaOH has a pH that can dissolve the BFFS and form sufficient C-A-S-H, formation of a large amount of crystal phase diminishes the strength of AA-BFFS.

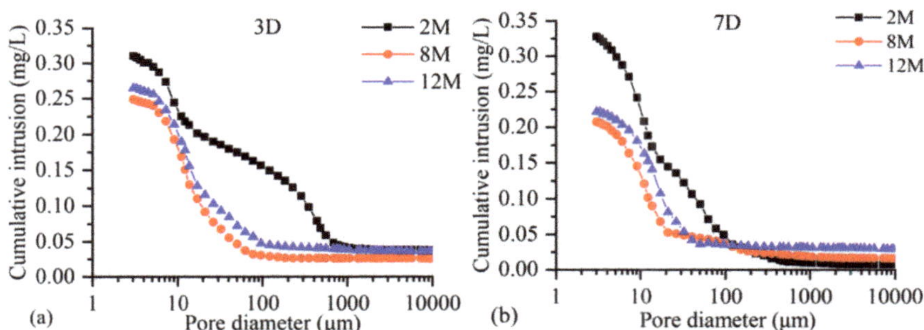

Figure 3. Cumulative pore volumes of alkali-activated BFFS paste at different ages: (**a**) 3 d; (**b**) 7 d.

Figure 3b shows the pore structure of AA-BFFS at 7 d, which is the same as at 3 d. Compared to the result at 3 d, the 2 M paste contains fewer pores from 100 to 1000 nm in size. The compressive strength of the 2 M paste at 7 d (17 MPa) is much greater than at 3 d (2.6 MPa), indicating that the pores between 100 and 1000 nm have a great influence on the strength of alkali-activated BFFS. The pore structures of 8 M and 12 M activated BFFS at 7 d are also improved. There are pores from 10 to 100 nm in size in the 7-d pore structure. The compressive strength of the 8 M (32.6 MPa) and 12 M (15.55 MPa) pastes at 7 d is greater than at 3 d (24.9 MPa and 11.8 MPa, respectively). The strength development of the 2 M paste is much greater than that of the 8 M and 12 M pastes, which indicates that the pores from 100 to 1000 nm have a greater impact on the compressive strength than the pores from 10 to 100 nm in AA-BFFS.

3.3. Phase Compositions of Hardened Pastes

XRD was used to analyzed the phase composition in order to understand the effect of reaction production on the pore structure. Figure 4 shows the XRD spectra of the hardened pastes hydrated for 1 d, 3 d, 7 d and 28 d. The main crystal product of AA-BFFS is hydrotalcite (PDF#89-0460), which is the same as that of alkali-activated BFS [35,36]. The XRD patterns show that as the reaction progresses, calcite ($CaCO_3$) and gypsum ($CaSO_4 \cdot 2H_2O$) are gradually dissolved and consumed; spinel ($MgO \cdot Al_2O_3$) did not participate in the reaction because it has a stable crystal structure, as reported in a previous study [36]. The main amorphous phase is C-A-S-H gel, which corresponds to the C-S-H(I) peaks in Figure 4 [36].

Comparing the XRD patterns of slag activated by different concentrations of sodium hydroxide, it was found that as the concentration increases, more crystals are formed in the product [37]. As spinel is stable in the pore solution of alkali-activated material, it can be used as an internal standard material for semiquantitative analysis. For 2 M activated BFFS, the rate of product generation was slow. Hydrotalcite was not observed in the XRD spectrum until 3 d (Figure 3a). The calcite in the 2 M XRD patterns was not consumed at 28 d, indicating that the low concentration of 2 M NaOH does not have sufficient alkalinity to dissolve BFFS and provide sufficient ions for crystal growth. Compared with paste activated by higher concentrations of NaOH, few C-S-H(I) peaks were observed in the XRD patterns. This can also be explained by the low concentration of NaOH. With an increase in age, the peak of hydrotalcite in the 2 M pattern did not change significantly, indicating that further reaction in the 2 M paste is minimal. The phase composition development of the 2 M paste explains the poor pore structure, and the low strength development of the 2 M mortar.

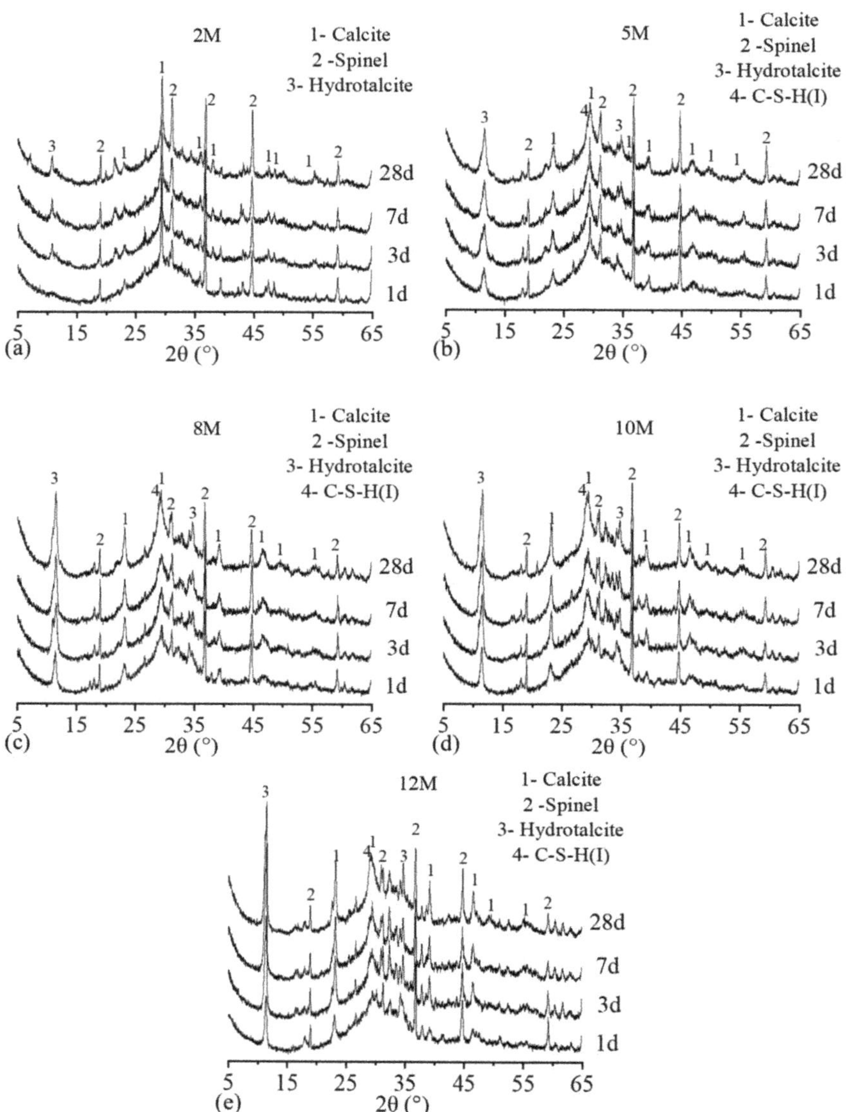

Figure 4. XRD spectra of hardened pastes at different ages: (**a**) 2 M, (**b**) 5 M. (**c**) 8 M. (**d**) 10 M and (**e**) 12 M.

When a higher concentration of NaOH (>2 M) was used to activate the BFFS, the hydrotalcite peak was clearly observed at 1 d. At the same age, the peaks of crystalline products and C-S-H(I) were higher with an increase in concentration, indicating that high alkalinity can promote product formation. High alkalinity accelerates the dissolution of BFFS, increasing the ion concentration in the solution [38,39] and promoting crystallization and growth of the products. Unlike the 2 M paste, the reaction continued to 28 d, as the peaks of hydrotalcite and C-S-H(I) are sharper than the peak of spinel (Figure 4b–e). This explains the strength and pore structure development. Comparing Figure 4c and Figure 4e, it is observed that the peaks of C-S-H(I) are similar in intensity according to the spinel peaks. The hydrotalcite peak of the 12 M paste (Figure 4c) is more intense than that of the 8 M paste (Figure 4e), indicating that with an increase in the NaOH concentration, more hydrotalcite was formed than C-A-S-H gel. C-A-S-H is the main product providing strength to alkali-activated material. The formation of hydrotalcite consumes more calcium and aluminum in the solution, reducing the Ca and Al content in the C-A-S-H gel, resulting in a low-degree C-A-S-H gel. Hydrotalcite and portlandite are preferred to enrich the particles and fine aggregates, forming an interfacial transition zone (ITZ) and reducing the compressive strength. As a high concentration of NaOH promotes generation of hydrotalcite and other crystals, the compressive strength of high-concentration NaOH is lower.

3.4. Isothermal Calorimetry

Isothermal calorimetry was measured to analyze the reaction mechanism and production of the reaction products. Figure 5 shows the isothermal calorimetry results for the alkali-activated pastes within 96 h. Isothermal calorimetry was performed to evaluate the effect of concentration on the kinetics of alkali-activated BFFS. The heat evolution curves of the alkali-activated BFFS are shown in Figure 5a. The evolution peak occurred in the early stage of the reaction (within 4 h). With an increase in the concentration, the evolution peak became more intense. The first exothermic peak of alkali-activated materials can be interpreted as the heat of wetting and the heat of dissolution according to [40]. The results show that a higher pH of the solution increases the dissolution of BFFS. This is consistent with the XRD results. A fast and intense evolution indicates fast dissolution of the BFFS, which provides sufficient ions in the concentration for crystallization. The 2 M paste shows a low heat evolution in poor formation of product and pore structure. The 5 M curves in Figure 5a show a broad shoulder peak from 3–4 h. This curve is widely accepted as an acceleration peak. The shape of this curve is similar to that for Portland cement, although the temporal occurrences are different [41].

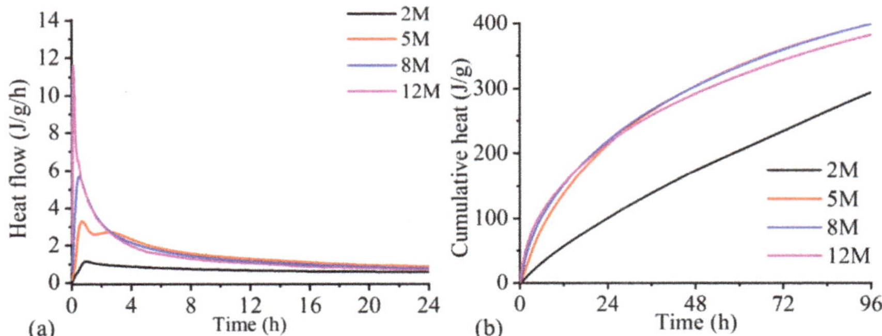

Figure 5. Isothermal calorimetry curves for paste: (**a**) heat flow and (**b**) cumulative heat.

Figure 5b shows the cumulative heat of the AA-BFFS. The total heat of the 2 M paste is significantly less than that of the others, indicating that the low pH of the 2 M NaOH reduces the reaction rate and the product formation of the paste (mainly C-A-S-H gel). This is direct evidence to explain the low strength and poor pore structure of the 2 M paste. However, for high-concentration pastes, the total heat within 72 h is similar, which is consistent with the XRD results. The cumulative heat and heat evolution indicate that the violent reaction at the early age hinders the later C-A-S-H production.

3.5. FT-IR Spectra of Hardened Pastes

The chemical bonding and the polymerization degree of C-A-S-H was analyzed by FT-IR. Figure 6 shows the FT-IR spectra of hardened pastes hydrated for 1 d, 3 d, 7 d and 28 d. Three obvious peaks are observed. The most important peak is at approximately 950–1000 cm^{-1}, corresponding to the Si–O asymmetric stretching vibrations (ν3) of Q2 units [41–43]. The peak at 870–875 cm^{-1} is related to the asymmetric stretching (v3) and out-of-plane bending (v2) modes of CO$_3^{2-}$ ions [41,44]. The weak peak at 650–700 cm^{-1} corresponds to the Al–O–Si bonds [43].

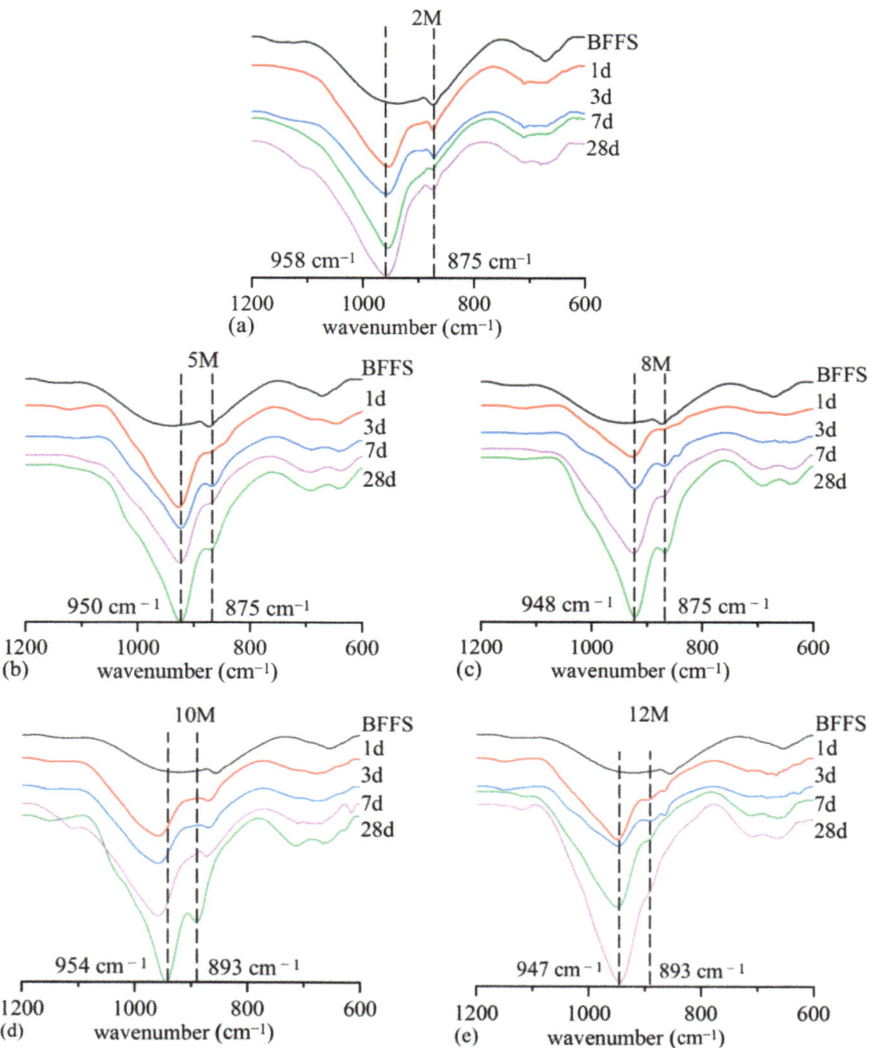

Figure 6. FT-IR spectra of hardened pastes at different ages: (**a**) 2 M, (**b**) 5 M, (**c**) 8 M, (**d**) 10 M and (**e**) 12 M.

The band at 875 cm^{-1} corresponds to calcite in the precursor. The shifting of this band can be interpreted as the dissolution of calcite and the formation of other unknown carbonates. Figure 5a shows that the CO$_3^{2-}$ bond of the 2 M paste remains at 28 d, indicating that calcite did not react in the 2 M paste. This is consistent with the results in Figure 4a. For the 5 M and 8 M concentrations, the CO$_3^{2-}$ bond broadens and becomes weaker, indicating that calcite dissolved. For the last two concentrations, the CO$_3^{2-}$ bond is weaker, disappears or shifts to other wavenumbers, indicating that calcite dissolved and transformed into other carbonates. The results indicate that higher concentrations promote crystal dissolution and new crystal formation, which is consistent with the XRD results.

The bands at 950–1000 cm^{-1} are different for different concentrations. At 28 d, with an increase in the concentration, the band at 950–1000 cm^{-1} first shifted toward a lower wavenumber and then back to a higher wavenumber. The lowest wavenumber was observed in 8 M activated BFFS (948 cm^{-1}, Figure 6c), and is consistent with the compressive strength. The shifting of the band from 950–1000 cm^{-1} is caused by different Si/Al ratios in the Si-O-T (Si, Al) [45]. A higher wavenumber corresponds to an "Si-rich" bond with a lower degree. A lower wavenumber corresponds to an "Al-rich" bond with a higher polymerization degree [35]. Thus, the wavenumber trend from low concentration to high concentration can be interpreted as the Al/Si ratio and polymerization degree of C-A-S-H gel first increasing and then decreasing when the concentration is greater than 8 M. This means the optimal concentration result in the higher polymerization degree of C-A-S-H gel. This explains the compressive strength and pore structure, which are consistent with the XRD patterns.

4. Conclusions

In this study, the properties of alkali-activated BBFS were investigated at different concentrations of NaOH (2 M, 5 M, 8 M, 10 M, 12 M). Based on the results, the following conclusions can be drawn.

- Optimal concentration on the compressive strength of mortars is found; 8 M NaOH shows the best performance. Both lower concentrations and higher concentrations cause strength degradation.
- The pore structure characterized by MIP indicates that at the low concentration (2M) more pore between 10 and 1000 nm is existed. This demonstrates the low compressive strength of 2M. The pore structure of 8M and 12M is similar which means the pore structure cannot explain the strength degradation at high concentration.
- The phase composition characterized by XRD reveals that less C-A-S-H gel is generated at low concentration and result in the poor pore structure. The low alkalinity of the 2M NaOH leads to lower reaction rate at the early age according to isothermal calorimetry result. Eventually, the amount of less C-A-S-H gel is generated.
- An increase in the concentration promotes the generation of C-A-S-H and hydrotalcite. According to XRD result, when the concentration is greater than 8 M, more hydrotalcite is formed compared to C-A-S-H. Isothermal calorimetry indicates that, the violent reaction at the early age hinders the later formation of the C-A-S-H. This causes the compressive strength degradation at high concentration.

FT-IR result indicates that the polymerization degree of C-A-S-H gel is consistent with the compressive strength, the optimal concentration of 8M shows the highest polymerization degree. The result indicates both low and high concentration reduce the polymerization degree of C-A-S-H which also do harm to the compressive strength.

Author Contributions: Z.H.: Conceptualization, Data curation, Formal analysis, Investigation, Methodology, Writing—original draft. Y.Z.: Investigation, Methodology. Y.C.: Conceptualization, Formal analysis, Funding acquisition, Supervision, Validation, Writing—review & editing. All authors have read and agreed to the published version of the manuscript.

Funding: This research was funded by [National Natural Science Foundation of China] grant number [No. 51822807].

Institutional Review Board Statement: Not applicable.

Informed Consent Statement: Informed consent was obtained from all subjects involved in the study.

Data Availability Statement: The data presented in this study are available within the manuscript.

Acknowledgments: The authors would like to acknowledge the National Natural Science Foundation of China (No. 51822807).

Conflicts of Interest: The authors declare no conflict of interest.

References

1. Index Mundi, Nickel: World Plant Production, by Country and Product. 2012. Available online: https://www.indexmundi.com/en/commodities/minerals/nickel/nickel_t12.html (accessed on 12 October 2021).
2. Vartiainen, A. Proceedings of the Twelfth International Ferroalloys Congress. In *The Twelfth International Ferroalloys Congress: Sustainable Future*; Outotec Oyj: Helsinki, Finland, 2010.
3. Cheng, M.M. Current development status, market analysis and prospect of ferronickel in China. *Expr. Inf. Min. Ind.* **2008**, *8*. (In Chinese)
4. Saha, A.K.; Khan, M.N.N.; Sarker, P.K. Value added utilization of by-product electric furnace ferronickel slag as construction materials: A review. *Resour. Conserv. Recy.* **2018**, *134*, 10–24. [CrossRef]
5. Wang, Q.; Huang, Z.; Wang, D. Influence of high-volume electric furnace nickel slag and phosphorous slag on the properties of massive concrete. *J. Therm. Anal. Calorim.* **2017**, *131*, 873–885. [CrossRef]
6. Huang, Y.; Wang, Q.; Shi, M. Characteristics and reactivity of ferronickel slag powder. *Constr. Build. Mater.* **2017**, *156*, 773–789. [CrossRef]
7. Rahman, M.A.; Sarker, P.K.; Shaikh, F.U.A.; Saha, A.K. Soundness and compressive strength of Portland cement blend-ed with ground granulated ferronickel slag. *Constr. Build. Mater.* **2017**, *140*, 194–202. [CrossRef]
8. Zhai, M.; Zhu, H.; Liang, G.; Wu, Q.; Zhang, C.; Hua, S.; Zhang, Z. Enhancing the recyclability of air-cooled high-magnesium ferronickel slag in cement-based materials: A study of assessing soundness through modifying method. *Constr. Build. Mater.* **2020**, *261*, 120523. [CrossRef]
9. Maragkos, I.; Giannopoulou, I.P.; Panias, D. Synthesis of ferronickel slag-based geopolymers. *Miner. Eng.* **2009**, *22*, 196–203. [CrossRef]
10. Saha, A.K.; Sarker, P. Sustainable use of ferronickel slag fine aggregate and fly ash in structural concrete: Mechanical properties and leaching study. *J. Clean. Prod.* **2017**, *162*, 438–448. [CrossRef]
11. Saha, A.K.; Sarker, P.K. Expansion due to alkali-silica reaction of ferronickel slagfine aggregate in OPC and blended cement mortars. *Constr. Build. Mater.* **2016**, *123*, 135–142. [CrossRef]
12. Zhang, Z.; Zhu, Y.; Yang, T.; Li, L.; Zhu, H.; Wang, H. Conversion of local industrial wastes into greener cement through geopolymer technology: A case study of high-magnesium nickel slag. *J. Clean. Prod.* **2017**, *141*, 463–471. [CrossRef]
13. Yang, T.; Wu, Q.; Zhu, H.; Zhang, Z. Geopolymer with improved thermal stability by incorporating high-magnesium nickel slag. *Constr. Build. Mater.* **2017**, *155*, 475–484. [CrossRef]
14. Provis, J.L.; van Deventer, J.S.J. (Eds.) *Alkali-activated Materials: State-of-the-Art Report, RILEM TC 224-AAM*; Springer/RILEM: Dordrecht, Germany, 2014.
15. Wang, D.; Wang, Q.; Huang, Z. New insights into the early reaction of NaOH-activated slag in the presence of CaSO4. *Compos. Pt. B-Eng.* **2020**, *198*, 108207. [CrossRef]
16. Luo, T.; Wang, Q.; Zhuang, S. Effects of ultra-fine ground granulated blast-furnace slag on initial setting time, fluidity and rheological properties of cement pastes. *Powder Technol.* **2019**, *345*, 54–63.
17. Dimas, D.D.; Giannopoulou, I.P.; Panias, D. Utilization of Alumina Red Mud for Synthesis of Inorganic Polymeric Materials. *Miner. Process. Extr. Met. Rev.* **2009**, *30*, 211–239. [CrossRef]
18. Gong, C.; Yang, N. Effect of phosphate on the hydration of alkali-activated red mud–slag cementitious material. *Cem. Concr. Res.* **2000**, *30*, 1013–1016. [CrossRef]
19. Kumar, A.; Kumar, S. Development of paving blocks from synergistic use of red mud and fly ash using geopolymerization. *Constr. Build. Mater.* **2013**, *38*, 865–871. [CrossRef]
20. Donatello, S.; Maltseva, O.; Fernandez-Jimenez, A.; Palomo, A. The Early Age Hydration Reactions of a Hybrid Cement Containing a Very High Content of Coal Bottom Ash. *J. Am. Ceram. Soc.* **2013**, *97*, 929–937. [CrossRef]
21. Kumar, S.; García-Triñanes, P.; Teixeira-Pinto, A.; Bao, M. Development of alkali activated cement from mechanically activated silico-manganese (SiMn) slag. *Cem. Concr. Compos.* **2013**, *40*, 7–13. [CrossRef]
22. Wang, D.; Wang, Q.; Xue, J. Reuse of hazardous electrolytic manganese residue: Detailed leaching characterization and novel application as a cementitious material. *Resour. Conserv. Recycl.* **2019**, *154*, 104645. [CrossRef]
23. Kalinkin, A.M.; Kumar, S.; Gurevich, B.I.; Alex, T.C.; Kalinkina, E.V.; Tyukavkina, V.V.; Kalinnikov, V.T.; Kumar, R. Geopolymerization behavior of Cu–Ni slag mechanically activated in air and in CO2 atmosphere. *Int. J. Miner. Process.* **2012**, *112–113*, 101–106. [CrossRef]
24. Onisei, S.; Pontikes, Y.; Van Gerven, T.; Angelopoulos, G.; Velea, T.; Predica, V.; Moldovan, P. Synthesis of inorganic polymers using fly ash and primary lead slag. *J. Hazard. Mater.* **2012**, *205–206*, 101–110. [CrossRef] [PubMed]
25. Provis, J.; Palomo, A.; Shi, C. Advances in understanding alkali-activated materials. *Cem. Concr. Res.* **2015**, *78*, 110–125. [CrossRef]
26. Shi, C.; Roy, D.; Krivenko, P. *Alkali-Activated Cements and Concretes*; CRC Press: Boca Raton, FL, USA, 2005.
27. Li, C.; Sun, H.; Li, L. A review: The comparison between alkali-activated slag (Si + Ca) and metakaolin (Si + Al) cements. *Cem. Concr. Res.* **2010**, *40*, 1341–1349. [CrossRef]
28. Wang, D.; Wang, Q.; Zhuang, S.; Yang, J. Evaluation of alkali-activated blast furnace ferronickel slag as a cementitious ma-terial: Reaction mechanism, engineering properties and leaching behaviors. *Constr. Build. Mater.* **2018**, *188*, 860–873. [CrossRef]

29. Tian, H.; Kong, X.; Su, T.; Wang, D. Comparative study of two PCE superplasticizers with varied charge density in Port-land cement and sulfoaluminate cement systems. *Cem. Concr. Res.* **2019**, *115*, 43–58. [CrossRef]
30. Scrivener, K.L.; Snellings, R.; Lothenbach, B. *A Practical Guide to Microstructural Analysis of Cementitious Materials*; CRC Press: Boca Raton, FL, USA, 2016.
31. De Filippis, U.; Prud'Homme, E.; Meille, S. Relation between activator ratio, hydration products and mechanical properties of alkali-activated slag. *Constr. Build. Mater.* **2020**, *266*, 120940. [CrossRef]
32. Bondar, D.; Lynsdale, C.; Milestone, N.B.; Hassani, N.; Ramezanianpour, A. Effect of type, form, and dosage of activators on strength of alkali-activated natural pozzolans. *Cem. Concr. Compos.* **2010**, *33*, 251–260. [CrossRef]
33. Xu, H.; van Deventer, J.S.J. The geo-polymerisation of alumino-silicate minerals. *Int. J. Miner. Process.* **2000**, *59*, 247–266. [CrossRef]
34. Wu, Z.; Shi, C.; Khayat, K.; Wan, S. Effects of different nanomaterials on hardening and performance of ultra-high strength concrete (UHSC). *Cem. Concr. Compos.* **2016**, *70*, 24–34. [CrossRef]
35. Puertas, F.; Torres-Carrasco, M. Use of glass waste as an activator in the preparation of alkali-activated slag. Mechanical strength and paste characterization. *Cem. Concr. Res.* **2014**, *57*, 95–104. [CrossRef]
36. Garcia-Lodeiro, I.; Palomo, A.; Fernández-Jiménez, A.; Macphee, D.E. Compatibility studies between N-A-S-H and C-A-S-H gels. Study in the ternary diagram Na_2O–CaO–Al_2O_3–SiO_2–H_2O. *Cem. Concr. Res.* **2011**, *41*, 923–931. [CrossRef]
37. Chithiraputhiran, S.; Neithalath, N. Isothermal reaction kinetics and temperature dependence of alkali activation of slag, fly ash and their blends. *Constr. Build. Mater.* **2013**, *45*, 233–242. [CrossRef]
38. Song, S.; Jennings, H.M. Pore solution chemistry of alkali-activated ground granulated blast-furnace slag. *Cem. Concr. Res.* **1999**, *29*, 159–170. [CrossRef]
39. Zuo, Y.; Nedeljković, M.; Ye, G. Pore solution composition of alkali-activated slag/fly ash pastes. *Cem. Concr. Res.* **2018**, *115*, 230–250. [CrossRef]
40. Yao, X.; Zhang, Z.; Zhu, H.; Chen, Y. Geopolymerization process of alkali–metakaolinite characterized by isothermal calorimetry. *Thermochim. Acta* **2009**, *493*, 49–54. [CrossRef]
41. Zhang, Z.; Wang, H.; Provis, J.L. Quantitative study of the reactivity of fly ash in geopolymerization by FTIR. *J. Sustain. Cem. Mater.* **2012**, *1*, 154–166. [CrossRef]
42. Yu, P.; Kirkpatrick, R.J.; Poe, B.; McMillan, P.F.; Cong, X. Structure of Calcium Silicate Hydrate (C-S-H): Near-, Mid-, and Far-Infrared Spectroscopy. *J. Am. Ceram. Soc.* **2004**, *82*, 742–748. [CrossRef]
43. Fang, Y.H.; Lu, Z.L.; Wang, Z.L. FT-IR Study on Early-Age Hydration of Alkali-Activated Slag Cement. *Key Eng. Mater.* **2011**, *492*, 429–432. [CrossRef]
44. Mollah, M.Y.A.; Lu, F.; Cocke, D.L. X-raydiffraction (XRD) fourier transform infrared spectroscopy (FT-IR) characterization Portland cement type-V. *Sci. Total Environ.* **1998**, *224*, 57. [CrossRef]
45. Hajimohammadi, A.; Provis, J.L.; van Deventer, J.S.J. The effect of silica availability on the mechanism of geopolymerisation. *Cem. Concr. Res.* **2011**, *41*, 210–216. [CrossRef]

Article

Effects of Different Content of Phosphorus Slag Composite Concrete: Heat Evolution, Sulphate-Corrosion Resistance and Volume Deformation

Kuisheng Liu [1] and Yong Cui [2,*]

[1] Beijing Urban Construction Group Co., Ltd., Beijing 100000, China; liuks@mail.bucg.com
[2] Department of Civil Engineering, Tsinghua University, Beijing 100084, China
* Correspondence: cuiyong20@mails.tsinghua.edu.cn; Tel.: +86-186-1082-8446

Abstract: Phosphorus slag (PS) and limestone (LS) composite (PLC) were prepared with a mass ratio of 1:1. The effects of the content of PLC and the water/binder ratio on the mechanical properties, durability and dry shrinkage of concrete were studied via compressive strength, electric flux, sulfate dry/wet cycle method, saturated drainage method, isothermal calorimeter, adiabatic temperature rise instrument and shrinkage deformation instrument. The results show that PLC can greatly reduce the adiabatic temperature rise of concrete. The adiabatic temperature rise is 55 °C with 33 wt.% PLC, 10 °C lower than that of the control sample. The addition in the content of PLC does not affect the long-term strength of concrete. When the water/binder ratio decreases by 0.1–0.15, the long-term strength of concrete with PLC increases by about 10%, compared with the control group. At the age of 360 days, the chloride permeability of L-11 (i.e., the content of PLC was 20%, the water/binder ratio was 0.418) and L-22 (i.e., the content of PLC was 33%, the water/binder ratio was 0.39) decrease to the "very low" grade. The strength loss rate of L-11 and L-22 after 150 sulfate dry/wet cycles is about 18.5% and 19%, respectively, which is 60% of the strength loss rate of the control sample. The drying shrinkage of L-11 and L-22 reduces by 4.7% and 9.5%, respectively, indicating that PLC can also reduce the drying shrinkage.

Keywords: phosphorus slag; limestone; concrete; sulphate-corrosion resistance; volume deformation

Citation: Liu, K.; Cui, Y. Effects of Different Content of Phosphorus Slag Composite Concrete: Heat Evolution, Sulphate-Corrosion Resistance and Volume Deformation. *Crystals* **2021**, *11*, 1293. https://doi.org/10.3390/cryst11111293

Academic Editors: Yifeng Ling, Chuanqing Fu, Peng Zhang, Peter Taylor, José L. García and Leonid Kustov

Received: 22 September 2021
Accepted: 20 October 2021
Published: 25 October 2021

Publisher's Note: MDPI stays neutral with regard to jurisdictional claims in published maps and institutional affiliations.

1. Introduction

Phosphorus slag (PS), the abbreviation of granulated electric furnace phosphorus slag powder, is an industrial waste residue containing calcium silicate. It is produced in the process of industrial production of yellow phosphorus preparation from phosphorus ore, silica and coke at a high temperature of about 1400 °C in an electric furnace. The obtained melt is then quenched by water to form granules, and ground to obtain PS powder [1,2]. The main mineral phase of PS is amorphous glass with a little crystalline phase, such as calcium phosphate, calcium orthosilicate and anorthite [3,4].

Barnesp [5] believed that the hydration activity of PS itself is very low. However, under the action of calcium hydroxide, the Ca^{2+}, $[AlO_4]^{5-}$, Al^{3+} and $[SiO_4]^{4-}$ in the glass of PS dissolve into solution to form C-A-S-H gel. In addition, P_2O_5 and fluoride in PS may react with $Ca(OH)_2$, resulting in the formation of the insoluble fluorohydroxyapatite. This precipitation wraps around the cement particles, therefore delaying the hardening and setting of the cement. On the other hand, when the content of PS is low, the formation of fluorohydroxyapatite is not enough to completely package around the cement particles, so that the hydration continues. With further hydration, the alkali concentration in the liquid gradually increases, while the osmotic pressure caused by the concentration difference enables H_2O or OH^- to penetrate into the coating of cement; then, the hydration continues and the strength increases. Li [6,7] et al. stated that with the addition of a certain amount of PS, the number of gel pores accumulates with the increase of hydration time.

Meanwhile, new C-A-S-H continuously polymerizes with the decomposing of glass in PS. The crosslinking filling of ettringite and C-A-S-H gel makes the structure of cement paste more compact, resulting in the reduction of the pore sized > 50 nm and the increase of the strength. It is worth noting that the addition of PS alone is an easy way to reduce the early strength and prolong the setting time of concrete, which is an urgent problem to be solved in engineering application.

Limestone powder (LS) is a common concrete admixture. Its main chemical composition is CaO with a small amount of SiO_2, Al_2O_3, MgO, Fe_2O_3, etc. The main mineral phase of limestone powder is crystalline $CaCO_3$ ($\geq$80%) with a small amount of quartz ($\leq$10%) [8,9].

Rizwan et al. [10] found that neither limestone powder nor fly ash can improve the performance of self-leveling mortar, while the fluidity, strength, microstructure, relative water absorption and early volume stability of self-leveling mortar mixed with limestone powder and fly ash composite are better than those prepared with other admixtures. Li et al. [11] found that when the ultra-fine limestone powder is added to replace cementitious materials, the compressive strength of concrete was unchanged with the same water/binder ratio, while the tensile strength, stiffness, and durability of concrete are improved. Hu et al. [12] found that the 28-day compressive strength of concrete mixed with 20% ultra-fine fly ash and 10% limestone powder is very close to that of pure cement concrete. The test results of Temiz, et al. [13] showed that under the condition of low water/binder ratio (<0.4), the three-component concrete mixed with limestone powder, cement, FA (fly ash), or GGBS (granulated blast furnace slag) can obtain similar or even higher long-term strength than the two-component mud concrete (cement and FA or GGBS). We can know that LS can act together with other volcanic ash materials to play the role of "1 + 1 > 2". However, it is rarely reported that LS and PS are used as admixtures in concrete at the same time. Whether PLC has an impact on durability and shrinkage performance is also a problem to be explored.

In this paper, phosphorus slag (PS) and limestone (LS) composite (PLC) were prepared with a mass ratio of 1:1. The effects of the content of PLC and the water/binder ratio on the mechanical properties, durability and dry shrinkage of concrete were studied via compressive strength, electric flux, sulfate dry/wet cycle method, saturated drainage method, isothermal calorimeter, adiabatic temperature rise instrument and shrinkage deformation instrument. The obtained results provide significance for theoretical and engineering guiding for the application of PLC in mass concrete and corrosion-resistant concrete.

2. Materials and Methods

2.1. Materials

The cement is P·O 42.5 ordinary Portland cement produced by Jinyu Group. Its basic performance is shown in Table 1, in accordance with the requirements of Chinese standard GB/T 175. The specific surface area and the chemical composition of PS and LS are shown in Table 2. The reaction formula of PS is shown as in Equation (1). The size of coarse aggregate is 5–25 mm, and other indexes meet the requirements of Chinese standard GB/T 14685. The fine aggregate is river sand, and the fineness modulus is 2.4. The water is tap water.

$$2Ca_3(PO_4)_2 + 10C + 6SiO_2 \rightarrow P_4 + 10CO + 6(CaO \cdot SiO_2) \tag{1}$$

Table 1. Properties of cement.

Cement Grade	Setting Time		Flexural Strength		Compressive Strength		Specific Surface Area	Requirement of Normal Consistency
	Initial Setting	Final Setting	3 d	28 d	3 d	28 d		
PO42.5	155	275	6.2	9.7	27.9	48.9	360	28%

Table 2. Chemical composition and specific surface area of phosphorus slag and limestone powder.

Composition	SiO_2	Al_2O_3	Fe_3O_4	CaO	MgO	K_2O	SO_3	P_2O_5	F	Loss
PS	38.53	5.65	0.05	44.2	1.24	1.46	3.12	3.13	1.72	0.58
LS	4.32	1.43	0.76	51.23	1.13	0.20	-	-	-	41.5

2.2. Sample Preparation

Table 3 shows the mix proportion of concrete with different content of PLC. The pure cement was taken as the control sample, and the mass ratio of PS to LS was 1:1. L-1 was using PLC to replace 20% of pure cement and the water/binder ratio was 0.425. L-11 was using PLC to replace 20% of pure cement, but the water/binder ratio was 0.418. The total amount of L-2 cementitious materials was 420 kg/m^3. PLC accounted for 33% of the total cementitious materials, and the water/binder ratio was 0.405. Meanwhile, the water cement ratio of L-22 was reduced to 0.39 with the same content of PLC. The workability of concrete was adjusted by a water reducing agent to keep consistent slump. The effects of the content of PLC and water/binder ratio on the mechanical properties, shrinkage and durability of concrete were studied.

Table 3. Mix proportion of concrete mixed with phosphorus slag powder and limestone powder (unit kg/m^3).

	Cement	PS	LS	Sand	Stone	Water
C	400	0	0	770	1010	170
L-1	320	40	40	770	1010	170
L-11	320	40	40	773	1010	167
L-2	280	70	70	770	1010	170
L-22	280	70	70	776	1010	164

2.3. Test Methods

(1)　Mechanical property

The compressive strength of concrete was tested in accordance with Chinese standard GB/T 50,081 (standard for test methods of mechanical properties of ordinary concrete). Cube specimens of 100 mm × 100 mm × 100 mm were used to measure the compressive strength of concrete for 7 d, 28 d, 90 d and 360 d after curing to a specific age in a standard curing room (20 °C and 95 RH%).

(2)　Chloride permeability

After curing to a specific age in a standard curing room (20 °C and 95 RH%), the 100 mm × 100 mm × 100 mm cube specimens were cut into 100 mm × 100 mm × 50 mm specimens. The chloride permeability of concrete at 28 d and 360 d was measured according to ASTM C1202 (standard test method for chloride ion permeability resistance of concrete). The instrument model is CABR-RCP9. After installing the test block, 0.3 mol/L NaOH solution and 3 wt.% NaCl solution were prepared immediately (ensuring the accuracy of solution concentration), followed by injecting NaOH solution at the positive extreme and NaCl solution at the negative extreme without any liquid leakage. The measurement time was 6 h, and the electric flux of each group of concrete was recorded.

(3)　Sulphate-corrosion Resistance

The sulphate-corrosion resistance of 100 mm × 100 mm × 100 mm cube specimens was tested by the dry/wet cycle method [14]. Each sample was dried in an oven at 80 ± 5 °C for 6 h, cooled it for 2 h, and then soaked in Na_2SO_4 solution with concentration of 5%. This process is one cycle (i.e., one dry wet cycle is completed in one day). At the same time, the samples with the same ratio were placed in the standard curing room for curing. After 120 and 150 cycles, the compressive strength of the specimens in the standard curing room was measured as S1 and the compressive strength of concrete after the dry wet cycle was measured as S2. The strength loss rate was calculated according to the following Equation (2):

$$\eta = (S1 - S2)/S1 \times 100\% \qquad (2)$$

(4) Connected porosity

The connected porosity of concrete was measured by the "water saturation drying" method [15]. A piece of 10 cm × 10 cm × 2 cm concrete sheet was cut and prepared, the volume V of the test piece was measured by a drainage method, and the mass m_1 of the test piece after vacuum water saturation was measured. The test piece was dried in an oven at 80 °C for 14 d, and the mass m_2 after drying was measured. The connected porosity was calculated according to the following Equation (3):

$$P = (m_1 - m_2)/\rho V \qquad (3)$$

where ρ is the density of water.

(5) Hydration heat release

The Calmetrix 8000 HPC isothermal calorimeter was used to determine the hydration heat release of the sample slurry [16]. The test method meets the requirements of ASTM C1702-2015. The temperature was set at 25 °C to determine the heat release rate and total heat release of C, L-11 and L-22 within 72 h.

(6) Adiabatic temperature rise

Concrete (C, L-11 and L-22) prepared according to the mix proportion was put into the concrete adiabatic temperature rise instrument to measure its adiabatic temperature rise. The amount of concrete used for each test was 40~45 L. The temperature control error was less than ±0.1 °C, and the minimum temperature resolution was 0.02 °C. The automatic data acquisition system collected data every 5 min, and guaranteed the concrete center temperature to be 0.1 °C higher than the concrete outer wall temperature during the measurement process.

(7) Shrinkage property

The shrinkage deformation of concrete (C, L-11 and L-22) was measured by contact method. $100 \times 100 \times 400$ mm³ concrete specimen was molded. After demolding, the specimen was cured at room temperature for 7 days, and it was covered with plastic film to prevent moisture loss. After curing for 7 days, the plastic film was removed, then the deformation length was measured by a length comparator with a dial indicator (accuracy 0.001 mm) and calibrated before each measurement. The results were the mean of three groups of concrete.

3. Results and Discussion

3.1. Compressive Strength

Figure 1 shows the compressive strength (a) 7 d (b) 28 d (c) 90 d (d) 360 d of concrete mixed with different content of PLC. The compressive strength of the control sample (C) was ~35 MPa at the age of 7 days. When 20% cement was replaced by PLC, the compressive strength of L-1 decreased to ~30 MPa. However, when reducing the water/binder ratio from 0.425 to 0.418, the compressive strength of L-11 increased from ~30 MPa to ~35 MPa. When 33% cement was replaced by PLC, the compressive strength of L-2 further reduced to ~25 MPa. By comparison, when lowing the water/binder ratio from 0.405 to 0.39, the compressive strength of L-22 increased from ~25 MPa to ~35 MPa accordingly, indicating that reduction of water can make up for the loss of the early strength of concrete caused by adding PLC to a certain extent. However, at the age of 90 days and 360 days, the compressive strength of concrete mixed with 20% and 33% PLC was all around 50 MPa, indicating that the addition of PLC does not affect the long-term strength of concrete. When the water/binder ratio decreased by 0.1–0.15, the long-term strength of concrete increased by ~10%, compared with the control sample.

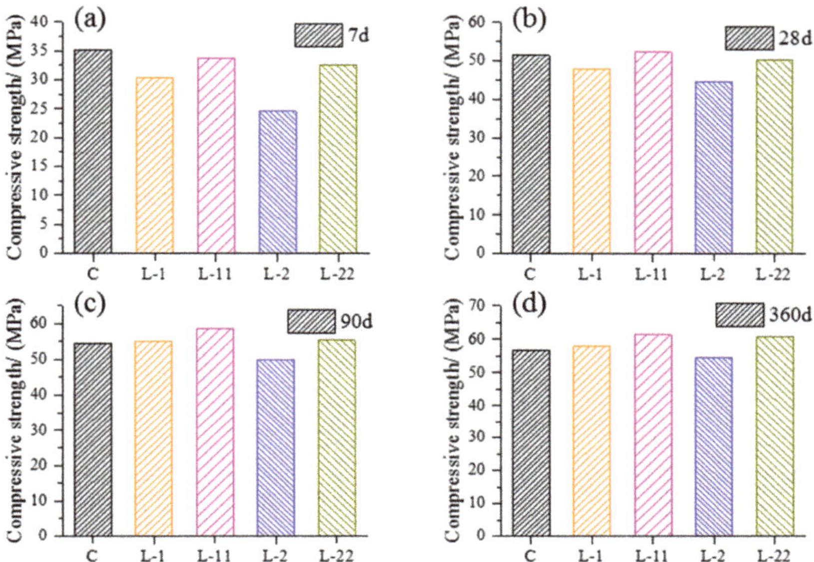

Figure 1. Compressive strength of concrete mixed with PLC: (**a**) 7 d (**b**) 28 d (**c**) 90 d (**d**) 360 d.

3.2. Chloride Permeability, Sulfate Attack and Connected Porosity

Figure 2 shows the chloride permeability (a) 28 d (b) 360 d of concrete mixed with different content of PLC. At the age of 28 days, the chloride permeability of the control sample (C), L-1, L-11 and L-22 was in the "low" grade, whereas the chloride permeability of L-2 (i.e., the content of PLC was 33%, and the water/binder ratio was 0.405) was in the "medium" grade. At the age of 360 days, the chloride permeability of L-11 (i.e., the content of PLC was 20%, the water/binder ratio was 0.418) and L-22 (i.e., the content of PLC was 33%, the water to binder ratio was 0.39) decreased to the "very low" grade. However, the chloride permeability of the control sample (C) and L-1 was still in the "low" grade.

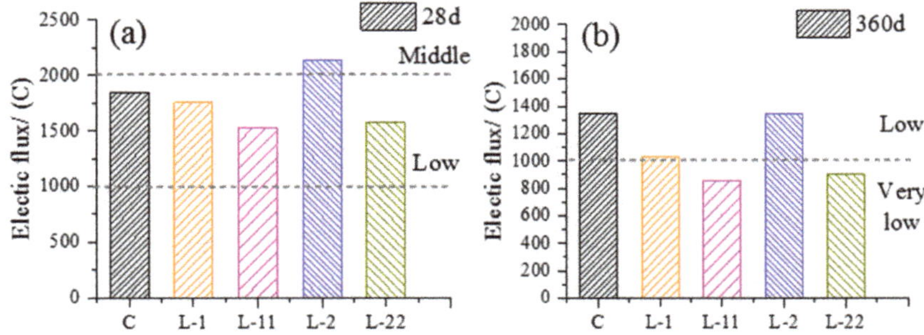

Figure 2. Chloride ion permeability of concrete mixed with PLC: (**a**) 28 d (**b**) 360 d.

Figure 3 shows the strength loss rate of concrete mixed with different content of PLC after sulfate immersion (a) 120 cycles (b) 150 cycles. The strength loss rate of the control sample (C) after 120 cycles was 17.5%, and its strength loss rate after 150 cycles was as high as 30.5%. The strength loss rates of L-11 and L-22 after 120 cycles were only one quarter of that of the control sample, that is, 3.8% and 3.6%, respectively. After 150 cycles, the strength loss rates of L-11 and L-22 were 18.5% and 19%, respectively, which was 60% of the strength loss rate of the control sample. This trend shows that PLC can effectively improve the sulfate resistance of concrete.

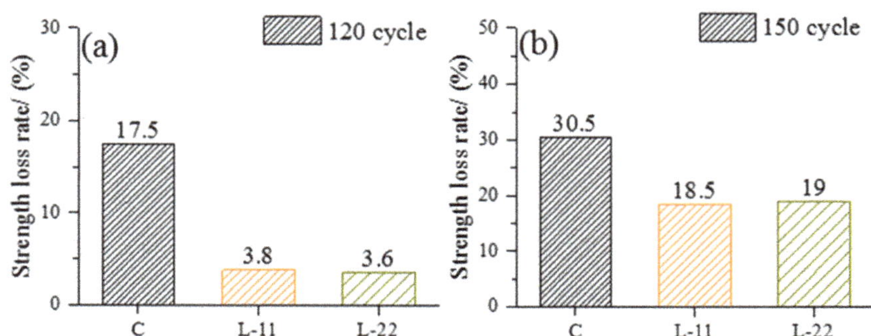

Figure 3. Strength loss rate of concrete mixed with different content of PLC after sulfate immersion: (**a**) 120 cycles (**b**) 150 cycles.

Figure 4 shows the connected porosity (a) 28 d (b) 360 d of concrete mixed with different content of PLC. The trend of chloride permeability can be explained by the trend of the connected porosity, that is, the lower the connected porosity, the better the chloride-attack resistance. At the age of 28 days, the connected porosity of the control sample (C) was 10.98. After adding 20% and 33% PLC, the connected porosity of L-1 and L-2 decreased to 10.56 and 10.85, respectively. The connected porosity of L-11 and L-22 decreased to 10.13 and 10.26, respectively, after a 0.1–0.15 reduction in water/binder ratio. Similarly, at the age of 360 days, the connected porosity of the samples showed the same trend, indicating that adding PLC or reducing the water/binder mass ratio can reduce the connected porosity of concrete, and the effect is the most positive when they are both performed at the same time.

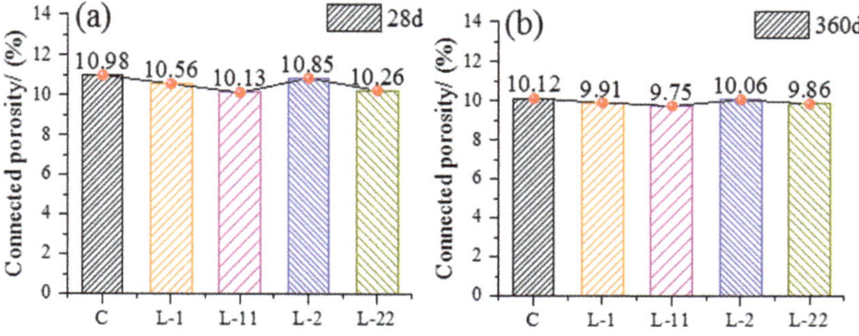

Figure 4. Connected porosity (**a**) 28 d (**b**) 360 d of concrete mixed with different content of PLC.

3.3. Hydration Heat and Adiabatic Temperature Rise

Figure 5 shows the (a) heat evolution and (b) total heat release of cement paste mixed with PLC. It can be seen from Figure 5a that the second exothermic peak of L-11 (i.e., the content of PLC was 20%, and the water/binder ratio was 0.418) was slightly earlier than that of the control sample, indicating that 20% of PLC can promote the reactivity of cement. However, the second exothermic peak of L-22 (i.e., the content of PLC was 33%, and the water/binder ratio was 0.39) was delayed, indicating that when the content of PLC is above certain threshold, the cement hydration will be weakened. It can be seen from Figure 5b that the total heat release of L-11 and L-22 decreased by 12% and 25%, respectively, indicating that PLC can significantly reduce the heat release of cement hydration. It can also be seen from Figure 6 that the adiabatic temperature rises of L-11 and L-22 were 60 °C and 55 °C, respectively, lower than that of the control sample (65 °C), which is consistent with the results of hydration heat release.

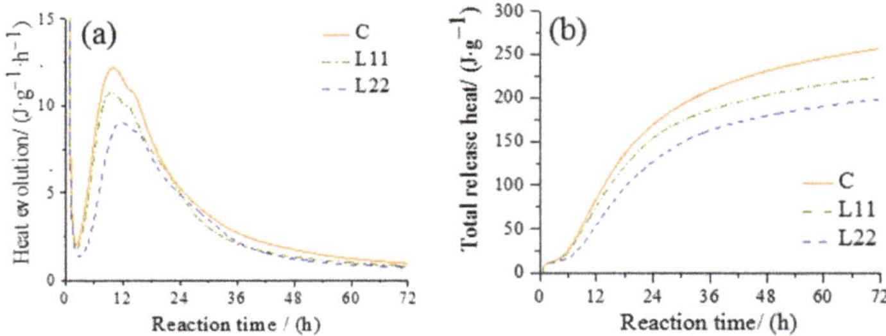

Figure 5. (**a**) Heat release rate (**b**) total heat release of cement paste mixed with PLC.

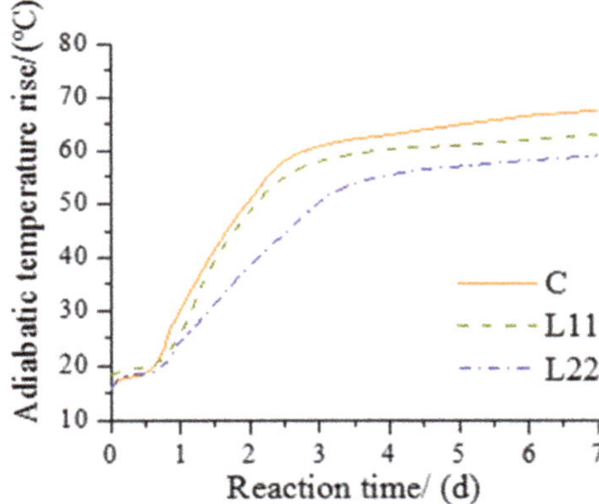

Figure 6. Adiabatic temperature rise of concrete mixed with PLC.

3.4. Drying Shrinkage

Figure 7 shows the drying shrinkage of concrete mixed with PLC. The drying shrinkage of both the control group and concrete mixed with PLC developed rapidly in the early stage. There is no obvious difference in the drying shrinkage curve between 1–50 days and their values were all around 350. From 50–150 days, the drying shrinkage (380) of concrete mixed with 33% PLC was less than that of the control group (420), and from 150–360 days, the drying shrinkage tended to be stable. The drying shrinkage rates of the control group C, L-11 (i.e., the content of PLC was 20%, and the water/binder ratio was 0.418) and L-22 (i.e., the content of PLC was 33%, and the water/binder ratio was 0.39) were 420, 400 and 380, respectively.

3.5. Reaction Mechanism

Figure 8 shows the reaction mechanism of PLC in concrete. When no PLC is added, as described in Equations (4)–(6), the reaction in concrete mainly produces C-A-S-H gel, ettringite and calcium hydroxide, among which calcium hydroxide is unfavorable for strength and durability. When the PLC is added, as described in Equations (7)–(10), the main reactions of PS in concrete are the pozzolanic reaction effect and filling effect. PS contains large amount of glass, which can react with calcium hydroxide produced by the hydration of cement to form C-A-S-H gel. These gels increase the compactness of concrete.

At the same time, the reduction in the content of calcium hydroxide can enhance the sulfate resistance of concrete. The main functions of limestone in concrete are nucleation effect and filling effect. Due to the heterogeneous nucleation of cement hydration, the surface of LS particles can become the attachment point of the crystal nucleus, which make the low-energy crystal nucleus and nucleation matrix (particle surface of LS) a replacement of the high-energy crystal nucleus and liquid interface. This reduces the nucleation barrier, promotes the hydration of cement and accelerates the strength development of concrete. Meanwhile, both PS and LS have the filling effect. This is because the fineness of PS and LS is less than that of cement, which can supplement the fine particles missing in cement and form continuous gradation in cementitious materials, filling the pores in concrete. These filling effects can improve the pore size distribution of concrete and reduce the connected porosity of concrete so as to improve the durability of concrete.

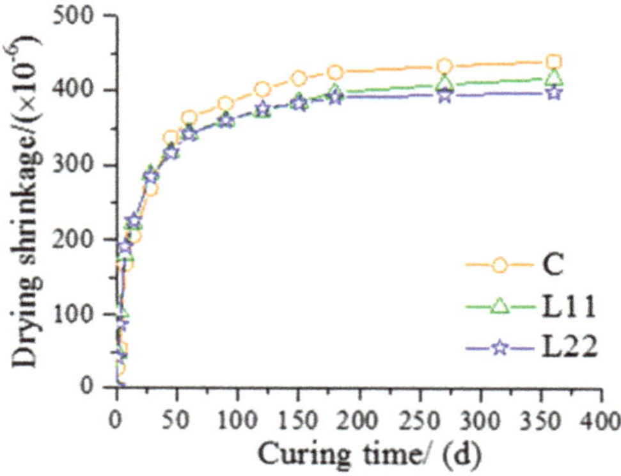

Figure 7. Drying shrinkage of concrete mixed with PLC.

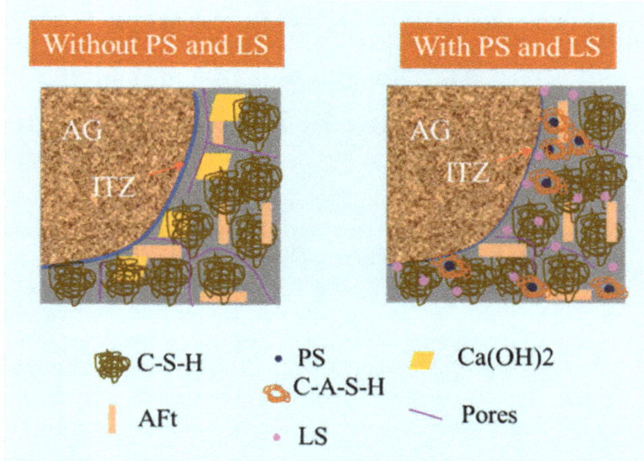

Figure 8. The reaction mechanism of PLC in concrete.

Without PS and LS:

$$C_3S + H_2O \rightarrow C\text{-}S\text{-}H + Ca(OH)_2 \tag{4}$$

$$C_2S + H_2O \rightarrow C\text{-}S\text{-}H + Ca(OH)_2 \tag{5}$$

$$C_3A + C\hat{S}H + H_2O \rightarrow AFt \tag{6}$$

With PS and LS:

$$C_3S + H_2O + LS \rightarrow C\text{-}S\text{-}H \text{ (LS nucleation)} + Ca(OH)_2 \tag{7}$$

$$C_2S + H_2O + LS \rightarrow C\text{-}S\text{-}H \text{ (LS nucleation)} + Ca(OH)_2 \tag{8}$$

$$SiO_2 \cdot Al_2O_3(PS) + Ca(OH)_2 + LS \rightarrow C\text{-}A\text{-}S\text{-}H \text{ (LS nucleation)} \tag{9}$$

$$C_3A + C\hat{S}H + H_2O \rightarrow AFt \tag{10}$$

4. Conclusions

Phosphorus slag (PS) and limestone (LS) composite (PLC) were prepared with a mass ratio of 1:1. The effects of the content of PLC and the water/binder ratio on the mechanical properties, durability and dry shrinkage of concrete were studied via compressive strength, electric flux, sulfate dry/wet cycle method, saturated drainage method, isothermal calorimeter, adiabatic temperature rise instrument and shrinkage deformation instrument.

(1) The adiabatic temperature rises of L-11 and L-22 are 60 °C and 55 °C, respectively, lower than that of the control sample C (65 °C). The addition of PLC does not affect the long-term strength of concrete. When the water/binder ratio decreases by 0.1–0.15, the long-term strength of concrete with PLC increases by about 10% compared with the control sample.

(2) At the age of 360 days, the chloride permeability of L-11 (i.e., the content of PLC was 20%, the water/binder ratio was 0.418) and L-22 (i.e., the content of PLC was 33%, the water/binder ratio was 0.39) decreases to the "very low" grade. After 150 cycles, the strength loss rates of L-11 and L-22 are about 18.5% and 19%, respectively, which is 60% of the strength loss rate of the control sample, indicating that PLC can effectively improve the sulfate resistance of concrete.

(3) The drying shrinkage of samples C, L-11 and L-22 are 420, 400 and 380, respectively. The addition of PLC can reduce the drying shrinkage. The obtained results provide theoretical and technical support for the application of PLC in mass concrete and corrosion-resistant concrete.

Author Contributions: Conceptualization, K.L. and Y.C.; methodology, K.L.; software, Y.C.; validation, K.L. and Y.C.; formal analysis, K.L.; investigation, K.L.; resources, Y.C.; data curation, Y.C.; writing—original draft preparation, K.L.; writing—review and editing, Y.C.; visualization, Y.C.; supervision, Y.C.; project administration, Y.C.; funding acquisition, Y.C. All authors have read and agreed to the published version of the manuscript.

Funding: This research was funded by the National Natural Science Foundation of China, grant number 51822807.

Institutional Review Board Statement: Not applicable.

Informed Consent Statement: Not applicable.

Data Availability Statement: All the data supporting reported results can be found in this manuscript.

Conflicts of Interest: The authors declare no conflict of interest.

References

1. Lebedeva, O.E.; Dubovichenko, A.E.; Kotsubinskaya, O.I.; Sarmurzina, A.G. Preparation of porous glasses from phosphorus slag. *J. Non-Crystal. Solids* **2000**, *277*, 10–14. [CrossRef]
2. Chen, J.S.; Zhao, B.; Wang, X.M.; Zhang, Q.L.; Wang, L. Cemented backfilling performance of yellow phosphorus slag. *Int. J. Miner. Metal. Mater.* **2010**, *17*, 121–126. [CrossRef]
3. Chen, X.; Fang, K.; Yang, H.; Peng, H. Hydration kinetics of phosphorus slag-cement paste. *J. Wuhan Univ. Techn.-Mater. Sci. Ed.* **2011**, *26*, 142–146. [CrossRef]
4. Ting, L.; Qiang, W.; Shiyu, Z. Effects of ultra-fine ground granulated blast-furnace slag on initial setting time, fluidity and rheological properties of cement pastes. *Powder Technol.* **2019**, *345*, 54–63. [CrossRef]

5.   Barnes, P.; Bensted, J. *Structure and Performance of Cement*; Wu, Z.; Wang, R., Translators; China Construction Industry Press: Beijing, China, 1991; pp. 248–262.

6.   Li, D.X.; Shen, J.L.; Mao, L.X.; Wu, X. The influence of admixture on the properties of phosphorous slag cement. *Cem. Con. Res.* **2000**, *30*, 1169–1173. [CrossRef]

7.   Li, D.X.; Shen, J.L.; Chen, L.; Wu, X. The influence of fast-setting/early-strength agent on high phosphorous slag content cement. *Cem. Con. Res.* **2001**, *31*, 19–24.

8.   Yan, P.; Mi, G.; Qiang, W. A comparison of early hydration properties of cement–steel slag binder and cement–limestone powder binder. *J. Therm. Analy. Calor.* **2014**, *115*, 193–200. [CrossRef]

9.   De Weerdt, K.; Haha, M.B.; Le Saout, G.; Kjellsen, K.O.; Justnes, H.; Lothenbach, B. Hydration mechanisms of ternary Portland cements containing limestone powder and fly ash. *Cem. Con. Res.* **2011**, *41*, 279–291. [CrossRef]

10.   Rizwan, S.A.; Bier, T.A. Blends of limestone powder and fly-ash enhance the response of self-compacting mortars. *Constr. Build. Mater.* **2012**, *27*, 398–403. [CrossRef]

11.   Li, L.G.; Kwan, A.K.H. Adding limestone fines as cementitious paste replacement to improve tensile strength, stiffness and durability of concrete. *Cem. Con. Comp.* **2015**, *60*, 17–24. [CrossRef]

12.   Hu, J.; Li, M.Y. The properties of high-strength concrete containing super-fine fly ash and limestone powder. *Appl. Mech. Mater.* **2014**, *477*, 926–930. [CrossRef]

13.   Temiz, H.; Kantarcı, F. Investigation of durability of CEM II B-M mortars and concrete with limestone powder, calcite powder and fly ash. *Constr. Build. Mater.* **2014**, *68*, 517–524. [CrossRef]

14.   Qiang, W.; Feng, J.J.; Yan, P.Y. An explanation for the negative effect of elevated temperature at early ages on the late-age strength of concrete. *J. Mater. Sci.* **2011**, *46*, 7279–7288.

15.   Zhang, Z.; Wang, Q.; Chen, H.; Zhou, Y. Influence of the initial moist curing time on the sulfate attack resistance of concretes with different binders. *Constr. Build. Mater.* **2017**, *144*, 541–551. [CrossRef]

16.   Wang, D.; Wang, Q.; Huang, Z. New insights into the early reaction of NaOH-activated slag in the presence of $CaSO_4$. *Comp. Part B Eng.* **2020**, *198*, 108207. [CrossRef]

crystals

Article

An Experimental Study of the Road Performance of Cement Stabilized Coal Gangue

Junfeng Guan [1], Meng Lu [1], Xianhua Yao [1,*], Qing Wang [2,*], Decai Wang [1], Biao Yang [1] and Huaizhong Liu [3]

1   School of Civil Engineering and Communication, North China University of Water Resources and Electric Power, Zhengzhou 450045, China; junfengguan@ncwu.edu.cn (J.G.); lumengmeng012@163.com (M.L.); wangdecai@ncwu.edu.cn (D.W.); Yangbiao1112@163.com (B.Y.)
2   School of Highway, Henan College of Transportation, Zhengzhou 450015, China
3   State Key Laboratory of Hydraulics and Mountain River Engineering, Sichuan University, Chengdu 610017, China; huaizhong.liu@scu.edu.cn
*   Correspondence: yaoxianhua@ncwu.edu.cn (X.Y.); wanglinwq2008@163.com (Q.W.)

Citation: Guan, J.; Lu, M.; Yao, X.; Wang, Q.; Wang, D.; Yang, B.; Liu, H. An Experimental Study of the Road Performance of Cement Stabilized Coal Gangue. *Crystals* **2021**, *11*, 993. https://doi.org/10.3390/cryst11080993

Academic Editors: Peter Taylor, Yifeng Ling, Chuanqing Fu and Peng Zhang

Received: 28 July 2021
Accepted: 19 August 2021
Published: 20 August 2021

Abstract: The research into the road performance of coal gangue is of great significance for the consumption of coal gangue and reducing pollution. In this paper, the coal gangues were prepared by separation and crushing processes, and their gradations were also optimized. Aiming to identify the possible problems of coal gangue as a pavement base, an unconfined compressive strength test, a splitting test, a freeze–thaw test, and a drying shrinkage test of cement stabilized gangue with varying cement amounts were carried out, and the test results were compared and analyzed. The test results showed that the maximum dry density and optimum moisture content (OMC) of the optimized cement stabilized gangue and cement stabilized macadam increased with cement content. The maximum dry density and OMC of cement stabilized macadam were larger than that of cement stabilized gangue with the same cement content. The optimized 7-day unconfined compressive strength of cement stabilized gangue can meet the requirements for a secondary and lower highway base and subbase. The OMC and cement content are the critical factors affecting the compressive strength loss rate of cement stabilized gangue after freeze–thaw cycles. The smaller the OMC of cement stabilized gangue and the larger the cement content, the lower the compressive strength loss rate. With an increase in cement content, the drying shrinkage strain of cement stabilized gangue increased. The results show that a cement content of 4% is optimal for the cement stabilized coal gangue, which can be used for the light traffic base and heavy traffic subbase of class II and below highways. It provides a basis, guide, and reference for the application of coal gangue materials in a high-grade highway base.

Keywords: coal gangue; gradation; cement content; unconfined compressive strength; freeze–thaw cycle

1. Introduction

Coal plays an essential role in the energy supply of China and many countries in the world. Coal gangue is produced during coal mining, accounting for almost 10–25% of total coal extraction [1–3]. A large number of coal gangues are piled up around mines. With continuous coal mining, more and more coal gangues are piled up, but the total utilization rate is low [4]. On the one hand, the accumulation of coal gangue occupies many valuable land resources and pollutes the soil, groundwater, and air in the accumulation place. On the other hand, a large amount of coal gangue accumulation may cause spontaneous combustion endangering the safety of the public and property. Therefore, there is an acute need to tackle this underutilization of coal gangue from a sustainability perspective [5–8].

As a suitable filling material, coal gangue was mainly used for filling subsidence areas and land reclamation earlier in many countries. In order to improve the utilization rate of coal gangue, researchers in some European countries started investigating the application of coal gangue in different fields, such as preparation of chemical materials, mining minerals, filling subgrade, and preparation of building materials [9–11]. In recent years, Li, Dong, and others [12–15] have analyzed the mechanical properties, from micro and macro perspectives, of cement mortar with calcined coal gangue replacing the cement content and have analyzed the specific factors affecting its properties. Many experts and scholars [16–20] have carried out a large number of tests to study the mechanical properties of coal gangue concrete and have analyzed the factors affecting its durability. Coal gangue material has been widely used in many civil engineering fields, such as for highway subgrade, a foundation cushion in construction engineering, etc. Many countries have taken the lead in applying coal gangue to engineering examples, such as the road network in northern France, the road network in the Ruhr region of Germany, and the railway stations of Gloucester and Croydon in Britain [21]. The application of coal gangue in highway engineering in China is gradually increasing with the advancement in transport infrastructure [22–24]. Many research findings show that the strength of coal gangue as a subgrade filler can fully meet the requirements of subgrade design [25–28]. At present, coal gangue material is mainly used as a subgrade filler and cushion, and the research into its use as a highway base is relatively scarce.

With the development of traffic infrastructure, the use of coal gangue in road engineering in China is progressively increasing. Relevant research and engineering applications show that the mechanical properties of coal gangue as a subgrade filler can fully meet the requirements of subgrade design [29,30]. He et al. [31] carried out screening tests, compaction tests, consolidation tests, permeability tests, bearing ratio tests, and direct shear tests to study the influence of soil on the engineering mechanical properties of coal gangue. The results indicated that the strength of soil gangue used as subgrade filler can fully meet the subgrade design requirements. Wu et al. [32] studied the strength and deformation characteristics of a coal gangue subgrade filler under different confining pressures, different gradations, and different compactness through large-scale triaxial tests with the method of artificial grading. The test results showed that the compactness of the coal gangue subgrade filler should not be less than 93%. Di et al. [33] conducted a preliminary study on the engineering properties of coal gangue through a compaction test and triaxial strength test. They systematically analyzed the variation model of shear strength parameters, maximum dry density, and optimum moisture content of coal gangue with coarse-grained material content. Geng et al. [34] carried out unidirectional frost heaving tests on common filling materials and coal gangue for high-speed railway subgrade under open system and closed system conditions. The test results showed that it was feasible to use coal gangue as a high-speed railway subgrade filler in permafrost regions. Zhou et al. [35] used lime fly ash stabilized coal gangue as a pavement base material and designed 15 different ratios. The test results confirmed that its strength meets the requirements of the expressway and first-class highway for the base and subbase. However, its freeze-thaw resistance is insufficient, so it is necessary to add cement to improve the durability of stabilized coal gangue. Therefore, the utilization and applicability of coal gangue are limited due to the constraints mentioned above. The utilization rate is not high, which necessitates exploring methods for improvement. The research into the frost resistance and drying shrinkage performance of coal gangue material in the base of high-grade highways is still sparse. The research into the application of coal gangue material in the base of high-grade highways is scarce, which does not allow the full utilization of coal gangue.

Given the existing problems in coal gangue applications as a highway base, this paper takes the coal gangue material produced in the Hebi area of Henan Province as its research subject. The optimum gradation of stabilized coal gangue was determined by using the power function model $y = ax^b$ [36], and cement was used to stabilize the coal gangue. Subsequently, through a compaction test, an unconfined compressive strength test, a freeze–thaw test, a splitting test, and a drying shrinkage test, the mechanical properties and durability of cement stabilized gangue were analyzed. The objective was to provide a basis, guide, and reference for the application of coal gangue materials in a high-grade highway base.

2. Raw Materials

2.1. Coal Gangue

The undisturbed coal gangue used in the test was provided by Hongchang Building Materials Co., Ltd. in Hebi City, Henan Province. The reserves of coal gangue in Hebi City are more than 20 million tons, and these are persistently accumulating. The undisturbed coal gangue was screened in the mining area, and the coal gangue with a particle size less than 30 mm was separated (Figure 1). The coal gangue with a particle size greater than 30 mm was selected for crushing and classification. In order to ensure the particle size of coal gangue, an impact crusher was used for crushing (Figure 2), and the particles greater than 30 mm were crushed and sieved. According to the particle size distribution of coal gangue, it was divided into four grades: machine-made sand, 4.75–9.5 mm, 9.5–19 mm, and 19–31.5 mm, as shown in Figure 3.

Gemini300 field emission scanning electron microscope (manufacturer: Carl Zeiss, Germany) was used to inspect the surface structure of concrete samples; Bruker D8 Advance X-ray diffractometer (manufacturer: Brooke Company, Germany) was used for X-ray diffraction analysis, and XRD data were analyzed with jade 6 software; The element distribution in the micro region of the material was analyzed qualitatively and quantitatively by SYMMETRYS EBSD energy chromatograph (manufacturer: Oxford Company, UK).The results of the scanning electron microscope (SEM), energy-dispersive X-ray spectroscopy (EDS) analysis, and X-ray diffraction (XRD) analysis are shown in Figure 4. The chemical composition and other physical and mechanical properties are shown in Tables 1–3. It can be seen that the basic physical and mechanical properties met the requirements for the class II highway base and subbase in the Chinese standard JTG/TF20-2015 "Technical guidelines for construction of highway roadbases" [37].

Figure 1. Separation of coal gangue.

Figure 2. Coal gangue crushing.

(a) Machine-made sand (0–4.75 mm)

(b) 4.75–9.5 (mm)

(c) 9.5–19 (mm)

(d) 19–31.5 (mm)

Figure 3. Sampling of coal gangue.

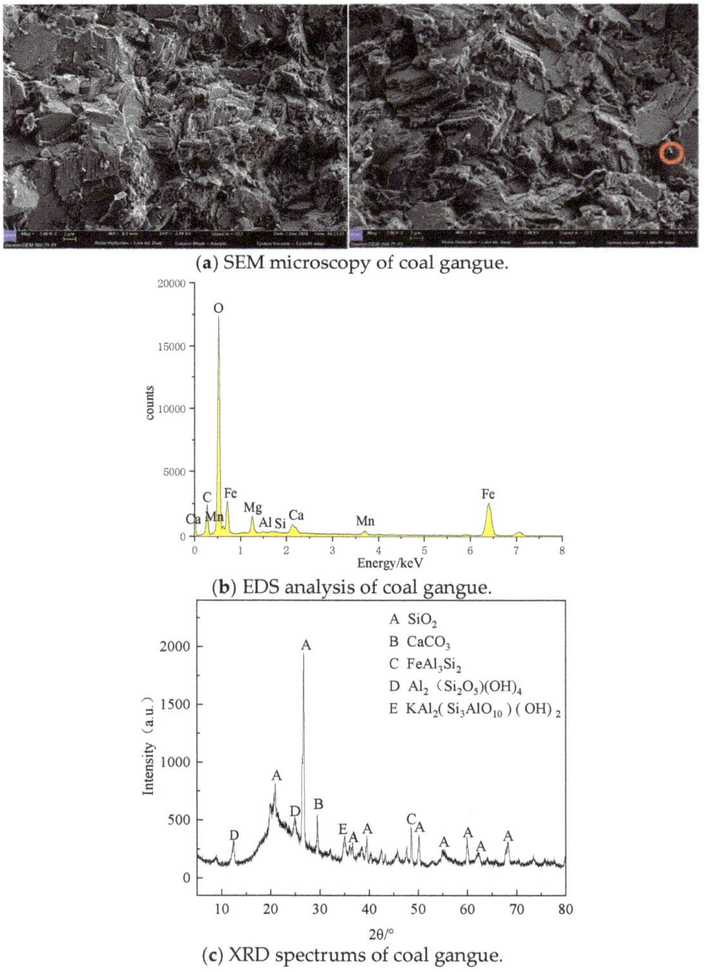

(a) SEM microscopy of coal gangue.

(b) EDS analysis of coal gangue.

(c) XRD spectrums of coal gangue.

Figure 4. Microstructure of coal gangue.

Table 1. Chemical elements of coal gangue (atomic fraction)/wt.%.

O	C	Fe	Mg	Ca	Al	Si	Mn
56.6	24.2	15.2	2.2	0.9	0.4	0.3	0.2

Table 2. Physical and mechanical properties of coal gangue coarse aggregate (wt.%).

Apparent Density/g.cm^{-3}	Bulk Density/g.cm^{-3}	Porosity/%	Crushing Value/%	Dust Content below 0.075 mm/%	Soft Rock Content/%
2.618	1.473	43.8	27.8	0.85	1.2

Table 3. Physical and mechanical properties of coal gangue fine aggregate (wt.%).

Apparent Density/g.cm^{-3}	Bulk Density/g.cm^{-3}	Porosity/%	Fineness Modulus	Water Absorption Rate/%
2.74	1.507	45	3.2	6.9

2.2. Crushed Stone

The gravel and machine-made sand used in the test were provided by Hongchang Building Materials Co., Ltd. in Hebi City, Henan Province. Based on the gradation/particle size distribution, the gravel and sand were classified into four grades (machine-made sand, 4.75–9.5 mm gravel, 9.5–19 mm gravel, and 19–1.5 mm gravel), as shown in Figure 5.

(**a**) Machine-made sand (0–4.75 mm) (**b**) 4.75–9.5 (mm) (**c**) 9.5–19 (mm) (**d**) 19–31.5 (mm)

Figure 5. Sampling of crushed stone.

The results of the scanning electron microscope (SEM), energy dispersive X-ray spectroscopy (EDS) analysis, and X-ray diffraction (XRD) analysis are shown in Figure 6. Its chemical composition and physical and mechanical properties are shown in Tables 4–6. It can be seen that its basic physical and mechanical properties can meet the requirements for the class II highway base and subbase in the Chinese standard JTG/TF20-2015 "Technical guidelines for construction of highway roadbases" [37].

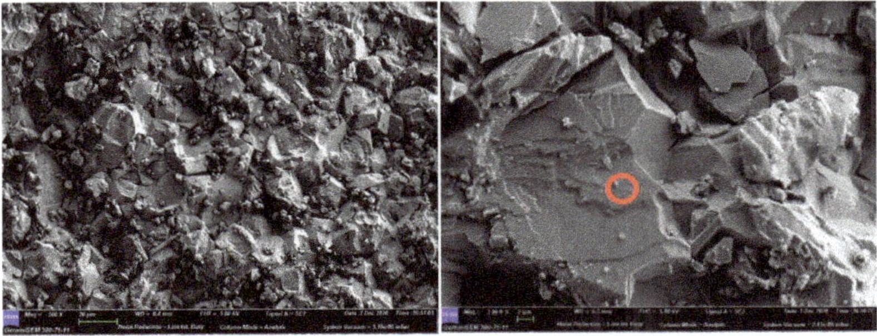

(**a**) SEM microscopy of crushed stone

Figure 6. *Cont.*

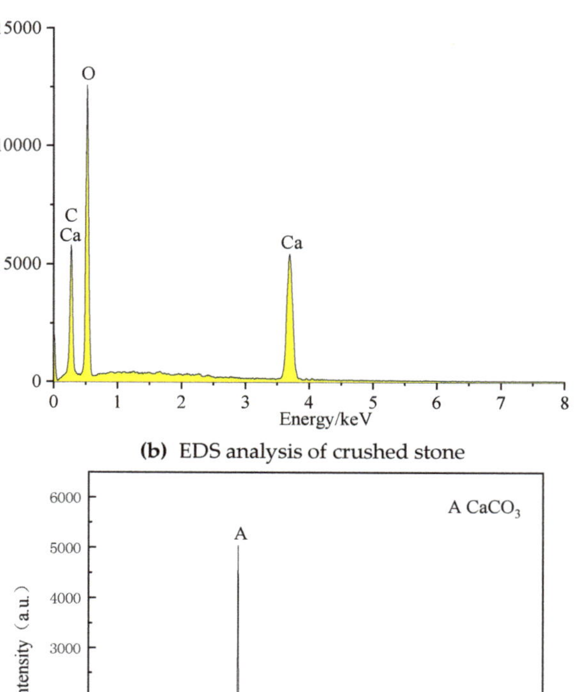

(b) EDS analysis of crushed stone

(c) XRD spectrums of crushed stone

Figure 6. Microstructure of crushed stone.

Table 4. Chemical elements of crushed stone (atomic fraction)/wt.%.

O	C	Ca
63.6	26.3	10.1

Table 5. Physical and mechanical properties of crushed stone coarse aggregate (wt.%).

Apparent Density/g.cm^{-3}	Bulk Density/g.cm^{-3}	Porosity/%	Crushing Value/%	Dust Content Below 0.075 mm/%	Soft Rock Content/%
2.650	1.523	42.5	25.2	0.81	1.2

Table 6. Physical and mechanical properties of crushed stone fine aggregate (wt.%).

Apparent Density/g.cm^{-3}	Bulk Density/g.cm^{-3}	Porosity/%	Fineness Modulus	Water Absorption Rate/%
2.74	1.504	45.1	3.33	3.5

2.3. Cement

Ordinary Portland cement (P·O·42.5) with a density of 3150 kg/m^3 was selected as cement. The properties of cement are shown in Table 7. It can be seen that the cement used in this test met the requirements of TG/TF30 "Technical guidelines for the construction of highway cement concrete pavements". It can be used as the admixture in cement stabilized gangue and cement stabilized macadam.

Table 7. Technical properties of cement.

Fineness/%	Setting Time (h)		Requirement of Normal Consistency/%	Flexural Strength (MPa)		Compressive Strength (MPa)	
	initial	final		3d	28d	3d	28d
3.1	3.6	6.5	28	6.9	9.4	27.6	45.8

3. Test Method

3.1. Gradation Optimization

Firstly, the raw materials of coal gangue and crushed stone with different particle sizes were sieved to determine the gradation curve. Further, the gradation of coal gangue was optimized according to JTG/TF20-2015 "Technical guidelines for construction of highway roadbases" [37] to meet the gradation requirements for a secondary highway base and subbase. The gravel was screened and synthesized to meet the standard gradation based on the original four particle size ranges.

3.2. Compaction Test

According to JTG E51-2009 "Test methods of materials stabilized with inorganic binders for highway engineering" [36], heavy compaction tests were carried out on cement stabilized gangue with a cement content of 3%, 4%, 5%, 6%, and 7% and cement stabilized macadam with a cement content of 4%, 5% and 6% (Figure 7).

(a) Material Preparation　　　　　　(b) Electric Compaction Instrument

Figure 7. Compaction test.

3.3. Unlateral Limited Compressive Strength Test

According to the compaction test results, the maximum dry density and optimum moisture content of cement stabilized gangue and cement stabilized macadam with different cement contents were determined. The specimens were formed according to the compaction degree of 96%. According to JTG E51-2009 "Test methods of materials stabilized with inorganic binders for highway engineering" [36], the unconfined compressive strength test of cement stabilized coal gangue and cement stabilized macadam was carried out for seven days. The static pressure was applied by the press, and 13 specimens (of size 150 mm × Φ150 mm) for each group were formed. After demolding, the specimens were cured in the standard curing room for seven days. On the last day of the curing period, the specimens were soaked in water for 24 h, and then the surface moisture was removed for subsequent testing of 7-day unconfined compressive strength (Figure 8).

(**a**) Specimen   (**b**) Experimental instruments

Figure 8. Unconfined compressive strength test.

3.4. Splitting Tensile Strength Test

According to JTG E51-2009 "Test methods of materials stabilized with inorganic binders for highway engineering" [36], the splitting tensile test was carried out on cement stabilized gangue with a cement content of 3%, 4%, and 5%. Thirteen specimens (of size 150 mm × Φ150 mm) in total were prepared. After demolding, the standard curing was carried out for 90 days according to the method of T0845-2009 [36]. During the test, the loading rate of the press was controlled at 1 mm/min. The splitting tensile test process is shown in Figure 9.

Figure 9. Splitting tensile test.

3.5. Freeze–Thaw Resistance Test

According to JTG E51-2009 "Test methods of materials stabilized with inorganic binders for highway engineering" [36], the freeze–thaw tests of cement stabilized gangue and cement stabilized macadam with a cement content of 4%, 5%, 6%, and 7% were carried out, respectively. Each group was statically pressed with a press to form 13 specimens (of size 150 mm × Φ150 mm). After demolding and marking, the specimens were put into a low-temperature chamber and frozen at −18 °C for 16 h.

After freezing for one cycle, they were weighed, and each was melted in a 20 °C water tank for 8 h. After melting, the surface of the specimen was dried, and the sample was weighed again. The above freezing and thawing cycles were repeated until five freeze–thaw cycles were completed, and the unconfined compressive strength was measured. The process of the freeze–thaw test and the appearance of specimens before and after freeze–thaw are shown in Figures 10 and 11, respectively.

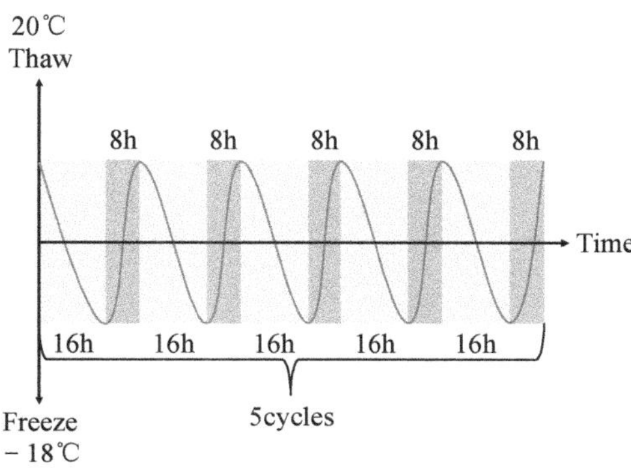

(**a**) Low temperature test chamber   (**b**) Freeze–thaw cycles

Figure 10. Freeze–thaw test process.

(**a**) Before freezing and thawing   (**b**) After freezing and thawing

Figure 11. Freeze–thaw test.

3.6. Drying Shrinkage Test

According to JTG E51-2009 "Test methods of materials stabilized with inorganic binders for highway engineering" [36], drying shrinkage tests were carried out on cement stabilized gangue with a cement content of 4%, 5%, and 6% (Figure 12). The mixture was prepared with the best moisture content obtained in the experiment, and the specimen was formed according to the compaction degree of 96%. The static pressure method was used to form the mixture, the speed was 2 kN/s, and the specimen size was 100 mm × 100 mm × 400 mm per beam. Each group had six specimens, out of which three specimens were used to measure the shrinkage deformation while the other three specimens were used to measure the drying shrinkage water loss ratio. The data was recorded every day during the first seven days and after every two days after 7-day age for 31 days.

(**a**) Specimens

(**b**) Conservation environment

Figure 12. Dry shrinkage test.

4. Test Results and Data Discussion

4.1. Gradation Optimization Design

First, the raw materials of coal gangue with different particle sizes were sieved. The sieving/screening results are shown in Table 8. According to Appendix A of JTG/TF20-2015 "Technical guidelines for construction of highway roadbases" [37], the gradation design of inorganic binder stabilized material was optimized. Three control points were selected: a nominal maximum particle size of 31.5 mm with a passing rate of 100%, a particle size of 4.75 mm with a passing rate of 40%, and a particle size of 0.075 mm with a passing rate of 2%. The power function model $y = ax^b$ was used to construct the gradation curve of coal gangue. Three control points were brought into the formula of power function model as follows:

$$y = ax^b \tag{1}$$

where: x is particle size (mm), y is the passing rate (%), and 'a' and 'b' are the coefficients ($a = 18.807$, $b = 0.4843$).

Therefore, the particle gradation of coal gangue after adjustment is shown in Table 9 and Figure 13 (c-c-2 refers to grade c-c-2 recommended in JTG/TF20-2015 "Technical guidelines for construction of highway roadbases" [37]). Therefore, the particle size distribution of coal gangue designed in this paper was well within the recommended c-c-2 grading range, which meets the gradation requirements for a secondary highway base and subbase [37]. The particle gradation of crushed stone is given in Table 10 and plotted in Figure 14.

Table 8. Screening results of coal gangue aggregates with different particle sizes.

Sieve Size/mm	31.5	26.5	19	16	13.2	9.5	4.75	2.36	1.18	0.6	0.3	0.15	0.075
0–4.75 mm	-	-	-	-	-	-	100	62	45.9	31.7	24.8	15.2	9.2
4.75–9.5 mm	-	-	-	-	100	96.8	0	-	-	-	-	-	-
9.5–19 mm	-	-	100	75.9	40	7.9	0.6	-	-	-	-	-	-
19–31.5 mm	100	71.9	6.1	3.1	2	-	-	-	-	-	-	-	-

Table 9. Coal gangue particle grading.

Sieve Size/mm	31.5	26.5	19	16	13.2	9.5	4.75	2.36	1.18	0.6	0.3	0.15	0.075
Passing (%)	100	92	78.3	72	65.6	56	40	24.1	14.6	9.0	5.4	3.3	2.0
Upper of C-C-2	100	100	87	82	75	66	50	36	26	19	14	10	7
Lower of C-C-2	100	90	73	65	58	47	30	19	12	8	5	3	2

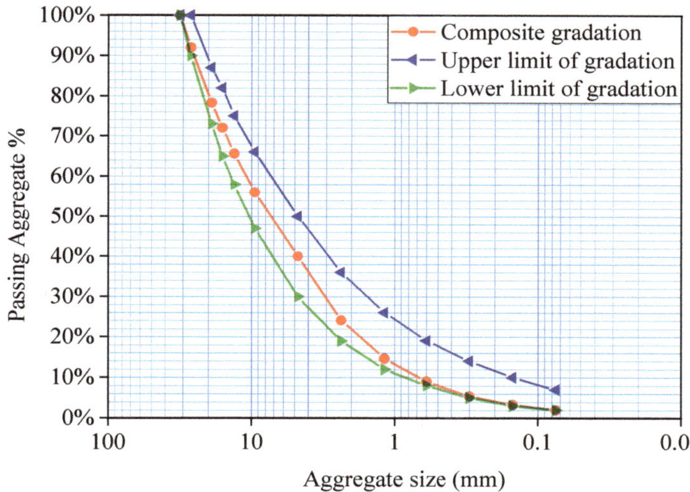

Figure 13. Grading curve of coal gangue.

Table 10. Crushed stone particle grading.

Sieve size/mm	31.5	26.5	19	16	13.2	9.5	4.75	2.36	0.6	0.075
Passing (%)	99.8	96.6	86	79.9	70	60.5	38.3	25.6	10.7	1.9
C-C-2	100	100–90	87–73	82–65	75–58	66–47	50–30	36–19	19–8	7–2

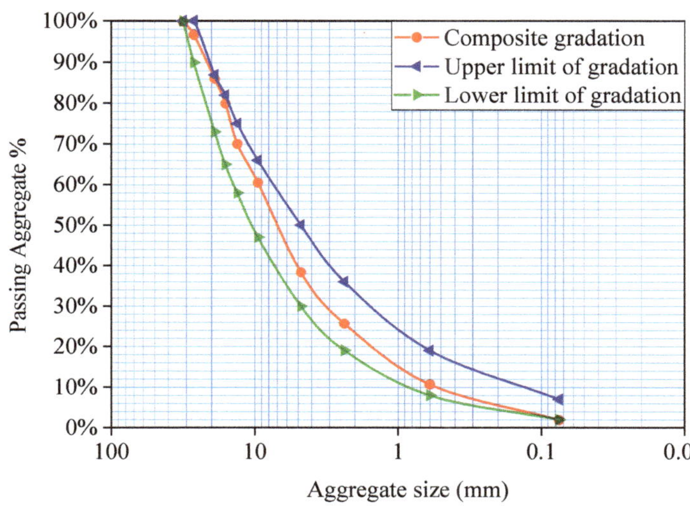

Figure 14. Grading curve of crushed stone.

4.2. Compaction Test

The maximum dry density and optimum moisture content of cement stabilized gangue cement stabilized macadam with different cement contents are shown in Figure 15. It can be seen from Figure 15a that the maximum dry density and optimal moisture content of cement stabilized coal gangue increased with the increase in cement content, and the change range is small. When the cement content was 7%, the maximum dry density and optimum moisture content of cement stabilized gangue were the highest, while with 3% cement content, these were the minimum. Compared with Figure 15a,b, under the condition of the same cement content, the maximum dry density of cement stabilized macadam was higher. The optimum moisture content was smaller than that of cement stabilized gangue.

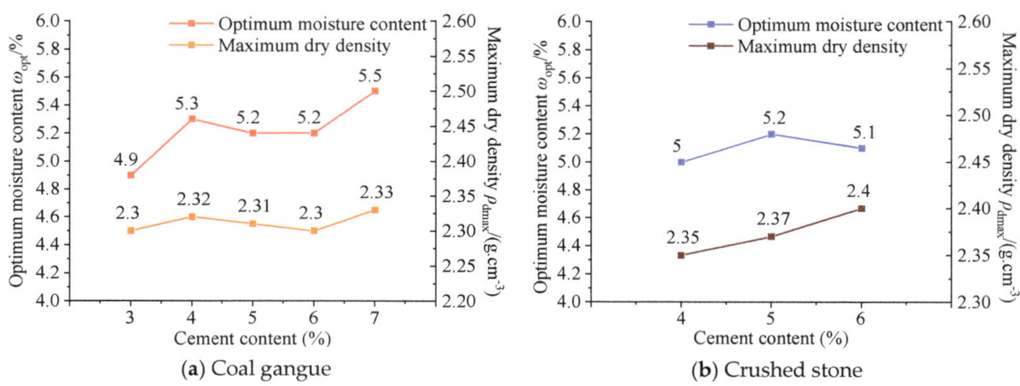

Figure 15. Test results of compaction test of mixture.

4.3. 7-Day Unconfined Compressive Strength

The representative value of 7-day unconfined compressive strength of cement stabilized macadam with 4% cement content was $R_d = 4.0$ MPa. The representative values of the 7-day unconfined compressive strength of cement stabilized gangue with different cement contents are shown in Figure 16. It is clear from Figure 16 that the representative value of unconfined compressive strength of cement stabilized gangue with 4% cement content was about 57.5% of the cement stabilized macadam with the same cement content. Compared with the 7-day unconfined compressive strength of 3% cement stabilized gangue, the 7-day unconfined compressive strength of cement stabilized gangue with 4%, 5%, 6%, and 7% cement increased by 0.2 MPa, 1.7 MPa, 1.9 MPa, and 2.1 MPa, respectively. Reasons may be that with the increase in cement content, the pores around the specimen are correspondingly reduced, which can make the structural shape more dense, so as to increase the strength of cement stabilized coal gangue [38]. The 7-day unconfined compressive strength of 7% cement stabilized gangue was higher than that of the 4% cement stabilized macadam. It can be seen from Figure 16 and Table 11 that when the cement content is the same, the 7-day test results of cement stabilized coal gangue in this paper were higher than those of other researchers in Table 11. In this paper, the 7-day unconfined compressive strength of cement stabilized coal gangue with 3% and 4% cement content can meet the requirements for a medium and light traffic base (2.0–4.0 MPa) and a heavy traffic subbase (2.0–4.0 MPa) of class II and below highways, respectively; other researchers needed the cement content of cement stabilized coal gangue to reach 5% or 6%, respectively.

Compared with the 7-day unconfined compressive strength requirements of cement stabilized material base and subbase for different highway grades in JTG/TF20-2015 ("Technical guidelines for construction of highway roadbases" [37]), in this paper, the 7-day unconfined compressive strength of cement stabilized coal gangue with cement content of 5% met the requirements for a heavy traffic base (3.0–5.0 MPa) and an extremely heavy and extra heavy traffic subbase (2.5–4.5 MPa) of class II and below highways. The 7-day unconfined compressive strength of cement stabilized coal gangue with a cement content of 6% and 7% met the requirements of an extremely heavy traffic base (3.0–5.0 MPa) and an extremely heavy traffic subbase of class II and below highways requirements for an extra heavy traffic base (4.0–6.0 MPa) and subbase (2.5–4.5 MPa), respectively.

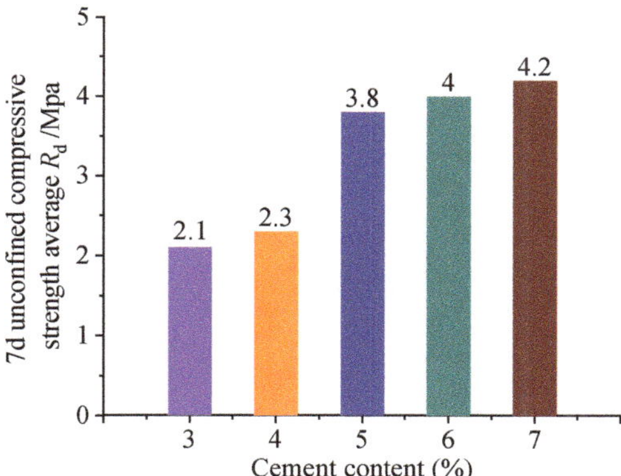

Figure 16. Test results of 7-day unconfined compressive strength.

Table 11. 7-day unconfined compressive strength test results of different researchers.

Researcher	Cao [39]	Yan [40]		Hu [41]	
Cement content/%	4	5	4	5	6
Compressive strength/MPa	1.92	2.69	1.81	2.34	2.92

4.4. Splitting Tensile Strength of Cement Stabilized Coal Gangue

The representative values of the 90-day splitting tensile strength of cement stabilized gangue with different cement contents are shown in Figure 17. It can be seen that the 90-day splitting tensile strength values of cement stabilized coal gangue with 3%, 4%, and 5% content were 0.65 MPa, 0.87 MPa, and 1.06 MPa. Compared with the 90-day splitting tensile strength of 3% cement stabilized gangue, the 7-day unconfined compressive strength of cement stabilized gangue with 4% and 5% cement increased by 0.22 MPa and 0.41 MPa, respectively. The results also indicated that the 90-day splitting tensile strength of cement stabilized coal gangue increased with cement content, and showed the same change law as the unrestricted compressive strength. The splitting tensile strength mainly reflects the cementing ability of cement stabilized aggregate inside aggregate. It mainly depends on the cohesion between aggregates. The C–S–H gel after hydration of cement is the source of this cohesive force [40]. Therefore, an increase in cement content can improve the splitting tensile strength of cement stabilized coal gangue.

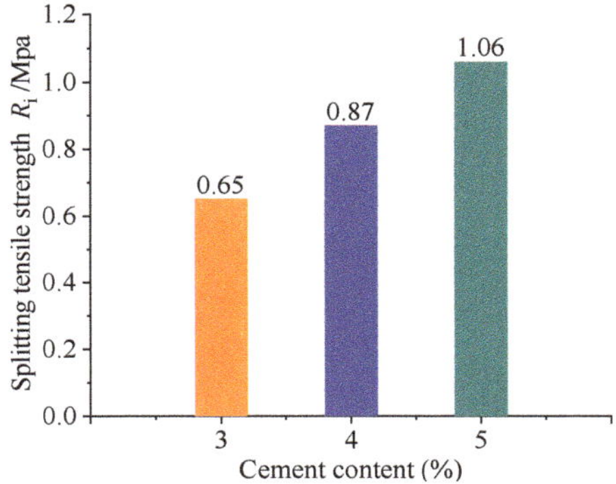

Figure 17. Test results of splitting tensile.

The splitting tensile strength of the cement stabilized macadam gangue mixture with different gradations of 5.5% cement content in [42] was 0.648 MPa and 0.723 MPa, respectively, and the splitting tensile strength of the water cement stabilized macadam gangue mixture with different gradations of 5.5% cement content in [43] were 0.749 MPa, 0.652 MPa, and 0.731 MPa. It was found that the splitting tensile strengths of cement stabilized coal gangue in this paper were higher than those of [42,43]. Reasons may be that the different gradations will lead to different maximum dry density and optimal water content of the sample. In this paper, the power function model $y = ax^b$ [36] was used. The corresponding maximum dry density of the optimized gradation sample was greater than the maximum dry density of [42,43], and the optimal water content was also less than the maximum dry density of [42,43] resulting in a more compact structure of cement stabilized coal gangue optimized in this paper. The recommended range of values for the splitting tensile strength of cement stabilized macadam is 0.4–0.6 MPa in the specifications [36]. Clearly, the 90-day splitting tensile strength of cement stabilized coal gangue with a cement content of 3% and 4% can meet the requirements of the specification.

4.5. Freeze–Thaw Resistance Test of Cement Stabilized Coal Gangue

Table 12 and Figure 18 show the frost resistance index and freeze-thaw compressive strength loss diagram of cement stabilized gangue with 4%, 5%, 6%, and 7% cement content. It can be seen from these results that the compressive strength loss BDR of the cement stabilized coal gangue with cement contents of 4%, 5%, 6% and 7% were greater than 75% after five freeze-thaw cycles, and the mass loss rate was less than 1%. Its compressive strength loss BDR was greater than the frost resistance requirements of lime fly ash stabilized materials in the heavy freezing area of Expressway and class I Highway (70%) [44], and the mass loss rate was less than the specification requirements (5%) [36]. The compressive strength loss of the cement stabilized gangue with 5% and 6% cement was the least, followed by the cement stabilized gangue with 7% cement which had a higher loss in compressive strength after freeze-thaw cycles. The compressive strength loss with 4% cement content was the highest. Reasons may be that the cement stabilized coal gangue with a cement content of 4% has a large number of pores due to the small cement content, and due to its high water content, the volume expansion of the water in the pores during the freezing process is large, which destroys the original pore structure and produces microcracks, resulting in greater strength loss than the other three cement contents [38]. Therefore, the cement content and moisture content are the main factors affecting the compressive strength loss of the cement stabilized coal gangue after freeze-thaw cycles.

Table 12. Frost resistance index.

Cement Content/%	4%	5%	6%	7%
BDR/%	75.16	94.33	92.26	89.29
Mass loss/%	0.07	0.26	0.81	0.24

4.6. Drying Shrinkage Test

It can be seen from Table 13 and Figure 19 that with the increase in cement content, the drying shrinkage strain of the cement stabilized gangue increased. At the same time, the hydration and setting of cement will cause volume shrinkage, which is positively correlated with cement content. Therefore, the larger the cement content, the larger the deformation of cement stabilized gangue, and correspondingly the larger the dry shrinkage strain. The changing trend was drastic at first, and then slowed down with the increase in age [38].

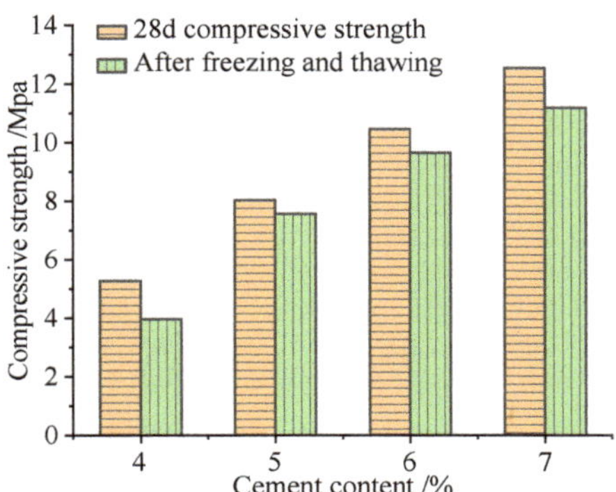

Figure 18. Freeze–thaw compressive strength loss diagram.

Table 13. Dry shrinkage strain value of cement stabilized gangue ($\times 10^{-6}$).

Cement Content/%	Age (d)			
	7	15	23	31
4	112.5	345	512.5	627.5
5	165	382.5	620	850
6	230	480	705	895

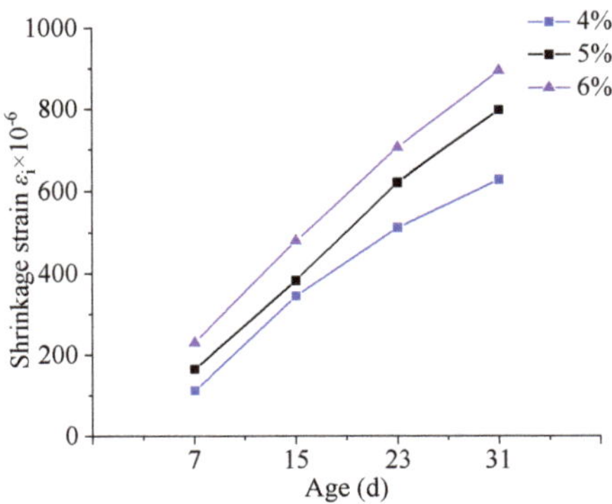

Figure 19. Relationship between dry shrinkage strain and age of cement stabilized gangue.

5. Conclusions

In this paper, according to JTG/TF20-2015, "Technical guidelines for construction of highway roadbases" [37], the gradation design of inorganic binder stabilized materials was optimized, and the power function model $y = ax^b$ was used to optimize the gradation of coal gangue. The optimized coal gangue gradation curve met the gradation requirements for the secondary highway base and subbase. It provides a basis, guide, and reference for the application of coal gangue materials in a high-grade highway base.

(1)   Coal gangue was prepared by a sorting and crushing process, and the grading of coal gangue raw materials was optimized by a power function model. The prepared cement stabilized coal gangue pavement base material with 4% cement content had a 7-day unconfined compressive strength of 2.3 MPa, a 90-day splitting tensile strength of 0.87 MPa, a frost resistance index BDR of 75.16%, a mass loss rate of 0.07%, and a 31-day dry shrinkage strain of 627.5×10^{-6}, which can be used for medium and light traffic bases and the heavy traffic subbase of class II and below highways.

(2) The test results show that the cement content and the grading of the coal gangue raw materials are the main factors affecting the 7-day unconfined compressive strength of the cement stabilized coal gangue. An increase in cement content can improve the 90-day splitting tensile strength of cement stabilized coal gangue. Cement content and moisture content are the key factors affecting the frost resistance index of cement stabilized coal gangue. The larger the cement content is, the larger the dry shrinkage strain of cement stabilized coal gangue is, and the shrinkage strain decreases first and then increases with an increase in age.

(3) In view of the possible problems of cement stabilized coal gangue as a pavement base, the comprehensive experimental analyses were carried out to verify the feasibility of cement stabilized coal gangue as a highway base and subbase.

Author Contributions: J.G. and X.Y. designed the experiments; M.L., Q.W., D.W., B.Y. and H.L. carried out the experiments; X.Y. and M.L. analyzed the experimental results; J.G. and M.L. reviewed, and edited the manuscript; J.G. received the funding. All authors have read and agreed to the published version of the manuscript.

Funding: This project was sponsored by the National Natural Science Foundation of China (51779095), the Program for Science & Technology Innovation Talents in Universities of Henan Province (20HASTIT013), Sichuan Univ, the State Key Lab Hydraul & Mt River Engn (SKHL2007), and the Innovation project of the 12th postgraduate of North China University of Water Resources and Electric Power (YK2020-11).

Data Availability Statement: All the relevant data and models used in the study have been provided in the form of figures and tables in the published article.

References

1. Bian, Z.; Dong, J.; Lei, S.; Leng, H.; Mu, S.; Wang, H. The iMpact of disposal and treatment of coal mining wastes on environment and farmland. *Environ. Geol.* **2009**, *58*, 625–634. [CrossRef]
2. Duan, X.; Xia, J.; Yang, J. Influence of coal gangue fine aggregate on microstructure of cement mortar and its action mechanism. *J. Build. Mater.* **2014**, *17*, 700–705.
3. Ma, J.; Yu, Z.M.; Shu, S.H.; Zeng, Z.H. University G. Environmental Hazards of Coal Gangue and the Control Measures. *Coal Eng.* **2015**, *47*, 70–73.
4. Liu, H.; Liu, Z. Recycling utilization patterns of coal mining waste in China. *Resour. Conserv. Recycl.* **2010**, *54*, 1331–1340.
5. Gong, C.; Yan, J.; Liu, J.; Huaijun, Y.U. Biology Migration and Distribution Characteristics of Trace Elements in Reconstructed Soil with Coal Gangue Filling. *Agric. Sci. Technol.* **2016**, *46*, 189–192.
6. Wang, S.; Luo, K.; Wang, X.; Sun, Y. Estimate of sulfur, arsenic, mercury, fluorine emissions due to spontaneous combustion of coal gangue: An important part of Chinese emission inventories. *Environ. Pollut.* **2016**, *209*, 107–113. [CrossRef]
7. Qin, J.; Zhao, R.; Chen, T.; Zi, Z.; Wu, J. Co-combustion of municipal solid waste and coal gangue in a circulating fluidized bed combustor. *Int. J. Coal Sci. Technol.* **2019**, *6*, 218–224. [CrossRef]
8. Mog, H.; Adam, M.J.; Ajalloeian, R.; Hajiannia, A. Preparation and application of alkali-activated materials based on waste glass and coal gangue: A review. *Constr. Build. Mater.* **2019**, *221*, 84–98.
9. Skaryńska, K.M. Reuse of coal mining wastes in civil engineering—Part 2: Utilization of minestone. *Waste Manag.* **1995**, *15*, 83–126. [CrossRef]
10. Marland, S.; Han, B.; Merchant, A. The effect of microwave radiation on coal grindability. *Fuel* **2000**, *79*, 1283–1288. [CrossRef]
11. Sripriya, R.; Rao, P.; Choudhury, B.R. Optimisation of operating variables of fine coal flotation using a combination of modified flotation parameters and statistical techniques. *Int. J. Miner. Process.* **2003**, *68*, 109–127. [CrossRef]
12. Li, W.X.; Wang, D.L.; Niu, J.S.; Ma, X.W. Preparation of Coal Gangue Cement Mortar. *Adv. Mater. Res.* **2013**, *684*, 159–162. [CrossRef]
13. Dong, Z.; Xia, J.; Fan, C.; Cao, J. Activity of calcined coal gangue fine aggregate and its effect on the mechanical behavior of cement mortar. *Constr. Build. Mater.* **2015**, *100*, 63–69. [CrossRef]
14. Qiu, Y.L.; Zhang, X.X.; Liu, K.P.; Hu, X.Y.; Guan, B.W. Research on Mechanical Behaviour of Cement Mortar with High-Volume Coal Gangue. *Adv. Mater. Res.* **2011**, *261–263*, 685–689. [CrossRef]
15. Qin, B.; Ji, Y.; Bai, Z.; Zhao, Q. Research on mechanical behavior of calcined coal gangue cement mortar with different chemical activators. In Proceedings of the 2017 3rd International Forum on Energy, Environment Science & Materials, Shenzhen, China, 25–26 November 2017; Atlantis Press: Amsterdam, The Netherlands, 2018; pp. 653–656.
16. Wang, H.; Amp, Y.V. Research on the Freeze-thaw Damage Law of Coal Gangue Concrete. *Fly Ash Compr. Util.* **2019**, *2*, 42–45.

17. Zheng, Y.; Zhang, P.; Cai, Y.; Jin, Z.; Moshtagh, E. Cracking resistance and mechanical properties of basalt fibers reinforced cement-stabilized macadam. *Compos. Part B Eng.* **2019**, *165*, 312–334. [CrossRef]
18. Huang, M.; Duan, J.; Wang, J. Research on Basic Mechanical Properties and Fracture Damage of Coal Gangue Concrete Subjected to Freeze-Thaw Cycles. *Adv. Mater. Sci. Eng.* **2021**, *7*, 1–12.
19. Chen, Y.; Niu, W.; Ding, Z. Research on Elastic Modulus of Gangue Concrete. In Proceedings of the 14th International Congress on the Chemistry of Cement ICCC, Beijing, China, 13–16 October 2015.
20. Guan, J.; Li, Q.; Wu, Z.; Zhao, S.; Dong, W.; Zhou, S. Minimum specimen size for fracture parameters of site-casting dam concrete. *Constr. Build. Mater.* **2015**, *93*, 973–982. [CrossRef]
21. Wang, D.; Zuo, Y.; Li, X.; Fan, D.; Cui, Y.; Wang, W. The mineralogy characteristics of coal shale and its utilization in building materials. *Block-Brick-Tile* **2006**, *6*, 17–23.
22. Ma, Q.; Liu, H. Analysis of Obstacle of Development and Application that Coal Gangue Used in China Road Engineering. *Adv. Mater. Res.* **2014**, *900*, 510–513. [CrossRef]
23. Li, L.; Long, G.; Bai, C.; Ma, K.; Wang, M.; Zhang, S. Utilization of Coal Gangue Aggregate for Railway Roadbed Construction in Practice. *Sustainability* **2020**, *12*, 4583. [CrossRef]
24. Yu, Q.; Miao, L.; Liu, S. Application Study and Practice of Coal Gangue Applied in Road Construction. *J. Highw. Transp. Res. Dev.* **2002**, *2*, 1–5.
25. Peng, Y.; Yang, J.S.; Liu, J.H. Mechanical Properties of Coal Gangue Used as Subgrade Filling. *Appl. Mech. Mater.* **2012**, *178–181*, 1226–1229. [CrossRef]
26. Peng, Y.; Yang, J.S.; Qing, H.J. Study of Pavement Characteristic of Coal Gangue Used as Roadbed. *Soil Eng. Found.* **2008**, *4*, 57–59.
27. Li, C. Study on Field Rolling Test of Coal Gangue Used as Expressway Subgrade Filling. *Value Eng.* **2017**, *36*, 153–155.
28. Chen, R.; Wang, P.; Liu, P.; Chen, W.; Kang, X.; Yang, W. Experimental study on soil-water characteristic curves of subgrade coal gangue filler. *Rock Soil Mech.* **2020**, *41*, 372–378.
29. Liu, C.; Zhou, C. Application of coal gangue in road construction. *China's Coal Ind.* **2001**, *9*, 46.
30. He, J.; Jin, M.; Yang, J. Study on mechanical properties and filling technology of coal gangue mixed with soil in road engineering. *China Civ. Eng. J.* **2008**, *41*, 87–93.
31. Xu, D. Construction of railway embankment using gangue. *Railw. Constr. Technol.* **2000**, *2*, 38–40.
32. Wu, J.; Gao, W.; Zhang, Z.; Tang, X.; Yi, M. Study on strength and deformation characteristics of gangue subgrade filler. *J. Railw. Sci. Eng.* **2021**, *18*, 885–891.
33. Di, G.; Li, H.; Li, L.; Zhang, F. Experimental study on mechanical properties of road coal gangue. *Railw. Eng.* **2011**, *8*, 127–129.
34. Geng, L.; Tang, H.; Luo, J.; Xiuli, D.U.; Ling, X.; Feng, J. Experimental Study on Frost-Heaving Characteristics of High Speed Railway Subgrade Fillers Mixed with Coal Gangue. *Railw. Eng.* **2019**, *59*, 41–45.
35. Zhou, M.; Li, Z.G.; Wu, Y.Q.; Zhang, X.F.; Ai, L. Experimental Research on Lime-Fly Ash-Cement Stabilized Coal Gangue Mixture. *J. Build. Mater.* **2010**, *13*, 213–217.
36. Ministry of Transport. *JTG E51—2009, Test Methods of Materials Stabilized with Inorganic Binders for Highway Engineering*; People's Communications Press: Beijing, China, 2009. (In Chinese)
37. Ministry of Transport. *JTG/TF 20—2015, Technical Guidelines for Construction of Highway Roadbases*; People's Communications Press: Beijing, China, 2015. (In Chinese)
38. Su, Y.H.; Wang, X.M.; Lv, C.; Gao, J. Strength and durability of low volume cement stabilized coal gangue. *J. Inn. Mong. Agric. Univ.* **2021**, *42*, 67–72.
39. Cao, D.; Ji, J.; Liu, Q.; He, Z.; Wang, H.; You, Z. Coal Gangue Applied to Low-Volume Roads in China. *Transp. Res. Rec.* **2018**, *2204*, 258–266. [CrossRef]
40. Yan, G.Y.; Zhou, M.K.; Chen, X.; Yu, G.; Liu, B.L.; Kang, Z. Study on Application of Coal Gangue Aggregate Pavement Base Material. *J. Wuhan Univ. Technol. Transp. Sci. Eng.* **2020**, *45*, 6.
41. Hu, P. Feasibility Research of Huaibei Washing Gangue Using in Base Course. *Mod. Transp. Technol.* **2019**, *16*, 1–5.
42. Guo, Y.X.; Li, C.; Li, M. Experimental study on cement stabilized macadam-gangue mixture in road base. *Int. J. Coal Prep. Util.* **2019**, *6*, 1–14. [CrossRef]
43. Li, M.; Li, C.; Guo, Y.X. Experimental Study on Performance of Cement Stabilized Macadam-Gangue Mixture. *Bull. Chin. Ceram. Soc.* **2019**, *276*, 200–206.
44. Ministry of Transport. *JTG D50-2017, Specifications for Design of Highway Asphalt Pavement*; People's Communications Press: Beijing, China, 2017. (In Chinese)

crystals

Article

Performance Evaluation of Sustainable Concrete Comprising Waste Polypropylene Food Tray Fibers and Palm Oil Fuel Ash Exposed to Sulfate and Acid Attacks

Hossein Mohammadhosseini [1,*], **Rayed Alyousef** [2,*], **Shek Poi Ngian** [1] and **Mahmood Md. Tahir** [1]

[1] Institute for Smart Infrastructure and Innovative Construction (ISIIC), School of Civil Engineering, Universiti Teknologi Malaysia (UTM), Skudai 81310, Malaysia; shekpoingian@utm.my (S.P.N.); mahmoodtahir@utm.my (M.M.T.)
[2] Department of Civil Engineering, Prince Sattam Bin Abdulaziz University, Al-Kharj 16273, Saudi Arabia
* Correspondence: mhossein@utm.my (H.M.); r.alyousef@psau.edu.sa (R.A.)

Abstract: Sulfate and acid attacks cause material degradation, which is a severe durability concern for cementitious materials. The performance of concrete composites comprising waste plastic food trays (WPFTs) as low-cost fibers and palm oil fuel ash (POFA) exposed to acid and sulfate solutions has been evaluated in an immersion period of 12 months. In this study, visual assessment, mass variation, compressive strength, and microstructural analyses are investigated. For ordinary Portland cement (OPC), six concrete mixtures, including 0–1% WPFT fibers with a length of 20 mm, were prepared. In addition, another six mixtures with similar fiber dosages were cast, with 30% POFA replacing OPC. It was discovered that adding WPFT fibers and POFA to concrete reduced its workability. POFA concrete mixes were found to have higher long-term compressive strength than OPC concrete mixes cured in water. As a result of the positive interaction between POFA and WPFT fibers, both the crack formation and spalling of concrete samples exposed to acid and sulfate solutions were reduced, as was the strength loss. The study's findings show that using WPFT fibers combined with POFA to develop a novel fiber-reinforced concrete subjected to chemical solutions is technically and environmentally feasible. WPFT fibers have a significant protective effect on concrete against chemical attacks.

Keywords: sustainability; concrete composites; durability; sulfate and acid attacks; WPFT fibers

Citation: Mohammadhosseini, H.; Alyousef, R.; Poi Ngian, S.; Tahir, M.M. Performance Evaluation of Sustainable Concrete Comprising Waste Polypropylene Food Tray Fibers and Palm Oil Fuel Ash Exposed to Sulfate and Acid Attacks. *Crystals* **2021**, *11*, 966. https://doi.org/10.3390/cryst11080966

Academic Editors: Yifeng Ling, Chuanqing Fu, Peng Zhang and Peter Taylor

Received: 31 July 2021
Accepted: 14 August 2021
Published: 16 August 2021

Publisher's Note: MDPI stays neutral with regard to jurisdictional claims in published maps and institutional affiliations.

1. Introduction

Over the past few decades, global recycling and energy recovery rates have steadily grown, consequently reducing contributions to landfill sites. Around the world, the rate of landfilling varies greatly. Plastics have become an essential and fundamental part of our lives in numerous shapes and types, and the use of plastic has progressively increased. Global plastic manufacturing reached about 288 million tons in 2012 [1]. Approximately half of this amount was spent on one-time consumable commercial products, which have contributed significantly to the rise of plastic-related waste. Furthermore, Wu and Montalvo [2] stated that plastics contribute an ever-increasing quantity to the solid waste stream, owing to their wide variety of uses. Besides, because computable information on waste plastic manufacturing is usually maintained in-house and accomplished on a business-to-business basis, it is rarely made public. According to Plastics Europe, European countries produce 25 million tons of waste plastic annually. Eriksen et al. [3] reported in 2014 that only 29.7% of waste was effectively recycled, with about 39.5% being used as reutilization and energy resources, and 30.8% being disposed of. Hearn and Ballard [4] stated that the generated waste plastics are non-biodegradable and can remain in nature for decades or centuries. Additionally, chemical reactions in some waste plastics can emit hazardous chemicals, contaminating the air, soil, and underground waters. Consequently, the generation of plastic wastes from any source is seen as a significant ecological problem.

Food packaging has changed substantially since the nineteenth century in response to global innovations and client needs. In this regard, Blanco [5] pointed out that over the last three decades, the utilization of plastics in various types and forms in food packaging has grown dramatically, as these materials are available in vast amounts with lower cost, transferability, good barrier properties, and prospective utility. Silvestre et al. [6] also stated that packing contributes to about 42% of the worldwide plastic sector, which has raised from about 5 million tons in the 1950s to over 100 million tons worldwide, with the packaging industry accounting for nearly 2% of GNP in developing countries. Despite all the benefits of polymeric food packaging stated above, Martino et al. [7] discovered that most of these polymers are conventionally created with high microbial resistance and have become unsustainable for the ecosystem. Because of the increased attention to decreasing the ecological challenges connected with plastic waste discarding and recycling, several researchers have been motivated to invent new materials that are less harmful to the environment [8–10]. Based on the location and accessible technologies and equipment, post-consumer waste plastics could be attained with various methods. The recycling process is complicated, according to Hubo et al. [11], since biological particles may contaminate waste plastic food trays. Kumar et al. [12] stated that the prevention of waste plastic generation in the first place, which is directly related to public awareness, is one of the preeminent techniques to avoid the accretion of a large number of waste plastics. These strategies and attempts are similar to those targeted at the appropriate and well-organized valorization of the enormous masses of plastic waste that are undoubtedly formed daily. Thus, according to Siddique et al. [13], the first-ever option that comes to mind is mostly the dumping and burning of such wastes.

According to Almeshal et al. [14], recovering plastic wastes from pre-and post-consumer sources is one of the most popular technological advances to recycle such waste, generate new raw materials, and end the loop of plastic waste; however, due to a deficiency of equipment and technology, most of the polypropylene food trays, including several impurities, are not appropriate to send for recycling and be used as secondary raw materials. Conventional approaches to discarding the massive volume of generated plastic wastes worldwide include burning and dumping [15,16]. Consequently, a reliable waste disposal strategy is required for this form of solid waste. A mechanical or chemical mechanism creates fresh raw materials during reprocessing. Eriksen et al. [3] reported that these unique raw materials could be used in production procedures to complete the circle or be utilized in other activities, such as buildings or similar construction industries. The recycling and reprocessing of solid wastes such as plastics are critical steps toward sustainable development in modern life [17]. Household plastic wastes, particularly polypropylene type of food trays, are a dispersed and contaminated resource, causing less reused plastic and reduced potential for closed-loop recovery. Lower physicomechanical properties and governing necessities for the chemical composition of reusable plastics can limit the possibility of closed-loop recycling [18].

Furthermore, one of the sectors where a considerable amount of plastic as waste material can be utilized is the construction industry. A great deal of research has been carried out in regard to plastic waste components as aggregates and fiber constituents in concrete mixtures [19]. Construction materials, like polymeric-based fibers, formed from raw plastic components are widely available. Researchers have been interested in reused plastics from several suppliers, and various investigations on the properties of concrete made from waste plastic materials and its potential ecological impacts have been conducted [20]. Concrete in various types has been recognized in construction industries for its simplicity of manufacturing, strength, and specific durability characteristics; however, concrete elements are frequently subjected to adverse circumstances, either naturally, like sulfate-rich coastal areas and soils, or artificially, including industrial wastewater and drainage, which impact the overall efficiency of hardened concrete. Sotiriadis et al. [21] stated that concrete degrades under the influence of such harsh conditions due to chemical reactions involving ion exchange. These compounds attack the C-S-H crystals of the cement matrix and the formation of soluble components such as ettringite and gypsum in the concrete matrix. Due to the creation of these new products, the microstructure of the matrix changes, causing the concrete to deteriorate [22].

Sulfate attack is a severe issue that shortens the service life of concrete. Sulfates are widespread in groundwaters when they occur in sulfate-bearing rocks with sulfate mineral-rich soils. Furthermore, as revealed by Bulatovic et al. [23], seawater contains a significant amount of sulfate. Generally, the solid forms of sulfate do not enter and attack the concrete components aggressively, but sulfate ions find their way into the previous matrix and chemically react with the cement constituents when dissolved in water. In mortar and concrete, a sulfate attack results in the creation of expanding gypsum and ettringite components. Consequently, the calcium aluminate hydrate of the cement paste is primarily influenced in this case [24,25]. Aside from sulfates, which can penetrate concrete and cause severe damage, acids are yet another chemical that could also lead concrete structures to deteriorate. Acid attack is not entirely resistant to cementitious materials [26]. Hadigheh et al. [27] stated that the cement hydrate's most sensitive component is $Ca(OH)_2$, but the acid could also affect the C-S-H gels in the matrix, depending on the types of acid and their concentration. Lu et al. [28] reported that acidic rains in which sulfuric acid is the primary element cause severe corrosion in concrete components. Furthermore, according to Mohammadhosseini and Tahir [29], the main consequence of sulfuric attack is gypsum, observable on the surface and linked to the volume expansion of concrete specimens that could induce tensile stress structural elements, resulting in cracking and spalling. Besides, Bankir and Sevim [30] pointed out that the accumulation of gypsum in reinforced concrete reduces the corrosion rate of rebar by sealing the outer surface of the samples if it is not washed out. Different chemical processes among cement paste and gypsum, such as the formation of calcium aluminate, can result in ettringite development with a higher quantity, leading to further specimen cracks and a reduction in the concrete compressive strength.

In addition to mechanical characteristics, fiber-reinforced concrete has shown satisfactory performance after contact with chemicals such as acid and sulfate attacks [31]. For instance, adding polypropylene (PP) fiber to concrete reduces the pre-yield toughness of reinforced concrete beams and the corrosion of steel bars [32]. Sulfate attacks on hardened concrete are among the most prevalent in terms of deterioration due to the chemical reaction among sulfate ions and the cement paste, which finally directs to the formation of ettringite and gypsum and therefore causes concrete to deteriorate [33]. Owing to the high surface free energy, the addition of PP fiber to concrete improves its resistance to magnesium sulfate and acid attacks, as well as good resistance to crack formation by minimizing cracks in the interfacial transition zones (ITZ) [34]. Consequently, adding fibers to concrete enhances its mechanical properties and improves its ability to resist chemical attacks. Bolat et al. [35] reported that several strategies could be used to develop the durability of concrete structures. For instance, admixtures improve the strength and show superior durability performance in concrete by generating a dense matrix by giving a well-graded particle size dissemination for reducing the entry of severe components into the concrete. These procedures have little effect on the ductility of concrete [36], where a study investigated the use of WPFT fiber and POFA ashes in the development of concrete composites and evaluated the combination of these materials on the performance of concrete under sulfate and acid attacks. Concrete specimens were immersed in water, 10% $MgSO_4$, and 5% H_2SO_4 solutions for 12 months, and concrete performance was assessed in terms of visual assessment, mass change, and variation in strength analysis. Moreover, the effects of chemical attacks on concrete microstructure were examined by scanning electron microscopy (SEM). In addition, the use of waste materials in concrete, such as plastic waste, contributes to the reduction and environmental protection of generated waste.

2. Experimental Program

2.1. Materials

In this study, ordinary Portland cement (type I) was utilized and met the ASTM C 150-07 standards. Also, raw POFA ashes from a palm oil mill in Johor, Malaysia, were collected. The ash was dried and sieved at a temperature of $100 \pm 5\,^{\circ}$C to remove bigger components and reduce the carbon content. For every 4 kg of raw ashes, fine particles lesser than 150 μm were processed for grinding in a Los Angeles abrasion device for around 2 h. The acquired POFA met the requirements of BS 3892 part 1 of 1992, and it may be classified as class C and F according to ASTM C618-15. Table 1 lists the physical characteristics and chemical compositions of the OPC and POFA.

Table 1. Physical properties and chemical composition of the OPC and POFA.

Material	Physical Properties		Chemical Composition (%)							
	Specific Gravity	Blaine Fineness	SiO_2	Al_2O_3	Fe_2O_3	CaO	MgO	K_2O	SO_3	LOI
OPC	3.15	3990	20.4	5.2	4.19	62.4	1.55	0.005	2.11	2.36
POFA	2.42	4930	62.6	4.65	8.12	5.7	3.52	9.05	1.16	6.25

In this work, clean and dry natural river sand with a maximum size of 4.75 mm was used as fine aggregate with a 2.3 fineness modulus and 2.6 specific gravity value and water absorption of 0.7%. Also, crushed granite passed through a 10 mm sieve with a 2.7 specific gravity and water absorption of 0.5% was utilized as a coarse aggregate. Potable water was used for mixing and curative reasons. A constant dosage of a polymeric superplasticizer was added to balance the flowability of fresh concrete. In addition, the fibers used in this research were fabricated from a polypropylene form of post-consumer waste food trays. The trays of different sizes and forms were recovered as post-consumer waste and washed with water to remove any contaminants that could impair the characteristics of the concrete, as shown in Figure 1a. The hand-cutting operation was performed with scissors to generate consistent sheets. To be employed in the main experiments as fibers, the unpolluted plastic strips from waste food trays were cut to a uniform width of 2 mm, thickness of 0.3 mm, and 20 mm length (Figure 1b,c). The fundamental features of the waste polypropylene food tray fibers are listed in Table 2.

Figure 1. (**a**) Post-consumer waste plastic food trays, (**b**) fabricated fibers and (**c**) WPFT fibers used in this study.

Table 2. Typical characteristics of the fabricated WPFT fibers used in this study.

Plastic Type	Strand Shape	Dimension (W × L) (mm)	Density Range (g/cm³)	Thickness (mm)	Tensile Strength (MPa)	Elongation (%)
Polypropylene	Rectangular	2 × 20	0.94	0.3	550	7–10

2.2. Formulation of the Experimental Prototype

Table 3 shows various formulations of the experimental prototype. The DOE concrete mix design approach was used in this research, using a water/binder (w/b) ratio of 0.49. In total, twelve mixes with varying fiber dosages were made, with a control batch (B1) being cast without the addition of WPFT fibers and POFA. Two groups of concrete mixtures, namely, OPC and POFA-based, were made with the WPFT fiber dosages of 0, 0.2, 0.4, 0.6, 0.8, and 1.0% in a total of 12 batches. In the POFA-based mixtures, POFA replaced OPC by 30%.

Table 3. The proportions of the different concrete mixtures.

Mix	Cement (kg/m^3)	POFA (%)	POFA (kg/m^3)	Water (kg/m^3)	Fine Aggregate (kg/m^3)	Coarse Aggregate (kg/m^3)	V_f (%)
B1	445	-	-	220	830	865	-
B2	445	-	-	220	830	865	0.2
B3	445	-	-	220	830	865	0.4
B4	445	-	-	220	830	865	0.6
B5	445	-	-	220	830	865	0.8
B6	445	-	-	220	830	865	1.0
B7	312	30	133	220	830	865	-
B8	312	30	133	220	830	865	0.2
B9	312	30	133	220	830	865	0.4
B10	312	30	133	220	830	865	0.6
B11	312	30	133	220	830	865	0.8
B12	312	30	133	220	830	865	1.0

2.3. Testing Methods

Concrete cubes of a 100 mm size were cast for the entire experimental work in accordance with the specifications of BS EN 12390-2: 2009 and BS EN 12390-3: 2009 at various curing periods up to 12 months. Chemical immersion experiments were utilized in this research work to investigate the performance of concrete composites comprising WPFT fibers and POFA against magnesium sulfate and sulfuric acid attacks. Using scanning electron microscopy (SEM), microstructural investigation of several concrete mixtures was undertaken to evaluate the morphological and degradation mechanisms. Overall, 36 concrete cubes of a 100 mm size were cast and coated with plastic to minimize fast vaporization for each test. The samples were then remolded after 24 hours and submerged in water for 28 days to cure. Afterward, the samples were withdrawn, wiped dry, and weighed to the nearest 0.1 g with an electronic weighing balance, recording the initial weight with the succeeding weight throughout the exposure period. Next, the concrete cubes were immersed in 10% $MgSO_4$ and 5% H_2SO_4 solutions for the sulfate and acid resistance tests.

The concrete sample cubes were entirely submerged in the chemical solutions for 12 months, and the pH values of the solutions changed quickly in both tests. Nevertheless, the pH for the sulfate resistance test was restricted to a concentrated value of 8.5 and 2.5 for the acid solution. The pH was controlled through refreshing the solutions and was then maintained constantly during the immersion duration. No standardized approach is currently in place to test concrete resistance against sulfuric acid and sulfate assaults. Furthermore, the testing techniques of ASTM C1012-15, ASTM C267-01-2012, and ASTM C452-15 guide testing the chemical resistance of mortar and polymer concrete. Consequently, following the standard specifications and the current literature, the concentration levels of 10% and 5% were selected in this study for the magnesium sulfate and sulfuric acid, respectively.

At the end of the exposure period (12 months), to evaluate the consequences of chemical attacks on the performance of the concrete samples, the samples were taken from the testing tanks and cleaned with plain water, then dried in a testing room at ambient temperature for about 30 minutes. After the specimens had sufficiently dried, visual inspection was carried out to observe the degradation degree for each mixture after 12 months of exposure in the $MgSO_4$ and H_2SO_4 solutions. The evaluation considered the specimen edges, texture, color, size, and geometry. After that, the average masses of specimens for each group were measured to calculate the variation in the mass by following Equation (1):

$$ML_t = \frac{M_t - M_i}{M_i} \times (100) \tag{1}$$

where M_t is the average mass of samples after 12 months exposure to chemical solutions (gr), M_i is the original mass earlier absorption (gr), and ML_t is the cumulative mass loss after 12 months of exposure.

An evaluation of the strength loss factor (SLF %) after 12 months of exploration was carried out to indicate the level of deterioration in concrete mixtures in terms of reduction in compressive strength, and the calculation was made with the following formula (Equation (2)):

$$SLF = \frac{F_{cw} - F_{cs}}{F_{cw}} \times 100\% \tag{2}$$

where SLF indicates the percentage of loss in compressive strength of concrete cubes exposed to H_2SO_4 and $MgSO_4$ solutions for 12 months, F_{cw} is the obtained strength values of concrete specimens after 12 months cured in plain water, and F_{cs} is the residual strength of the cubic samples immersed in H_2SO_4 and $MgSO_4$ solutions for the same period.

3. Results and Discussion

3.1. Workability

A slump test was carried out to investigate the effects of WPFT fibers on the fluidity of concrete mixtures and the results are presented in Figure 2. WPFT fiber addition led to a reduction in the slump values, with the highest slump value being 185 mm for the OPC control mix. Slump values of 140 mm and 95 mm were noted for mixes comprising 0.2% and 0.4% WPFT fibers, respectively. According to the ACI 211-02 standard requirements, these concrete mixtures have fluid and plastic workability and can be used in structural members with congested and regular reinforcing. A minimum slump value of 35 mm was measured for the mixture containing 1% WPFT fibers. The results also demonstrate that the POFA mixes had less workability than the OPC-based mixes. Increases in the dosage of fibers had a comparable consequence on the flowability of the POFA mixtures. The slump of mixes with fiber dosages of 0, 0.2, 0.4, 0.6, 0.8, and 1%, for example, were measured as 165, 115 mm, 70 mm, 55 mm, 40 mm, and 25 mm. Previous research has shown that the PP types of waste plastic fiber harm the workability of concrete [37]. This could be attributed to the fact that fibers with a large surface area absorb more cement paste, which causes the viscosity to rise, resulting in a lower slump value. Furthermore, the high fiber content and wide surface area of short fibers results in reduced workability [38].

3.2. Water-cured Compressive Strength

Cubic compressive strength testing for the concrete samples cured in plain water at various curing periods was carried out and the results are illustrated in Figure 3. The obtained results show a decrease in the compressive strength values as the fiber volume fractions increased. Nevertheless, the reduction in strength was minor, and the strength values were within the acceptable range for structural applications. After 365 days of curing in plain water, the strengths of OPC mixtures reinforced with 0, 0.2, 0.4, 0.6, 0.8, and 1% WPFT fibers were 48.5, 47.2, 44.8, 43.2, 39.7, and 37.4 MPa, accordingly. Generally, the modulus of elasticity of PP types of fibers ranged between 3.5–4.9 GPa, and these comparatively low values classified the PP fiber as a soft material, and the matrix was treated as a soft composite, and, therefore, caused a drop in the compressive strength of reinforced concrete specimens [39,40].

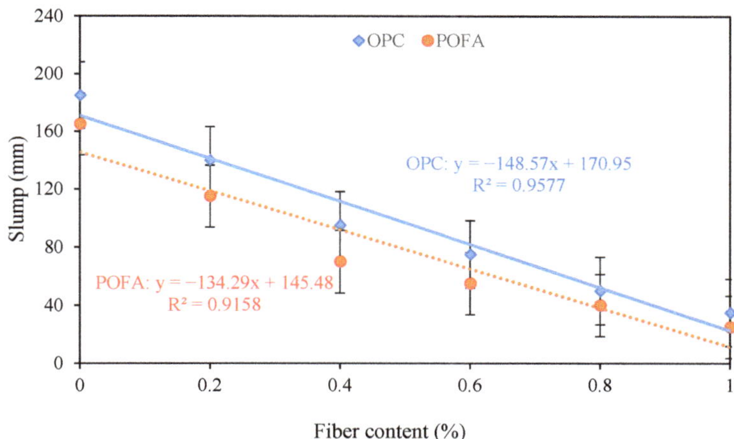

Figure 2. Effects of the WPFT fibers at various dosages on the slump of fresh concrete mixtures.

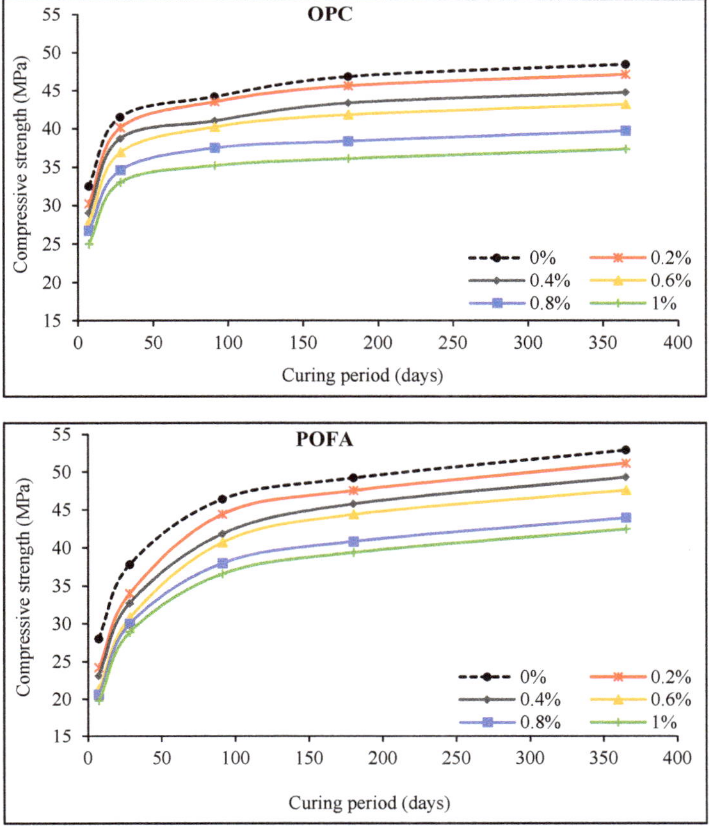

Figure 3. The variation in the water-cured compressive strength of OPC and POFA concrete mixtures containing WPFT fibers.

Furthermore, replacing 30% of OPC with POFA resulted in greater compressive strength values, particularly after 12 months of curing in water. The strength value for the plain POFA mix was recorded as 52.9 MPa, which is approximately 10% greater than the 48.5 MPa compressive strength value observed for the plain OPC mixture without any fibers. The testing results showed that the POFA mixes obtained higher strength values at extended curing periods than the OPC mixes. According to Zeyad et al. [41], the higher strength of POFA mixtures could be attributed to pozzolanic activity at ultimate ages, leading to increased concrete strength by creating supplementary C-S-H gels. POFA mixtures incorporating WPFT fibers showed a similar trend to the OPC mixtures, where the inclusion of fibers and rises in fiber volume fractions resulted in slight decreases in compressive strength values [42]. After 365 days of immersion in water, the compressive strength values of POFA mixes containing 0, 0.2, 0.4, 0.6, 0.8, and 1% WPFT fiber were 52.9, 51.2, 49.3, 47.6, 43.9, and 42.4 MPa, respectively, as shown in Figure 3 These values were higher than those recorded for the OPC mixtures with the exact fiber dosages.

3.3. Sulfate Attack

3.3.1. Visual Assessment

The performance of cubic concrete samples immersed in the $MgSO_4$ solution was visually inspected monthly in this investigation. The degradation scale used to assess the damage is shown in Table 4. Figure 4 depicts the appearances of specimens after 12 months of exposure, while Figure 3 depicts the deterioration degree of the OPC and POFA specimens in a $MgSO_4$ solution for the entire exposure period. The plain OPC mix without WPFT fibers was the only mix that showed initial signs of deterioration, such as small cracks along the corners of specimens. POFA-based specimens, on the other hand, showed the first marks of degradation after three months of immersion. As shown in Figure 5, the deterioration of all specimens, with and without fibers, accelerated over time. The evolution of the OPC-based mixtures was faster than that of the samples containing 30% POFA. The specimens containing a higher dose of WPFT fibers were less severely deteriorated after 12 months of exposure than the plain concrete mixes.

Since $CaCO_3$ and $Ca(OH)_2$ are the most vulnerable elements of the cement hydrate, degradation was more intense in the OPC specimens, which have higher contents of these components. This finding suggests that sulfate-induced degradation is exacerbated by the low resistance of $CaCO_3$ and $Ca(OH)_2$ to sulfate assault, resulting in the formation of calcium sulfate ($CaSO_4$) [21]. The small cracks and slight deterioration that occurred at the corners of samples were the initial evidence of attack in all cases, which appears as small cracks, slow expansion of samples, and the spalling of edges. A soft white substance was also applied to the interior surfaces of the fractures. The presence of fibers was discovered to influence the degree of deterioration, as well as the progression. As can be seen, the deterioration of the specimens studied progressed more slowly in combinations reinforced with WPFT fibers, and greater fiber volume fractions showed stronger resilience. It is also worth noting that the final degree of deterioration in POFA-based combinations was lower than in the OPC mixtures for all specimens.

3.3.2. Mass Gain

Figure 6 depicts the variation in the masses of the concrete cube samples when immersed in a $MgSO_4$ solution for 12 months, revealing that all mixtures gained mass after the end of the exposure period due to sulfate particle absorption by the specimens [21]. The results revealed rises in masses by 1.3 and 0.8% for the control mixes of OPC and POFA, respectively, after 12 months of exposure. The addition of WPFT fiber to the mixture, on the other hand, drastically reduced the mass increase of the specimens. The addition of fiber in the mixes appears to have contributed to the reduction in mass gain, as shown in Figure 6. For fiber contents of 0.2, 0.4, 0.6, 0.8, and 1%, the mass gains of OPC fibrous mixes were 0.48, 0.56, 0.96, 1.27, and 1.57%, respectively. The results also demonstrated that adding 30% POFA to the fibrous composites resulted in lesser mass gains when compared to the OPC mixes when submerged in a sulfate solution.

Figure 4. The appearance of concrete specimens exposed to a $MgSO_4$ solution for 12 months.

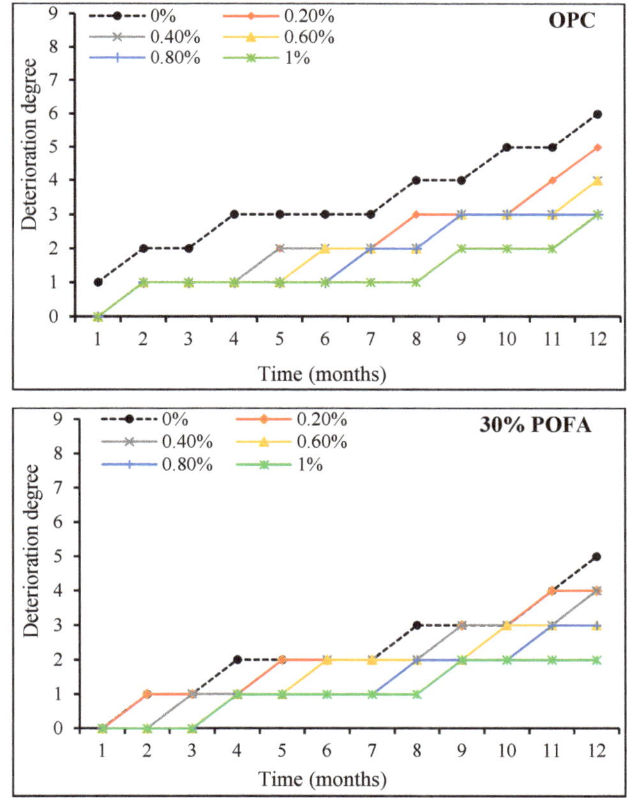

Figure 5. Deterioration degree of concrete specimens reinforced with WPFT fibers exposed to $MgSO_4$ solutions.

Table 4. Assumed degree of deterioration and the detected damages.

Deterioration Scale	Observed Damages
0	No noticeable deterioration
1	Slight deterioration at edges in the form of microcracks
2	Deterioration at edges and corners
3	Cracking at corners and along the edges of samples
4	Severe cracks and expansion along the edges
5	Widespread cracks and sample expansion
6	Further expansion and side deterioration of samples
7	Wide expansion and spalling
8	Extensive deterioration and washout
9	Sample breakdown

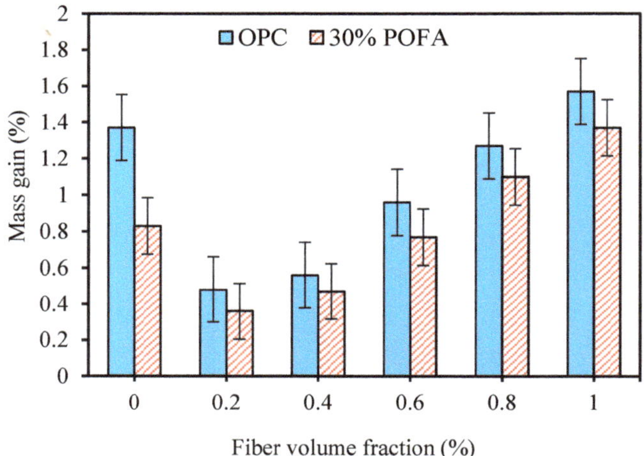

Figure 6. The effects of the WPFT fibers on the mass gain of concrete samples exposed to a $MgSO_4$ solution.

The variations in the masses of POFA-based mixes were recorded as 0.36, 0.47, 0.77, 1.1, and 1.35% for the similar fiber levels, respectively, which indicates a gain in mass for all specimens. The development of a grid structure by fibers, which restricts particle penetration and disruption into specimens, might be attributed to the drop in the mass gain of specimens reinforced with WPFT fiber. As a result of the irregular pore arrangement of the POFA concrete and the quantity of calcium hydroxide contained in the matrix, the use of POFA in mixtures is favorable when in direct contact with a sulfate attack. Microcracks are the primary entry points for hazardous chemical ions, such as sulfate ions, into the concrete interior. The exposure of concrete specimens, either plain concrete or fiber-reinforced concrete, followed several steps, as shown in Figure 7, including: (1) initial exposure, (2) chemical reactions, (3) stress development in the concrete specimens, (4) crack formation, (5) crack development and infiltration, and (6) substantial damage in various forms. It can be seen that there were no noteworthy changes in the process of sulfate attacks amongst the plain mix and mixes reinforced with fibers, with gypsum and ettringite development producing expansion tension to cause concrete degradation [24]. The bridging action of the fiber, which arrests microcracks, reduces the permeability and spalling of the concrete significantly. Furthermore, the fibers prevent the passage of disturbance particles into specimens by arresting microcracks induced by excessive stress [25]. Consequently, the concrete mixtures comprising fibers gained less mass than conventional plain concrete.

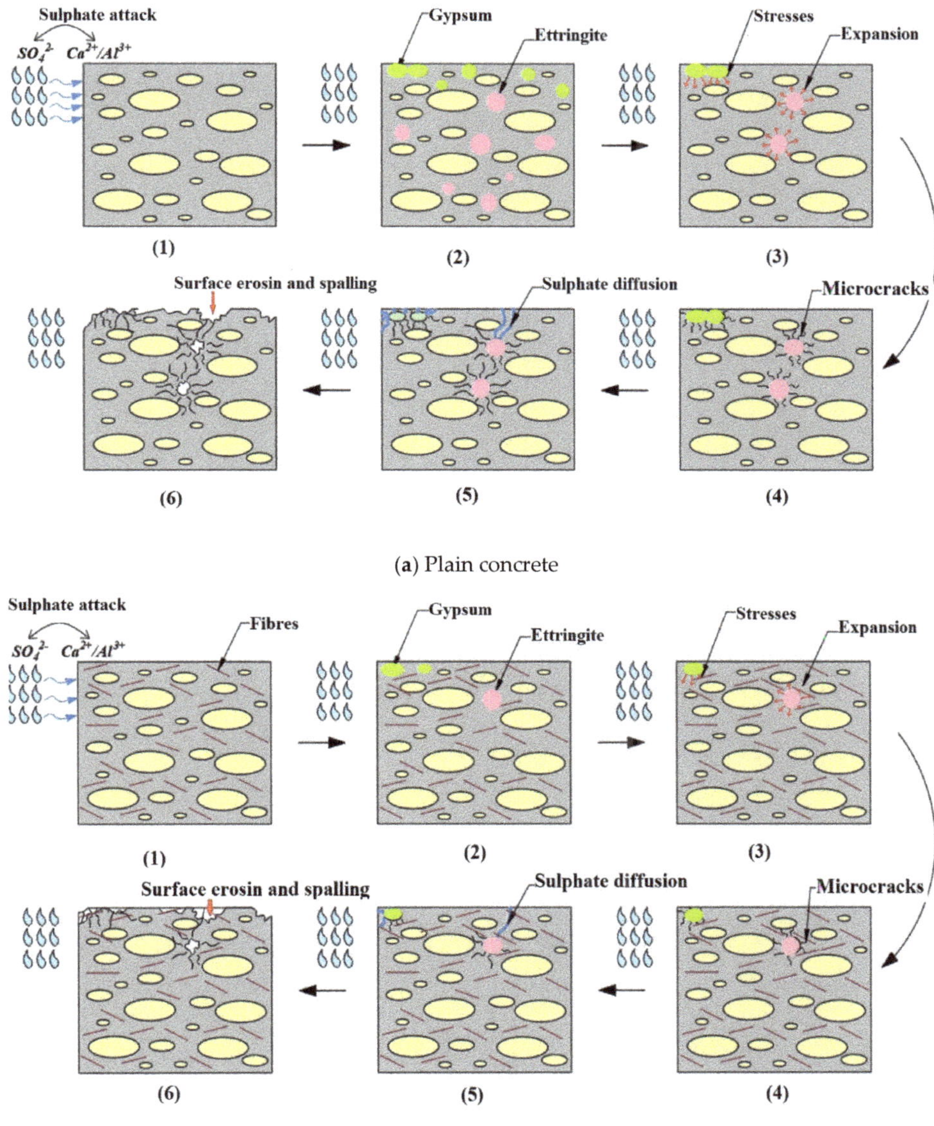

Figure 7. Diagrammatic representation of sulfate attack against (**a**) plain concrete and (**b**) fiber-reinforced concrete.

3.3.3. Compressive Strength Loss

Generally, concrete performance often degrades when exposed to a chemical attack and extends the immersion period, owing to physical destruction and undesirable chemical breakdown of the matrix. The inclusion of WPFT fibers improved the resistance to magnesium sulfate assault, as seen in Figure 8. After 12 months of exposure, the compressive strength of concrete mixtures exposed to sulfate attacks was evaluated, and the attained outcomes were associated with the water-cured mixes. The strength loss was described as the variation in strength among water-cured mixes and those exposed to $MgSO_4$. The data in Figure 8 demonstrate that all of the blends had lost strength. When the OPC mixes were associated with those of POFA mixes, the loss was found to be greater. For example, in the OPC mixes, strength losses of 9.5, 6.3, 5.2, 9.6, 12.5, and 14.6 MPa were detected for fiber dosages of 0, 0.2, 0.4, 0.6, 0.8, and 1%, correspondingly, whereas the values of 7.5, 5.7, 3.6, 8.4, 10.7, and 12.5 MPa were recorded as strength losses for the same fiber dosages, respectively, in the POFA mixtures. It was observed that the obtained values are lower in POFA mixtures than those of OPC, which indicates the better performance of POFA concrete when exposed to chemical attacks.

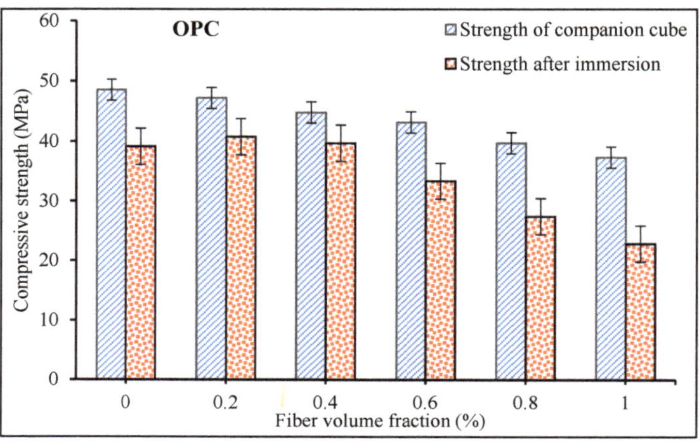

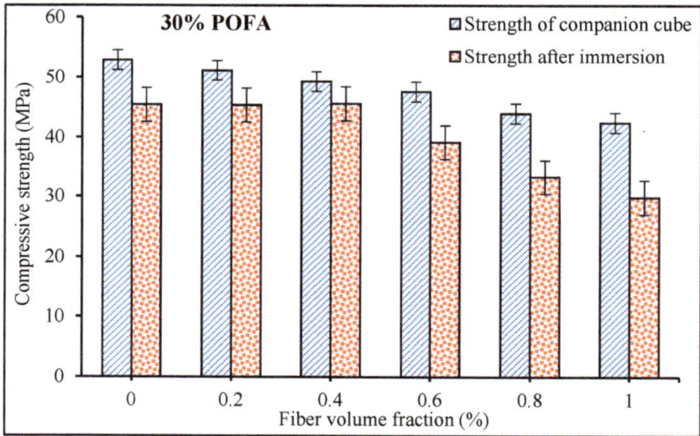

Figure 8. The variation in the 365-day compressive strength of concrete mixtures containing WPFT fibers exposed to a $MgSO_4$ solution.

The tensile expansion strain from the development of ettringite and gypsum in matrix induces cracking and reduces the strength characteristics of concrete because of its low tensile strength. Furthermore, the magnesium attack causes calcium compounds to be released from C-S-H gels, reducing matrix stiffness and causing the specimens to break down [26]. By microscopic investigation, Honglei et al. [43] reported the formation of gypsum with double layers on the surface of the concrete samples, with only a slight amount of ettringite and monosulfate. Magnesium sulfate attack is characterized by a loss of strength and adhesion rather than cracking and expansion [44]. Furthermore,

WPFT fibers inhibited the creation of tiny cracks and failure of specimens due to their bridging effect.

The strength loss factor (SLF), following Equation (2), was used to indicate the decline of compressive strength of concrete mixtures containing WPFT fibers after sulfate attack. Figure 9 illustrates the SLF after 12 months of exposure to the $MgSO_4$ solution. Sulfate reactions altered all of the mixes and resulted in strength reduction. Under the same circumstances, the SLF values were much higher in the OPC fibrous composites, but the POFA specimen containing 0.4% WPFT fibers had the lowest SLFs; however, a further rise in the dosage of fibers resulted in a minor surge in the SLF values. There has been no research conducted so far on the combined effect of POFA and WPFT fibers on the sulfate resistance of concrete. Nevertheless, the study results are comparable to those of Behfarnia and Farshadfar [34] when considering the effective combination of PP fibers and pozzolanic ashes in concrete bare to chemical attacks. They reported that the combination of PP fibers and pozzolanic ashes has a significant effect on improving concrete resistance against sulfate attacks.

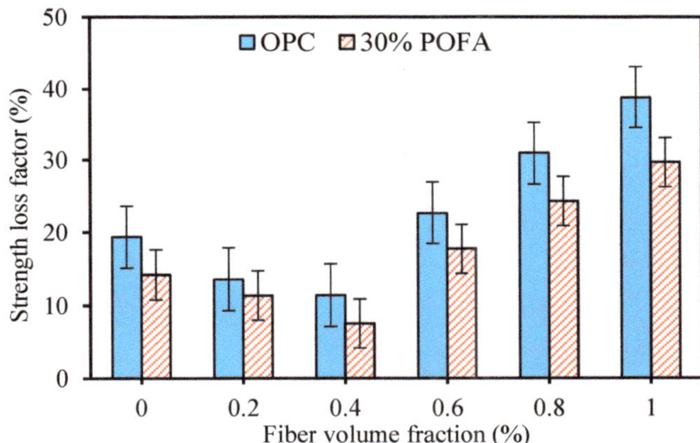

Figure 9. The effects of WPFT fibers on the strength loss factor of concrete exposed to a $MgSO_4$ solution.

3.4. Sulfate Attack

3.4.1. Visual Assessment

The cubic samples submerged in a H_2SO_4 solution were visually inspected monthly for 12 months, similar to the sulfate resistance test. Figure 10 reveals the appearance of concrete specimens with and without WPFT fibers when exposed to the acid solution after 12 months. The initial stage in measuring the deteriorating consequence of acidic attack is commonly visual assessment through analyzing the color change, subsurface cracks, the deposition of additional components on the surface, and the spalling of specimens. As illustrated in Figure 11, the appearance and color of the concrete specimens provides a comprehensive indicator of the acid's effect on the material, regardless of whether the color reflects the original or acid-attacked surface. It was detected that the addition of WPFT fibers and POFA in concrete mixtures results in slight changes in the shape and texture of concrete samples exposed to acid, although both the plain and fiber-reinforced OPC samples were fragmented and distorted after the 12 months of exposure.

The damage shown in Figure 12 was measured using the degradation scale provided in Table 4. As illustrated in Figure 12, all specimens, with and without fibers, deteriorated faster over time. The evolution of the OPC-based mixtures was faster than that of the 30% POFA samples. For example, after 12 months of exposure, the plain OPC mix had a deterioration degree of 8, whereas the plain POFA mix had a deterioration degree of 6. For the entire testing period, the degree of degradation decreased as the fiber volume percentage increased. It is worth noting that a larger fiber content exhibited an excellently increased acid resistance and lower deterioration level. When exposed to acid, concrete specimens containing WPFT fibers showed less spalling than the control specimens without fibers.

3.4.2. Mass Loss

The fiber reinforcement had a considerable influence on mass loss, as demonstrated by the results. This could be ascribed to the fiber bridging action, which reduced concrete components from spalling and improved performance in an acidic environment. The mass loss vs. fiber volume fraction test results are shown in Figure 13. At the end of 12 months of exposure to H_2SO_4 solutions, all concrete specimens showed the same mass reduction behavior. The plain OPC specimens were found with a comparatively higher rate of deterioration in acidic solutions. As OPC contains about 62.5% CaO, it results in faster chemical reactions with acid and forms increased levels of gypsum and ettringite, which are washed out quickly and result in a higher mass rate loss spalling.

Figure 10. The appearance of various concrete specimens exposed to a H_2SO_4 solution for 12 months.

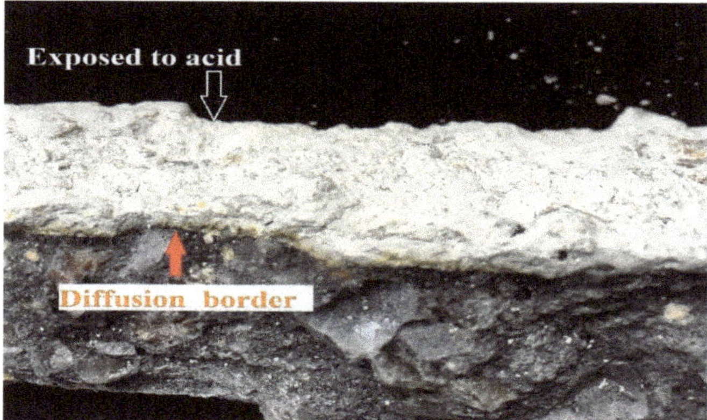

Figure 11. The effects of H_2SO_4 exposure on the texture and color of the concrete matrix.

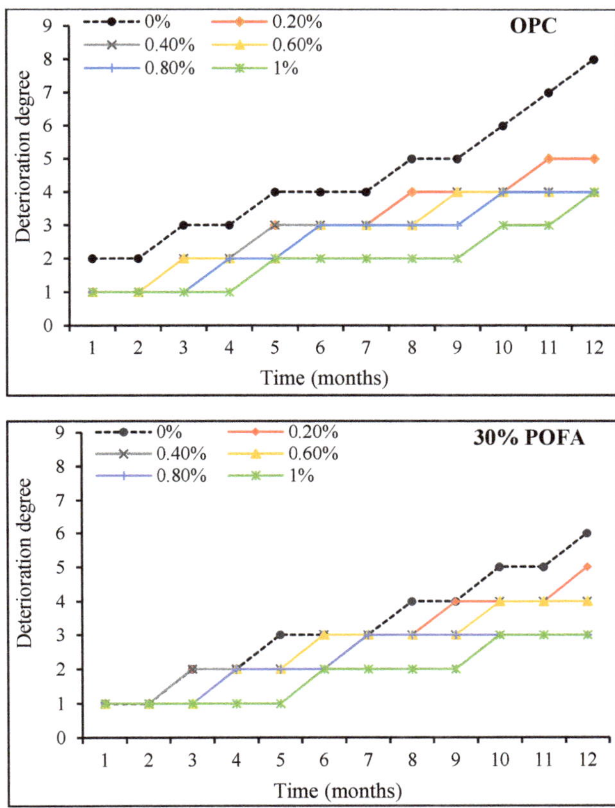

Figure 12. Deterioration degree of concrete specimens comprising WPFT fibers exposed to H_2SO_4 solutions.

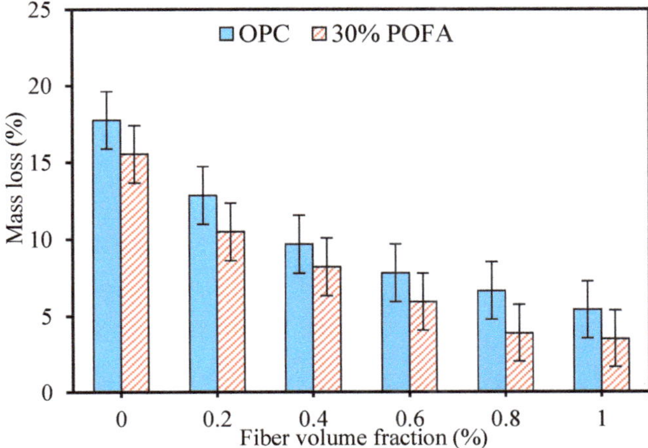

Figure 13. The effects of the WPFT fibers on the mass variation of the concrete specimens after 12 months of exposure in a H_2SO_4 solution.

In contrast, the CaO content is only about 5% in POFA and provides it good acid resistance as a material in concrete production [22]. The reinforcement of concrete with WPFT fibers reduced the rate of mass loss in the OPC mixtures. In the OPC mixes reinforced with 0, 0.2, 0.4, 0.6, 0.8, and 1% WPFT fibers, mass loss values of 17.5, 12.7, 9.7, 7.8, 6.6, and 5.4% were recorded, respectively, indicating that the levels of mass loss in fibrous mixtures was much lower than those in plain mixtures. Acid penetrated the concrete and dissolved the cement paste binder, producing soft and soluble gypsum that subsequently interacted with $Ca(OH)_2$ to form ettringite, leading to mass losses. The presence of fibers helped the concrete constituents to join via a bridging action, resulting in less breakage and deterioration of the matrix. Another harmful consequence of acid attack was the reactions among aluminate and calcium components in the OPC, leading to the formation of soluble products. These exceptionally expansive and soluble products caused internal pressures in the matrix, and microcracks and voids then formed and initiated the deterioration of the concrete and a loss in the strength of the concrete components [21].

It is worth noting that the mass loss rates in the 30% POFA mixes were far lower than the OPC specimens. This might be due to a lower concentration of reactive compounds, like CaO, which slowed the degradation process and prevented ettringite's creation, which would have caused the concrete to expand [23]. The replacement of cement with POFA resulted in a more remarkable performance for the concrete. For the exact dosages of WPFT fibers, the mass losses of the POFA mixes were 15.6, 10.5, 8.3, 5.8, 3.5, and 3.2%, respectively. According to the findings, the incorporation of WPFT fibers and POFA led to the enhancement of concrete against acid attack and reduced spalling and mass loss.

3.4.3. Strength Loss

Figure 14 shows a comparison of the strength values among water-cured samples and specimens submerged in a H_2SO_4 solution for 365 days. The extensive immersion in the acid solution was thought to be the cause of the strength reduction in all mixes. It was discovered that as the amount of WPFT fibers in the concrete increased, the strength loss reduced, whereas the rate of strength loss was higher in the plain concrete mix. When comparing the OPC specimens to 30% POFA mixes, the loss was shown to be greater. For OPC mixes comprising 0, 0.2, 0.4, 0.6, 0.8, and 1% WPFT fibers, values of 31.4, 27.5, 24.7, 23.4, 21.7, and 20.8 MPa were recorded as strength losses, respectively. For the same amounts of fiber, the losses in the strength of POFA mixes were recorded as 28.4, 25.4, 22.8, 21.7, 20.5, and 19.9 MPa.

Due to the good pozzolanic activity of POFA and a lower CaO content, the combined effect of WPFT fiber and 30% POFA caused a minor rate of strength loss. This is because OPC's CaO content reacts with POFA's reactive SiO_2 during the hydration process, resulting the formation of additional products, which helps to enhance the acid resistance of the matrix [22]. The use of POFA, on the other hand, has permitted the production of concrete that is very resistant to acid attacks. The presence of WPFT fibers reduced the creation of fine cracks and specimen spalling due to their linking effect. As such, the reinforced specimens immersed in the acid solution maintained their ductile behavior with a similar mode of failure, almost identical to that of the water-cured materials under compressive loads. Nevertheless, the control specimens without any fibers were completely distorted, and their failure modes were not the same as typical of water-cured concrete. In addition, as shown in Figure 15, the obtained SLF values of OPC mixtures were slightly higher than those noted for POFA mixes after 12 months of exposure to the acid solution. This behavior can be explained as follows: Adding 30% POFA to concrete specimens resulted in more $Ca(OH)_2$ being consumed throughout the hydration procedure, resulting in secondary C-S-H gels in the matrix, which reduced the micro-sized void density in the matrix [25]. The lower SLF values of reinforced POFA mixes signify the positive interaction of POFA and WPFT fiber in enhancing concrete performance against acid attacks.

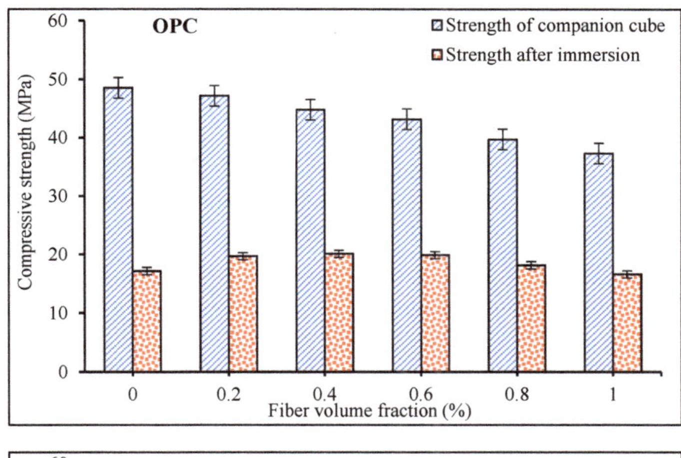

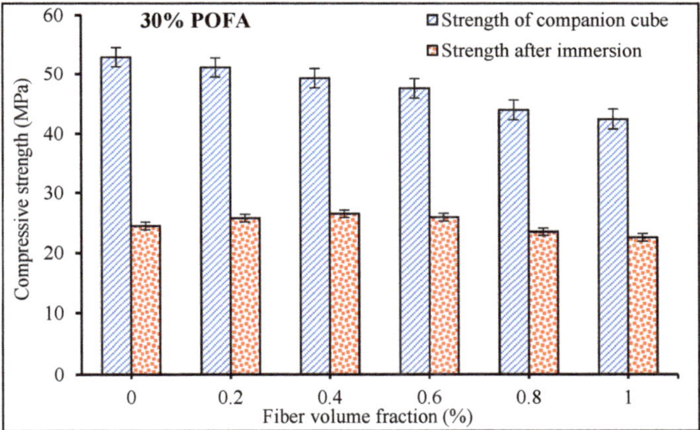

Figure 14. The variation in the 365-day compressive strength of concrete mixtures containing WPFT fibers exposed to a H_2SO_4 solution.

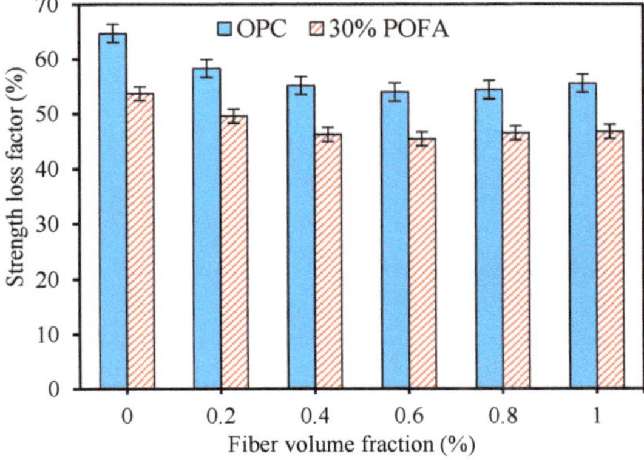

Figure 15. The effects of WPFT fibers on the strength loss factor of concrete mixtures exposed to a H_2SO_4 solution.

3.5. Scanning Electron Microscopy Analysis

Chemical attacks are primarily manifested by losing strength and adherence rather than cracking and expansion. Figure 16 shows scanning electron microscopy (SEM) images of water-cured concrete samples and those immersed in $MgSO_4$ and H_2SO_4 solutions for 365 days. Figure 16a shows the homogeneous spreading of C-S-H gel for water-cured specimens after 365 days. The POFA-based mixes had enhanced performance under chemical attacks due to the consumption of $Ca(OH)_2$ in high amounts, which is a soluble component in OPC during the pozzolanic reactions between POFA and OPC particles along the exposure period [22]. The SEM results of conventional OPC and POFA concrete mixtures treated with a $MgSO_4$ solution are shown in Figure 16b. The SEM image demonstrates how the morphology of the matrix changed when immersed in a sulfate solution. The spherical pores in the OPC concrete mix gradually filled with more precipitated particles due to sulfate exposure. It can be detected that the pore volume of the OPC matrix varied during the exposure period, which affects both the mechanical properties and the durability of the concrete. Consequently, due to the pozzolanic activity of POFA and the development of extra C-S-H gel, the majority of the voids in the mixtures containing 30% POFA filled up with these products, especially at later ages, leaving less space for the freshly harmful particles formed during the exposure period in the sulfate [41]. Accordingly, the strength and durability of concrete improved with the addition POFA as partial cement replacement [45,46].

Besides, the microstructures of the OPC and POFA samples after 12 months of exposure to a severe H_2SO_4 solution are shown in Figure 16c. The production of gypsum at a high level can be seen in the OPC specimens. In the POFA matrix, the generation of gypsum was marginally lower. This was attributed to the existence of a greater amount of $Ca(OH)_2$ particles in the OPC, which caused the formation of a soluble product such as gypsum in the matrix. Generally, gypsum with a weak structure is the main product of a chemical reaction between sulfuric acid and cement particles, which causes the specimen to expand and split [36]. During the hydration process, the $Ca(OH)_2$ in the POFA matrix was consumed, generating C-S-H gels, which have a greater ability to resist acid attacks. Besides, the reactions between other components of OPC, such as calcium aluminate and gypsum, may result in the creation of ettringite, which expands the specimens and contributes to the formation of more cracks [47,48].

(a)

Figure 16. *Cont.*

Figure 16. SEM images of plain OPC and POFA matrices at 12 months of exposure in (**a**) water, (**b**) $MgSO_4$, (**c**) and H_2SO_4 solutions.

4. Conclusions

The long-term performance of concrete comprising WPFT fibers and POFA was investigated against sulfate and acid attacks after 365 days. Concrete specimens were exposed to 10% $MgSO_4$ and 5% H_2SO_4 solutions. The performance of the concrete mixtures was assessed with the help of visual inspection, mass change, residual strength, and microstructural analysis. In the light of attained outcomes, the following conclusions can be drawn:

- Adding WPFT fibers into the concrete mixture resulted in a reduction in workability. The addition of 1% fibers resulted in a slump value of 35 mm, which is comparatively lower than that of the 185 mm which was obtained for plain concrete mix.
- The compressive strength reduced as the fiber dosage increased. Initially, the strength improvement of POFA mixes was essentially identical to that of OPC concrete mixes. POFA-based concrete mixtures had higher compressive strength than OPC concrete mixtures after extended curing times. At 365 days, the obtained compressive strength values for POFA concrete mixes ranged from 42 MPa to 52.9 MPa, which is higher than the values reported for OPC mixes. POFA's pozzolanic activity, which produces additional hydration products, can account for the higher strength values in POFA mixes.
- After being exposed to acid, the mass variation of plain and reinforced concrete specimens was negative, whereas the mass variation of sulfate-treated samples was

positive. The rate of mass change in mixtures comprising WPFT fiber and POFA, on the other hand, was lower than that in the OPC combinations. Adding WPFT fibers to concrete exposed to chemical attacks was a good technique for controlling mass changes. The mass loss of POFA concrete specimens reinforced with 1% fibers after 12 months of exposure in the H_2SO_4 solution was recorded as 3.5%, which is comparatively lower than that of the 15.5% recorded for the plain concrete without fibers.

- Adding WPFT fibers and POFA to concrete specimens improved their sulfate resistance. Due to the bridging action, the presence of WPFT fibers reduced expansion and crack development, whereas the control specimen expanded due to the high calcium oxide content in the sulfate solution. Consequently, a concrete mixture containing 0.4% WPFT fibers and 30% POFA exposed to sulfate has excellent potential for reducing compressive strength loss to only 3.7 MPa.
- Concrete specimens containing POFA and WPFT fibers showed a respectable level of acid resistance. Concrete's acid resistance rose as the fiber volume percentage grew. It was discovered that concrete comprising 1% WPFT fibers and 30% POFA had a compressive strength loss of 19.9 MPa, which is comparatively lesser than that of the 31.4 MPa noted for the control mix.
- Due to the increased CaO content in ordinary Portland cement, the deterioration of OPC-based concrete mixtures was more severe and quicker than that of the POFA mixtures exposed to aggressive environments.
- SEM analysis of the pastes treated with 10% $MgSO_4$ and 5% H_2SO_4 solutions revealed how the specimens changed in these harsh conditions. As a result of exposure, the pores of the OPC matrix eventually filled up with other precipitated elements. Furthermore, in the POFA mixtures, most voids were filled with these products due to POFA's pozzolanic action and additional C-S-H gel, especially at later ages, which developed the performance of the concrete.
- Concrete made from waste plastic food tray fibers and palm oil fuel could be commercialized and used to develop industrial structures and other similar uses where high resistance to harsh environments is required.

Author Contributions: All authors contributed to the paper evenly. Conceptualization, H.M. and R.A.; methodology, H.M. and R.A.; software H.M., S.P.N. and R.A.; validation, H.M. and R.A.; formal analysis, H.M. and R.A.; investigation, H.M. and R.A.; resources, H.M. and R.A.; data curation, H.M., S.P.N. and R.A.; writing—original draft preparation, H.M., M.M.T. and R.A.; writing—review and editing, H.M. and R.A.; visualization, H.M., S.P.N. and R.A.; supervision, M.M.T.; project administration, R.A. and S.P.N.; funding acquisition, H.M. and R.A. All authors have read and agreed to the published version of the manuscript.

Funding: This research was funded by University of Technology Malaysia under Postdoctoral fellowship grants.

Data Availability Statement: The data presented in this study are available on request from the corresponding author. The data are not publicly available due to the size of the research.

Acknowledgments: The authors would like to acknowledge the financial and technical supports received from Universiti Teknologi Malaysia and Prince Sattam Bin Abdulaziz University (Saudi Arabia).

Conflicts of Interest: The authors declare no conflict of interest.

References

1. Gu, L.; Ozbakkaloglu, T. Use of recycled plastics in concrete: A critical review. *Waste Manag.* **2016**, *51*, 19–42. [CrossRef]
2. Wu, S.; Montalvo, L. Repurposing waste plastics into cleaner asphalt pavement materials: A critical literature review. *J. Clean. Prod.* **2020**, *280*, 124355. [CrossRef]
3. Eriksen, M.; Christiansen, J.; Daugaard, A.E.; Astrup, T. Closing the loop for PET, PE and PP waste from households: Influence of material properties and product design for plastic recycling. *Waste Manag.* **2019**, *96*, 75–85. [CrossRef]
4. Hearn, G.; Ballard, J. The use of electrostatic techniques for the identification and sorting of waste packaging materials. *Resour. Conserv. Recycl.* **2005**, *44*, 91–98. [CrossRef]

5. Blanco, I. Lifetime prediction of food and beverage packaging wastes. *J. Therm. Anal. Calorim.* **2016**, *125*, 809–816. [CrossRef]
6. Silvestre, C.; Duraccio, D.; Cimmino, S. Food packaging based on polymer nanomaterials. *Prog. Polym. Sci.* **2011**, *36*, 1766–1782. [CrossRef]
7. Martino, V.; Ruseckaite, R.; Jiménez, A. Thermal and mechanical characterisation of plasticised poly (L-lactide-co-D, L-lactide) films for food packaging. *J. Therm. Anal. Calorim.* **2006**, *86*, 707–712. [CrossRef]
8. Mohammadhosseini, H.; Yatim, J.M.; Sam, A.R.M.; Awal, A.A. Durability performance of green concrete composites containing waste carpet fibers and palm oil fuel ash. *J. Clean. Prod.* **2017**, *144*, 448–458. [CrossRef]
9. Alqahtani, F.K.; Abotaleb, I.S.; ElMenshawy, M. Life cycle cost analysis of lightweight green concrete utilizing recycled plastic aggregates. *J. Build. Eng.* **2021**, *40*, 102670. [CrossRef]
10. Mohammadhosseini, H.; Yatim, J.M. Microstructure and residual properties of green concrete composites incorporating waste carpet fibers and palm oil fuel ash at elevated temperatures. *J. Clean. Prod.* **2017**, *144*, 8–21. [CrossRef]
11. Hubo, S.; Leite, L.; Martins, C.; Ragaert, K. Evaluation of post-industrial and post-consumer polyolefin-based polymer waste streams for injection moulding. In Proceedings of the 6th Polymers & Mould Innovations International Conference, Guimaraes, Portugal, 10–12 September 2014; pp. 201–206.
12. Kumar, S.; Panda, A.K.; Singh, R. A review on tertiary recycling of high-density polyethylene to fuel. *Resour. Conserv. Recycl.* **2011**, *55*, 893–910. [CrossRef]
13. Siddique, R.; Khatib, J.; Kaur, I. Use of recycled plastic in concrete: A review. *Waste Manag.* **2008**, *28*, 1835–1852. [CrossRef] [PubMed]
14. Almeshal, I.; Tayeh, B.A.; Alyousef, R.; Alabduljabbar, H.; Mohamed, A.M.; Alaskar, A. Use of recycled plastic as fine aggregate in cementitious composites: A review. *Constr. Build. Mater.* **2020**, *253*, 119146. [CrossRef]
15. Alyousef, R.; Mohammadhosseini, H.; Alrshoudi, F.; Alabduljabbar, H.; Mohamed, A.M. Enhanced performance of concrete composites comprising waste metalised polypropylene fibres exposed to aggressive environments. *Crystals.* **2020**, *10*, 696. [CrossRef]
16. Mohammadhosseini, H.; Alyousef, R.; Tahir, M.M. Towards Sustainable Concrete Composites through Waste Valorisation of Plastic Food Trays as Low-Cost Fibrous Materials. *Sustainability.* **2021**, *13*, 2073. [CrossRef]
17. Alrshoudi, F.; Mohammadhosseini, H.; Alyousef, R.; Alabduljabbar, H.; Mustafa Mohamed, A. The impact resistance and deformation performance of novel pre-packed aggregate concrete reinforced with waste polypropylene fibres. *Crystals.* **2020**, *10*, 788. [CrossRef]
18. Mohammadhosseini, H.; Ngian, S.P.; Alyousef, R.; Tahir, M.M. Synergistic effects of waste plastic food tray as low-cost fibrous materials and palm oil fuel ash on transport properties and drying shrinkage of concrete. *J. Build. Eng.* **2021**, *42*, 102826. [CrossRef]
19. Jain, A.; Siddique, S.; Gupta, T.; Jain, S.; Sharma, R.K.; Chaudhary, S. Evaluation of concrete containing waste plastic shredded fibers: Ductility properties. *Struct. Concr.* **2021**, *22*, 566–575. [CrossRef]
20. Colangelo, F.; Cioffi, R.; Liguori, B.; Iucolano, F. Recycled polyolefins waste as aggregates for lightweight concrete. *Compos. Part B Eng.* **2016**, *106*, 234–241. [CrossRef]
21. Sotiriadis, K.; Nikolopoulou, E.; Tsivilis, S. Sulfate resistance of limestone cement concrete exposed to combined chloride and sulfate environment at low temperature. *Cem. Concr. Compos.* **2012**, *34*, 903–910. [CrossRef]
22. Mohammadhosseini, H.; Tahir, M.M.; Sam, A.R.M.; Lim, N.H.A.S.; Samadi, M. Enhanced performance for aggressive environments of green concrete composites reinforced with waste carpet fibers and palm oil fuel ash. *J. Clean. Prod.* **2018**, *185*, 252–265. [CrossRef]
23. Bulatović, V.; Melešev, M.; Radeka, M.; Radonjanin, V.; Lukić, I. Evaluation of sulfate resistance of concrete with recycled and natural aggregates. *Constr. Build. Mater.* **2017**, *152*, 614–631. [CrossRef]
24. Wu, F.; Yu, Q.; Liu, C.; Brouwers, H.J.H.; Wang, L.; Liu, D. Effect of fibre type and content on performance of bio-based concrete containing heat-treated apricot shell. *Mater. Struct.* **2020**, *53*, 1–16. [CrossRef]
25. Neville, A. The confused world of sulfate attack on concrete. *Cem. Concr. Res.* **2004**, *34*, 1275–1296. [CrossRef]
26. Hill, J.; Byars, E.; Sharp, J.; Lynsdale, C.; Cripps, J.; Zhou, Q. An experimental study of combined acid and sulfate attack of concrete. *Cem. Concr. Compos.* **2003**, *25*, 997–1003. [CrossRef]
27. Hadigheh, S.A.; Gravina, R.; Smith, S. Effect of acid attack on FRP-to-concrete bonded interfaces. *Constr. Build. Mater.* **2017**, *152*, 285–303. [CrossRef]
28. Lu, C.; Zhou, Q.; Wang, W.; Wei, S.; Wang, C. Freeze-thaw resistance of recycled aggregate concrete damaged by simulated acid rain. *J. Clean. Prod.* **2021**, *280*, 124396. [CrossRef]
29. Mohammadhosseini, H.; Tahir, M.M. Durability performance of concrete incorporating waste metalized plastic fibres and palm oil fuel ash. *Constr. Build. Mater.* **2018**, *180*, 92–102. [CrossRef]
30. Bankir, M.B.; Sevim, U.K. Performance optimization of hybrid fiber concretes against acid and sulfate attack. *J. Build. Eng.* **2020**, *32*, 101443. [CrossRef]
31. Meng, C.; Li, W.; Cai, L.; Shi, X.; Jiang, C. Experimental research on durability of high-performance synthetic fibers reinforced concrete: Resistance to sulfate attack and freezing-thawing. *Constr. Build. Mater.* **2020**, *262*, 120055. [CrossRef]
32. Guo, L.; Wu, Y.; Xu, F.; Song, X.; Ye, J.; Duan, P.; Zhang, Z. Sulfate resistance of hybrid fiber reinforced metakaolin geopolymer composites. *Compos. Part B Eng.* **2020**, *183*, 107689. [CrossRef]

33. Mohammadhosseini, H.; Alyousef, R.; Lim, N.H.A.S.; Tahir, M.M.; Alabduljabbar, H.; Mohamed, A.M.; Samadi, M. Waste metalized film food packaging as low cost and ecofriendly fibrous materials in the production of sustainable and green concrete composites. *J. Clean. Prod.* **2020**, *258*, 120726. [CrossRef]

34. Behfarnia, K.; Farshadfar, O. The effects of pozzolanic binders and polypropylene fibers on durability of SCC to magnesium sulfate attack. *Constr. Build. Mater.* **2013**, *38*, 64–71. [CrossRef]

35. Bolat, H.; Şimşek, O.; Çullu, M.; Durmuş, G.; Can, Ö. The effects of macro synthetic fiber reinforcement use on physical and mechanical properties of concrete. *Compos. Part B Eng.* **2014**, *61*, 191–198. [CrossRef]

36. Guo, X.; Xiong, G. Resistance of fiber-reinforced fly ash-steel slag based geopolymer mortar to sulfate attack and drying-wetting cycles. *Constr. Build. Mater.* **2021**, *269*, 121326. [CrossRef]

37. Turk, K.; Bassurucu, M.; Bitkin, R.E. Workability, strength and flexural toughness properties of hybrid steel fiber reinforced SCC with high-volume fiber. *Constr. Build. Mater.* **2021**, *266*, 120944. [CrossRef]

38. Li, K.; Yang, C.; Huang, W.; Zhao, Y.; Wang, Y.; Pan, Y.; Xu, F. Effects of hybrid fibers on workability, mechanical, and time-dependent properties of high strength fiber-reinforced self-consolidating concrete. *Constr. Build. Mater.* **2021**, *277*, 122325. [CrossRef]

39. Bentegri, I.; Boukendakdji, O.; Kadri, E.-H.; Ngo, T.; Soualhi, H. Rheological and tribological behaviors of polypropylene fiber reinforced concrete. *Constr. Build. Mater.* **2020**, *261*, 119962. [CrossRef]

40. Mohammadhosseini, H.; Tahir, M.M.; Alaskar, A.; Alabduljabbar, H.; Alyousef, R. Enhancement of strength and transport properties of a novel preplaced aggregate fiber reinforced concrete by adding waste polypropylene carpet fibers. *J. Build. Eng.* **2020**, *27*, 101003. [CrossRef]

41. Alrshoudi, F.; Mohammadhosseini, H.; Alyousef, R.; Alghamdi, H.; Alharbi, Y.R.; Alsaif, A. Sustainable use of waste polypropylene fibers and palm oil fuel ash in the production of novel prepacked aggregate fiber-reinforced concrete. *Sustainability.* **2020**, *12*, 4871. [CrossRef]

42. Nili, M.; Afroughsabet, V. The effects of silica fume and polypropylene fibers on the impact resistance and mechanical properties of concrete. *Constr. Build. Mater.* **2010**, *24*, 927–933. [CrossRef]

43. Honglei, C.; Zuquan, J.; Penggang, W.; Jianhong, W.; Jian, L. Comprehensive resistance of fair-faced concrete suffering from sulfate attack under marine environments. *Constr. Build. Mater.* **2021**, *277*, 122312. [CrossRef]

44. Li, Y.; Yang, X.; Lou, P.; Wang, R.; Li, Y.; Si, Z. Sulfate attack resistance of recycled aggregate concrete with NaOH-solution-treated crumb rubber. *Constr. Build. Mater.* **2021**, *287*, 123044. [CrossRef]

45. Alrshoudi, F.; Mohammadhosseini, H.; Tahir, M.M.; Alyousef, R.; Alghamdi, H.; Alharbi, Y.; Alsaif, A. Drying shrinkage and creep properties of prepacked aggregate concrete reinforced with waste polypropylene fibers. *J. Build. Eng.* **2020**, *32*, 101522. [CrossRef]

46. Mohammadhosseini, H.; Lim, N.H.A.S.; Tahir, M.M.; Alyousef, R.; Samadi, M.; Alabduljabbar, H.; Mohamed, A.M. Effects of waste ceramic as cement and fine aggregate on durability performance of sustainable mortar. *Arab. J. Scie. Eng.* **2020**, *45*, 3623–3634. [CrossRef]

47. Mohammadhosseini, H.; Alrshoudi, F.; Tahir, M.M.; Alyousef, R.; Alghamdi, H.; Alharbi, Y.R.; Alsaif, A. Durability and thermal properties of prepacked aggregate concrete reinforced with waste polypropylene fibers. *J. Build. Eng.* **2020**, *32*, 101723. [CrossRef]

48. Mohammadhosseini, H.; Alrshoudi, F.; Tahir, M.M.; Alyousef, R.; Alghamdi, H.; Alharbi, Y.R.; Alsaif, A. Performance evaluation of novel prepacked aggregate concrete reinforced with waste polypropylene fibers at elevated temperatures. *Constr. Build. Mater.* **2020**, *259*, 120418. [CrossRef]

Article

Effect of Ultrafine Metakaolin on the Properties of Mortar and Concrete

Shengli Zhang [1], Yuqi Zhou [2], Jianwei Sun [3],* and Fanghui Han [4]

1   Beijing Urban Construction Engineering Co., Ltd., Beijing 100071, China; shenglizhang@mail.buce.cn
2   China Construction First Group Construction and Development Co., Ltd., Beijing 100102, China; zhouyuqi@chinaonebuild.com
3   School of Civil Engineering, Qingdao University of Technology, Qingdao 266033, China
4   School of Civil and Resource Engineering, University of Science and Technology Beijing, Beijing 100083, China; hanfanghui@ustb.edu.cn
*   Correspondence: jianwei2019@mail.tsinghua.edu.cn

Abstract: This study investigated the influence of ultrafine metakaolin replacing cement as a cementitious material on the properties of concrete and mortar. Two substitution levels of ultrafine metakaolin (9% and 15% by mass) were chosen. The reference samples were plain cement concrete sample and silica fume concrete sample with the same metakaolin substitution rates and superplasticizer contents. The results indicate that simultaneously adding ultrafine metakaolin and a certain amount of polycarboxylate superplasticizer can effectively ensure the workability of concrete. Additionally, the effect of adding ultrafine metakaolin on the workability is better than that of adding silica fume. Adding ultrafine metakaolin or silica fume can effectively increase the compressive strength, splitting tensile strength, resistance to chloride ion penetration and freeze–thaw properties of concrete due to improved pore structure. The sulphate attack resistance of mortar can be improved more obviously by simultaneously adding ultrafine metakaolin and prolonging the initial moisture curing time.

Keywords: ultrafine metakaolin; silica fume; strength; durability

Citation: Zhang, S.; Zhou, Y.; Sun, J.; Han, F. Effect of Ultrafine Metakaolin on the Properties of Mortar and Concrete. *Crystals* **2021**, *11*, 665. https://doi.org/10.3390/cryst11060665

Academic Editors: Yifeng Ling, Chuanqing Fu, Peng Zhang and Peter Taylor

Received: 21 May 2021
Accepted: 8 June 2021
Published: 10 June 2021

Publisher's Note: MDPI stays neutral with regard to jurisdictional claims in published maps and institutional affiliations.

1. Introduction

Concrete made with Portland cement has been available for nearly two hundred years. In the past two centuries, with the development of science and technology, the composition and performance of Portland cement have undergone great changes, and construction technology using concrete has also undergone earth-shaking changes [1,2]. The two changes above have led to complete changes in the composition, preparation methods and performance of modern concrete. Mineral admixtures such as supplementary cementitious materials are an indispensable part of modern concrete [3,4]. The use of high-activity admixtures such as slag and low-activity or inert admixtures such as limestone powder effectively reduces energy and resource consumption by decreasing the amount of cement used in the concrete production process and ensures the green and sustainable development of the concrete industry [3,4]. More importantly, the use of admixtures reduces the dependence of modern concrete strength on cement strength, improves the rheological properties of the mixture, and increases the durability of concrete structures.

At present, the mineral admixtures used in the concrete preparation process worldwide are mainly fly ash, slag, and limestone powder [5–8]. However, with the development of the concrete industry, the stock of high-quality admixtures has decreased, leading to gradually rising prices. Therefore, the available high-quality and economical mineral admixtures will become increasingly scarce and unable to meet the needs of the national construction market. On the premise that the annual production of admixtures cannot be changed, the main technical means to offset the scarcity of high-quality admixtures is to increase their reactivity to reduce the dosage. Mechanical grinding is the most direct method used to improve the reactivity of powders. Ultrafine grinding involves the application of strong mechanical forces that distort the crystal structure of the mineral, rapidly resulting in lattice dislocations, defects, and recrystallization, which improves the reaction activity [9–12]. On the other hand, ultrafine grinding increases the specific surface area of the admixture, thereby enhancing micro aggregation and nucleation effects in the cement matrix [13–15]. In addition, ultra-high-performance concrete (UHPC) has attracted increasing attention due to it meeting the requirements for special concrete structures such as super high-rise buildings and long-span bridges [16–18]. Ultra-high-activity mineral admixtures, such as ultrafine powders, are indispensable for the preparation of UHPC [19–21]. Therefore, it is necessary to study the action mechanism and application effects of ultrafine admixtures in concrete, whether from the perspective of improving the overall quality of admixtures, solving the problem of the scarcity of high-quality admixtures, or meeting the needs of special engineering and ensuring the supply of UHPC.

Metakaolin is an amorphous aluminum silicate formed by the calcination of kaolin clay at temperatures ranging from 500 °C to 800 °C [22,23]. In the process of heating, most octahedral alumina is converted into more active tetra-coordinated and penta-coordinated units [24,25]. When the crystal structure is completely or partially broken, or the bonds between the kaolinite layers are broken, kaolin undergoes a phase change and finally forms metakaolin with poor crystallinity [25,26]. As its molecular arrangement is irregular, metakaolin is in a metastable thermodynamic state and has gelling properties under appropriate excitation [23–26]. Metakaolin has high pozzolanic activity, which means that it can react with $Ca(OH)_2$ and produce C-S-H gel and alumina-containing phases, including C_4AH_{13}, C_2ASH_8, and C_3AH_6, at ambient temperature [23]. Therefore, metakaolin is mainly used as a mineral admixture in cement and concrete. Compared to silica fume and fly ash, metakaolin has a very high reactivity level [23]. Previous studies have shown that metakaolin can increase the mechanical strength of concrete to varying degrees, depending mainly on the replacement rate of metakaolin, the water/binder ratio, and the age at testing [22,23,27]. Remarkably, metakaolin has a positive effect on reducing drying shrinkage and improving durability [22,23,27–30].

In current practical engineering applications, 99.9% of metakaolin particles are less than 16 µm [23]. The mean particle size is generally 3 µm, which is significantly smaller than that of cement particles but not as fine as silica fume [23]. Ultrafine metakaolin with a specific surface area greater than 20,000 m^2/kg can be obtained by grinding. The pozzolanic activity of metakaolin is related to the particle size. Theoretically, it is practical to grind metakaolin so that it can play a greater role in concrete. In this paper, ultrafine metakaolin obtained by grinding was used as a supplementary cementitious material in concrete and mortar. The macroscopic properties, including the mechanical strength and durability of concrete and mortar, were investigated. Plain cement concrete and silica fume concrete were employed as reference samples.

2. Materials and Methods

2.1. Raw Materials

In this paper, Portland cement with a strength grade of 42.5 from Beijing Jinyu Group Co., Ltd. in China was used as the cementitious material. Ultrafine metakaolin and silica fume with similar specific surface areas were used as mineral admixtures. The chemical compositions of the raw materials are presented in Table 1. The total content of SiO_2 and Al_2O_3 in ultrafine metakaolin is more than 99%. The particle size distribution of ultrafine metakaolin is shown in Figure 1a. The X-ray diffraction (XRD) patterns of ultrafine metakaolin are presented in Figure 1b. Figure 1b shows that the amorphous phase is the main constituent, and SiO_2 crystals are the main crystalline mineral phase in ultrafine metakaolin. The microstructure of ultrafine metakaolin is shown in Figure 1c. Ultrafine metakaolin is broken into small angular particles. Fine aggregates of mortar and concrete were ISO standard sand and river sand with a fineness modulus of 2.9, respectively.

Crushed limestone with a continuous size gradation from 4.75 mm to 20 mm was used as coarse aggregate. A polycarboxylate superplasticizer with a water reduction rate of 20% was used to adjust the flowability of fresh mortar and concrete.

Table 1. Main chemical compositions of raw materials/%.

	CaO	SiO_2	Al_2O_3	Fe_2O_3	MgO	SO_3	Na_2Oeq *	LOI
Cement	54.86	21.10	6.33	4.22	2.60	2.66	0.53	2.03
Ultrafine Metakaolin	0.13	52.72	46.29	0.28	0.15	0.08	0.12	0.64
Silica Fume	-	94.67	-	-	-	-	0.06	2.51

* $Na_2O_{eq} = Na_2O + 0.658K_2O$; LOI: loss on ignition.

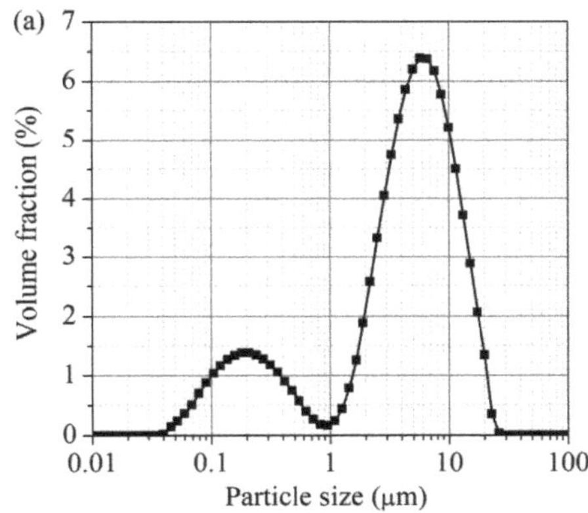

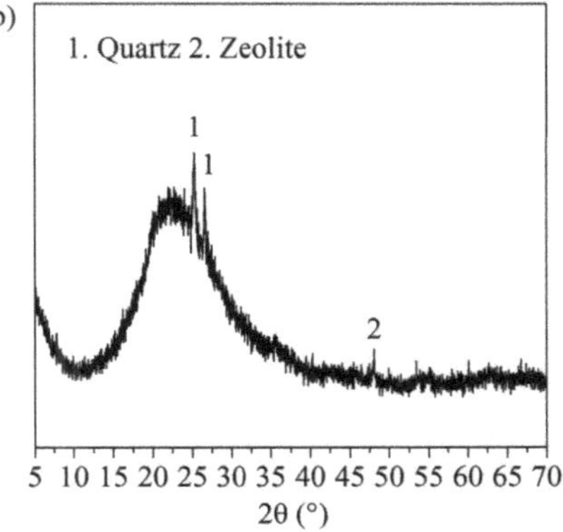

Figure 1. Characteristics of ultrafine metakaolin used in this study: (**a**) particle size distribution; (**b**) XRD analysis; (**c**) micromorphology.

2.2. Mix Proportions

Table 2 shows the mix proportions of concrete. The total amount of binder was 400 kg/m^3. A water/binder ratio of 0.4 and a sand ratio of 0.44 were selected. Two substitution rates of ultrafine metakaolin (9% and 15% by mass) were used, corresponding to sample M1 and sample M2. The plain cement concrete sample (sample C), concrete sample containing 9% silica fume (sample S1) and concrete sample containing 15% silica fume (sample S2) were regarded as the reference samples. The superplasticizer dosages by mass percent of the total cementitious materials in samples C, M1 and M2 were 0.8%, 0.85% and 0.95%, respectively. The fresh concrete was poured into different sizes of molds, including 100 mm × 100 mm × 100 mm and 100 mm × 100 mm × 400 mm. Only three mix proportions (samples C, M1 and M2) were used for the mortar test. The water/binder ratio was the same as concrete (0.4), and the sand/binder ratio was 3.0. Fresh mortar was poured into steel molds with dimensions of 40 mm × 40 mm × 160 mm for sulphate attack test. After 24 h, all samples were unmolded and cured under scheduled regimes.

Table 2. Mix proportions of concretes/$kg \cdot m^{-3}$.

Sample	Cement	Ultrafine Metakaolin	Silica Fume	Fine Aggregate	Coarse Aggregate	Water	Superplasticizer
C	400	0	0	854	1086	160	3.2
M1	364	36	0	854	1086	160	3.4
S1	364	0	36	854	1086	160	3.4
M2	340	60	0	854	1086	160	3.8
S2	340	0	60	854	1086	160	3.8

2.3. Curing Conditions and Test Methods

In the fresh state, the workability of fresh concrete was determined according to Chinese National Standard GB/T 50080-2016. The fresh concrete was poured into the slump bucket (upper 100 mm, lower 200 mm, and height 300 mm). After each pouring, a tamper bar was used to hammer it evenly 25 times. After tamping, the bucket was pulled up, and the concrete collapsed due to its own weight. The height value after the slump was recorded. The slump value and loss ratio were calculated by the height difference between peak height and slump height. In this study, different methods were used to cure concrete and mortar. Concrete was cured under standard curing conditions at a constant temperature (20 ± 2 °C) and relative humidity (95 ± 1%). The compressive strength of concrete after 1 d, 3 d, 7 d, 28 d and 90 d and the splitting tensile strength after 28 d and 90 d were measured according to China National Standards GB/T 50081-2011. Tests of compressive strength and splitting tensile strength were performed after casting at a certain age by using three specimens for each test. The chloride ion penetrability resistance and freeze–thaw resistance of concrete were determined by the American Society of Testing Materials Standard ASTM C1202 and Chinese National Standard GB/T 50082-2009, respectively. For the chloride ion penetrability resistance test, the cube specimen of 100 mm × 100 mm × 100 mm was used to cut one specimen of 50 mm × 100 mm × 100 from the middle. The charge passed in 6 h was used to evaluate the chloride ion penetrability resistance. Three test blocks were tested for each group of specimens. For the freeze–thaw resistance test, 100 mm × 100 mm × 100 mm cube blocks were used. The mass change and dynamic elasticity modulus were tested after 300 cycles. The average value for six specimens was obtained to ensure the accuracy of the test. For the connected porosity test, the cut specimen of 50 mm × 100 mm × 100 mm was used. The connected porosity was measured by the vacuum saturation–drying method. The connected porosity P was determined using the Equation (1):

$$P = (m_1 - m_2)/m_1 \times 100\%, \tag{1}$$

where m_1 is the mass of concrete that was completely saturated with water by the vacuum saturation method for 3 d and m_2 represents the mass of concrete that was dried at 40 °C for 14 d.

Two curing conditions for mortar were used: 3 d and 7 d initial moist curing. After initial moist curing, all mortars were placed in a natural environment. The compressive strength was tested after 28 d, and then the mortar was semi-immersed in a solution containing 10% sodium sulphate (by mass) for 28 d, 56 d and 90 d. The concentration of the sodium sulphate solution was maintained by periodically replacing the solution. Meanwhile, the reference specimens cured in water for the same lengths of time were tested for compressive strength and flexural strength. Therefore, the sulphate attack resistance was evaluated by the relative compressive strength and flexural strength loss for the same curing time.

3. Results and Discussion

3.1. Workability

Table 3 presents the values of slump and slump loss for all mixtures with different dosages of ultrafine metakaolin. No segregation or bleeding was observed during mixture experiments. Obviously, the workability of fresh concrete and replacement rate of ultrafine metakaolin are nonlinearly related in Table 3. Compared to plain cement concrete, adding 9% ultrafine metakaolin decreases workability, while adding 15% ultrafine metakaolin has little influence on workability. The addition of 9% ultrafine metakaolin results in a smaller average particle size and larger specific surface area of the composite binder, which leads to the availability of less free water in the concrete matrix. Therefore, sample M1 has poorer workability. However, sample M2 has the highest content of superplasticizer compared to sample C, which seriously weakens the water absorption effect of ultrafine metakaolin particles. Compared to silica fume concrete with the same replacement rate and superplasticizer content, ultrafine metakaolin concrete has a higher slump value. This indicates that the compatibility of the superplasticizer and ultrafine metakaolin concrete is better. After 0.5 h, the slump loss values show the same change tendency as the slump. However, the slump loss ratio is significantly changed. Compared to plain cement concrete, adding 9% ultrafine mineral admixtures and 6.25% polycarboxylate superplasticizer has little influence on the slump loss ratio. The addition of 15% ultrafine mineral admixtures and 18.75% polycarboxylate superplasticizer obviously decreases the slump loss ratio. Adding ultrafine metakaolin has a more effective impact.

Table 3. Slump and slump loss of fresh concretes.

	C	M1	S1	M2	S2
0 h (mm)	235	224	213	233	221
0.5 h (mm)	180	172	162	186	172
Loss ratio (%)	23.40	23.21	23.94	20.17	22.17

3.2. Mechanical Strength

The compressive strength and splitting tensile strength of all concrete are shown in Figure 2. Overall, adding mineral admixtures increases the compressive strength of concretes at all ages, as shown in Figure 2a. It is more obvious at the later ages. However, different admixture dosages have various effects on ultrafine metakaolin concrete. The compressive strength of sample M1 is slightly higher than that of sample M2 at 1 d, and it is obviously higher than that of sample M2 at 3 d and 7 d. At 1 d, the early compressive strengths of all concretes are approximately 15 MPa. At 3 d, the compressive strength of plain cement concrete is 30 MPa, while the maximum strength is nearly 40 MPa (sample M1). At 7 d, the compressive strengths of all concretes increase by approximately 10 MPa. However, the opposite trend occurs at 28 d and 90 d. At 28 d, the compressive strengths of sample C and sample M1 increase by approximately 10 MPa, but the compressive strength of sample M2 reaches 62 MPa, increasing by 17 MPa. Therefore, adding 15% ultrafine metakaolin can increase the strength of concrete to meet the strength requirements of C60. The compressive strength of the composite concrete increases slowly from 28 d to 90 d. Thus, sample M2 has the highest compressive strength at late ages. Compared to plain cement concrete, adding 15% ultrafine metakaolin increases the compressive strength at 28 d and 90 d by 24% and 20%, respectively. There is not much difference in the compressive strengths of ultrafine metakaolin concrete and silica fume concrete with the same replacement rate and superplasticizer content at all ages. Remarkably, the 7 d compressive strengths of all concretes reached nearly 72–83% of the 28-d strength, which suggests that a high superplasticizer content has no obvious negative effect on the early strength of concrete.

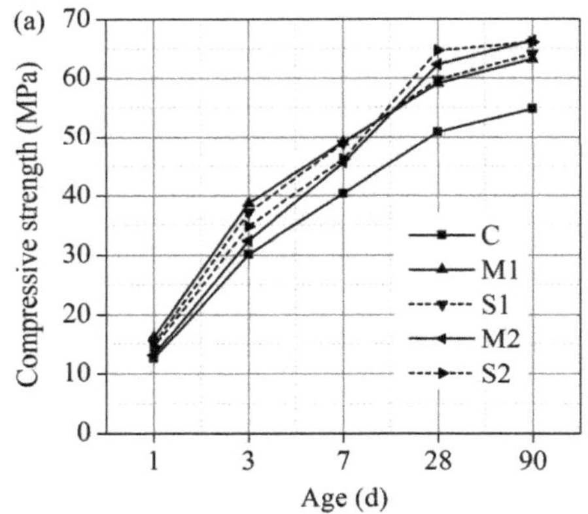

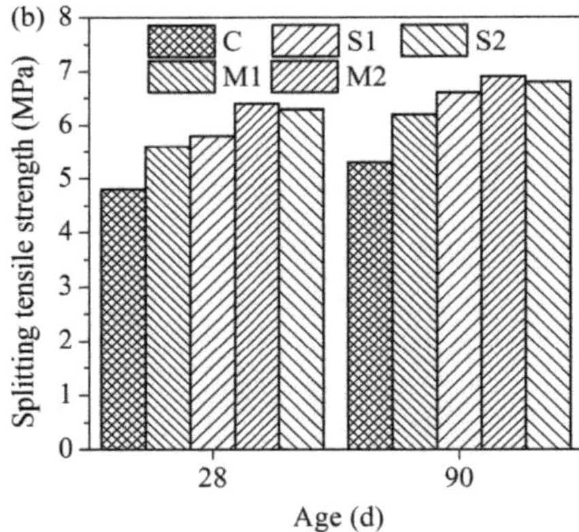

Figure 2. Mechanical strength of concrete: (**a**) compressive strength; (**b**) splitting tensile strength.

Figure 2b shows the same influence of ultrafine mineral admixtures on the splitting tensile strength and compressive strength. The results show that the splitting tensile strength of concrete increases as the ultrafine admixture content increases. There is little difference in the splitting tensile strength of ultrafine metakaolin concrete and silica fume concrete with the same replacement rate and superplasticizer content at 28 d and 90 d. Adding 9% ultrafine metakaolin increases the splitting tensile strength by approximately 17% at 28 d and 90 d. Adding 15% ultrafine metakaolin increases the splitting tensile strength at 28 d and 90 d by approximately 33% and 30%, respectively. The maximum splitting tensile strength is close to 7 MPa at 90 d (samples M2 and S2).

3.3. Connected Porosity

The connected porosity of concrete is an index used to measure the transport capacity of water and solution erosion, which is closely related to permeability. The connected porosities of all concretes at 28 d are shown in Figure 3. Remarkably, the 28-d connected porosity values of all concretes are in the 11–14% range. As shown in Figure 3, the connected porosity of concrete obviously decreases with increasing ultrafine metakaolin content. The connected porosity of silica fume concrete is relatively lower than that of ultrafine metakaolin concrete at the same replacement rate and superplasticizer content. This indicates that the addition of ultrafine mineral admixtures refines the pore structure of concrete and makes the matrix denser, which improves the durability of concrete. This result is consistent with the findings of Erhan et al. [22], who found that ultrafine metakaolin substantially enhanced the pore structure of concrete and reduced the presence of harmful large pores, especially at a high replacement level. The beneficial effect of adding silica fume is slightly better than that of adding ultrafine metakaolin.

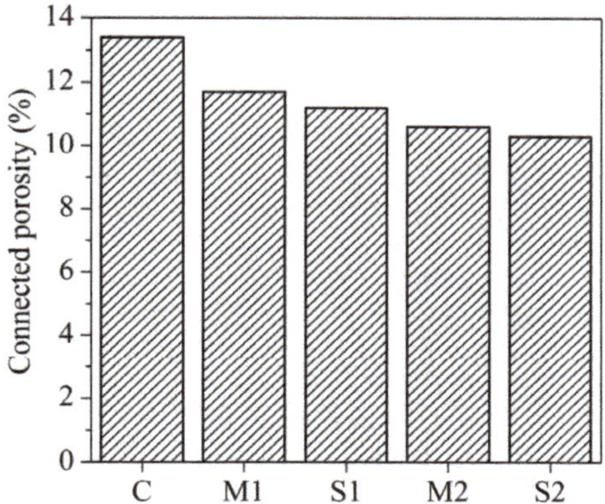

Figure 3. Connected porosity of concrete.

3.4. Chloride Ion Penetrability of Concrete

According to ASTM C1202, the penetrability grades of concrete at 28 d and 90 d are shown in Figure 4. Apparently, Figure 4 shows that the penetrability grades of samples C, M1 and M2 at 28 d are "moderate", "low" and "very low", respectively. At 90 d, although the charge passed by all concretes decreases due to lower connected porosity, there is no change in penetrability levels. Meanwhile, the penetrability grades of ultrafine metakaolin concrete and silica fume concrete with the same replacement rate show little difference at 28 d and 90 d for the same superplasticizer content. This result is attributed to similar pore structures. Therefore, adding ultrafine mineral admixtures can improve the resistance to chloride ion penetration of concretes at 28 d and 90 d. Ultrafine metakaolin has the same effect as silica fume.

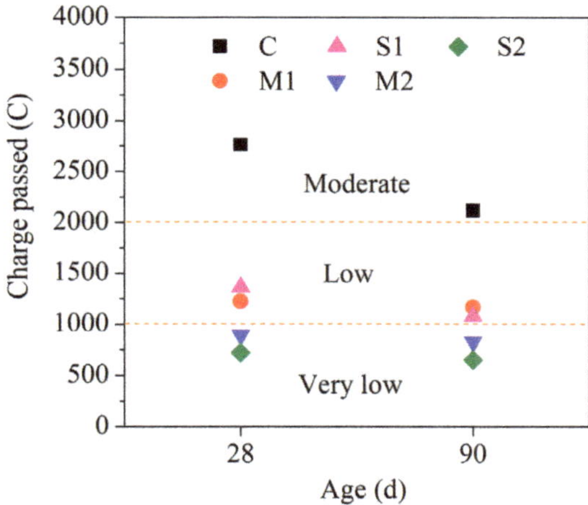

Figure 4. Chloride ion penetrability of concrete.

3.5. Freeze–Thaw Resistance

Freeze–thaw damage is the main factor affecting the instability of concrete structures in cold areas, which seriously threatens the safety and service life of concrete structures. The relative dynamic elasticity modulus and mass loss of all concretes after 300 freeze–thaw cycles are presented in Figure 5a,b, respectively. Figure 5a shows that the relative dynamic elasticity moduli of samples M1–S2 are all 81–86% after 300 cycles; therefore, sample M2 has the best freeze–thaw resistance, and its relative dynamic elasticity modulus is 85.5%. However, the relative dynamic elasticity modulus of plain cement concrete is only 62.8%, which is much less than those of composite concretes. Figure 5b shows that the mass loss of plain cement concrete exceeds 5%. However, the mass loss of samples M1–S2 is very small, less than 4% after 300 cycles in Figure 5b. Therefore, ultrafine metakaolin concrete and silica fume concrete can meet the frost resistance requirements in cold areas; they have the lowest frost resistance grade of F300. With increasing ultrafine metakaolin or silica fume content, the relative dynamic elasticity modulus of concrete tends to increase, and the mass loss decreases obviously. Meanwhile, it can hardly observe surface denudation of samples M1, S1, M2 and S2, and the surface damage layers of these samples above are very thin. Thus, composite concrete has excellent apparent performance. Therefore, adding ultrafine metakaolin or silica fume has a positive influence on the freeze–thaw resistance of concrete.

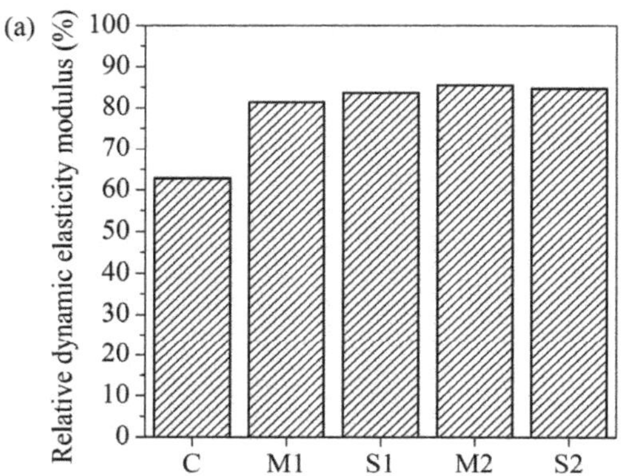
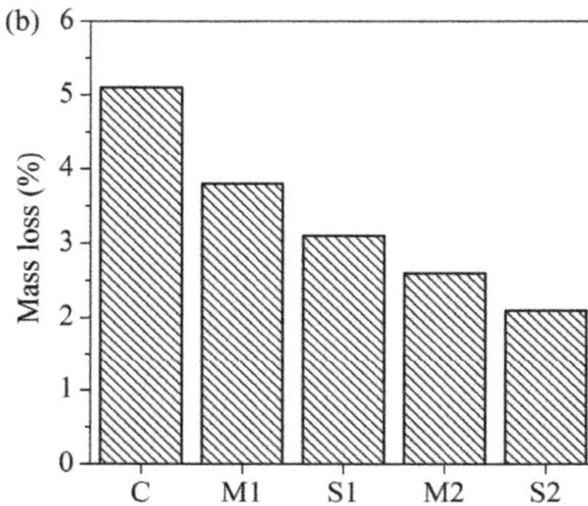

Figure 5. Results of fast freeze-thaw cycle tests of concretes after 300 cycles: (**a**) relative dynamic elasticity modulus; (**b**) mass loss.

3.6. Sulphate Attack

Figure 6 shows the influence of different initial moisture curing times on the 28-d compressive strength. The 28-d compressive strengths of samples C, M1 and M2 for 3-d initial moisture curing are 55.3 MPa, 59.1 MPa and 67.9 MPa, respectively. Compared to sample C, the growth rates of samples M1 and M2 are 6.9% and 22.8%, respectively. The 28-d compressive strengths of samples C, M1 and M2 for 7-d initial moisture curing are 57.4 MPa, 64.7 MPa and 70.8 MPa, respectively. Adding 9% and 15% ultrafine metakaolin increases the 28-d compressive strength by 12.7% and 23.3%, respectively. The strength growth rate of the 28-d compressive strength significantly increases with increasing ultrafine metakaolin content. The growth rate for 7-d initial moisture curing is higher than the growth rate for 3-d initial moisture curing. Meanwhile, compared to that for 3-d initial moisture curing, the 28-d compressive strength of mortar for 7-d initial moisture curing increases. The growth rates of samples C, M1 and M2 are 3.8%, 9.5% and 4.3%, respectively. The growth rates of samples M1 and M2 are higher than that of sample C. Therefore, prolonging the initial moisture curing time is more favorable for ultrafine metakaolin concrete than plain cement concrete.

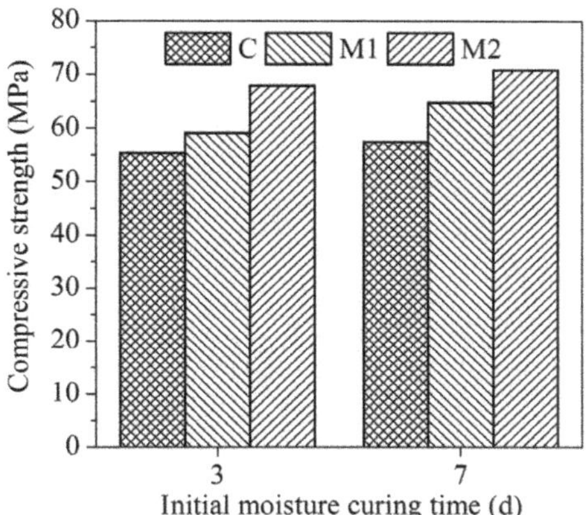

Figure 6. The 28-d compressive strength with different initial moisture curing time.

The visual appearance of samples C, M1 and M2 exposed to sulphate attack for different ages is shown in Figure 7a–c, respectively. The numbers marked in Figure 7 refer to the initial moisture curing time. Many salt crystals precipitate from the upper parts of the mortars and tend to increase with prolonged semi-immersion time. In the process of semi-immersion, no obvious cracks appear on the surfaces of all mortars. The relative compressive strengths of the mortars after different semi-immersion times are shown in Figure 8a. When the mortars were initially moisture cured for 3 d, the relative compressive strength of all mortars after 28 d of semi-immersion are less than 100%. However, the relative compressive strengths for certain mortars after 56 d and 90 d of semi-immersion are more than 100%, especially for plain cement mortar. When the mortars are initially moisture cured for 7 d, the relative compressive strengths of almost all mortars after semi-immersion are less than 100%. Therefore, from the results for the relative compressive strengths, prolonging the initial moisture curing time has an adverse effect on the sulphate attack resistance. This is unreasonable. Previous studies have shown that prolongation of the initial moisture curing time is beneficial to cement hydration and pore structure development, and sulphate attack resistance should not decrease with prolongation of the initial moisture curing time [23,27]. Therefore, the compressive strength cannot be used as an index to evaluate sulphate attack resistance of rectangular mortars in semi-immersion tests. The flexural strength losses of mortars after different semi-immersion times are shown in Figure 8b. Apparently, the flexural strength losses of mortars decrease with increasing ultrafine metakaolin at any semi-immersion age. When the semi-immersion time is the same, prolonging the initial moisture curing time has little impact on the flexural strength loss of samples C and M1. However, compared to the flexural strength loss for 3-d initial moisture curing, the flexural strength loss of sample M2 for 7-d initial moisture curing is relatively low. The addition of ultrafine metakaolin significantly improves the sulphate attack resistance of mortar, and the sulphate attack resistance also significantly improves with increasing ultrafine metakaolin. Therefore, in the semi-immersion test, it is reasonable to evaluate the sulphate attack resistance of mortar by flexural strength loss.

Figure 7. Visual appearance of mortar exposed to sulphate attack: (**a**) C; (**b**) M1; (**c**) M2.

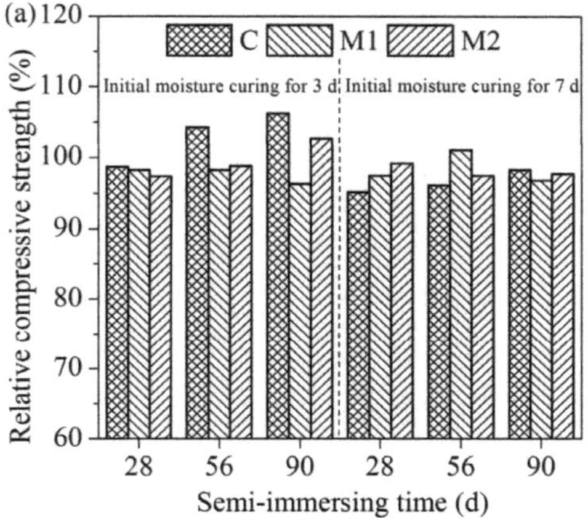
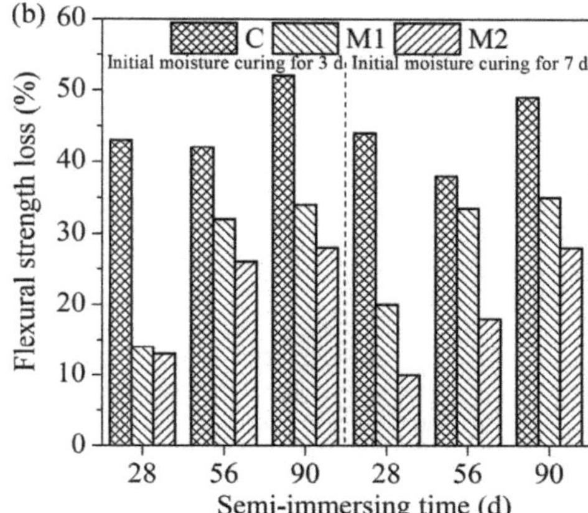

Figure 8. Results of sulphate attack: (**a**) relative compressive strength; (**b**) flexural strength loss.

4. Conclusions

(1) Simultaneously adding ultrafine metakaolin and a certain amount of polycarboxylate superplasticizer can effectively ensure the workability of concrete. The effect of adding ultrafine metakaolin is better than that of adding silica fume in the same case.

(2) The compressive strength and splitting tensile strength increase as the ultrafine admixture content increases. Adding 15% ultrafine metakaolin increases the compressive strength and splitting tensile strength at 28 d by approximately 24% and 33%, respectively. The effect of mixing the same amounts of ultrafine metakaolin and silica fume on the mechanical strength is not significant.

(3) Adding ultrafine metakaolin obviously reduces the connected porosity of concrete, which effectively improves its resistance to chloride ion penetration and freeze–thaw cycles. Silica fume has the same effect as ultrafine metakaolin.

(4) Prolonging the initial moisture curing time is more favorable for ultrafine metakaolin concrete than plain cement concrete in terms of compressive strength. Simultaneously adding ultrafine metakaolin and prolonging the initial moisture curing time can significantly improve the sulphate attack resistance of mortar.

Author Contributions: Conceptualization, S.Z.; methodology, Y.Z. and F.H.; software, S.Z. and Y.Z.; validation, J.S. and Y.Z.; writing—original draft preparation, J.S.; writing—review and editing, J.S. and F.H. All authors have read and agreed to the published version of the manuscript.

Funding: This research was funded by the National Natural Science Foundation of China (No. 51908033) and Beijing Natural Science Foundation (No. 8204067).

Institutional Review Board Statement: Not applicable.

Informed Consent Statement: Not applicable.

Data Availability Statement: Data are contained within the article.

Conflicts of Interest: The authors declare no conflict of interest.

References

1. Abedalqader, A.; Shatarat, N.; Ashteyat, A.; Katkhuda, H. Influence of temperature on mechanical properties of recycled asphalt pavement aggregate and recycled coarse aggregate concrete. *Constr. Build. Mater.* **2021**, *269*, 121285. [CrossRef]

2. Mei, S.; Wang, Y. Viscoelasticity: A new perspective on correlation between concrete creep and damping. *Constr. Build. Mater.* **2020**, *265*, 120557. [CrossRef]

3. Wang, D.; Wang, Q.; Huang, Z. Reuse of copper slag as a supplementary cementitious material: Reactivity and safety. *Resour. Conserv. Recycl.* **2020**, *162*, 105037. [CrossRef]

4. Wang, D.; Wang, Q.; Huang, Z. New insights into the early reaction of NaOH-activated slag in the presence of $CaSO_4$. *Compos. Part B Eng.* **2020**, *198*, 108207. [CrossRef]

5. Sun, J.; Chen, Z. Influences of limestone powder on the resistance of concretes to the chloride ion penetration and sulfate attack. *Powder Technol.* **2018**, *338*, 725–733. [CrossRef]

6. Sun, J.; Zhang, Z.; Hou, G. Utilization of fly ash microsphere powder as a mineral admixture of cement: Effects on early hydration and microstructure at different curing temperatures. *Powder Technol.* **2020**, *375*, 262–270. [CrossRef]

7. Wang, Q.; Feng, J.J.; Yan, P.Y. An explanation for the negative effect of elevated temperature at early ages on the late-age strength of concrete. *J. Mater. Sci.* **2011**, *46*, 7279–7288. [CrossRef]

8. Wang, Q.; Wang, D.; Chen, H. The role of fly ash microsphere in the microstructure and macroscopic properties of high-strength concrete. *Cem. Concr. Compos.* **2017**, *83*, 125–137. [CrossRef]

9. Yao, G.; Liu, Q.; Wang, J.; Wu, P.; Lyu, X. Effect of mechanical grinding on pozzolanic activity and hydration properties of siliceous gold ore tailings. *J. Clean. Prod.* **2019**, *217*, 12–21. [CrossRef]

10. Yao, G.; Cui, T.; Zhang, J.; Wang, J.; Lyu, X. Effects of mechanical grinding on pozzolanic activity and hydration properties of quartz. *Adv. Powder Technol.* **2020**, *31*, 4500–4509. [CrossRef]

11. Ting, L.; Qiang, W.; Shiyu, Z. Effects of ultra-fine ground granulated blast-furnace slag on initial setting time, fluidity and rheological properties of cement pastes. *Powder Technol.* **2019**, *345*, 54–63. [CrossRef]

12. Zhao, Y.; Gao, J.; Liu, C.; Chen, X.; Xu, Z. The particle-size effect of waste clay brick powder on its pozzolanic activity and properties of blended cement. *J. Clean. Prod.* **2020**, *242*, 118521. [CrossRef]

13. Irassar, E.F.; Bonavetti, V.L.; Castellano, C.C.; Trezza, M.A.; Rahhal, V.F.; Cordoba, G.; Lemma, R. Calcined illite-chlorite shale as supplementary cementing material: Thermal treatment, grinding, color and pozzolanic activity. *Appl. Clay Sci.* **2019**, *179*, 105143. [CrossRef]

14. Li, D.; Sun, R.; Wang, D.; Ren, C.; Fang, K. Study on the pozzolanic activity of ultrafine circulating fluidized-bed fly ash prepared by jet mill. *Fuel* **2021**, *291*, 120220. [CrossRef]

15. Yao, G.; Wang, Z.; Yao, J.; Cong, X.; Anning, C.; Lyu, X. Pozzolanic activity and hydration properties of feldspar after mechanical activation. *Powder Technol.* **2021**, *383*, 167–174. [CrossRef]

16. Ahmed, T.; Elchalakani, M.; Karrech, A.; Mohamed Ali, M.S.; Guo, L. Development of ECO-UHPC with very-low-C_3A cement and ground granulated blast-furnace slag. *Constr. Build. Mater.* **2021**, *284*, 122787. [CrossRef]

17. Herald Lessly, S.; Lakshmana Kumar, S.; Raj Jawahar, R.; Prabhu, L. Durability properties of modified ultra-high performance concrete with varying cement content and curing regime. *Mater. Today Proc.* **2020**. [CrossRef]

18. Ding, M.; Yu, R.; Feng, Y.; Wang, S.; Zhou, F.; Shui, Z.; Gao, X.; He, Y.; Chen, L. Possibility and advantages of producing an ultra-high performance concrete (UHPC) with ultra-low cement content. *Constr. Build. Mater.* **2021**, *273*, 122023. [CrossRef]

19. Lv, L.S.; Wang, J.Y.; Xiao, R.C.; Fang, M.S.; Tan, Y. Chloride ion transport properties in microcracked ultra-high performance concrete in the marine environment. *Constr. Build. Mater.* **2021**, *291*, 123310. [CrossRef]

20. Lu, Z.; Feng, Z.; Yao, D.; Li, X.; Ji, H. Freeze-thaw resistance of Ultra-High performance concrete: Dependence on concrete composition. *Constr. Build. Mater.* **2021**, *293*, 123523. [CrossRef]

21. Dixit, A.; Verma, A.; Pang, S.D. Dual waste utilization in ultra-high performance concrete using biochar and marine clay. *Cem. Concr. Compos.* **2021**, *120*, 104049. [CrossRef]

22. Güneyisi, E.; Gesoğlu, M.; Mermerdaş, K. Improving strength, drying shrinkage, and pore structure of concrete using metakaolin. *Mater. Struct. Constr.* **2008**, *41*, 937–949. [CrossRef]

23. Siddique, R.; Klaus, J. Influence of metakaolin on the properties of mortar and concrete: A review. *Appl. Clay Sci.* **2009**, *43*, 392–400. [CrossRef]

24. Zhan, P.-M.; He, Z.-H.; Ma, Z.-M.; Liang, C.-F.; Zhang, X.-X.; Abreham, A.A.; Shi, J.-Y. Utilization of nano-metakaolin in concrete: A review. *J. Build. Eng.* **2020**, *30*, 101259. [CrossRef]

25. Raheem, A.A.; Abdulwahab, R.; Kareem, M.A. Incorporation of metakaolin and nanosilica in blended cement mortar and concrete—A review. *J. Clean. Prod.* **2021**, *290*, 125852. [CrossRef]

26. Sabir, B.; Wild, S.; Bai, J. Metakaolin and calcined clays as pozzolans for concrete: A review. *Cem. Concr. Compos.* **2001**, *23*, 441–454. [CrossRef]

27. Elahi, M.M.A.; Shearer, C.R.; Naser Rashid Reza, A.; Saha, A.K.; Khan, M.N.N.; Hossain, M.M.; Sarker, P.K. Improving the sulfate attack resistance of concrete by using supplementary cementitious materials (SCMs): A review. *Constr. Build. Mater.* **2021**, *281*, 122628. [CrossRef]

28. Panesar, D.K.; Zhang, R. Performance comparison of cement replacing materials in concrete: Limestone fillers and supplementary cementing materials—A review. *Constr. Build. Mater.* **2020**, *251*, 118866. [CrossRef]

29. Al Menhosh, A.; Wang, Y.; Wang, Y.; Augusthus-Nelson, L. Long term durability properties of concrete modified with metakaolin and polymer admixture. *Constr. Build. Mater.* **2018**, *172*, 41–51. [CrossRef]

30. Hossain, M.M.; Karim, M.R.; Hasan, M.; Hossain, M.K.; Zain, M.F.M. Durability of mortar and concrete made up of pozzolans as a partial replacement of cement: A review. *Constr. Build. Mater.* **2016**, *116*, 128–140. [CrossRef]

Article

Experimental Study on the Effect of Compound Activator on the Mechanical Properties of Steel Slag Cement Mortar

Junfeng Guan [1], Yulong Zhang [1], Xianhua Yao [1,*], Lielie Li [1], Lei Zhang [2] and Jinhua Yi [3]

[1] School of Civil Engineering and Communication, North China University of Water Resources and Electric Power, Zhengzhou 450045, China; shuaipipi88@126.com (J.G.); 18838089801@163.com (Y.Z.); 1000-lilili@163.com (L.L.)

[2] Institute of Engineering Mechanics, Yellow River Institute of Hydraulic Research, Zhengzhou 450045, China; hkyzhanglei@163.com

[3] Institute of Road and Bridge Engineering, Hunan Communication Polytechnic, Changsha 410132, China; y371926012@163.com

* Correspondence: yaoxianhua@ncwu.edu.cn

Abstract: In this study, activator, metakaolin, and silica fume were used as a compound activator to improve the activity of steel slag powder. The influence of activator, steel slag powder, metakaolin, and silica fume on the resulting strength of steel slag cement mortar was investigated by orthogonal experiments. For four weight fractions of steel slag powder (10%, 20%, 30%, 40%), the experimental results indicate that the compressive strength of mortar can reach up to more than 85% of the control group while the flexural strength can reach up to more than 90% of the flexural strength of the control group. Through orthogonal analysis, it is determined that the activator is the primary factor influencing the mortar strength. According to the result of orthogonal analysis, the optimal dosages of activator, steel slag powder, metakaolin, and silica fume are suggested. The GM $(0, N)$ prediction model of compressive strength and flexural strength was established, and the compressive strength and flexural strength of mortar with the optimal dosage combinations were predicted. The prediction results show that by using the optimal dosage combination, the mortar strength can reach the level of P·O·42.5 cement. Considering the different strength and cost requirements of cementitious materials in practical engineering, the economic benefits of replacing cement with steel slag powder activated by compound activator in various proportions and equal amounts were presented. The results show that the method proposed in this study can reduce the cost of cementitious materials.

Keywords: steel slag powder; compound activator; mortar strength; orthogonal experiment; GM $(0, N)$ model

Citation: Guan, J.; Zhang, Y.; Yao, X.; Li, L.; Zhang, L.; Yi, J. Experimental Study on the Effect of Compound Activator on the Mechanical Properties of Steel Slag Cement Mortar. *Crystals* **2021**, *11*, 658. https://doi.org/10.3390/cryst11060658

Academic Editors: Chuanqing Fu, Peng Zhang, Peter Taylor and Yifeng Ling

Received: 19 May 2021
Accepted: 6 June 2021
Published: 10 June 2021

Publisher's Note: MDPI stays neutral with regard to jurisdictional claims in published maps and institutional affiliations.

Highlights:

- A method for using a compound activator to improve the activity of steel slag powder is proposed.
- The optimal dosage combination of activator, steel slag, metakaolin, and silica fume is suggested.
- The economic benefit analysis is carried out on the steel slag powder activated by the compound activator to replace part of the cement.

1. Introduction

Steel slag is a byproduct of steel production, which accounts for about 15% of the mass of steel production [1–3]. In China, the generation of steel slag is huge, whereas the total utilization rate is low [4]. The accumulation of steel slag not only takes up a lot of land but also pollutes the surrounding environment [5]. Therefore, it is imperative to improve the utilization of steel slag [6–8]. The composition of steel slag is similar to that of cement, and as such, it has the potential of replacing cement as a cementitious material. If steel slag can be effectively used in the cement industry, it will benefit the solid waste utilization, energy conservation, and environmental protection [9–13]. However, the inherent low activity of steel slag restricts its application in the cement industry [14–16].

In order to address this issue, researchers have been employing different methods to improve the activity of steel slag. Commonly used activation methods include physical activation, chemical activation, thermal activation, and steel slag restructuring. The physical activation method increases the specific surface area of the steel slag by grinding the steel slag into ultra-fine powder by using a ball mill, thereby increasing the hydration rate. Zhu et al. [17] used a grinding aid mixed with sulfonate, alcohol, and metaphosphoric to grind steel slag, which increased the early hydration rate of steel slag. Altun et al. [18] ground steel slag to 4000 cm^2/g and 4700 cm^2/g specific surface area and used 30% of it to replace Portland cement in mortar preparation which led to the 28-days compressive strength of mortar to be 38.5 MPa and 45.8 MPa, respectively. Chemical activation enhances the activity of steel slag by changing the mineral formation process, primarily including alkali activation and acid activation [19–21]. Peng et al. [22] used water glass as a steel slag activator, and the 28-days compressive strength of the mortar with 40% steel slag dosage reached 51.4 MPa. Sun et al. [23] used water glass to activate the steel slag activity, and the results showed that the pore structure of hardened cement paste was more compact than that of the steel slag paste activated by sodium silicate, while the compressive strength of alkali–activated steel slag hardened pastes was only 30–40% of the strength of ordinary cement pastes. Huo et al. [20,24] used phosphoric acid and formic acid to activate the activity of steel slag, and research result showed that the compressive strength of the activated steel slag pastes significantly improves at 3 days and 7 days ages. Zhang et al. [25] used water glass, industrial residues, and a mixture of sodium hydroxide, calcium oxide, and alum as the compound activator for steel slag; and the 28 days compressive strength of steel slag blended cement was reported as 47.7 MPa. Du et al. [26] used dihydrate gypsum and silica fume as steel slag compound activator in a ratio of 1:4, which greatly improved the strength of steel slag cementitious materials at the early and late ages. Thermal activation can depolymerize the vitreous phase in the steel slag, thereby increasing the activity. Lin et al. [27] investigated the effect of using thermally activated steel slag-fly ash-gypsum system (the autoclave temperature was 100 °C). It was shown that the 28 days compressive strength reached 46.8 MPa and 43.5 MPa for the pretreatment material dosages of 35% and 40%, respectively. Steel slag reconstruction is the addition of different materials to adjust the chemical composition of steel slag to cause a chemical reaction at high temperature for absorbing free calcium oxide to generate reactive substances such as dicalcium silicate, tricalcium silicate, tricalcium aluminate, which improved the activity of steel slag. Kang et al. [28] mixed basic oxygen furnace steel slag and electric arc furnace steel slag in an appropriate ratio and reheated these at a high temperature in the laboratory, which significantly improved the activity of steel slag. Yin et al. [29] reconstructed the steel slag by reducing FeO_x to improve the hydraulic activity of the steel slag, and the result confirmed that the activity index was 92% when direct reduction slag replaced the cement with a mass ratio of 30%. Zhao et al. [30] reconstructed the steel slag by adding electric furnace slag and fly ash to the converter steel slag. The compressive strength of the paste with 30% reconstructed steel slag dosage could reach 99.9% of that for the pure cement paste. Many studies have shown that the physical activation takes a long time and has little effect on the activity of steel slag at a later age. Also, the cost of the chemical activator is much higher than the other methods. Whereas steel slag reconstruction can improve its activity to a certain degree, it still lags behind the cement clinker. Other methods can also improve the activity of steel slag, but these are still in the experimental stage [31,32].

There are many researches on chemical alkaline activation methods, but alkaline activation has high cost. Therefore, the compound activator composed of neutral materials (some salts such as sodium sulfate, sodium aluminate, and some mineral admixtures such as silica fume) was used to activate the activity of steel slag powder in this paper, and the optimal dosage of each component of compound activator in cementitious material was determined through experiments. A grey prediction model GM $(0, N)$ was established to predict the strength of steel slag cement mortar with the optimal dosage of each component. The effectiveness of the proposed activation method was verified by the test results and the model prediction results. Furthermore, considering the different requirements of engineering for cementitious materials, economic benefit analysis is carried out for different mix proportions to check whether the proposed method will reduce the cost of cementitious materials and the extent of reduction to provide basis for engineering.

2. Experimental and Methods

2.1. Raw Materials

Seven different kinds of raw materials were used in the experimental program of this study: cement, sand, steel slag powder, activator, metakaolin, silica fume, and water. The cement was P·O·42.5 Portland cement with an apparent density of 3150 kg/m^3. The sand used in the study was ISO standard sand. The steel slag powder was produced by Taiyuan Iron & Steel Group Co., Ltd. in Taiyuan, China, with a specific surface area of 450 m^2/kg. The activator was composed of various materials, and the corresponding components and mass percentages are shown in Table 1. Water quenched slag and stone powder in the activator can accelerate the hydration reaction. In addition, the particle size of water quenched slag is larger than that of steel slag powder, so the water quenched slag can be combined with steel slag powder to optimize the grading. Metakaolin was obtained by calcining kaolin at 700 °C for 24 h. Silica fume was purchased from the market with a density of 2.3 g/cm^3. Ordinary potable water was used as well. The chemical compositions of cement, steel slag powder, metakaolin, and silica fume are shown in Table 2, while the particle size distribution is given in Figure 1. The actual samples of cement, steel slag powder, activator, metakaolin, and silica fume are shown in Figure 2.

Table 1. Components of activator (wt %).

Silica Fume	Sodium Aluminate	Sodium Sulfate	Sodium Tripolyphosphate	Water Quenched Slag	Desulfurization Gypsum	Stone Powder
6.25	0.63	1.25	0.62	21.25	3.13	66.87

Table 2. Main chemical compositions of materials (wt %).

Composition	SiO$_2$	Al$_2$O$_3$	Fe$_2$O$_3$	CaO	SO$_3$	MgO	Na$_2$O	MnO	K$_2$O
Cement	19.31	5.86	3.15	60.33	4.42	3.03	0.12	-	1.13
Steel slag powder	14.24	1.94	19.69	46.19	-	10.06	-	1.36	-
Metakaolin	55.36	35.46	1.83	0.54	-	0.03	0.04	-	0.26
Silica fume	91.33	0.85	0.57	0.47	0.47	1.55	0.42	-	1.38

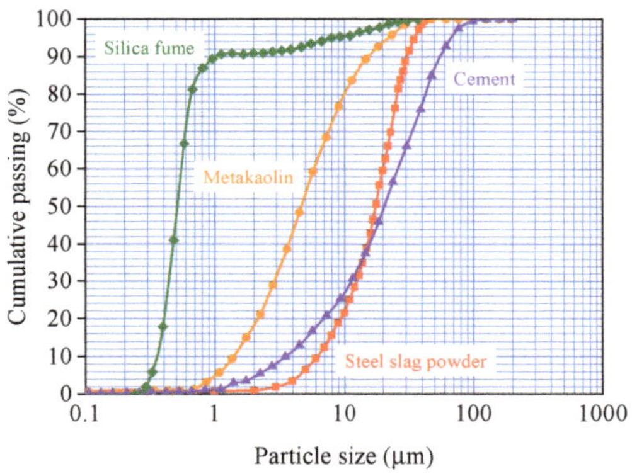

Figure 1. Particle size distribution of materials.

(a) Cement (b) Steel slag power

(c) Activator (d) Metakaolin (e) Silica fume

Figure 2. Samples of cementitious material components in this study (**a**) cement; (**b**) steel slag powder; (**c**) activator; (**d**) metakaolin and (**e**) silica fume.

2.2. Mix Proportions

The experimental design method is often chosen according to the experimental purpose [33–35]. In order to study the influence of various factors on the mortar strength, the orthogonal method was adopted for designing the experimental mix proportions. Activator, steel slag powder, metakaolin and silica fume were taken as four factors. The dosage of each factor was set at several levels according to the mass percentage of cementitious material. Due to the low activity of steel slag powder itself, the activation method is difficult to have a good influence on the strength of steel slag cement mortar with excessive steel slag powder dosage, so the dosage of steel slag powder is set to 10%, 20%, 30%, and 40%, a total of four levels. Al_2O_3 in metakaolin can accelerate the hydration reaction of SiO_2 in cementitious materials, so as to improve the strength of steel slag cement mortar. A small amount of metakaolin can have positive effect on the strength of steel slag cement mortar, so the dosage of metakaolin is set at four levels: 5%, 10%, 15%, and 20%. The silica fume can improve the strength of steel slag cement mortar, but excessive dosage will lead to high cost, so the silica fume dosage is set at 4 levels: 2%, 4%, 6%, and 8%. Level details of factors are shown in Table 3. According to the orthogonal design method, sixteen groups of mix proportions were designed. Another group of pure cement mortar specimens was also designed as the control group for comparative study. The mix proportions are given in Table 4.

Table 3. Level details of factors (wt %).

Dosage Level	Factor			
	Activator	Steel Slag Powder	Metakaolin	Silica Fume
1	5	10	5	2
2	10	20	10	4
3	15	30	15	6
4	20	40	20	8

Table 4. Mix proportions (g).

Experimental Group	Activator	Steel Slag Powder	Metakaolin	Silica Fume	Cement	Sand	Water
G0	0	0	0	0	450	1350	225
G1	22.5	45	22.5	9	351	1350	225
G2	22.5	90	45	18	274.5	1350	225
G3	22.5	135	67.5	27	198	1350	225
G4	22.5	180	90	36	121.5	1350	225
G5	45	45	45	27	288	1350	225
G6	45	90	22.5	36	256.5	1350	225
G7	45	135	90	9	171	1350	225
G8	45	180	67.5	18	139.5	1350	225
G9	67.5	45	67.5	36	234	1350	225
G10	67.5	90	90	27	175.5	1350	225
G11	67.5	135	22.5	18	207	1350	225
G12	67.5	180	45	9	148.5	1350	225
G13	90	45	90	18	207	1350	225
G14	90	90	67.5	9	193.5	1350	225
G15	90	135	45	36	144	1350	225
G16	90	180	22.5	27	130.5	1350	225

2.3. Test Methods

The mortar was mechanically mixed by a mixer. Before mixing, the activator and steel slag powder was mixed uniformly. Then, metakaolin, silica fume, and cement were gradually added to obtain the cementitious materials mix, and the mixing was further continued. The mixing is done according to the following steps. First, water was added to the mixing pot followed by the addition of all the cementitious materials. The machine was started and the materials in the pot were slowly mixed at low speed for 30 s. The sand was uniformly added after 30 s, and the mixing was continued at high speed for an additional 30 s after the sand was fully added. The mixing was stopped for 90 s, and a rubber scraper was used to scrape off the paste on the blade and the wall of the mixing pot. Then, it was further mixed at high speed for 60 s. The detailed mixing procedure is shown in Figure 3. The mortar should be formed immediately after preparation. The specimens were cast in three connected test molds, each having a size of 40 mm × 40 mm × 160 mm. The mortar should be put into the test molds in two layers, and a vibrating table should be used for vibrating the layers. Each time it was vibrated 60 times. Then, the mortar molds were put into a curing box with humidity of 95% and a temperature of 20 °C for curing. After 24 h, the demolding was done, and the test block was cured for 180 days under standard curing conditions. Then, the mechanical properties of mortar were tested according to GB/T 17671-1999 [36]. The flexural strength testing machine is an electric flexural testing machine model DKZ-5000 with 5 kN measuring range, manufactured by Wuxi Jianyi Instrument & Machinery Co., Ltd. in Wuxi, China. Load control was selected as the test procedure of flexural strength, and the loading rate was 50 N/s. The compressive strength testing device is Hualong universal testing machine model WAW-600 with 600 kN measuring range, manufactured by Shanghai Hualong Test Instrumens Co., Ltd. in Shanghai, China. Load control was selected as the test procedure of compressive strength, and the loading rate was 2500 N/s.

Figure 3. Mixing progress of steel slag mortar.

2.4. GM (0, N) Prediction Model

In practice, we often encounter problems such as the lack of data, grey characteristics of the data itself, and the need to consider the correlation between predictive variables and multiple factors [37,38]. Grey system theory provides a method to solve such problems. GM (0, N) and GM (1, N) are common multi-factor grey prediction models [39,40]. GM (1, N) is a little complicated because it involves the first-order differentiation, whereas the GM (0, N) model is relatively simple to establish as it has high prediction accuracy [41,42]. The establishment of the GM (0, N) model is as follows [42]:

Take $X_1^{(0)} = \left(x_1^{(0)}(1), x_1^{(0)}(2), \cdots, x_1^{(0)}(n) \right)$ as the system characteristic data sequence, and

$$X_2^{(0)} = (x_2^{(0)}(1), x_2^{(0)}(2), \cdots, x_2^{(0)}(n))$$
$$X_3^{(0)} = (x_3^{(0)}(1), x_3^{(0)}(2), \cdots, x_3^{(0)}(n))$$
$$\cdots$$
$$X_N^{(0)} = (x_N^{(0)}(1), x_N^{(0)}(2), \cdots, x_N^{(0)}(n))$$

as the relative factors data sequences. $X_i^{(1)}$ is the 1-AGO sequence of $X_i^{(0)}$ ($i = 1, 2, \cdots, N$), then call

$$x_1^{(1)}(k) = \sum_{i=2}^{N} b_i x_i^{(1)}(k) + a \tag{1}$$

as the GM (0, N) model.

$$B = \begin{bmatrix} x_2^{(1)}(2) & x_3^{(1)}(2) & \cdots & x_N^{(1)}(2) & 1 \\ x_2^{(1)}(3) & x_3^{(1)}(3) & \cdots & x_N^{(1)}(3) & 1 \\ \vdots & \vdots & & \vdots & \vdots \\ x_2^{(1)}(n) & x_3^{(1)}(n) & \cdots & x_N^{(1)}(n) & 1 \end{bmatrix}, Y = \begin{bmatrix} x_1^{(1)}(2) \\ x_1^{(1)}(3) \\ \vdots \\ x_1^{(1)}(n) \end{bmatrix}$$

Then the least-squares estimate of the parameter column $\hat{b} = [b_2, b_3, \cdots, b_N, a]^T$ is

$$\hat{b} = (B^{\mathrm{T}}B)^{-1} B^{\mathrm{T}} Y \tag{2}$$

3. Analysis of Results

The 180 days compressive strength and flexural strength test results of the control group G0 (pure cement mortar specimens) are 48.0 MPa and 9.2 MPa, respectively. The orthogonal experimental results are shown in Table 5.

Table 5. Orthogonal experimental results of mechanical properties.

Experimental Group	Activator (%)	Steel Slag Powder (%)	Metakaolin (%)	Silica Fume (%)	Compressive Strength (MPa)	Flexural Strength (MPa)
G1	5	10	5	2	43.4	8.2
G2	5	20	10	4	37.2	7.8
G3	5	30	15	6	45.6	8.3
G4	5	40	20	8	44.6	7.9
G5	10	10	10	6	44.2	8.8
G6	10	20	5	8	40.6	8.7
G7	10	30	20	2	29.8	7.4
G8	10	40	15	4	28.4	7.6
G9	15	10	15	8	47.9	8.2
G10	15	20	20	6	24.3	3.0
G11	15	30	5	4	31.1	3.7
G12	15	40	10	2	29.0	8.4
G13	20	10	20	4	16.0	4.0
G14	20	20	15	2	14.5	2.5
G15	20	30	10	8	12.5	2.4
G16	20	40	5	6	11.5	2.2

The compressive strength test results are shown in Table 5 and Figure 4a. When the steel slag powder dosage is 10%, the mix proportion G9 has the highest compressive strength (47.9 MPa). The dosages of activator, metakaolin, and silica fume are 15%, 15%, 8%, respectively. When the steel slag powder dosage is 20%, the compressive strength of the mix proportion G6 is the highest (40.6 MPa), and the corresponding dosages of activator, metakaolin, and silica fume are 10%, 5%, and 8%, respectively. When the steel slag powder dosage is 30%, the mix proportion G3 has the highest compressive strength (45.6 MPa), and the corresponding dosages of activator, metakaolin, and silica fume are 5%, 15%, and 6%, respectively. Furthermore, when the steel slag powder dosage is 40%, the mix proportion G4 has the highest compressive strength (44.6 MPa), and the corresponding dosages of activator, metakaolin, and silica fume are 5%, 20%, and 8%, respectively. As shown in Figure 4a, with the change of dosage of steel slag powder (10–40%), the highest compressive strength at each dosage level is more than 85% of the compressive strength of the control group. As shown in Figure 4a, the compressive strength of steel slag cement mortar shows an overall trend of decline from G1 to G16, while the dosage of activator (5–20%) is gradually increasing. The reason why the strength of steel slag cement mortar decreases may be that the content of stone powder increases with the increase of the activator dosage, and excessive stone powder will have negative effect on the strength of steel slag cement mortar. G9 is the mix proportion with the highest compressive strength among 16 mix proportions. The reasons may be divided into two aspects. On the one hand, the dosage of steel slag powder (10%) is relatively low, and the negative effect of low activity of steel slag powder on steel slag cement mortar is relatively small. On the other hand, it may be related to the relatively large dosage of metakaolin (15%) and silica fume (8%). SiO_2 in metakaolin and silica fume hydrate to form calcium silicate, which improves the compressive strength. In addition, it is difficult to identify the main cause in the condition of many factors, and further analysis is needed to determine the cause.

The flexural strength test results are given in Table 5 and Figure 4b. When the steel slag powder dosage is 10%, the mix proportion G5 has the highest flexural strength (8.8 MPa), and the corresponding dosages of activator, metakaolin, and silica fume, are 10%, 10%, and 6%, respectively. When the steel slag powder dosage is 20%, the mix proportion G6 has the highest flexural strength (8.7 MPa), and the dosages of activator, metakaolin, and silica fume are 10%, 5%, and 8%, respectively. At 30% dosage of steel slag powder, the mix proportion G3 has the highest flexural strength (8.3 MPa), and the corresponding dosages of activator, metakaolin, and silica fume are 5%, 15%, and 6%, respectively. When the steel slag powder dosage is 40%, the mix proportion G12 has the highest flexural strength (8.4 MPa), and the corresponding dosages of activator, metakaolin, and silica fume are 15%, 10%, and 2%, respectively. As shown in Figure 4b, with the change of steel slag powder dosage (10–40%), the highest flexural strength at each level of dosage is greater than 90% of the flexural strength of the control group. As shown in Figure 4b, there is little difference in the flexural strength of steel slag cement mortar of G1–G9, and the flexural strength of G10, G11, G13, G14, G15, and G16 is relatively low, while the flexural strength of G12 is relatively high. Similar to the compressive strength, it is difficult to find out the influence law of each factor only from the test results, and further analysis is needed to determine the cause.

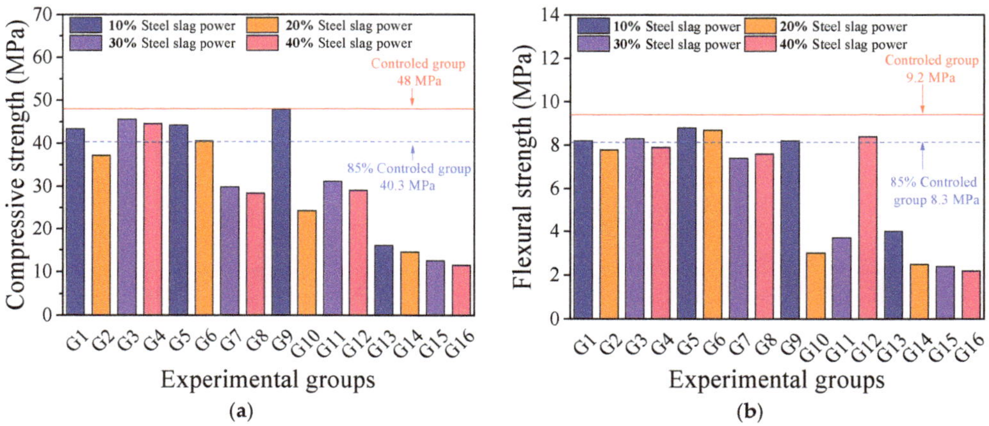

Figure 4. Strength of steel slag cement mortar with different mix proportions: (**a**) Compressive strength of experimental groups; (**b**) Flexural strength of experimental groups.

4. Orthogonal Analysis

The analysis methods of orthogonal experimental results include range analysis and variance analysis. Range analysis is a method to analyze the influence of each factor on the system value in a multi-factor system. It can not only determine the primary factor influencing the system value but also gives the optimal level combination of several factors. Variance analysis is also a method of multi-factor system analysis, which can not only estimate the error size but also represents the significant degree of influence of each factor on the system value [43]. The above two methods are used to analyze the experimental results.

4.1. Compressive Strength Analysis

The results of the compressive strength range analysis are shown in Table 6. According to the range (R) of compressive strength caused by the change of dosage level, the factors are in order as follows: activator (29.0 MPa) > steel slag powder (9.5 MPa) > silicon powder (8.2 MPa) > metakaolin (5.4 MPa). Activator dosage has the most significant influence on the experimental results of compressive strength. It indicates that the activator is the primary influencing factor of compressive strength. The results of compressive strength variance analysis are shown in Table 7. Activator, steel slag powder, metakaolin, and silica fume have a very profound effect on the compressive strength. The influence degree of each factor on compressive strength can be further characterized according to the value of '*F*' of each factor. The greater the value of '*F*' of each factor, the more significant the influence is. Therefore, the order of the various factors according to the influence degree of compressive strength from large to small is the activator, steel slag powder, silica fume, and metakaolin. The results of range analysis and variance analysis indicate that the activator had the greatest influence on the compressive strength and is the primary factor influencing the compressive strength.

The relationship curve between the compressive strength and the dosage levels of each influencing factor is shown in Figure 5. The influence of the activator dosages on the compressive strength is shown in Figure 5a. The compressive strength decreases gradually with the increase of the activator dosage, and it is at maximum when the activator dosage is 5%. The influence of steel slag powder dosage on the compressive strength is shown in Figure 5b. The compressive strength decreases gradually with the increase of the steel slag powder dosage, and it reaches the maximum when the steel slag powder dosage is 10%. The influence of metakaolin dosage on compressive strength is shown in Figure 5c. The compressive strength first increases and then decreases with the increase of metakaolin dosage. The compressive strength reaches the maximum when the metakaolin dosage is 15%. The influence of silica fume dosage on compressive strength is shown in Figure 5d. The compressive strength first decreases and then increases with the increase of silica fume dosage. The compressive strength reaches the maximum value when the activator content is 8%. The compressive strength of steel slag cement mortar with 10% steel slag powder dosage is significantly different from that with 20%, 30%, and 40% steel slag powder dosage. The reason is that the activity of steel slag powder is low, and it is difficult to achieve a higher compressive strength even if the activation method is adopted. With the increase of the activator dosage, the content of stone powder also increases, and the negative effect caused by excessive stone powder is greater than the activation effect of activator on steel slag. When the dosage of metakaolin is 20%, the compressive strength is lower than the compressive strength corresponding to dosages of 5–15%, so the dosage should not be exceed 15%. With the increase of silica fume dosage (4–8%), the compressive strength of steel slag cement mortar increases greatly. The dosages corresponding to the maximum compressive strength are taken as the optimal dosage. Therefore, for the compressive strength of mortar, the optimal dosage combination of the four factors is activator 5%, steel slag powder 10%, metakaolin 15%, and silica fume 8%.

Table 6. Range analysis of compressive strength.

Index	Compressive Strength (MPa)			
	Activator	Steel Slag Powder	Metakaolin	Silica Fume
K_1	170.8	151.4	126.6	116.6
K_2	142.9	116.6	122.9	112.7
K_3	132.3	119.0	136.3	125.6
K_4	54.6	113.5	114.7	145.6
k_1	**42.7**	**37.9**	31.7	29.2
k_2	35.7	29.2	30.7	28.2
k_3	33.1	29.8	**34.1**	31.4
k_4	13.7	28.4	28.7	36.4
MAX	42.7	37.9	34.1	**36.4**
MIN	13.7	28.4	28.7	28.2
R	29.0	9.5	5.4	8.2

Note: K_i is the sum of multiple test results at a certain level, k_i is the mean value of multiple test results at a certain level, and R is the range of mean value of test results at different levels.

Table 7. Variance analysis of compressive strength.

Material	SS	d_f	MS	F	Significant Degree
Activator	5562.2	3	1854.1	154.6	**
Steel slag powder	703.0	3	234.3	19.5	**
Metakaolin	181.0	3	60.3	5.0	**
Silica fume	485.8	3	161.9	13.5	**
Se_1	300.9	3		$F_{0.01}$ (3, 35) = 4.4	
Se_2	118.8	32		$F_{0.05}$ (3, 35) = 2.9	
Se	419.7	35	12.0	$F_{0.2}$ (3, 35) = 1.6	
Sum	7232.9	47			

Note: SS is the sum of squares; d_f is the degrees of freedom; MS is the mean square; Se_1 is the System error; Se_2 is the experiment error; Se is the overall error; F is the MS/Se; ** is very significant ($F > F_{0.01}$ (3, 35)).

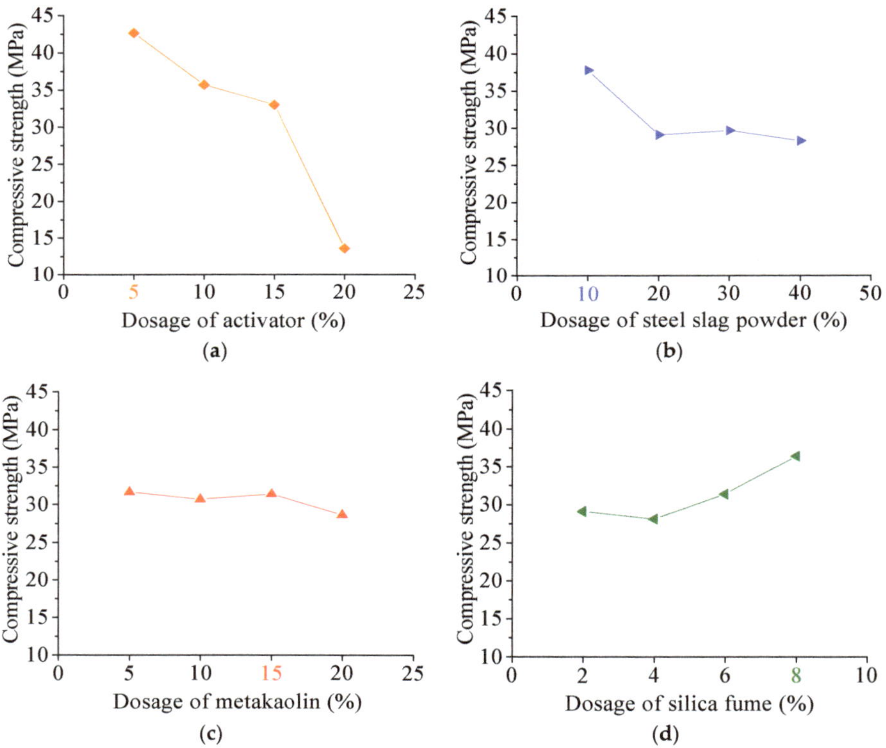

Figure 5. Relationship between compressive strength and dosage levels of each factor: (**a**) activator; (**b**) steel slag powder; (**c**) metakaolin and (**d**) Silica fume.

4.2. Analysis of Flexural Strength

The results of the flexural strength range analysis are shown in Table 8. According to the range (R) of flexural strength caused by the change of dosage level, the factors are in order as follows: activator (5.4 MPa)> steel slag powder (1.8 MPa) > metakaolin (1.3 MPa) > silica fume (1.2 MPa). Activator dosage has the most significant influence on the experimental results of flexural strength. It indicates that the activator is the primary influencing factor of flexural strength. The results of the variance analysis of flexural strength are shown in Table 9. Activator, steel slag powder, metakaolin, and silica fume all have a very significant influence on flexural strength. The influence degree of each factor on flexural strength can be further distinguished according to the value of '*F*' of each factor. Therefore, the order of the different factors according to the influence degree of the flexural strength from large to small is the activator, steel slag powder, metakaolin, and silica fume. The results of range analysis and variance analysis indicate the activator had the greatest influence on the flexural strength and is the primary factor affecting the flexural strength.

The relationship curve between the flexural strength and the dosage levels of each factor is shown in Figure 6. The influence of the activator dosages on the flexural strength is shown in Figure 6a. The flexural strength first increases and then decreases with the increase of the activator dosage. The flexural strength reaches the maximum when the activator dosage is 10%. The influence of steel slag powder dosage on the flexural strength, shown in Figure 6b, indicates that the flexural strength decreases gradually with the increase of the steel slag powder dosage. The flexural strength reaches the maximum when the steel slag powder dosage level is 10%. The influence of metakaolin dosage on flexural strength is shown in Figure 6c. The flexural strength first increases and then decreases with the increase of metakaolin dosage. The flexural strength reaches the maximum when the metakaolin dosage is 15%. The influence of silica fume content on flexural strength is shown in Figure 6d. The flexural strength decreases first and then increases with the increase of silica fume dosage, while it reaches the maximum value when the activator content is 8%. Similar to the compressive strength, the steel slag cement mortar with smaller dosage (5%, 10%) of activator has a larger flexural strength, while the steel slag cement mortar with larger dosage (15%, 20%) of activator has smaller flexural strength. When the dosages of metakaolin and silica fume were 10% and 8%, respectively, the corresponding flexural strength reached the maximum, which was very close to 15% and 8% of optimal dosage in compressive strength analysis. In addition, the strength corresponding to 40% dosage of steel slag powder is only smaller than the strength corresponding to 10% dosage, and the strength corresponding to 2% dosage of silica fume is only smaller than the strength corresponding to 8% dosage, as shown in Figure 6. This indirectly explains the reason why G12 has a large flexural strength, which is the result of the combined action of many factors. The dosages corresponding to the maximum flexural strength are taken as the optimal dosages. Therefore, for the flexural strength of mortar, the optimal dosage combination of the four factors is activator 10%, steel slag powder 10%, metakaolin 10%, and silica fume 8%.

Table 8. Range analysis of flexural strength.

Index	Flexural Strength (MPa)			
	Activator	Steel Slag Powder	Metakaolin	Silica Fume
K_1	32.2	29.1	22.8	26.4
K_2	32.5	22.0	27.4	23.0
K_3	23.2	21.8	26.6	22.4
K_4	11.1	26.1	22.2	27.2
k_1	8.0	7.3	5.7	6.6
k_2	8.1	5.5	6.8	5.7
k_3	5.8	5.4	6.7	5.6
k_4	2.8	6.5	5.6	6.8
MAX	8.1	7.3	6.8	6.8
MIN	2.8	5.4	5.6	5.6
R	5.4	1.8	1.3	1.2

Note: K_i is the sum of multiple test results at a certain level, k_i is the mean value of multiple test results at a certain level, and R is the range of mean value of test results at different levels.

Table 9. Variance analysis of flexural strength.

Material	SS	d_f	MS	F	Significant Degree
Activator	228.7	3	76.2	51.1	**
Steel slag powder	28.1	3	9.4	6.3	**
Metakaolin	15.6	3	5.2	3.5	*
Silica fume	13.3	3	4.4	3.0	*
Se_1	28.9	3		$F_{0.01}$ (3, 35) = 4.4	
Se_2	23.3	32		$F_{0.05}$ (3, 35) = 2.9	
Se	52.2	35	1.5	$F_{0.2}$ (3, 35) = 1.6	
Sum	314.6	47			

Note: SS is the sum of squares; d_f is the degrees of freedom; MS is the mean square; Se_1 is the System error; Se_2 is the experiment error; Se is the overall error; F is the MS/Se; ** is very significant ($F > F_{0.01}$ (3, 35)); * is significant ($F > F_{0.05}$ (3, 35)).

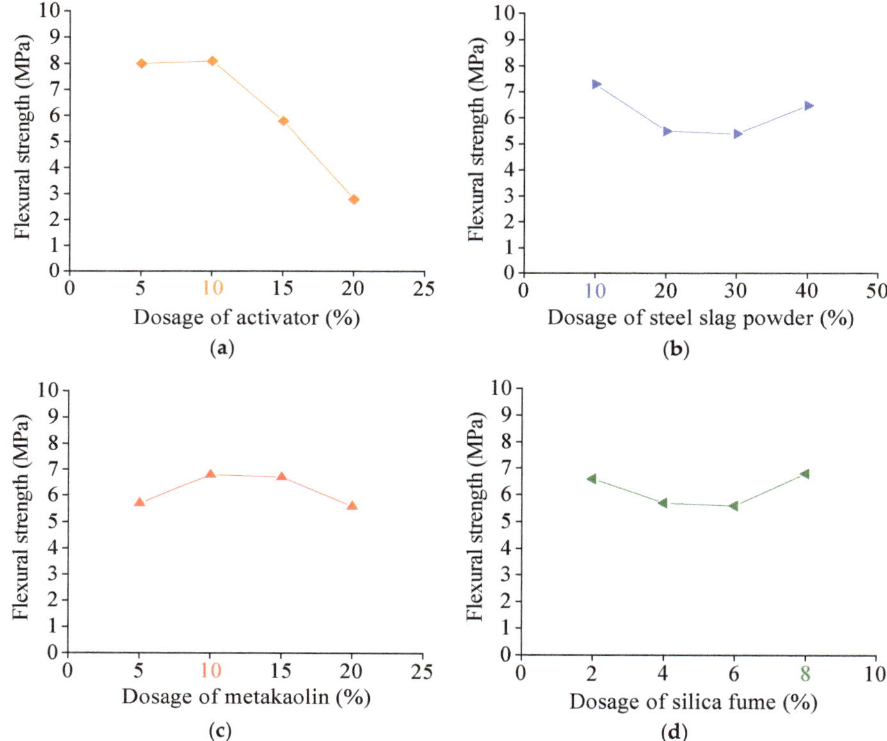

Figure 6. Relationship between flexural strength and dosage levels of each factor: (**a**) activator; (**b**) steel slag powder; (**c**) metakaolin and (**d**) Silica fume.

Through orthogonal analysis, it is determined that the activator is the primary factor affecting the strength of steel slag cement mortar, and the optimal dosages of each factor corresponding to the compressive strength and the flexure strength are obtained. It provides a basis for scientific research and engineering application. At the same time, it was found that excessive content of stone powder in the activator had a negative effect on the strength of steel slag cement mortar.

5. Prediction of Mortar Strength

The principle of orthogonal design method is to perform the overall evaluation of the system by uniformly sampling the level combinations of multiple factors. Therefore, the dosage level combinations of multiple factors adopted in the experiment may not necessarily take into account the optimal dosage level combination. The GM (0, N) model can be established to predict the system characteristic data of the optimal level combination obtained by the range analysis method. The data selection used to establish the model has a certain influence on the prediction accuracy of GM (0, N) model. Appropriately reducing the number of data used in the model establishment, according to the level value range of the primary factor of the system characteristic data, can improve the accuracy of the prediction model. The results of orthogonal analysis show that the activator is the main factor affecting the strength of mortar while the optimal dosages of the activator for compressive strength and flexural strength are lower 5% and 10%, respectively. Therefore, the GM (0, N) model was established by selecting the data of the first eight experimental groups for improving the prediction accuracy.

5.1. Prediction of Compressive Strength

Taking the compressive strength of the first eight groups as $X_1^{(0)}$, that is the system characteristic data sequence, and the dosage of activator, steel slag powder, metakaolin, and silica fume as the relative factors data sequence $X_2^{(0)}$, $X_3^{(0)}$, $X_4^{(0)}$, $X_5^{(0)}$, then

$$X_1^{(0)} = (43.4\,,37.2\,,45.6\,,44.6\,,44.2\,,40.6\,,29.8\,,28.4\,)$$
$$X_2^{(0)} = (5,5,5,5,10,10,10,10)$$
$$X_3^{(0)} = (10,20,30,40,10,20,30,40)$$
$$X_4^{(0)} = (5,10,15,20,10,5,20,15)$$
$$X_5^{(0)} = (2,4,6,8,6,8,2,4)$$

The data sequence is superimposed at once, and the parameter column satisfying the least-square estimation is obtained through Equation (2).

$$\hat{b} = [-0.541, -0.395, 1.361, 4.557, 41.867]^T \tag{3}$$

Thus, the GM (0, 5) model of compressive strength is obtained:

$$x_1^{(0)}(k) = -0.541x_2^{(1)}(k) - 0.395x_3^{(1)}(k) - 1.361x_4^{(1)}(k) - 4.557x_5^{(1)}(k) + 41.867 \tag{4}$$

As shown in Table 10, the average relative simulation error of GM (0, 5) model of compressive strength is 5.9%, while the accuracy is above 94%, which is a good prediction accuracy. When the optimal dosage combination of compressive strength was substituted into Equation (4), the prediction value of compressive strength was 55.6 MPa. The optimal dosage combination of the flexural strength was substituted into Equation (4) to obtain the prediction value of compressive strength of 51.5 MPa. The results show that the compressive strength of the two mixtures reaches the level of P·O·42.5 Portland cement.

Table 10. Simulation error check of GM (0, 5) model for compressive strength.

Number	Actual Value	Simulated Value	Residual	Relative Simulation Error
2	37.188	39.830	−2.642	0.071
3	45.583	38.612	6.971	0.153
4	44.625	50.581	−5.956	0.133
5	44.188	42.412	1.775	0.040
6	40.625	40.771	−0.146	0.004
7	29.750	29.894	−0.144	0.005
8	28.354	28.253	0.101	0.004
Mean				0.059

5.2. Prediction of Flexural Strength

Taking the flexural strength of the first eight groups as $X_1^{(0)}$, that is, the system characteristic data sequence, and the dosage of activator, steel slag powder, metakaolin, and silica fume as the relative factors data sequence $X_2^{(0)}$, $X_3^{(0)}$, $X_4^{(0)}$, $X_5^{(0)}$, then, the data sequence $X_i^{(1)}$ is superimposed at once to obtain the parameter column satisfying the least-squares estimation through Equation (2).

$$\hat{b} = [0.362, -0.019, 0.158, 0.591, 7.737]^T \tag{5}$$

Thus, the GM (0, 5) prediction model of flexural strength is obtained as

$$x_1^{(0)}(k) = 0.362x_2^{(1)}(k) - 0.019x_3^{(1)}(k) + 0.158x_4^{(1)}(k) + 0.591x_5^{(1)}(k) + 7.737 \tag{6}$$

As shown in Table 11, the average relative simulation error of GM (0, 5) model of flexural strength is 5.3%, while the corresponding accuracy is above 94%, which is a good prediction accuracy. When the optimal dosage combination of the flexural strength was substituted into Equation (6), the prediction value of the flexural strength is obtained as 9.7 MPa, reaching the level of P·O·42.5 Portland cement. By substituting the optimal dosage combination of compressive strength into Equation (6), the prediction value of flexural strength is 8.7 MPa, which is consistent with the strength of the reference group (9.2 MPa).

Table 11. Simulation error check of GM (0, 5) model for flexural strength.

Number	Actual Value	Simulated Value	Residual	Relative Simulation Error
2	7.766	8.203	−0.437	0.056
3	8.344	7.156	1.188	0.142
4	7.906	8.938	−1.032	0.130
5	8.844	8.556	0.288	0.033
6	8.734	8.758	−0.024	0.003
7	7.359	7.392	−0.033	0.004
8	7.586	7.594	−0.008	0.001
Mean				0.053

6. Economic Benefit Analysis

The cement industry is a material and energy-intensive industry, which not only consumes a lot of natural energy but also pollutes the environment. In practical engineering, cement as a cementitious material has a huge cost, while the cost of steel slag, metakaolin, and activator is relatively low. Activator, steel slag powder, metakaolin, and silica fume were used as cementitious materials in equal amounts instead of cement, and their economic benefits were evaluated. Through market research, the prices of activator, steel slag powder, metakaolin, silica fume, and ordinary silicate P·O·42.5 Portland cement are 170 RMB/ton, 100 RMB/ton, 400 RMB/ton, 1000 RMB/ton, and 450 RMB/ton, respectively. In practical engineering, the strength and cost requirements of cementitious materials are different according to different working conditions. Thus, it is imperative to provide the economic benefit analysis of various dosage combinations for appropriate binder selection. Based on the analysis of the above test results, the experimental group has a total of four steel slag powder dosage levels (10%, 20%, 30%, and 40%). The results of the economic benefit analysis based on the compressive strength and flexural strength are shown in Tables 12 and 13, respectively.

Table 12. Economic benefit analysis on the combination with highest compressive strength in each steel slag powder dosage level.

Dosage Combination of Binding Materials (%)					Cost (RMB/ton)		Reduction Rate (%)
Steel Slag Powder	Activator	Metakaolin	Silica Fume	Cement	Binging Materials in Study	Cement	
10	15	15	8	52	409.5		9.00
20	10	5	8	57	393.5		12.56
30	5	15	6	44	356.5	450	20.78
40	5	20	8	27	330		26.67
Max							26.67
Min							9.00

Table 13. Economic benefit analysis on the combination with highest flexural strength in each steel slag powder dosage level.

Dosage Combination of Binding Materials (%)					Cost (RMB/ton)		Reduction Rate (%)
Steel Slag Powder	Activator	Metakaolin	Silica Fume	Cement	Binging Materials in Study	Cement	
10	10	10	6	64	415		7.78
20	10	5	8	57	393.5		12.56
30	5	15	6	44	356.5	450	20.78
40	15	10	2	33	274		39.11
Max							39.11
Min							7.78

The results of the economic effect analysis indicate that the cost after cement replacement can be reduced by 9.00–26.67% compared to that before the replacement when the compressive strength is used as the benchmark for analysis. When the flexural strength is analyzed, it is seen that the cost after the cement replacement can be reduced by 7.78–39.11% compared to that before the replacement. For practical engineering, it is assumed that 10,000 m^3 of concrete with the required strength of 42.5 MPa, the amount of cementitious material per cubic meter of concrete is 0.4 ton. Using the method proposed in this study, the cost of cement per cubic meter of concrete can save 31–156 RMB. Thus, the total cost of cement replacement can be saved by at least 310,000 RMB and up to the maximum of 1560,000 RMB, which is a significant economic impact on the project.

7. Conclusions

A method of activating the activity of steel slag powder with neutral material is proposed. The validity of the proposed method is verified by experiments. Through orthogonal analysis, the optimal dosage combination of various components in the compound activator is determined. The grey prediction model is established to predict the strength of steel slag cement mortar under the optimal dosage combination of various factors. Considering the different requirements of cementitious materials in engineering, the economic benefits of several mix proportions are analyzed. The conclusions drawn from this study are appended below.

1/The experimental results show that with the change of steel slag powder dosage (10%, 20%, 30%, 40%), the compressive strength of mortar is affected. The highest strength of each dosage can reach more than 85% of the compressive strength of the control group. Similarly, the highest flexural strength can reach more than 90% of the flexural strength of the control group.

2/Through orthogonal analysis, it is ascertained that the activator is the primary factor influencing the strength of the steel slag cement mortar, and the optimal dosage combination of the compressive strength of the mortar is obtained as activator 5%, steel slag powder 10%, metakaolin 15%, and silica fume as 8%, while the optimal dosage combination of flexural strength is determined as activator 10%, steel slag powder 10%, metakaolin 10%, and silica fume 8%.

3/GM (0, 5) prediction models for compressive strength and flexural strength were established, respectively. The compressive strength and flexural strength of mortar were predicted. The prediction results of the compressive strength and flexural strength for the optimal dosage combination of compressive strength are 55.6 MPa and 8.7 MPa, respectively. The compressive strength and flexural strength at the optimal dosage combination for flexural strength are predicted to be 51.5 MPa and 9.7 MPa, respectively, which reach the strength level of P·O·42.5 Portland cement.

4/The research conducted on economic benefit analysis for multiple dosage combinations showed that the method proposed in this study can lower environmental pollution and reduce the project cost to a greater extent on the basis of meeting project requirements.

Author Contributions: J.G. and X.Y. designed the experiments. Y.Z., L.L., L.Z. and J.Y. carried out the experiments. X.Y. and Y.Z. analyzed the experimental results. J.G. and Y.Z. reviewed, and edited the manuscript. J.G. received the funding. All authors have read and agreed to the published version of the manuscript.

Funding: This research was funded by National Natural Science Foundation of China, grant number 51779095, Program for Science & Technology Innovation Talents in Universities of Henan Province, grant number 20HASTIT013, and Sichuan Univ, State Key Lab Hydraul & Mt River Engn, grant number SKHL2007. The APC was funded by Program for Science & Technology Innovation Talents in Universities of Henan Province, grant number 20HASTIT013.

Data Availability Statement: All the relevant data and models used in the study have been provided in the form of figures and tables in the published article.

Acknowledgments: This project was sponsored by National Natural Science Foundation of China (51779095), Program for Science & Technology Innovation Talents in Universities of Henan Province (20HASTIT013), Sichuan Univ, State Key Lab Hydraul & Mt River Engn (SKHL2007).

Conflicts of Interest: The authors declare no conflict of interest to this work.

References

1. Pang, B.; Zhou, Z.H.; Hou, P.K. Autogenous and engineered healing mechanisms of carbonated steel slag aggregate in concrete. *Constr. Build. Mater.* **2016**, *107*, 191–202. [CrossRef]
2. Furlani, E.; Tonello, G.; Maschio, S. Recycling of steel slag and glass cullet from energy saving lamps by fast firing production of ceramics. *Waste Manag.* **2010**, *30*, 1714–1719. [CrossRef]
3. Das, B.; Prakash, S.; Reddy, P. An overview of utilization of slag and sludge from steel industrie. *Resour. Conserv. Recycl.* **2007**, *50*, 40–57. [CrossRef]
4. Xiang, X.D.; Xi, J.C.; Li, C.H. Preparation and application of the cement-free steel slag cementitious material. *Constr. Build. Mater.* **2016**, *114*, 874–879. [CrossRef]
5. Guo, Y.; Xie, J.; Zheng, W. Effects of steel slag as fine aggregate on static and impact behaviours of concrete. *Constr. Build. Mater.* **2018**, *192*, 194–201. [CrossRef]
6. Saxena, S.; Tembhurkar, A.R. Impact of use of steel slag as coarse aggregate and wastewater on fresh and hardened properties of concrete. *Constr. Build. Mater.* **2018**, *165*, 126–137. [CrossRef]
7. Cardoso, C.; Camões, A.; Eires, R.; Mota, A.; Araújo, J.; Castro, F.; Carvalho, J. Using foundry slag of ferrous metals as fine aggregate for concrete. *Resour. Conserv. Recycl.* **2018**, *138*, 130–141. [CrossRef]
8. Martinho, F.C.; Picado-Santos, L.G.; Capitao, S.D. Influence of recycled concrete and steel slag aggregates on warm-mix asphalt properties. *Constr. Build. Mater.* **2018**, *185*, 684–696. [CrossRef]
9. Zhang, P.; Zheng, Y.X.; Wang, K.J. A review on properties of fresh and hardened geopolymer mortar. *Compos. Part B Eng.* **2018**, *152*, 79–95. [CrossRef]
10. Sheen, Y.N.; Wang, H.Y.; Sun, T.H. A study of engineering properties of cement mortar with stainless steel oxidizing slag and reducing slag resource materials. *Constr. Build. Mater.* **2013**, *40*, 239–245. [CrossRef]
11. Calmon, J.L.; Tristão, F.A.; Giacometti, M.; Meneguelli, M.; Moratti, M.; Teixeira, J.E. Effects of BOF steel slag and other cementitious materials on the rheological properties of self-compacting cement pastes. *Constr. Build. Mater.* **2013**, *40*, 1046–1053. [CrossRef]
12. Li, D.X.; Fu, X.H.; Wu, X.Q.; Tang, M.S. Durability study of steel slag cement. *Cem. Concr. Res.* **1997**, *27*, 983–987.

13. Li, Y.X.; Chen, Y.M.; Wei, J.X. A study on the relationship between porosity of the cement paste withmineral additives and compressive strength of mortar based on this paste. *Cem. Concr. Res.* **2006**, *36*, 1740–1743. [CrossRef]
14. Wu, X.; Hong, Z.; Hou, X. Study on steel slag and fly ash composite Portland cement. *Cem. Concr. Res.* **1999**, *29*, 1103–1106.
15. Hu, S.G. Research on Hydration of Steel Slag Cement Activated with Waterglass. *J. Wuhan Univ. Technol. Mater. Sci.* **2001**, *16*, 37–40.
16. Hu, S.G.; He, Y.J.; Lu, L.N. Effect of fine steel slag powder on the early hydration process of Portland cement. *J. Wuhan Univ. Technol. Mater. Sci.* **2006**, *21*, 147–149.
17. Zhu, X.; Hou, H.B.; Huang, X.Q. Enhance hydration properties of steel slag using grinding aids by mechano chemical effect. *Constr. Build. Mater.* **2012**, *29*, 476–481. [CrossRef]
18. Altun, İ.A.; Yılmaz, İ. Study on steel furnace slags with high MgO as additive in Portland cement. *Cem. Concr. Res.* **2002**, *32*, 1247–1249. [CrossRef]
19. Zhang, P.; Wang, K.X.; Wang, J. Mechanical properties and prediction of fracture parameters of geopolymer/alkali-activated mortar modified with PVA fiber and nano-SiO$_2$. *Ceram. Int.* **2020**, *46*, 20027–20037. [CrossRef]
20. Huo, B.B.; Li, B.L.; Chen, C. Surface etching and early age hydration mechanisms of steel slag powder with formic acid. *Constr. Build. Mater.* **2021**, *280*, 1–11. [CrossRef]
21. Zhang, P.; Wang, K.X.; Li, Q.F. Fabrication and engineering properties of concretes based on geopolymers/alkali-activated binders—A review. *J. Clean. Prod.* **2020**, *258*, 1–22. [CrossRef]
22. Peng, X.Q.; Liu, C.; Li, S. Research on the setting and hardening performance of Alkali-activated steel slag-slag based cementitious materials. *J. Hunan Univ. (Nat. Sci.)* **2015**, *42*, 47–52. (In Chinese)
23. Sun, J.W.; Zhang, Z.Q.; Zhuang, S.Y. Hydration properties and microstructure characteristics of alkali–activated steel slag. *Constr. Build. Mater.* **2020**, *241*, 1–9. [CrossRef]
24. Huo, B.B.; Li, B.L.; Huang, S.Y. Hydration and soundness properties of phosphoric acid modified steel slag powder. *Constr. Build. Mater.* **2020**, *254*, 1–9. [CrossRef]
25. Zhang, T.S.; Liu, F.T.; Liu, S.Q. Factors influencing the properties of a steel slag composite cement. *Adv. Cem. Res.* **2008**, *20*, 145–150. [CrossRef]
26. Du, J.; Liu, J.X. Compound effect of dehydrate gypsum and silica fume on strength of steel slag-cement binding materials. *J. Civ. Archit. Environ. Eng.* **2013**, *35*, 131–136. (In Chinese)
27. Lin, Z.S.; Tao, H.Z. Research for Increasing the Activation of Steel Slag and Fly Ash. *J. Wuhan Univ. Technol.* **2001**, *2*, 4–7. (In Chinese)
28. Kang, L.; Zhang, Y.J.; Zhang, L. Preparation, characterization and photocatalytic activity of novel CeO2 loaded porous alkali-activated steel slag-based binding material. *Int. J. Hydrog. Energy* **2017**, *42*, 17341–17349. [CrossRef]
29. Yin, S.H.; Guo, H.; Yu, Q.J. Steel Slag Reconstruction Experiment by Reduction FeOx and Its Mineral Composition. *J. Chin. Ceram. Soc.* **2013**, *41*, 966–971. (In Chinese)
30. Zhao, H.J.; Yu, Q.J.; Wei, J.X. Effect of Composition and Temperature on Structure and Early Hydration Activity of Modified Steel Slag. *J. Build. Mater.* **2012**, *15*, 399–405. (In Chinese)
31. Wei, R.; Li, H.; Zhang, W. Mechanism and Recent Development of Steel Slag Activating Activity. *Mater. Rep.* **2014**, *28*, 105–108+128. (In Chinese)
32. Yi, L.S.; Wen, J. Research Status and Progress on the Steel Slag Activity Excitation Technology. *Bull. Chin. Ceram. Soc.* **2013**, *32*, 2057–2062. (In Chinese)
33. Guan, J.F.; Yuan, P.; Hu, X.Z. Statistical analysis of concrete fracture using normal distribution pertinent to maximum aggregate size. *Theor. Appl. Fract. Mech.* **2019**, *101*, 236–253. [CrossRef]
34. Yao, X.H.; Li, L.L.; Guan, J.F. Initial cracking strength and initial fracture toughness from three-point-bending and wedge splitting concrete specimens. *Fatigue Fract. Eng. Mater. Struct.* **2021**, *44*, 601–621. [CrossRef]
35. Yao, X.H.; Qin, P.Q.; Guan, J.F. Residual mechanical properties and constitutive model of high-strength seismic steel bars through different cooling rates. *Materials* **2021**, *14*, 469. [CrossRef] [PubMed]
36. *GB/T 17671-1999, Method of Testing Cements-Determination of Strength*; China Standards Press: Beijing, China, 1999. (In Chinese)
37. Guan, J.F.; Hu, X.Z.; Yao, X.H. Fracture of 0.1 and 2 m long mortar beams under three-point-bending. *Mater. Des.* **2017**, *133*, 363–375. [CrossRef]
38. Miah, M.J.; Patoary, M.; Paul, S.C. Enhancement of Mechanical Properties and Porosity of Concrete Using Steel Slag Coarse Aggregate. *Materials* **2020**, *13*, 2865. [CrossRef]
39. Chen, S.F. The Study of Sexual Assault under Age Factors Affection via GM (*0, N*) and Grey Structure Modeling. *J. Grey Syst.* **2015**, *18*, 173–179.
40. Ren, J. GM (*1, N*) method for the prediction of anaerobic digestion system and sensitivity analysis of influential factors. *Bioresour. Technol.* **2018**, *247*, 1258–1261. [CrossRef]

41. Yu, Y.F.; Zhu, S.Y.; Wang, H. A Forecasting Method of Road Freight Volume Based on GM (0, *N*). *J. Wuhan Univ. Technol. (Traffic Sci. Eng.)* **2010**, *34*, 93–96. (In Chinese)
42. Liu, S.F. *Grey System Theory and Application*; Science Press: Beijing, China, 2014. (In Chinese)
43. Zhang, G.X. *Chinese Quality Engineer Handbook*; Enterprise Management Publishing House: Beijing, China, 2002. (In Chinese)

Article

Experimental Evaluation of Untreated and Pretreated Crumb Rubber Used in Concrete

Hamad Hassan Awan [1], Muhammad Faisal Javed [2,*], Adnan Yousaf [2], Fahid Aslam [3], Hisham Alabduljabbar [3] and Amir Mosavi [4,5,6,*]

1   National Institute of Transportation (SCEE), National University of Science and Technology, Islamabad 44000, Pakistan; hhawan.tn18@nit.nust.edu.pk
2   Department of Civil Engineering, COMSATS University Islamabad, Abbottabad Campus, Abbottabad 22060, Pakistan; engrayousaf@cuiatd.edu.pk
3   Department of Civil Engineering, College of Engineering in Al-Kharj, Prince Sattam Bin Abdulaziz University, Al-Kharj 11942, Saudi Arabia; f.aslam@psau.edu.sa (F.A.); h.alabduljabbar@psau.edu.sa (H.A.)
4   Faculty of Civil Engineering, Technische Universitat Dresden, 01069 Dresden, Germany
5   John von Neumann Faculty of Informatics, Obuda University, 1034 Budapest, Hungary
6   Department of Informatics, J. Selye University, 94501 Komarno, Slovakia
*   Correspondence: arbab_faisal@yahoo.com (M.F.J.); amir.mosavi@mailbox.tu-dresden.de (A.M.)

Citation: Awan, H.H.; Javed, M.F.; Yousaf, A.; Aslam, F.; Alabduljabbar, H.; Mosavi, A. Experimental Evaluation of Untreated and Pretreated Crumb Rubber Used in Concrete. *Crystals* 2021, *11*, 558. https://doi.org/10.3390/cryst11050558

Academic Editors: Yifeng Ling, Chuanqing Fu, Peng Zhang and Peter Taylor

Received: 8 April 2021
Accepted: 6 May 2021
Published: 17 May 2021

Publisher's Note: MDPI stays neutral with regard to jurisdictional claims in published maps and institutional affiliations.

Abstract: The present research aims at evaluating the mechanical performance of untreated and treated crumb rubber concrete (CRC). The study was also conducted to reduce the loss in mechanical properties of CRC. In this study, sand was replaced with crumb rubber (CR) with 0%, 5%, 10%, 15%, and 20% by volume. CR was treated with NaOH, lime, and common detergent for 24 h. Furthermore, water treatment was also carried out. All these treatments were done to enhance the mechanical properties of concrete that are affected by adding CR. The properties that were evaluated are compressive strength, indirect tensile strength, unit weight, ultrasonic pulse velocity, and water absorption. Compressive strength was assessed after 7 and 28 days of curing. The mechanical properties were decreased by increasing the percentage of the CR. The properties were improved after the treatment of CR. Lime treatment was found to be the best treatment of all four treatments followed by NaOH treatment and water treatment. Detergent treatment was found to be the worse treatment of all four methods of treatment. Despite increasing the strength it contributed to strength loss.

Keywords: crumb rubber concrete; crumb rubber; NaOH treatment; lime treatment; water treatment; detergent treatment; concrete; compressive strength; materials; mechanical properties

1. Introduction

With the rapid growth in industrialization, solid waste is also increasing at an alarming rate. It has become essential for the construction industry to find and apply new technologies to reduce waste produced by the industries and incorporate it in conventional concrete [1–3]. Among many other solid wastes, crumb rubber (CR) is perhaps one of the most challenging solid waste materials to cope with. CR is made by shredding tires having a size between 0.075 mm and 4.75 mm [4]. It is estimated that nearly 1 billion tires are generated every year, ending their serviceable life and out of this, about 50%, without any treatment goes to garbage or landfills. By 2030, it is estimated, there would be about 5 billion tires that will be disposed of [5]. About 300 million tons are generated in the USA, 10 million tons in Turkey and Iran, and in the European Community, it is about 3.4 million tons [6]. In order to avoid the negative and harmful ecological and environmental effects caused by waste tire disposal, a significant body has promoted its use in concrete. The major part of wasted tires is landfilled, globally. This rapid accumulation of tire waste has catastrophic ecological and environmental consequences, causing serious threats to human health (e.g., soil contamination, fire, and pests) [7,8]. There is a great potential in the construction industry to accumulate a larger part of the rubber by utilizing CR as a partial replacement of natural aggregates in concrete which results in a type of concrete named, crumb rubber concrete (CRC) [9,10]. Introducing CR into concrete increases energy dissipation, impact resistance [11], drying shrinkage [12], water absorption [13], ductility [14], damping ratio, durability, and toughness [1,10,15]. However, it has been found by many researchers that by increasing the percentage of CR the compressive strength of the concrete decreases [16–20].

The utilization of admixtures in concrete has led to some changes in concrete mix design, but in return, the resulted concrete is more durable and stronger [21,22]. Adhesion of rubber particles is the main cause of strength loss [23,24]. In order to recover strength loss, many researchers have tried diverse methods and techniques. Some researchers have concentrated on the treatment of CR in order to improve adhesion, some have used additives to recover the strength loss, and some have used a combination of treatment of CR with additives to recover strength loss by adding CR [25]. Metakaolin, fly ash and silica fume are the most common additives used in CRC [6,10,22,26–30] and treatments; NaOH treatment is the most common treatment [10,31,32]. Some researchers have increased the cement content [10] and some have used different water to cement ratios [16,33] in order to mitigate the strength losses. Eshmaiel Ganjian et al. [34], investigated the performance of concrete by incorporating discarded waste tire replacing fine (200–850 mm) aggregate with 5%, 7.5%, and 10% by weight. They found a 10–23% drop in the compressive strength of the concrete while the drop in tensile strength was found to be 30–60%. Eldin and Senouci [35] found a reduction of 85% in compressive strength when coarse aggregate was replaced while with the replacement of fine aggregate, there was a 65% drop in the compressive strength. Güneyisi et al. [36] investigated the performance of CRC with the addition of silica fume. They used CR for fine aggregates and tire chips for coarse aggregates. They found a 77% reduction in the unit weight of the normal weight concrete with 50% replacement of rubber as the total aggregate volume. They also found that silica fume was beneficial and helped to recover strength loss. They also proposed that rubber should not be used above 25% of the total volume of aggregate. Rezaifar et al. [6] found that loss in compressive strength of CRC with 10, 20, and 30% CR replacement was 17, 34, and 51% respectively. Rezaifar et al. [6], used metakaoline in conjunction with CR which lowered the strength loss to about 22%. They also found a decrease in the unit weight of the CRC. Mohammadi et al. [31], recovered a 25% of strength loss by treating CR with NaOH for periods of 20 min, 24 h, and 7 days. They found 24 h to be the optimum period. Onuaguluchi and Panesar [17] investigated the performance of CRC with pre-coated CR in conjunction with silica fume. They found an increase in compressive strength of 29% with 5% CR and silica fume and an increase in compressive strength of 14% with 10% CR and silica fume. Youssf et al. [10], investigated the performance of concrete by treatment using silica fume, NaOH (sodium hydroxide) solution, and cement content. They found 0.5 hours of treatment of rubber, 350 kg/m^3 cement content, and 0% silica fume replacing cement by weight as the best alternatives.

In this study, the focus is on the surface treatment of the CR to mitigate the strength loss of concrete due to the addition of CR rather than focusing on additives, admixtures, or increasing the cement content in the CRC. This study aims at finding new, better, and cheaper methods of surface treatments of CR to recover the strength loss of concrete by adding CR. This research study will contribute in better understanding the relationship between surface treatments of CR and mechanical properties of the CRC. This will pave the way to explore and advance the treatments' techniques in order to achieve the best results relating to losses in mechanical properties of CRC. The outcomes of this research study can be helpful in the practical application of using CR in conventional concrete.

2. Experimental Program

A total of 189 cylinders of 150 mm × 300 mm for 21 mixes (as shown in Table 1) were made for the research study. For each mix, nine cylinders were made. Three for assessing compressive strength at 7 days, three for assessing compressive strength at 28 days, and three for assessing the indirect tensile the strength at 28 days. Out of the 21 mixes, 1 mix was used for controlled mix, 4 mixes were for untreated CRC, and 16 mixes were for treated CRC as shown in Table 1. About 15 additional cylinders of 150 mm × 300 mm were made to evaluate the compressive strength of untreated CRC at 28 days after placing the concrete specimen in the oven at 200 °C for a time period of 6 hours. The specimen were then allowed to cool down at room temperature.

Table 1. Mix design.

Mix Type	Treatment	CR %	w/c Ratio	Net Water (kg/m^3)	Cement (kg/m^3)	Fine Aggregate		Coarse Aggregate
						Sand (kg/m^3)	Crumb Rubber (kg/m^3)	Gravel (kg/m^3)
C0 (CM)	-	0	0.50	203.97	400	661.82	0	1074.18
CU5	-	5	0.50	203.96	400	628.73	20.35	1074.18
CU10	-	10	0.50	203.95	400	595.64	40.70	1074.18
CU15	-	15	0.50	203.94	400	562.55	61.05	1074.18
CU20	-	20	0.50	203.93	400	529.46	81.40	1074.18
CN5	NaOH	5	0.50	203.96	400	628.73	20.35	1074.18
CN10	NaOH	10	0.50	203.95	400	595.64	40.70	1074.18
CN15	NaOH	15	0.50	203.94	400	562.55	61.05	1074.18
CN20	NaOH	20	0.50	203.93	400	529.46	81.40	1074.18
CL5	Lime	5	0.50	203.96	400	628.73	20.35	1074.18
CL10	Lime	10	0.50	203.95	400	595.64	40.70	1074.18
CL15	Lime	15	0.50	203.94	400	562.55	61.05	1074.18
CL20	Lime	20	0.50	203.93	400	529.46	81.40	1074.18
CW5	Water	5	0.50	203.96	400	628.73	20.35	1074.18
CW10	Water	10	0.50	203.95	400	595.64	40.70	1074.18
CW15	Water	15	0.50	203.94	400	562.55	61.05	1074.18
CW20	Water	20	0.50	203.93	400	529.46	81.40	1074.18
CD5	Detergent	5	0.50	203.96	400	628.73	20.35	1074.18
CD10	Detergent	10	0.50	203.95	400	595.64	40.70	1074.18
CD15	Detergent	15	0.50	203.94	400	562.55	61.05	1074.18
CD20	Detergent	20	0.50	203.93	400	529.46	81.40	1074.18

2.1. Concrete Materials and Properties

Ordinary Portland Cement (OPC) in compliance with ASTM C150 Type I, from Bestway cement factory was used. The sand was used as fine aggregate, crushed stone was used as coarse aggregate, and CR was used as a replacement of sand by volume ranging from 0–20%. Water used for the entire research project was ordinary tap water available. The fineness modulus of fine aggregate was found to be 2.71. The specific gravity, water absorption, and moisture content of the sand were 2.6, 1.71%, and 0.809% respectively. The coarse aggregates with a maximum size of 22.5 mm were used. The specific gravity, water absorption, and moisture content of coarse aggregate were 2.63, 0.431%, and 1.696% respectively. The specific gravity, water absorption, and moisture content of CR were found to be 1.599, 0.035%, and 0.085% respectively. The sieve analysis of CR, fine and coarse aggregates are shown in Table 2.

NaOH and lime were obtained from the local markets. Lime was in powdered form while NaOH was available in bottles of 1 kg in solid granular form. The detergent used in this research study was locally available detergent used for washing clothes. CR was collected from a CR supplier. CR was in ground form with particle size ranging from 4.75–0.075 mm in size.

Table 2. Sieve analysis of aggregates.

Sieve Size (mm)	0.075	0.15	0.3	0.6	1.18	2.36	4.75	9.5	12.5	19
Sand Passed (%)	0	3.8	11.9	36	77	94.8	99.4	100	100	100
Crumb Rubber (%)	-	0	3.6	23	43.6	68	99.7	100	100	100
Gravel Passed (%)	-	-	-	-	-	-	-	28.62	67.12	92.02

2.2. Mix Proportions

Mix design of controlled concrete was prepared according to British Standard (BS) i.e., in per cubic meter of concrete as presented in Table 1. Controlled concrete was designed for compressive strength of 21.7 MPa. Controlled concrete was the concrete having 0% CR. Water to cement ratio (0.5), cement content (400 kg/m^3), and coarse aggregate (1074.18 kg/m^3) were not changed throughout the study. In this study CM stands for controlled mix concrete, CN for NaOH treated, CL for lime treated, CW stands for water treated, and CD stands for detergent-treated CRC. Whereas 5, 10, 15, and 20 represent percentages of sand replaced with CR by volume.

2.3. Treatment of Crumb Rubber

Researchers have tried treatments of CR to improve the adhesion properties in order to improve the strength of concrete [37–40]. In this research project, four different types of treatments were used to treat the CR's surface namely lime treatment, NaOH treatment, detergent treatment, and water treatment in order to make the surface rougher and improve the interface adhesion of rubber/cement.

In the present study, 10% concentrated solutions of NaOH, detergent, and lime were made to treat the CR. Untreated CR was washed and then submerged in the solutions for 24 h. The time of treatment was taken on the basis of contact of CR with the solutions which was 24 h. After the time has elapsed, CR was extracted from the solutions and washed again to decrease the pH values as it may cause adverse effects on the concrete [10]. Water treatment was carried out by boiling the water and then submerging CR into it for a time period of 10 minutes. Then the water was allowed to cool and the CR was removed from the water. This treatment was done to remove zinc stearate layers on CR [41].

2.4. Specimen Preparation

The concrete batches were mixed in the laboratory with the help of a mixer in accordance with ASTM C192/C192M [42]. After uniform mixing, each specimen was cast in 150 × 300 mm cylinders and compacted with the help of a rod vibrator. After casting, the specimen were left at room temperature 24 ± 3 °C for a time period of 24 h. The specimen were then withdrawn from the molds and kept for curing in the tank until the time of testing at a temperature of 24 ± 3 °C in accordance with ASTM C192/C192M [42].

2.5. Testing Methods

Slump test was conducted according to ASTM C143 [43] and compaction factor test was conducted following IS: 1199–1959 [44]. The compressive strength of each mix was determined according to ASTM C 109M [45] and C 39 [46]. Testing was carried out at the curing age of 7 and 28 days. An indirect tensile strength test was conducted following AS 1012.10 [47] at a constant loading rate of 1.5 ± 0.15 MPa/min at 28 days of curing. Ultrasonic pulse velocity test was also performed according to ASTM C597 [48] at 28 days of curing. The water absorption test was performed in accordance with ASTM C642 [49] specifications. The weight of the concrete cylinders was obtained and divided by the volume of molds. The unit weight of concrete for all cylinders was assessed at 7 and 28 days.

3. Results and Discussions

In this section effect of untreated CR, NaOH treated CR, lime-treated CR, detergent-treated CR, and water-treated CR on water absorption, slump, compressive strength, and indirect tensile strength of concrete are discussed. Experimental results of all concrete mixes are shown in Table 3.

Table 3. Test results.

Mix Code	Treatment	CR %	Slump (mm)	Compressive Strength (MPa)		Indirect Tensile Strength (MPa)	Water Absorption (%)		Unit Weight (kg/m^3)		UPV (km/s)
				7 days	28 days		7 days	28 days	7 days	28 days	
CM	-	0	50	11.88	23.00	7.30	2.3	6.6	2331	2464	4.45
CU5	Untreated	5	50	11.00	20.63	7.04	1.8	6.2	2344	2364	4.41
CU10	Untreated	10	75	9.56	15.37	5.33	2.8	6.8	2121	2128	4.41
CU15	Untreated	15	150	6.62	11.45	3.60	3.6	7.4	1999	2021	4.40
CU20	Untreated	20	180	4.21	8.71	1.15	4.7	8.7	1882	1892	4.38
CN5	NaOH	5	65	11.82	21.40	7.19	4.7	3.7	2370	2382	4.41
CN10	NaOH	10	90	10.10	16.52	5.68	5.8	4.1	2117	2132	4.40
CN15	NaOH	15	150	7.25	12.25	3.95	8.0	4.9	1989	2014	4.39
CN20	NaOH	20	165	5.63	9.31	1.76	8.6	5.3	1869	1881	4.37
CL5	Lime	5	50	12.01	21.56	7.33	1.1	3.1	2341	2375	4.42
CL10	Lime	10	75	10.37	16.67	5.91	2.3	3.9	2101	2110	4.41
CL15	Lime	15	165	7.43	12.48	4.05	3.3	3.9	1903	2021	4.40
CL20	Lime	20	180	5.87	9.78	1.82	4.5	4.1	1875	1882	4.39
CW5	Water	5	50	11.23	21.05	7.08	4.9	4.6	2353	2363	4.40
CW10	Water	10	75	9.91	16.43	5.41	6.2	5.5	2111	2124	4.40
CW15	Water	15	150	7.10	12.01	3.66	8.2	6.7	1993	2015	4.39
CW20	Water	20	180	4.52	9.25	1.17	10.2	7.3	1874	1890	4.37
CD5	Detergent	5	50	10.95	20.57	7.00	4.6	4.0	2352	2360	4.39
CD10	Detergent	10	90	9.40	15.32	5.29	5.6	4.3	2109	2121	4.38
CD15	Detergent	15	165	6.40	11.43	3.53	6.5	5.3	1992	2015	4.38
CD20	Detergent	20	180	4.15	8.71	1.14	7.9	6.5	1881	1892	4.36

3.1. Slump and Compaction Factor

The slump of freshly mixed concrete for replacement levels of 0%, 5%, 10%, 15%, and 20% determined (as shown in Figure 1) with the maximum slump of 180 mm was recorded for CU 20, CL20, CW20, and CD20. The minimum slump of 50 mm was recorded for CM, CU5, CL5, CW5, and CD5. On average an increase of 52% slump was recorded for every increment of 5% in CR replacement. A total of 250% increase in a slump was recorded with 20% of replacement of sand with CR from that of controlled concrete. Albano et al. [50] used CR (0.59 and 0.29 mm) as fine aggregate and found a decrease in a slump. Bignozzi and Sandrolini [51] replaced the sand with CR of two sizes 0.5 to 2 mm and 0.05 to 0.7 mm and found no significant change in the behavior of fresh concrete. However, Onuaguluchi and Panesar [17] replaced the sand with CR and found a significant increase in the slump.

A 14% increase in compaction factor was recorded with 20% replacement of sand by CR. On average there was a 3.3% increase in compaction factor for every increment of 5% replacement with CR.

The increase in a slump and compaction factor in this study was due to the addition of poorly graded CR in the mixes with a high fineness modulus of 3.62 as compared to sand which had the fineness modulus of 2.77. With the increase in fineness modulus of concrete aggregates, the workability of CRC also increased.

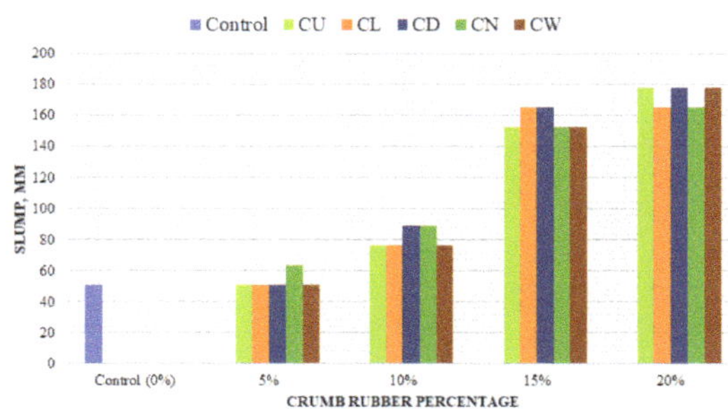

Figure 1. Slump test results.

The compaction factor of freshly prepared concrete at replacement levels of 0%, 5%, 10%, 15%, and 20% was determined as shown in Figure 2. Compaction factor increases with the increase in percentage levels of CR.

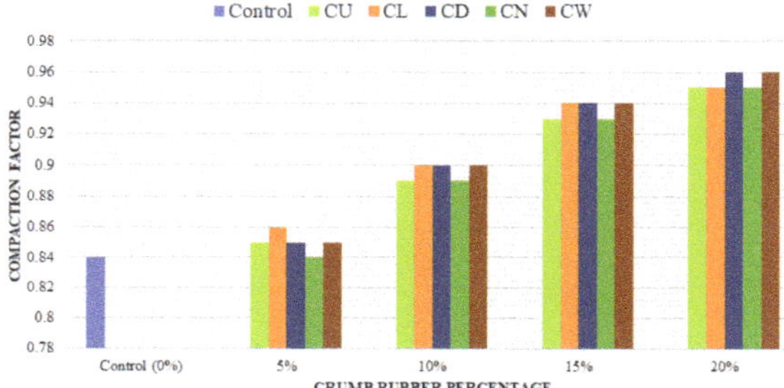

Figure 2. Compaction factor results.

3.2. Water Absorption

Water absorption tells us about the porosity and pore structure of the concrete. Sadek and El-Attar [52] found that water absorption is affected by CR when replaced with fine or coarse aggregates. However, they found that in the case of coarser rubber the increase in water absorption is greater as compared to the finer rubber aggregates. Water absorption was increased by increasing the percentage of the CR and was decreased by increasing the curing ages (as shown in Figures 3 and 4). The lowest absorption percentage was recorded at 1.15% for CL5 at 7 days and for 28 days it was 3.1% for CL5. The highest absorption percentage was recorded 10.23% for CW20 at 7 days and 8.69% for CU20 at 28 days. This increase in the water absorption was due to a decrease in unit weight and increase in porosity of CRC due to an increase in the percentage of CR.

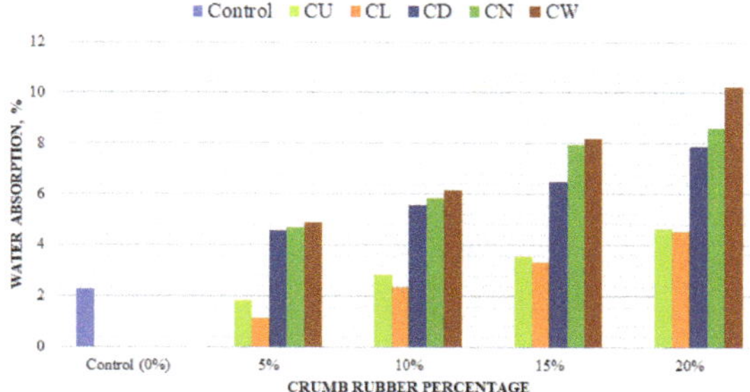

Figure 3. Water absorption at 7 days.

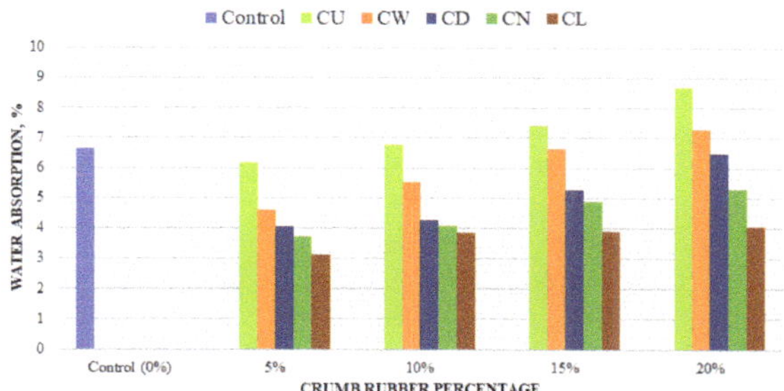

Figure 4. Water absorption at 28 days.

3.3. Density

As percentage levels of replacement of CR increased, the density was decreased. Corinaldesi et al. [53], has also found a decrease in density with the introduction of rubber particles. However, as the curing age increased, the density increased. The control mix showed an increase in the density from 7 to 28 days i.e., 2331 kg/m^3 to 2464 kg/m^3. The lowest amount of density recorded for CRC at 7 days was 1869 kg/m^3 and for 28 days it was 1881 kg/m^3 as shown in Figures 5 and 6 respectively. The increase in density as the curing period increased was due to the presence of water which helped internal curing. The water was available for the hydration of cementitious materials in concrete. The decrease in density as the replacement level increases was due to the low specific gravity of CR.

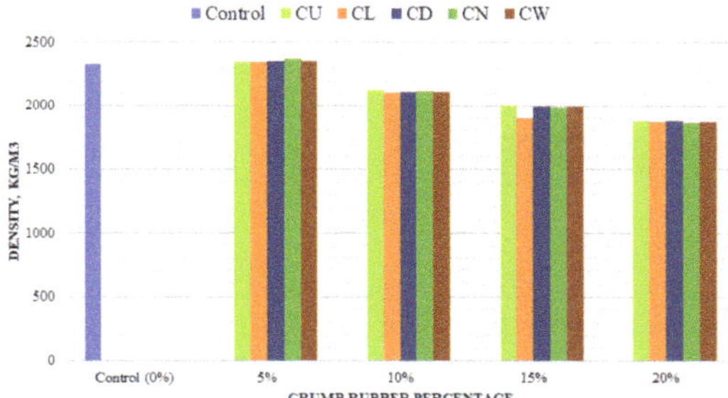

Figure 5. Density of concrete at 7 days.

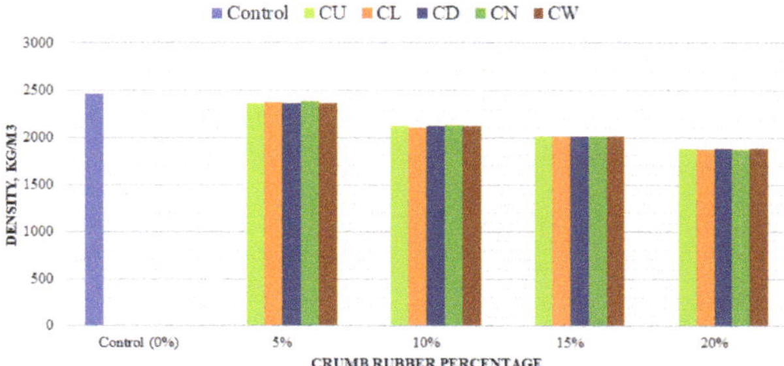

Figure 6. Density of concrete at 28 days.

3.4. Ultrasonic Pulse Velocity (UPV)

As the replacement level was increased there was a decrease in UPV values. Turgut and Yesilata [54] used CRs with sizes ranging from 4.75 mm (No. 4 Sieve) to 0.075 mm (No. 200 Sieve). They also found a decrease in UPV values with the increase in the percentage of CR. Salhi et. al [55] found a correlation between compressive strength and UPV to be good. The highest value of UPV was recorded for CL5 and it was 4.42 km/s which is a 0.67% decrease from that of controlled concrete. The lowest value of UPV was recorded for CD20 and it was 4.36 km/s which is a 2.02% decrease from that of controlled concrete. The UPV and density of the concrete share a direct relation. In this study, with the increase of CR, the density of the concrete was decreased as shown in Figure 7. It means the more the CR in the concrete; the more would be the cracks, pores, capillaries attributing to the enhancement of interfacial transition zone (ITZ) [56]. Due to the presence of pores, crack, and capillaries the values of UPV were decreased with the increase in the percentage of CR because it needs compact mass for the velocity of compression waves to travel.

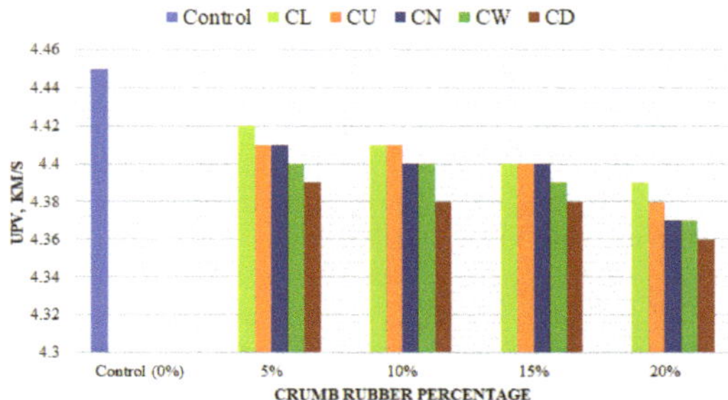

Figure 7. UPV results.

3.5. Compressive Strength

It is evident from many research studies that by increasing the percentage of CR the compressive strength of the concrete decreases [16–20,57,58]. There was a loss of 7.41% compressive strength at 7 days with 5% replacement, a loss of 19.53% with 10% replacement, 44.28% with 15% replacement, and a loss of 64.56% with 20% replacement of sand with CR in untreated CRC (Figure 8). At 7 days of curing, lime treatment managed to recover 9.96% of the strength loss, NaOH treatment recovered 7.54% of strength loss, and water treatment recovered 5.09% of strength loss at 7 days of curing. Detergent treatment did not recover strength loss however it decreased the strength further to 1.72% at 7 days of curing. At 28 days of curing 10.30% loss of compressive strength at 5% replacement, 33.17% at 10% replacement, 50.22% strength loss at 15% replacement, and 62.13% loss at 20% replacement of sand with CR were seen (Figure 9). At 28 days of curing, lime treatment managed to recover 8.56% of the strength loss, NaOH treatment recovered 6.27% of strength loss, and water treatment recovered 5.01% of strength loss at 28 days of curing. Detergent treatment did not recover strength loss, however it decreased the strength further to 0.20% at 28 days of curing. Figure 10 shows the comparison of strength loss recovered at 7 and 28 days respectively. It shows that the strength loss recovered or deteriorated for all treatments was greater at 7 days than 28 days except for water treatment.

3.6. Compressive Strength after Heating

Liang et al. [59] found a significant decrease in compressive strength of concrete after a rise in temperature. A greater drop in compressive strength of concrete samples was recorded at 28 days after placing them in the oven at 200 °C (Figure 11) as compared to the compressive strength at normal temperature (24 ± 3). Replacement of sand with CR showed very poor results when CRC was heated in the oven at a temperature of 200 °C. At 5% replacement level there was a loss of 61.38% in compressive strength, at 10% replacement, it increased to 87.13%, at 15% replacement, it further increased to 90.73%, and at 20% replacement level it reached 95.37%. This huge strength loss was due to the low softening point of CR which lies between 180 and 250 °C.

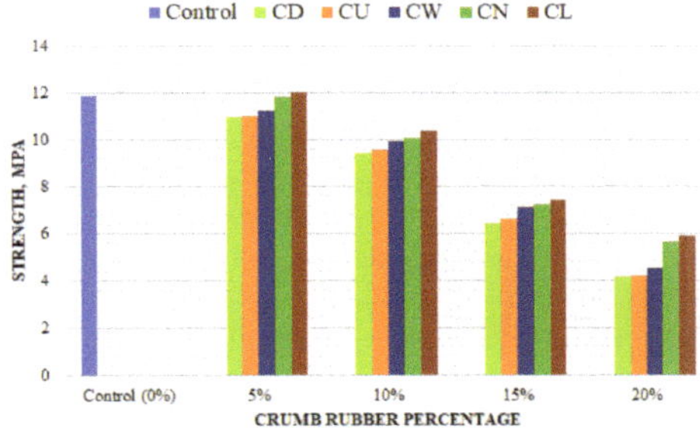

Figure 8. Compressive strength at 7 days.

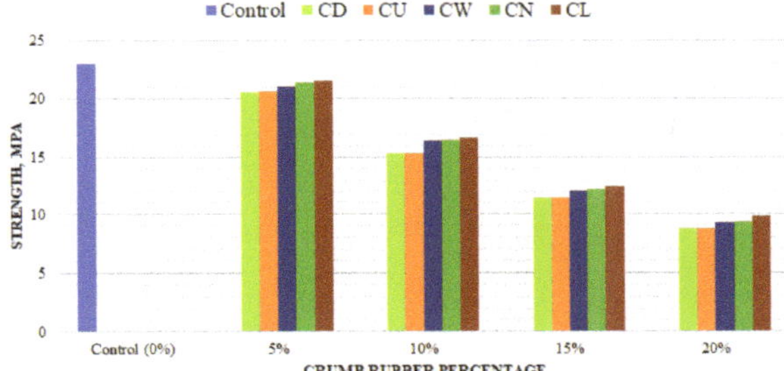

Figure 9. Compressive strength at 28 days.

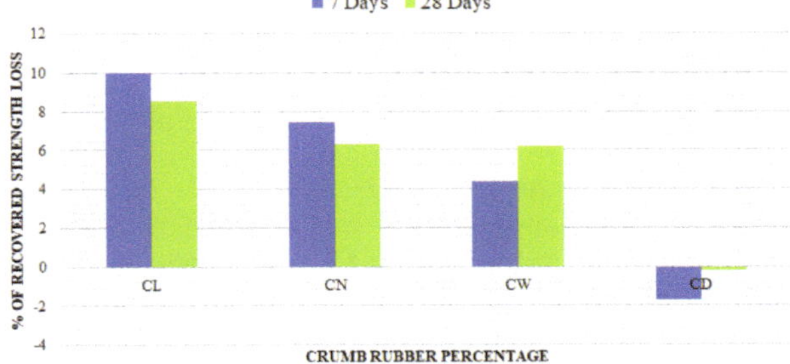

Figure 10. Percentage of recovered strength loss at 7 and 28 days.

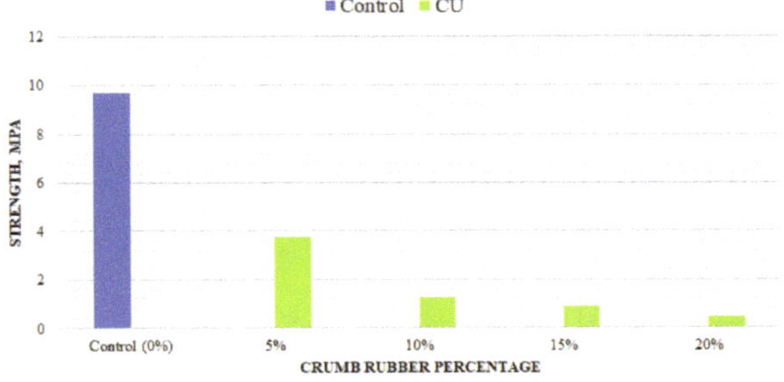

Figure 11. Compressive strength (after heating) at 28 days.

3.7. Indirect Tensile Strength

Indirect tensile strength was found to be following the same pattern of compressive strength. At 28 days of curing there was a loss of 3.56% in indirect tensile strength at 5% replacement, 24.25% loss at 10% replacement, 50.68% loss at 15% replacement, and 84.25% loss of indirect tensile strength at 20% replacement level of sand with CR (Figure 12). Batayneh et al. [8], also found that with the increase in CR, there is a loss in tensile strength of concrete. Lime treatment managed to recover 9.16% of strength loss, NaOH treatment recovered 6.14% of strength loss, and water treatment recovered 1.37% of strength loss. Detergent as in all cases reduced the tensile strength to a further 1.03%. The reduction in indirect tensile strength might be due to the weak bonding between CR and cement. The ITZ acted as a micro-crack between the two materials. This weak ITZ accelerated the reduction in tensile strength [60].

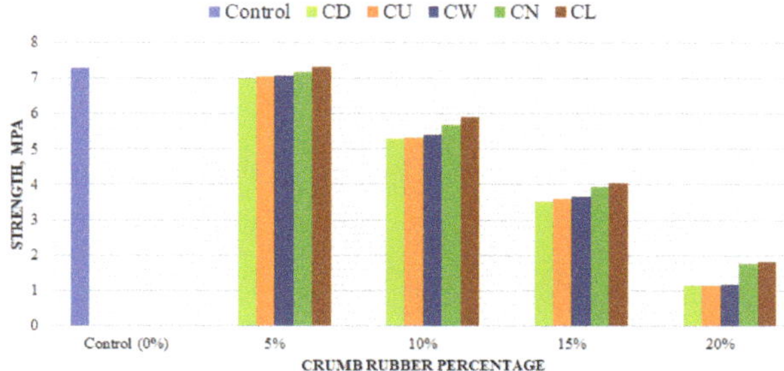

Figure 12. Indirect tensile strength at 28 days.

3.8. Scanning Electron Microscopy (SEM)

In order to check the morphology of CR, SEM was conducted on treated and untreated samples. SEM can render information on surface structure, chemical composition, crystalline structure, and electrical behavior of the top [61]. As the focus of this research was on surface treatments, SEM helped to look at the physical effects of surface treatments besides experimental results.

From Figures 13–17, it is visible that the surface of lime-treated CR is rougher than the remaining three giving the best results in the case of a compression test. After lime the surface of NaOH-treated CR is relatively rougher than water-treated and detergent-treated samples giving the second-best results. The surfaces of water-treated and detergent-treated CR were relatively slightly rougher than the untreated CR.

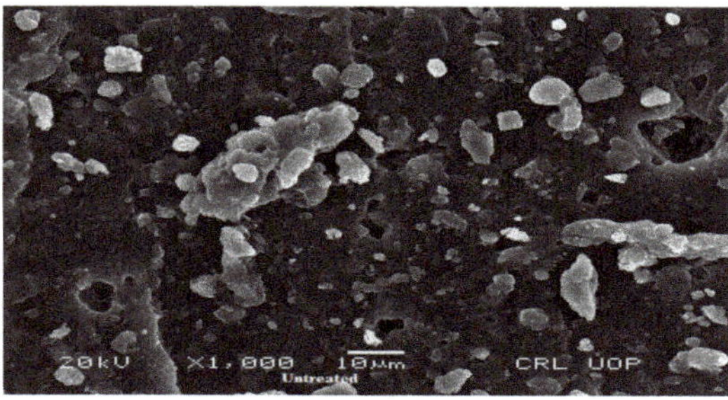

Figure 13. Untreated CR.

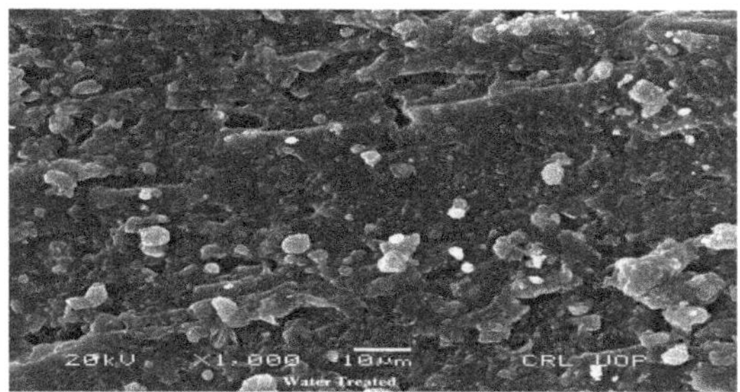

Figure 14. Water-treated CR.

Figure 15. Lime-treated CR.

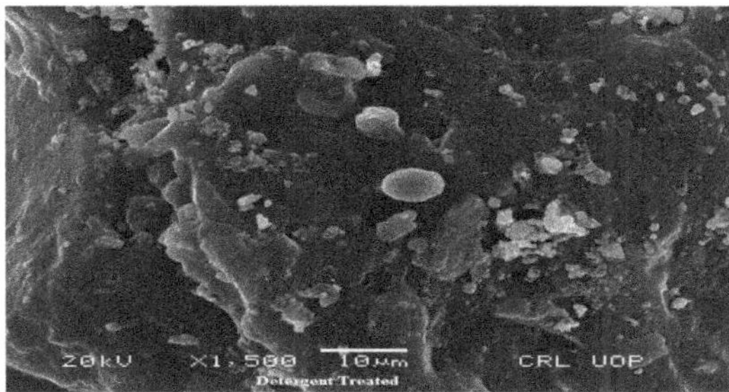

Figure 16. Detergent-treated CR.

Figure 17. NaOH-treated CR.

4. Conclusions

This study was conducted to evaluate the performance of untreated and treated CRC. CR was treated with lime, NaOH, detergent, and water. The fresh and hard properties of concrete were evaluated. Based on the experimental results of the research study, the following conclusions are drawn:

- A 250% increase in slump and 14% increase in compaction factor were recorded with 20% replacement of sand with CR.
- Water absorption increased with the addition of CR and a maximum of 10.23% water absorption was recorded at 7 days for 20% replacement of sand and it decreased as the curing period increased and recorded 8.69% as the maximum value at 28 days.
- The density of concrete dropped to 1869 kg/m^3 and 1881 kg/m^3 for 7 and 28 days respectively for 20% replacement. Based on its lightweight properties CR concrete can be used in stone backing, interior construction, false facades, and nailing concrete.
- Lime treatment was found to be the best treatment of all four treatments followed by NaOH treatment and water treatment. Lime treatment recovered a compressive strength of 10.30% at 28 days and 9.16% of tensile strength at 28 days.
- Detergent treatment was found to be the worse treatment of all four treatment methods. Despite of increasing the strength it contributed to compressive strength loss of 1.70% at 7 days and 0.20% at 28 days and a loss of 1.03% for indirect tensile strength at 28 days.
- CRC is not suitable for heat applications as it dropped 95.37% and 61% of its compressive strength with 20% and 5% replacement of sand, respectively.

Author Contributions: H.H.A., conceptualization, data analysis, writing original draft preparation; M.F.J., formal analysis and modeling, conceptualization, data analysis, writing original draft preparation; A.Y., supervision, review and editing; F.A., investigation and review; H.A., methodology and review and editing; A.M., review and editing, supervision. All authors have read and agreed to the published version of the manuscript.

Funding: This research received no external funding.

Institutional Review Board Statement: Not applicable.

Informed Consent Statement: Not applicable.

Data Availability Statement: The data used in this study is provided in the manuscript.

Acknowledgments: The support from the project GINOP-2.2.1-18-2018-00015 is acknowledged.

Conflicts of Interest: The authors declare no conflict of interest.

References

1. Xu, J.; Yao, Z.; Yang, G.; Han, Q. Research on crumb rubber concrete: From a multi-scale review. *Constr. Build. Mater.* **2020**, *232*, 117282. [CrossRef]
2. Gayana, B.C.; Chandar, K.R. Sustainable use of mine waste and tailings with suitable admixture as aggregates in concrete pavements—A review. *Adv. Concr. Constr.* **2018**, *6*, 221–243. [CrossRef]
3. Djebien, R.; Belachia, M.; Hebhoub, H. Effect of marble waste fines on rheological and hardened properties of sand concrete. *Struct. Eng. Mech.* **2015**, *53*, 1241–1251. [CrossRef]
4. Issa, C.A.; Salem, G. Utilization of recycled crumb rubber as fine aggregates in concrete mix design. *Constr. Build. Mater.* **2013**, *42*, 48–52. [CrossRef]
5. Thomas, B.S.; Gupta, R.C. A comprehensive review on the applications of waste tire rubber in cement concrete. *Renew. Sustain. Energy Rev.* **2016**, *54*, 1323–1333. [CrossRef]
6. Rezaifar, O.; Hasanzadeh, M.; Gholhaki, M. Concrete made with hybrid blends of crumb rubber and metakaolin: Optimization using Response Surface Method. *Constr. Build. Mater.* **2016**, *123*, 59–68. [CrossRef]
7. Rodríguez-Fernández, I.; Baheri, F.T.; Cavalli, M.C.; Poulikakos, L.D.; Bueno, M. Microstructure analysis and mechanical performance of crumb rubber modified asphalt concrete using the dry process. *Constr. Build. Mater.* **2020**, *259*, 119662. [CrossRef]
8. Batayneh, M.K.; Marie, I.; Asi, I. Promoting the use of crumb rubber concrete in developing countries. *Waste Manag.* **2008**, *28*, 2171–2176. [CrossRef]
9. Youssf, O.; Mills, J.E.; Benn, T.; Zhuge, Y.; Ma, X.; Roychand, R.; Gravina, R. Development of Crumb Rubber Concrete for Practical Application in the Residential Construction Sector–Design and Processing. *Constr. Build. Mater.* **2020**, *260*, 119813. [CrossRef]

10. Youssf, O.; Mills, J.E.; Hassanli, R. Assessment of the mechanical performance of crumb rubber concrete. *Constr. Build. Mater.* **2016**, *125*, 175–183. [CrossRef]
11. Atahan, A.O.; Yücel, A. Crumb rubber in concrete: Static and dynamic evaluation. *Constr. Build. Mater.* **2012**, *36*, 617–622. [CrossRef]
12. Li, D.; Zhuge, Y.; Gravina, R.; Benn, T.; Mills, J.E. Creep and drying shrinkage behaviour of crumb rubber concrete (CRC). *Aust. J. Civ. Eng.* **2020**, *18*, 187–204. [CrossRef]
13. Bravo, M.; de Brito, J. Concrete made with used tyre aggregate: Durability-related performance. *J. Clean. Prod.* **2012**, *25*, 42–50. [CrossRef]
14. Saberian, M.; Shi, L.; Sidiq, A.; Li, J.; Setunge, S.; Li, C.-Q. Recycled concrete aggregate mixed with crumb rubber under elevated temperature. *Constr. Build. Mater.* **2019**, *222*, 119–129. [CrossRef]
15. Bilondi, M.P.; Marandi, S.; Ghasemi, F. Effect of recycled glass powder on asphalt concrete modification. *Struct. Eng. Mech.* **2016**, *59*, 373–385. [CrossRef]
16. Mohammadi, I.; Khabbaz, H. Shrinkage performance of Crumb Rubber Concrete (CRC) prepared by water-soaking treatment method for rigid pavements. *Cem. Concr. Compos.* **2015**, *62*, 106–116. [CrossRef]
17. Onuaguluchi, O.; Panesar, D.K. Hardened properties of concrete mixtures containing pre-coated crumb rubber and silica fume. *J. Clean. Prod.* **2014**, *82*, 125–131. [CrossRef]
18. Atahan, A.O.; Sevim, U.K. Testing and comparison of concrete barriers containing shredded waste tire chips. *Mater. Lett.* **2008**, *62*, 3754–3757. [CrossRef]
19. Li, G.; Garrick, G.; Eggers, J.; Abadie, C.; Stubblefield, M.A.; Pang, S.-S. Waste tire fiber modified concrete. *Compos. Part B Eng.* **2004**, *35*, 305–312. [CrossRef]
20. Son, K.S.; Hajirasouliha, I.; Pilakoutas, K. Strength and deformability of waste tyre rubber-filled reinforced concrete columns. *Constr. Build. Mater.* **2011**, *25*, 218–226. [CrossRef]
21. Vakhshouri, B.; Nejadi, S. Self-compacting light-weight concrete; mix design and proportions. *Struct. Eng. Mech.* **2016**, *58*, 143–161. [CrossRef]
22. Guneyisi, E.; Gesoglu, M.; Mermerdas, K.; Ipek, S. Experimental investigation on durability performance of rubberized concrete. *Adv. Concr. Constr.* **2014**, *2*, 193–207. [CrossRef]
23. Youssf, O.; ElGawady, M.A.; Mills, J.E.; Ma, X. An experimental investigation of crumb rubber concrete confined by fibre reinforced polymer tubes. *Constr. Build. Mater.* **2014**, *53*, 522–532. [CrossRef]
24. Hofstetter, K.; Eberhardsteiner, J.; Mang, H. Efficient treatment of rubber friction problems in industrial applications. *Struct. Eng. Mech.* **2006**, *22*, 517–539. [CrossRef]
25. Pelisser, F.; Zavarise, N.; Longo, T.A.; Bernardin, A.M. Concrete made with recycled tire rubber: Effect of alkaline activation and silica fume addition. *J. Clean. Prod.* **2011**, *19*, 757–763. [CrossRef]
26. Williams, K.C.; Partheeban, P. An experimental and numerical approach in strength prediction of reclaimed rubber concrete. *Adv. Concr. Constr.* **2018**, *6*, 87–102. [CrossRef]
27. Nas, M.; Kurbetci, Ş. Durability properties of concrete containing metakaolin. *Adv. Concr. Constr.* **2018**, *6*, 159–175. [CrossRef]
28. Zhang, P.; Gao, J.-X.; Dai, X.-B.; Zhang, T.-H.; Wang, J. Fracture behavior of fly ash concrete containing silica fume. *Struct. Eng. Mech.* **2016**, *59*, 261–275. [CrossRef]
29. Karthikeyan, B.; Dhinakaran, G. Strength and durability studies on high strength concrete using ceramic waste powder. *Struct. Eng. Mech.* **2017**, *61*, 171–181. [CrossRef]
30. Golewski, G.L. Determination of fracture toughness in concretes containing siliceous fly ash during mode III loading. *Struct. Eng. Mech.* **2017**, *62*, 1–9. [CrossRef]
31. Mohammadi, I.; Khabbaz, H.; Vessalas, K. Enhancing mechanical performance of rubberised concrete pavements with sodium hydroxide treatment. *Mater. Struct.* **2015**, *49*, 813–827. [CrossRef]
32. Eldin, N.N.; Senouci, A.B. Rubber-tire particles as concrete aggregate. *J. Mater. Civ. Eng.* **1993**, *5*, 478–496. [CrossRef]
33. Thomas, B.S.; Gupta, R.C.; Kalla, P.; Cseteneyi, L. Strength, abrasion and permeation characteristics of cement concrete containing discarded rubber fine aggregates. *Constr. Build. Mater.* **2014**, *59*, 204–212. [CrossRef]
34. Ganjian, E.; Khorami, M.; Maghsoudi, A.A. Scrap-tyre-rubber replacement for aggregate and filler in concrete. *Constr. Build. Mater.* **2009**, *23*, 1828–1836. [CrossRef]
35. Eldin, N.N.; Senouci, A.B. Measurement and prediction of the strength of rubberized concrete. *Cem. Concr. Compos.* **1994**, *16*, 287–298. [CrossRef]
36. Güneyisi, E.; Gesoğlu, M.; Özturan, T. Properties of rubberized concretes containing silica fume. *Cem. Concr. Res.* **2004**, *34*, 2309–2317. [CrossRef]
37. Chou, L.-H.; Lin, C.-N.; Lu, C.-K.; Lee, C.-H.; Lee, M.-T. Improving rubber concrete by waste organic sulfur compounds. *Waste Manag. Res.* **2010**, *28*, 29–35. [CrossRef]
38. Tian, S.; Zhang, T.; Li, Y. Research on modifier and modified process for rubber-particle used in rubberized concrete for road. *Adv. Mater. Res.* **2011**, *243*, 4125–4130. [CrossRef]
39. Liu, H.; Wang, X.; Jiao, Y.; Sha, T. Experimental investigation of the mechanical and durability properties of crumb rubber concrete. *Materials* **2016**, *9*, 172. [CrossRef] [PubMed]
40. Li, Y.; Wang, M.; Li, Z. Physical and mechanical properties of Crumb Rubber Mortar (CRM) with interfacial modifiers. *J. Wuhan Univ. Technol. Sci. Ed.* **2010**, *25*, 845–848. [CrossRef]
41. Pacheco-Torgal, F.; Ding, Y.; Jalali, S. Properties and durability of concrete containing polymeric wastes (tyre rubber and polyethylene terephthalate bottles): An overview. *Constr. Build. Mater.* **2012**, *30*, 714–724. [CrossRef]

42. *Standard Practice for Making and Curing Concrete Test Specimens in the Laboratory*; ASTM C192/C192M-19; ASTM International: West Conshohocken, PA, USA, 2019.

43. *Standard Test Method for Slump of Hydraulic-Cement Concrete*; ASTM C143/C143M-15a; ASTM International: West Conshohocken, PA, USA, 2015.

44. *Indian Standard Methods of Sampling and Analysis of Concrete*; IS: 1199–1959; Bureau of Indian Standards: Old Delhi, India, 1959.

45. *Standard Test Method for Compressive Strength of Hydraulic Cement Mortars (Using 2-in. or [50 mm] Cube Specimens)*; ASTM C109/C109M-20b; ASTM International: West Conshohocken, PA, USA, 2016.

46. *Standard Test Method for Compressive Strength of Cylindrical Concrete Specimens*; ASTM C39/C39M-17a; ASTM Standard: West Conshohocken, PA, USA, 2017.

47. *Methods of Testing Concrete-Determination of Indirect Tensile Strength of Concrete Cylinders, Standards Australia*; AS 1012.10; Australian Standard: Sydney, Australia, 2000.

48. *Standard Test Method for Pulse Velocity through Concrete*; ASTM C597-16; ASTM International: West Conshohocken, PA, USA, 2016.

49. *Standard Test Method for Density, Absorption, and Voids in Hardened Concrete*; ASTM C642-13; ASTM International: West Conshohocken, PA, USA, 2013.

50. Albano, C.; Camacho, N.; Reyes, J.; Feliu, J.; Hernández, M. Influence of scrap rubber addition to Portland I concrete composites: Destructive and non-destructive testing. *Compos. Struct.* **2005**, *71*, 439–446. [CrossRef]

51. Bignozzi, M.; Sandrolini, F. Tyre rubber waste recycling in self-compacting concrete. *Cem. Concr. Res.* **2006**, *36*, 735–739. [CrossRef]

52. Sadek, D.M.; El-Attar, M.M. Structural behavior of rubberized masonry walls. *J. Clean. Prod.* **2015**, *89*, 174–186. [CrossRef]

53. Corinaldesi, V.; Mazzoli, A.; Moriconi, G. Mechanical behaviour and thermal conductivity of mortars containing waste rubber particles. *Mater. Des.* **2011**, *32*, 1646–1650. [CrossRef]

54. Turgut, P.; Yesilata, B. Physico-mechanical and thermal performances of newly developed rubber-added bricks. *Energy Build.* **2008**, *40*, 679–688. [CrossRef]

55. Salhi, M.; Ghrici, M.; Li, A.; Bilir, T. Effect of curing treatments on the material properties of hardened self-compacting concrete. *Adv. Concr. Constr.* **2017**, *5*, 359–375. [CrossRef]

56. Wang, J.; Guo, Z.; Yuan, Q.; Zhang, P.; Fang, H. Effects of ages on the ITZ microstructure of crumb rubber concrete. *Constr. Build. Mater.* **2020**, *254*, 119329. [CrossRef]

57. Padhi, S.; Panda, K. Fresh and hardened properties of rubberized concrete using fine rubber and silpozz. *Adv. Concr. Constr.* **2016**, *4*, 49–69. [CrossRef]

58. Solanki, P.; Dash, B. Mechanical properties of concrete containing recycled materials. *Adv. Concr. Constr.* **2016**, *4*, 207–220. [CrossRef]

59. Liang, J.F.; Wang, E.; He, C.F.; Hu, P. Mechanical behavior of recycled fine aggregate concrete after high temperature. *Struct. Eng. Mech.* **2018**, *65*, 343–348. [CrossRef]

60. Sofi, A. Effect of waste tyre rubber on mechanical and durability properties of concrete—A review. *Ain Shams Eng. J.* **2018**, *9*, 2691–2700. [CrossRef]

61. Barrentine, L.B. *An Introduction to Design of Experiments: A Simplified Approach*; ASQ Quality Press: Milwaukee, WI, USA, 1999.

Article

Effect of Recycled Coarse Aggregate and Bagasse Ash on Two-Stage Concrete

Muhammad Faisal Javed [1,*], **Afaq Ahmad Durrani** [2], **Sardar Kashif Ur Rehman** [1], **Fahid Aslam** [3], **Hisham Alabduljabbar** [3] and **Amir Mosavi** [4,5,6,7,*]

[1] Department of Civil Engineering, COMSATS University Islamabad, Abbottabad Campus, Khyber Pakhtunkhwa 22060, Pakistan; skashif@cuiatd.edu.pk

[2] Department of Civil Engineering, University of Louisiana at Lafayette, Lafayette, LA 70503, USA; C00480150@louisiana.edu

[3] Department of Civil Engineering, College of Engineering in Al-Kharj, Prince Sattam Bin Abdulaziz University, Al-Kharj 11942, Saudi Arabia; f.aslam@psau.edu.sa (F.A.); h.alabduljabbar@psau.edu.sa (H.A.)

[4] Faculty of Civil Engineering, Technische Universitat Dresden, 01069 Dresden, Germany

[5] Faculty Department of Informatics, J. Selye University, 94501 Komarno, Slovakia

[6] Faculty Information Systems, University of Siegen, Kohlbettstraße 15, 57072 Siegen, Germany

[7] Faculty John von Neumann Faculty of Informatics, Obuda University, 1034 Budapest, Hungary

* Correspondence: arbab_faisal@yahoo.com (M.F.J.); amir.mosavi@mailbox.tu-dresden.de (A.M.)

Citation: Javed, M.F.; Durrani, A.A.; Kashif Ur Rehman, S.; Aslam, F.; Alabduljabbar, H.; Mosavi, A. Effect of Recycled Coarse Aggregate and Bagasse Ash on Two-Stage Concrete. *Crystals* **2021**, *11*, 556. https://doi.org/10.3390/cryst11050556

Academic Editor: Yifeng Ling

Received: 16 April 2021
Accepted: 13 May 2021
Published: 16 May 2021

Abstract: Numerous research studies have been conducted to improve the weak properties of recycled aggregate as a construction material over the last few decades. In two-stage concrete (TSC), coarse aggregates are placed in formwork, and then grout is injected with high pressure to fill up the voids between the coarse aggregates. In this experimental research, TSC was made with 100% recycled coarse aggregate (RCA). Ten percent and twenty percent bagasse ash was used as a fractional substitution of cement along with the RCA. Conventional concrete with 100% natural coarse aggregate (NCA) and 100% RCA was made to determine compressive strength only. Compressive strength reduction in the TSC was 14.36% when 100% RCA was used. Tensile strength in the TSC decreased when 100% RCA was used. The increase in compressive strength was 8.47% when 20% bagasse ash was used compared to the TSC mix that had 100% RCA. The compressive strength of the TSC at 250 °C was also determined to find the reduction in strength at high temperature. Moreover, the compressive and tensile strength of the TSC that had RCA was improved by the addition of bagasse ash.

Keywords: bagasse ash; sustainable concrete; sustainable development; mechanical properties; natural coarse aggregate; recycled coarse aggregate; two-stage concrete; materials design; green concrete; recycled concrete

1. Introduction

The most extensively used building material around the globe is concrete [1]. Normal concrete is made by mixing coarse aggregate, fine aggregate, cement, water, and admixture in a mixer according to ASTM C192-02. In TSC, coarse aggregates are set in formwork, and then grout or mortar is infused through a pipe with high pressure from the bottom to the top. The grout fills the pores between the coarse aggregates [2]. Maximum density is achieved without mechanical compaction or using a vibrator [3]. In contrast to conventional concrete, compressive stresses in TSC are first passed through the body of coarse aggregates due to point-to-point contact and, after deformation, they then pass through the hardened mortar [2]. TSC is environmentally friendly concrete as the coarse aggregates are not mixed in a mixer, which reduces the consumption of energy. The grout is made on-site, which eliminates the usage of transit trucks, which in turn results in decreasing pollution and reduces cost [4]. The cost of TSC is 40% less than that of traditional concrete because it consists of about 70% coarse aggregate, consuming about 30% less cement [5]. Due to point-to-point contact of the coarse aggregates, the modulus of elasticity of TSC is very high [6]. Drying shrinkage in TSC is lower than in normal concrete, which results in less volume change [7]. TSC is mainly used where a low heat of hydration is required, such as mass concreting in tunnels, and underwater construction and repair [6].

Around 25 billion tons of concrete is produced annually, making it the largest consumer of Earth's natural resources, which are water, natural aggregates (gravel and crushed rock), and sand [8]. Around 12.6 billion tons of natural aggregate is used annually [9]. Due to urbanization, a lot of construction and demolition waste (CDW) is produced. CDW has a serious impact on our environment as it creates pollution when it is disposed of in landfills. Over 1 billion tons of CDW is produced annually [9]. To protect our environment from depleting virgin aggregate resources, recycled coarse aggregates (RCA) have been used to produce concrete [10]. RCA consists of natural aggregates and adhered mortar. Concrete obtained from demolished buildings is crushed to obtain RCA. Due to increased absorption capacity, 5% more water is required for concrete made with RCA to acquire a similar workability as that of normal concrete [11]. Due to traverse cracks in RCA, increased porosity, less information regarding the interfacial transition zone (ITZ) between RCA and cement paste, impurities, adhered mortar, and inferior quality, this makes the use of RCA as structural concrete difficult [12]. RCA has two ITZs, one with new cement paste and the other with old mortar. Therefore, the structure of concrete made with RCA demonstrates a more complex structure than that of concrete made with natural aggregates. Old ITZ has micro-cracks that decrease the strength of the concrete and also increase its water consumption [13]. RCA may contain impurities such as organic matter, sulphates, carbonates, and chlorides [14]. When Sheen et al. [15] made concrete using RCA, it was observed that the compressive strength of the concrete with RCA decreased because of fine particles. The compressive strength of the concrete with RCA can be increased by adding natural admixtures [16]. The permeability of the concrete can be increased by increasing the amount of the RCA [17]. The shape and texture of RCA depends on the crusher plant, which directly affects the workability of the concrete [18]. Pakistan is an agricultural country. Sugar cane is the second largest cash crop of Pakistan, which accounts for 3.6% of gross domestic product [19]. Bagasse is a by-product of the sugar industry and is used as an energy source for sugar production in this industry [20]. Sugarcane contains about 25% bagasse. Fourteen million tons of bagasse is produced in Pakistan annually. Bagasse is also used in the paper industry. When bagasse is burnt for energy purposes, it produces 3% ash, which is dumped in landfills [21]. Waste obtained from agriculture and some other industries can be used as a partial replacement of cement in concrete [22]. One of the ways to decrease the negative environmental impact of concrete is to use mineral admixtures as a partial replacement of cement, which will not only reduce pollution but will also decrease the cost and improve the strength of the concrete [23].

The objective of this research is to evaluate the mechanical properties of TSC made with NCA and RCA. In order to achieve the objective, four mixes of TSC were prepared. One mix was prepared with 100% NCA, while in the other three mixes, NCA was replaced with 100% RCA. Bagasse ash was used as a partial replacement of the cement at 10% and 20% in two mixes of the TSC that had RCA. Compressive strength, compressive strength at 250 °C, tensile strength, and mass loss at 250 °C were then determined. In addition, two mixes of conventional concrete were also prepared. One mix was prepared with 100% NCA, while the other mix was prepared with 100% RCA. The results of the compressive strength of the conventional concrete were then compared to that of the TSC.

2. Materials and Methods

2.1. Materials

2.1.1. Cement and Bagasse Ash

Ordinary Portland cement (ASTM Type-I) from the Cherat cement factory was used for the preparation of the TSC. Twenty-eight-day strength of cement was up to 69 MPa. The fineness of the cement was 93.15%. The surface area of the cement was 2137 cm^2/gm.

Bagasse ash was brought from Premier sugar mill, Mardan, Pakistan. It was then ground/crushed in PCSIR, Peshawar, Pakistan. It was passed through sieve#200. The specific gravity of bagasse ash was 1.35. The surface area of the bagasse ash was 2840.7 cm^2/gm. The chemical composition of the bagasse is shown in Table 1 and was used according to ASTM C618 criteria.

Table 1. Chemical composition of bagasse ash.

Chemical Composition	Bagasse Ash	ASTM C618 Requirement
Silicon dioxide (SiO_2)	64.81%	
Aluminum oxide (Al_2O_3)	11.36%	Minimum 70%
Ferric oxide (Fe_2O_3)	1.53%	
$SiO_2 + Al_2O_3 + Fe_2O_3$	77.70%	77.70% > 70%
Magnesium oxide (MgO)	3.49%	
Calcium oxide (CaO)	11.13%	
Sodium oxide (Na_2O)	1.14%	
Manganese oxide (MnO)	0.09%	
Potassium oxide (K_2O)	3.99%	
Phosphorous pentoxide (P_2O_5)	1.99%	
Titanium dioxide (TiO_2)	0.47%	

2.1.2. Aggregates

Coarse and fine aggregates were obtained from a quarry near COMSATS University Islamabad, Abbottabad Campus, Pakistan. The concrete waste from a demolished building was obtained from an empty plot near Daewoo terminal, Abbottabad, Pakistan. The demolished concrete waste was crushed with the help of a crusher to obtain the RCA. Twenty-five millimeters was the maximum size of both the natural and recycled aggregates. Table 2 shows the physical properties of both the natural and recycled aggregates.

Table 2. Physical properties of the natural, recycled coarse, and fine aggregates.

Physical Properties	Natural Coarse Aggregates	Recycled Coarse Aggregates	Fine Aggregate
Water absorption	1.85%	7.59%	1.1%
Specific gravity	2.75	2.62	2.35
Impact value	14.72%	22.23%	-
Fineness modulus	2.04	2.19	2.96
Density	1532.3 kg/m^3	1399.2 kg/m^3	1610 kg/m^3

2.1.3. Superplasticizer

Ultra-Super Plast 470 was used throughout the casting of the TSC. The Ultra-Super Plast 470 was procured from Ultra Chemicals, Peshawar, Pakistan. The product was light brown, and the specific gravity was 1.15.

2.2. Mixture Proportions

Six groups of concrete mixtures were prepared in total as shown in Table 3. Two mixtures were prepared with the conventional method. Four different mixtures of TSC were made. All six groups were prepared with a ratio of 1:1:2.7 (cement: fine aggregate: coarse aggregate). Control mix-N was prepared using the conventional method and contained 100% natural coarse aggregate (NCA). Control mix-R was prepared using the conventional method and contained 100% RCA. Control mix-TSCI was a TSC and was made with 100% NCA. Control mix-TSCII was a TSC and was made with 100% RCA. The fifth mix was a TSC and was prepared with 10% bagasse ash as a fractional substitution of the cement and 100% RCA. The sixth mix was a TSC and was made with 20% bagasse ash as a fractional substitution of the cement and 100% RCA. The water-to-cement ratio (W/C) used in this experimental research was 0.5. This ratio was used for all six of the mixes. Superplasticizer was added at a rate of 1% by the weight of the cement in TSC10BA and TSC20BA.

Table 3. Mix types with identification based on replacement ratio.

Mix Types	Concrete Mix Proportion
CM-N	Control mix made with the conventional method (100% NCA)
CM-R	Control mix made with the conventional method (100% RCA)
CM-TSCI	Control mix made with TSC (100% NCA and 100% cement)
CM-TSCII	Control mix made with TSC (100% RCA and 100% cement)
TSC10BA	TSC with 100% RCA and 10% cement replaced by bagasse ash
TSC20 BA	TSC with 100% RCA and 20% cement replaced by bagasse ash

2.3. Specimen Casting and Curing

CM-N and CM-R were normal concrete mixes. The RCA used during casting of the normal concrete and the TSC were prewet. Fine aggregate and coarse aggregates were first dry mixed for one minute in a mixer. Cement was then added and dry mixed for a further minute. Finally, water was added, and the total mixing time was five minutes. Cylindrical molds of 6 inches (152.4 mm) in diameter and 12 inches (254.8) in height were used for casting as per ASTM C 31/C 31M-03. For the TSC, a pipe with a diameter of 1 inch (25.4 mm) and a height of 2 m (2000 mm) was placed in the middle of a mold. Then, a mold was filled with coarse aggregates. Following this, grout was injected from the top via a pipe. The grout was injected with the help of pressure created by gravity. As the voids in the coarse aggregate were filled, the pipe was raised. After the appearance of grout at the top of a mold, the pipe was removed from the mold. This procedure was used for all the specimens and was also used by Abdelgader [24]. After 24 h, the specimens were taken out of the molds and were kept in a water tank. Ninety specimens were prepared in total; CM-N and CM-R had three specimens each, while each mix of the TSC had twenty-one specimens.

3. Results and Discussions

3.1. Compressive Strength

The compressive strength of all the concrete mixes is shown in Figure 1. Each compressive strength is an average of three measurements. The compressive strength of CM-N and CM-R was determined at day 28 of the curing process, whilst the compressive strength of all the TSC mixes was determined at days 7, 28, and 56 of the curing process. The figures show that the compressive strength of CM-TSCI is at its highest among all mixes at days 7, 28, and 56 of the curing process. The compressive strength of CM-TSCI is 3.32% more than that of CM-N at 28 days Thus, the compressive strength of the TSC is greater than that of the conventional concrete. The increase in strength is due to the point-to-point contact of coarse aggregates in the TSC; stresses in the TSC are first passed through the skeleton of the coarse aggregates and then through the grout, while in normal concrete, stresses are passed through a non-homogenous matrix. Due to this phenomenon, the crack mechanism and the ultimate strength of the TSC were different [25].

The compressive strength of CM-TSCI was at its highest at 56 days of curing. When 100% RCA is used, compressive strength is decreased. The compressive strength of CM-TSCII is reduced by 14.14% when compared with CM-TSCI at 28 days, while the compressive strength of CM-R is reduced by 22.13% when compared with CM-N at 28 days [26–28]. Noritaka Morohashi [29] evaluated the compressive strength of TSC by replacing 30% of the NCA with RCA in TSC, which resulted in a reduction in compressive strength of 12.6% at 28 days. RCA has a porous structure due to the adhered mortar and absorbs more water, which makes it weaker than NCA and causes a reduction in compressive strength [30]. This adhered mortar also decreases the density of RCA. Tavakoli and Soroushian [31] reported that the compressive strength of concrete made with RCA is less than that of concrete made with NCA when the same w/c ratio is used. Li [32] observed in his study that the percentage of RCA in concrete has an inverse relation with compressive strength.

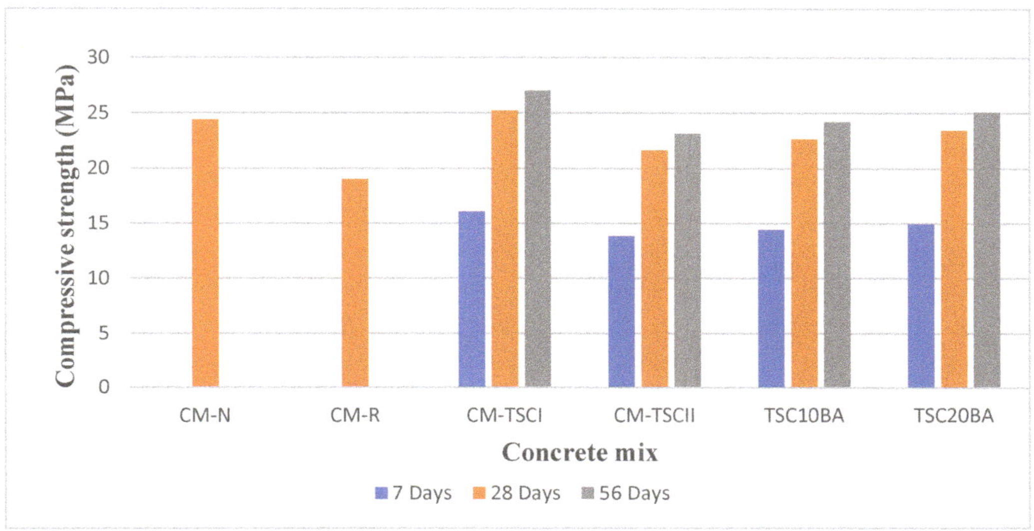

Figure 1. Compressive strength of the concrete mixes at 7, 28, and 56 days.

The compressive strength of the TSC mixes that had 100% RCA was improved by the addition of bagasse ash as a partial substitution of cement. The compressive strength of TSC10BA and TSC20BA was increased by 4.57% and 8.47%, respectively, when compared with CM-TSCII at 56 days of curing. The increase in strength is due to the pozzolanic reaction between calcium hydroxide and bagasse ash, which forms a calcium silicate hydrate gel [33]. Figure 2 shows the development of compressive strength of all four mixes of the TSC. Figure 2 clearly shows that the increase in compressive strength was rapid until day 7. CM-TSCII, TSC10BA, and TSC20BA show a very similar compressive strength at day 7. From day 7 to day 28, CM-TSCI shows more of an increase in compressive strength than that of the other mixes. From day 28 to day 56, the increase in compressive strength was low for all four of the mixes of the TSC. Kou, Poon, and Agrela [12] reported that there was an increase in compressive strength in RCA concrete mixes when mineral admixture was used. This is mainly due to the porous nature of the RCA; mineral admixture penetrates into the pores of the RCA, which improves the ITZ bond strength between the cement paste and the RCA. Another reason for the increase in compressive strength is the filling of cracks present in the RCA with hydration products.

3.2. Compressive Strength at 250 °C

The compressive strength of the TSC mixes at 20 °C and 250 °C is demonstrated in Table 4. The compressive strength at 250 °C is compared with the compressive strength of the mixes at 20 °C to determine the decrease in compressive strength due to high temperature. All mixes of the TSC demonstrated a loss of strength at 250 °C. The compressive strength loss in all of the mixes was less than 5.2%. The highest strength loss was 5.13% in TSC10BA and the lowest strength loss was 3.53% in CM-TSCI. The decrease in compressive strength at high temperature was mainly due to a loss of water and dehydration of the calcium silicate hydrate. Maanser et al. reported a 4% decrease in compressive strength at 200 °C [34]. The decrease in compressive strength in the mixes that had RCA was slightly more than that of CM-TSCI. This is mainly due to the weak interfacial bond between the RCA and the hardened paste in the concrete matrix. Otherwise, there is no significant difference in the strength loss between the TSC that has NCA and RCA.

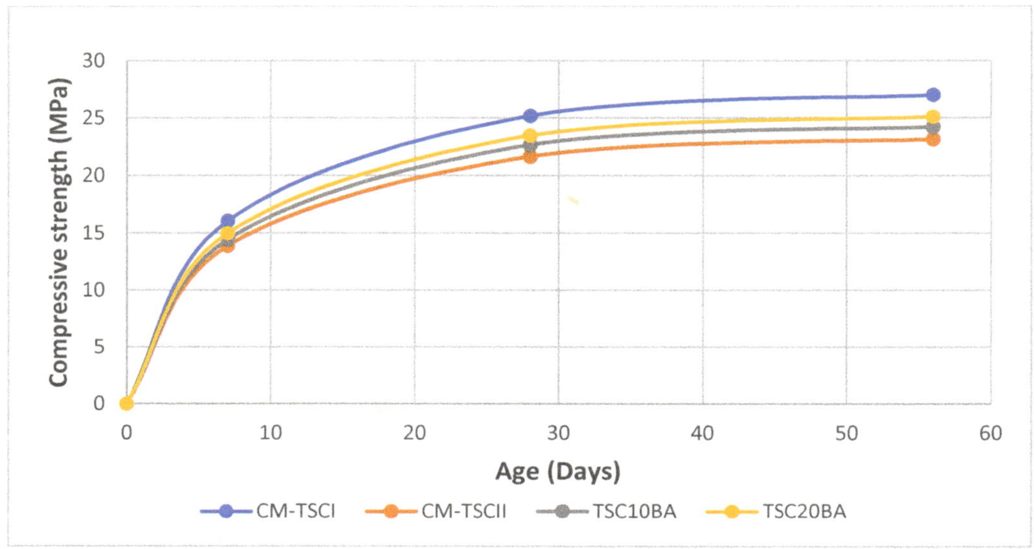

Figure 2. Development of compressive strength of TSC mixes.

Table 4. Mean compressive strength with standard deviation.

Mix Type	Days	Mean Compressive Strength (MPa)	SD
CM-N	7	-	-
	28	24.38	0.093
	56	-	-
CM-R	7	-	-
	28	18.99	0.098
	56	-	-
CM-TSCI	7	16.06	0.089
	28	25.19	0.095
	56	27.02	0.081
CM-TSCII	7	13.85	0.087
	28	21.63	0.098
	56	23.14	0.099
TSC10BA	7	14.41	0.076
	28	22.62	0.083
	56	24.2	0.095
TSC20BA	7	14.96	0.079
	28	23.44	0.082
	56	25.11	0.093

3.3. Tensile Strength

The results of the split tensile strength of the TSC at any given curing age are demonstrated in Figure 3. A split tensile test was only conducted on the TSC samples. Each value is an average of three measurements. It is evident from Figure 3 that the tensile strength of the TSC mixes that have RCA is less than that of the TSC mix that has NCA. The tensile strength of CM-TSCI is highest among all the mixes of the TSC throughout all curing days. There was a rapid decrease in tensile strength when 100% RCA was used. CM-TSCII tensile strength was reduced by 26.84% when compared with CM-TSCI at 56 days of curing. Tensile strength increased with the addition of bagasse ash as a partial substitution of the cement in TSC mixes that have 100% RCA. Tensile strength showed similar behavior as that of compressive strength. Lee and Choi [35] and Padmini, Ramamurthy, and Mathews [26] reported that the tensile strength of the concrete that had RCA was less than that of the concrete that had NCA. The tensile strength of TSC10BA and TSC20BA was increased by 22.1% and 26.86%, respectively, when compared with that of CM-TSCII at 56 days of curing.

The TSC mixes prepared with bagasse ash and 100% RCA showed better results in tensile strength as compared to the TSC mix that had only 100% RCA.

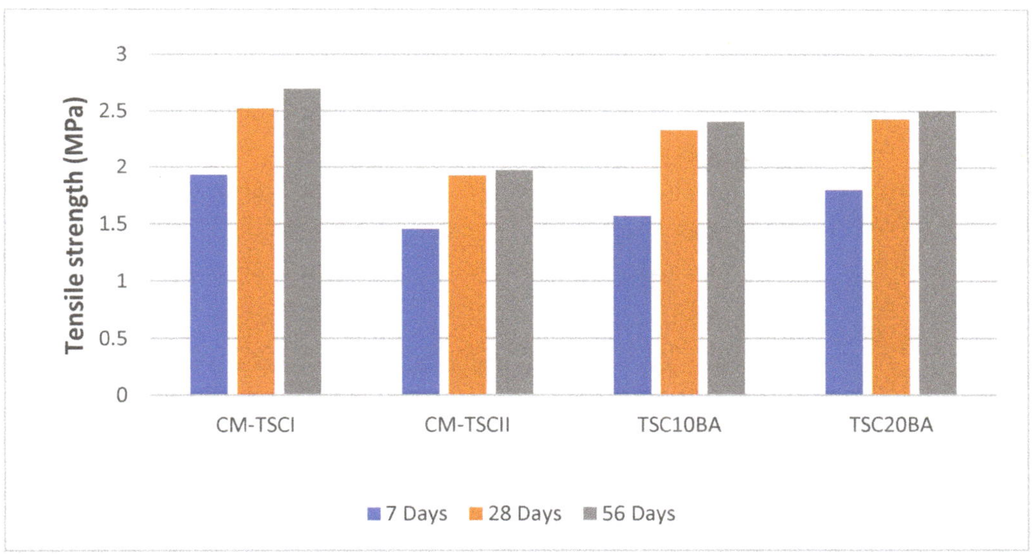

Figure 3. Tensile strength of concrete mixes at 7, 28, and 56 days.

3.4. Mass Loss at 250 °C

All the TSC mixes exhibited a mass loss due to the evacuation of free water due to an increase in temperature from 20 to 250 °C, which is demonstrated in Table 5. The decrease in mass loss is expressed in percentage form. The maximum mass loss was found in CM-TSCII, which was roughly 4.35%. The lowest mass loss was found in TSC10BA, which was 2.6%. The water present in concrete comes in three forms, which are the free, adsorbed, and bonded forms. This water escapes at high temperature and causes a mass loss. From 20 to 150 °C, a small mass loss occurs, while from 150 to 300 °C, there is an increase in mass loss. This mass loss is mainly due to dehydration of C-S-H [36]. A lesser mass loss is mainly due to the superplasticizer in TSC10BA and TSC20BA, which is due to the effect of resistances. Mean tensile strength is shown in Table 6 while percentage mass loss is shown in Table 7.

Table 5. Compressive strength at 20 °C and 250 °C.

Mix Types	Compressive Strength at 20 °C (MPa)	Compressive Strength at 250 °C (MPa)	% Decrease
CM-TSCI	25.18	24.29	3.53
CM-TSCII	21.63	20.67	4.43
TSC10BA	22.63	21.47	5.13
TSC20BA	23.45	22.33	4.77

Table 6. Mean tensile strength with standard deviation.

Mix Type	Days	Mean Compressive Strength (MPa)	SD
CM-TSCI	7	1.92	0.065
	28	2.51	0.078
	56	2.69	0.073
CM-TSCII	7	1.45	0.086
	28	1.92	0.087
	56	1.97	0.084
TSC10BA	7	1.57	0.092
	28	2.33	0.077
	56	2.41	0.088
TSC20BA	7	1.79	0.073
	28	2.42	0.067
	56	2.5	0.081

Table 7. Mass loss of the TSC.

Mix Types	Mass at 20 °C (kg)	Mass at 250 °C (kg)	% Decrease
CM-TSCI	13.24	12.69	4.15%
CM-TSCII	12.39	11.85	4.35%
TSC10BA	12.27	11.95	2.6%
TSC20BA	12.35	12.01	2.75%

4. Conclusions

The use of recycled coarse aggregate and bagasse ash in two-stage concrete would be beneficial for sustainable and environmentally friendly construction. The following conclusions are deduced from the experimental investigation:

1. Compressive strength of the samples made with the two-stage concrete method showed 3% more strength than that of the conventional concrete that had the same ratios, demonstrating the superior nature of the two-stage concrete method.
2. Compressive strength was decreased both in the TSC and the conventional concrete when RCA was used. However, compressive strength was reduced in the conventional concrete by 22%, compared to the TSC, which demonstrated a 20% increase in compressive strength.
3. A maximum increase in compressive strength was achieved when 20% bagasse ash was used. This increase in strength was found to be 8.47% at 56 days.
4. Minimum reduction in compressive strength (3.57%) at 250 °C was achieved when natural aggregates were used. However, the reduction in compressive strength in mixes that had RCA was also very minimal (maximum 5%).
5. Every mix that was used experienced mass loss at 250 °C. The maximum mass loss was 4.35% in the two-stage concrete that contained all natural aggregates.

Author Contributions: M.F.J., conceptualization, data analysis, writing original draft preparation; A.A.D., formal analysis and modeling, conceptualization, data analysis, writing original draft preparation; S.K.U.R., supervision, review and editing; F.A., investigation and review; H.A., methodology and review and editing; A.M., review and editing, supervision. All authors have read and agreed to the published version of the manuscript.

Funding: This research received no external funding.

Institutional Review Board Statement: Not Applicable.

Informed Consent Statement: Not Applicable.

Acknowledgments: We are grateful to the Department of Civil Engineering, COMSATS University Islamabad, Abbottabad Campus for their support in this research.

Conflicts of Interest: The authors declare no conflict of interest.

References

1. Meyer, C. Concrete materials and sustainable development in the USA. *Struct. Eng. Int.* **2004**, *14*, 203–207. [CrossRef]
2. Nowek, A.; Kaszubski, P.; Abdelgader, H.S.; Górski, J. Effect of admixtures on fresh grout and two-stage (pre-placed aggregate) concrete. *Struct. Concr.* **2007**, *8*, 17–23. [CrossRef]
3. Littlejohn, G. Grouted pre-placed aggregate concrete. In *Proceeding of Concrete in Ground*; The Concrete Society: London, UK, 1984; p. 13.
4. O'Malley, J.; Abdelgader, H.S. Investigation into viability of using two-stage (pre-placed aggregate) concrete in Irish setting. *Front. Archit. Civ. Eng. China* **2010**, *4*, 127–132. [CrossRef]
5. Abdelgader, H. Experimental–Mathematical Procedure of Designing the Two-Stage Concrete. Ph.D. Thesis, Technical University of Gdańsk, Gdańsk, Poland, 1995.
6. Abdelgader, H.S.; Elgalhud, A. Effect of grout proportions on strength of two-stage concrete. *Struct. Concr.* **2008**, *9*, 163–170. [CrossRef]
7. Abdelgader, H.S.; Górski, J. Influence of grout proportions on modulus of elasticity of two-stage concrete. *Mag. Concr. Res.* **2002**, *54*, 251–255. [CrossRef]
8. Assi, L.N.; Carter, K.; Deaver, E.; Ziehl, P. Review of availability of source materials for geopolymer/sustainable concrete. *J. Clean. Prod.* **2020**, *263*, 121477. [CrossRef]
9. Mehta, P.K. Greening of the concrete industry for sustainable development. *Concr. Int.* **2002**, *24*, 23–28.
10. Hansen, T.C. *Recycling of Demolished Concrete and Masonry*; CRC Press: Boca Raton, FL, USA, 1992.
11. Etxeberria, M.; Vázquez, E.; Marí, A.; Barra, M. Influence of amount of recycled coarse aggregates and production process on properties of recycled aggregate concrete. *Cem. Concr. Res.* **2007**, *37*, 735–742. [CrossRef]
12. Kou, S.-C.; Poon, C.-S.; Agrela, F. Comparisons of natural and recycled aggregate concretes prepared with the addition of different mineral admixtures. *Cem. Concr. Compos.* **2011**, *33*, 788–795. [CrossRef]
13. Tam, V.W.; Gao, X.; Tam, C.M. Microstructural analysis of recycled aggregate concrete produced from two-stage mixing approach. *Cem. Concr. Res.* **2005**, *35*, 1195–1203. [CrossRef]
14. Alengaram, U.J.; Salam, A.; Jumaat, M.Z.; Jaafar, F.F.; Saad, H.B. Properties of high-workability concrete with recycled concrete aggregate. *Mater. Res.* **2011**, *14*, 248–255.
15. Sheen, Y.-N.; Wang, H.-Y.; Juang, Y.-P.; Le, D.-H. Assessment on the engineering properties of ready-mixed concrete using recycled aggregates. *Constr. Build. Mater.* **2013**, *45*, 298–305. [CrossRef]
16. Corinaldesi, V.; Moriconi, G. Influence of mineral additions on the performance of 100% recycled aggregate concrete. *Constr. Build. Mater.* **2009**, *23*, 2869–2876. [CrossRef]
17. Gholamreza, F.; Razaqpur, O.; Isgor, B.; Abbas, A.; Fournier, B.; Simon, F. Anovel method for proportioning structural concrete mixes made with recycled concrete aggregate. *Concr. Int. Mag. ACI* **2009**.
18. Rashwan, S.; AbouRizk, S. Research on an alternative method for reclaiming leftover concrete. *Concr. Int.* **1997**, *19*.
19. Qureshi, M.A.; Afghan, S. Sugarcane cultivation in Pakistan. *Sugar Book Pub. Pak. Soc. Sugar Technol.* **2005**, 1–11.
20. Xie, Z.; Xi, Y. Hardening mechanisms of an alkaline-activated class F fly ash. *Cem. Concr. Res.* **2001**, *31*, 1245–1249. [CrossRef]
21. Amin, N.-U. Use of bagasse ash in concrete and its impact on the strength and chloride resistivity. *J. Mater. Civ. Eng.* **2010**, *23*, 717–720. [CrossRef]
22. Hansen, T.C. Recycled aggregates and recycled aggregate concrete second state-of-the-art report developments 1945–1985. *Mater. Struct.* **1986**, *19*, 201–246. [CrossRef]
23. Wild, S.; Khatib, J.M.; Jones, A. Relative strength, pozzolanic activity and cement hydration in superplasticised metakaolin concrete. *Cem. Concr. Res.* **1996**, *26*, 1537–1544. [CrossRef]
24. Abdelgader, H. Effect of the quantity of sand on the compressive strength of two-stage concrete. *Mag. Concr. Res.* **1996**, *48*, 353–360. [CrossRef]
25. Abdul Awal, A.S. Failure mechanism of prepacked concrete. *J. Struct. Eng.* **1988**, *114*, 727–732. [CrossRef]
26. Padmini, A.; Ramamurthy, K.; Mathews, M. Influence of parent concrete on the properties of recycled aggregate concrete. *Constr. Build. Mater.* **2009**, *23*, 829–836. [CrossRef]

27. Xiao, J.; Li, J.; Zhang, C. Mechanical properties of recycled aggregate concrete under uniaxial loading. *Cem. Concr. Res.* **2005**, *35*, 1187–1194. [CrossRef]

28. Katz, A. Properties of concrete made with recycled aggregate from partially hydrated old concrete. *Cem. Concr. Res.* **2003**, *33*, 703–711. [CrossRef]

29. Morohashi, N.; Abdelgader, H.S. Concrete with recycled aggregates-two-stage production method. *Concr. Plant Int.* **2013**, 34–42.

30. Katz, A. Treatments for the improvement of recycled aggregate. *J. Mater. Civ. Eng.* **2004**, *16*, 597–603. [CrossRef]

31. Tavakoli, M.; Soroushian, P. Strengths of recycled aggregate concrete made using field-demolished concrete as aggregate. *Mater. J.* **1996**, *93*, 178–181.

32. Li, X. Recycling and reuse of waste concrete in China: Part I. material behaviour of recycled aggregate concrete. *Resour. Conserv. Recycl.* **2008**, *53*, 36–44. [CrossRef]

33. Martirena-Hernández, J.; Betancourt-Rodriguez, S.; Middendorf, B.; Rubio, A.; Martínez-Fernández, L.; López, I.M.; González-López, R. Pozzolanic properties of residues of sugar industries (second part). *Mater. Construcción* **2001**, *51*, 67–72. [CrossRef]

34. Tebbal, N.; Rahmouni, Z.; Belouadah, M. Influence of an addition on the mechanical behavior of high performance concretesa subjected to high temperatures. In Proceedings of the 30th Meeting AUGC-IBPSA Chambéry, Savoie, France, 6–8 June 2012.

35. Lee, G.; Choi, H. Study on interfacial transition zone properties of recycled aggregate by micro-hardness test. *Constr. Build. Mater.* **2013**, *40*, 455–460. [CrossRef]

36. Maanser, A.; Benouis, A.; Ferhoune, N. Effect of high temperature on strength and mass loss of admixtured concretes. *Constr. Build. Mater.* **2018**, *166*, 916–921. [CrossRef]

Article

Prediction of Compressive Strength of Rice Husk Ash Concrete through Different Machine Learning Processes

Ammar Iqtidar [1,*], Niaz Bahadur Khan [2], Sardar Kashif-ur-Rehman [1], Muhmmad Faisal Javed [1], Fahid Aslam [3], Rayed Alyousef [3], Hisham Alabduljabbar [3] and Amir Mosavi [4,5,*]

1 Department of Civil Engineering, Abbottabad Campus, COMSATS University Islamabad, Abbottabad 22060, Pakistan; skashif@cuiatd.edu.pk (S.K.-u.-R.); arbabfaisal@cuiatd.edu.pk (M.F.J.)
2 School of Mechanical & Manufacturing Engineering, National University of Sciences & Technology, Islamabad 44000, Pakistan; niaz.bahadur@smme.nust.edu.pk
3 Department of Civil Engineering, College of Engineering in Al-Kharj, Prince Sattam Bin Abdulaziz University, Al-Kharj 11942, Saudi Arabia; f.aslam@psau.edu.sa (F.A.); r.alyousef@psau.edu.sa (R.A.); h.alabduljabbar@psau.edu.sa (H.A.)
4 Faculty of Civil Engineering, Technische Universitat Dresden, 01069 Dresden, Germany
5 John von Neumann Faculty of Informatics, Obuda University, 1034 Budapest, Hungary
* Correspondence: FA16-BCV-016@cuiatd.edu.pk (A.I.); amir.mosavi@mailbox.tu-dresden.de (A.M.)

Citation: Iqtidar, A.; Bahadur Khan, N.; Kashif-ur-Rehman, S.; Faisal Javed, M.; Aslam, F.; Alyousef, R.; Alabduljabbar, H.; Mosavi, A. Prediction of Compressive Strength of Rice Husk Ash Concrete through Different Machine Learning Processes. *Crystals* **2021**, *11*, 352. https://doi.org/10.3390/cryst11040352

Academic Editors: Shujun Zhang and Yifeng Ling

Received: 13 February 2021
Accepted: 22 March 2021
Published: 29 March 2021

Abstract: Cement is among the major contributors to the global carbon dioxide emissions. Thus, sustainable alternatives to the conventional cement are essential for producing greener concrete structures. Rice husk ash has shown promising characteristics to be a sustainable option for further research and investigation. Since the experimental work required for assessing its properties is both time consuming and complex, machine learning can be used to successfully predict the properties of concrete containing rice husk ash. A total of 192 data points are used in this study to assess the compressive strength of rice husk ash blended concrete. Input parameters include age, amount of cement, rice husk ash, super plasticizer, water, and aggregates. Four soft computing and machine learning methods, i.e., artificial neural networks (ANN), adaptive neuro-fuzzy inference system (ANFIS), multiple nonlinear regression (NLR), and linear regression are employed in this research. Sensitivity analysis, parametric analysis, and correlation factor (R^2) are used to evaluate the obtained results. The ANN and ANFIS outperformed other methods.

Keywords: rice husk ash; sustainable concrete; artificial neural networks; multiple linear regression; eco-friendly concrete; green concrete; sustainable development; artificial intelligence; data science; machine learning

1. Introduction

The world is making progress by leaps and bounds. New technologies and innovations are being introduced every day in every field. These advancements have altered the course of human history. One of the main aspects that has played a crucial role in shaping modern human civilization is infrastructure. From caves, mankind has started to live in strong and pleasing dwellings made by their own creative and innovative minds. Still, today infrastructure is considered to be the main element for progress in any country. The construction material that is used in abundance throughout the world for the construction of infrastructure is cement. However, along with the advantages of cement there are also certain adverse effects. Cement is said to be responsible for seven percent of the total carbon dioxide emissions worldwide [1]. It produces carbon dioxide while reacting when water is added to it. Secondly, a high temperature is required during the production of cement [2]. This high temperature is achieved by burning fossil fuels which increase the carbon footprint of cement. Our planet earth is suffering from problems of grave danger. Environmental deterioration and global warming are some of these alarming issues. If not controlled in due time, these problems will push the earth to the brink of extinction. One of the major causes of environmental degradation and global warming is said to be the emission of carbon dioxide from different products and processes [3,4]. Since cement is a crucial contributor to the total carbon dioxide emissions of the world, the importance of infrastructure cannot be undermined. It must be replaced with some other material that has a smaller carbon footprint as well as possessing the same or better properties than cement.

The materials that replicate the properties of ordinary Portland cement (OPC) are known as secondary cementitious materials (SCMs). They have smaller carbon dioxide emission rates [5]. SCMs are generally waste materials and byproducts of different industries. These materials become sources of various types of pollution if not discarded or utilized properly. SCMs can be used in different proportions and combinations to replicate the desired properties of OPC. Some of the SCMs are fly ash (FA), corn cob ash (CCA), sugarcane bagasse ash (SCBA), rice husk ash (RHA), ground granulated blast furnace slag (GGBFS), etc [6–9]. RHA is one of the SCMs obtained from the agricultural waste of rice crop. Rice grains are covered in rice husks (RH) which are used as a fuel to boil paddy in rice mills. RHA is obtained after utilizing rice husks as fuel. It contains more than 90 percent silica and can be used successfully as an SCM to synthesize concrete [10]. An illustration of the chemical composition of RHA is shown in Figure 1 [11]. Ameri et al. [12] conducted a research on concrete containing RHA. It was found that concrete containing RHA showed a vigorous increase in early compressive strength. However, by increasing the RHA content by more than 15 percent, the compressive strength was decreased. This is attributed to the excess amount of silica present in RHA which remains unreacted. The compressive strength of concrete with RHA as an SCM was 9, 12, 13, and 16 percent higher than that of control mix. Similarly, Chao Lung et al. [13] incorporated RHA in concrete and concluded that concrete containing RHA showed a strength 1.2 to 1.5 times greater than that of the control mix. Chindaprasirt et al. [14] tested the concrete containing RHA for sulphate attack resistance and reported that concrete containing RHA proved to be highly effective against sulphate attack. It was reported by Thomas et al. [15] in a review paper that concrete containing RHA has a dense microstructure, so it can be used to reduce the water absorption of concrete by up to 30 percent. Rattanachu et al. [16] conducted research in which grounded RHA was used with steel reinforcements. It was observed that the use of RHA in the presence of steel resisted the corrosion of steel due to the fine structure of RHA. Thus, several studies have been made on environmental impact of RHA. They are reported in Table 1:

Table 1. Environmental impact of rice husk ash (RHA).

Material in Which RHA Is Used	Results	Reference
Concrete	Utilization of RHA results in reduction of global warming potential (GWP)	[17]
Mortar	Use of RHA results in reduction of harmful environmental impacts	[18]
Concrete	RHA aids in reducing carbon footprint of concrete	[19]
Concrete blocks	Utilization of RHA shows positive environmental results	[20]

Hence, RHA can be utilized successfully as a cementitious material. RHA does not produce excessive amount of carbon dioxide. It can be used as a structural concrete. Not only does it contribute towards the strength of the concrete but also towards the long term durability properties of concrete [21].

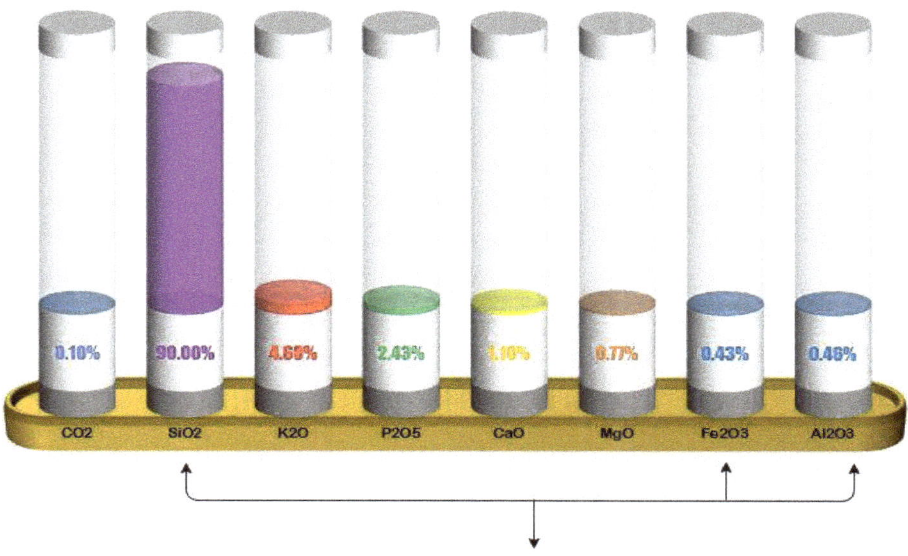

Figure 1. Chemical composition of RHA.

The rate of environmental deterioration does not allow one to spend an extensive amount of time on research and development of RHA blended concrete (RBC). Consequently, extensive lab works cannot be carried out on RBC. Along with that there is always an uncertainty regarding the mix design of RBC. This is due to the hygroscopic nature of RHA. Therefore, to predict the properties of different SCMs, artificial intelligence (AI) is being used throughout the globe. AI is used by different researchers to assess and predict the strength of concrete mixes. Table 2 lists the different previous studies conducted on SCMs to predict different properties. Different techniques such as artificial neural networks (ANN), LR, adaptive neuro-fuzzy inference system (ANFIS), and MNLR are used to successfully model and predict different properties of materials [22,23].

As AI research depends on mathematical modelling and parameters, it is a complex programming work and needs great optimization and care. Therefore, four programming techniques are being used to predict the compressive strength of RHA-based concrete in this research. These techniques are ANFIS, ANN, MNLR, and LR. To achieve the targeted accuracy and to cater the complexity of programming these four techniques will be compared with each other. A vast database of peer reviewed literature is used to model the prediction of compressive strength.

Table 2. Some recent studies using AI.

Material Used	No. of Data Points	Property Predicted	Modelling Technique Used	Reference
SCBA	65	Compressive strength	GEP, Multiple Linear Regression (MLR), Multiple Nonlinear Regression (MNLR)	[22]
Silica fume (SF) and zeolite	18	Compressive Strength	ANN	[23]
Recycled concrete aggregate	17	Compressive strength	ANN, Response Surface Methodology (RSM)	[24]
Recycled rubber concrete	72	Compressive strength	ANN, MNLR, ANFIS, Support vector machine (SVM)	[25]
Cellular concrete	99	Compressive strength	Backpropagation Neural Network (BPNN)	[26]
Fly ash (FA) and blast furnace slag (BFS)	135	Compressive strength	ANN	[27]
Foamed concrete	91	Compressive strength	Extreme Learning Machine (ELM)	[28]
Recycled aggregates	74	Compressive strength	ANN Convolutional Neural Network	[29]
Rubberized concrete	112	Compressive strength	ANN	[30]
Steel fiber added lightweight concrete	126	Compressive strength	ANN	[31]
Fiber reinforced polymer concrete (FRP)	98	Shear strength	ANN	[32]
FRP	84	Shear strength	ANN	[33]
High strength concrete (HSC)	187	Compressive strength	ANN	[34]

2. Data Collection

To predict the compressive strength (CS) of RHA, mathematical models are developed using a dataset of 192 data points from the vast literature review and existing studies on machine learning [12,13,35–39]. These data were collected through google Scopus. The constituents of concrete include RHA, OPC, aggregates, super plasticizer (SP), and water. The type of cement and curing methodology used in all the mixes is same. The CS of cubic specimens is converted into CS of cylinders by using 0.8 as a factor (as per BS 1881: Part 120:1983). The only output parameter in this study is compressive strength. The input parameters consist of main variables such as percentage of SP, curing age (CA), quantity of water used (W), amount of OPC (OPCP), quantity of aggregates (AGG), and amount of RHA (RHAP). Moreover, the description of collected data and its statistics are given in Table 3.

Table 3. Statistical analysis of input data.

Parameters	Mean	Standard Deviation	Kurtosis	Skewness	Minimum	Maximum
Input parameters						
Age (days)	34.57	33.52	−1.02	0.75	1	90
Plain cement (kg/m^3)	409.02	105.47	3.66	1.55	249	783
RHA (kg/m^3)	62.33	41.55	0.07	0.44	0	171
Water (kg/m^3)	193.54	31.93	−0.74	−0.42	120	238
Super plasticizer (kg/m^3)	3.34	3.52	−0.82	0.69	0	11.25
Aggregates (kg/m^3)	1621.51	267.77	−0.27	−0.74	1040	1970
Response						
Experimental compressive strength (MPa)	48.14	17.54	0.75	0.83	16	104.1

3. Methodology

The methodology section provides a brief detail about the approaches being made to determine the CS of concrete mathematically as shown in Figure 2. First, the AI processes used in this research are explained. The results obtained from AI data processing techniques are assessed for validity by different statistical parameters.

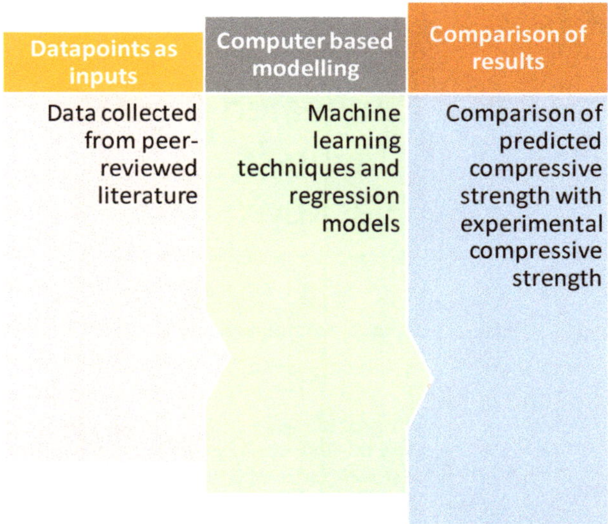

Figure 2. Adopted methodology for study.

3.1. Modeling Techniques

Machine learning-based modeling has been used in the past to predict the different mechanical properties of materials [40–42]. These types of modeling techniques can be utilized to develop models for prediction of a property of material. They do not need any knowledge of the rudimentary experimental processes. This section of paper provides a brief introduction of the predictive models used in this study. These models are as follow:

3.1.1. ANN

ANN is an artificial data analyzing technique. It is inspired by the learning capability of human brain. The most widely used type of ANN is feedforward back propagation (FFBP). As evident from Figure 3, an FFBP consists of at least three layers, namely, the input layer, hidden layer, and output layer. The nodes of these layers are connected in a proper sequence along with some weights. The input layer nodes do not perform any function on input data. Their function is to just receive the data from outside. It is a hidden layer where data are biased, weighted, and summed up. These processed data are then sent out to the output layers [43,44].

There are basically two types of FFBP, namely, single layer perceptron (SLP) and multiple layers perceptron (MLP). Both types of FFBP have their own advantages and disadvantages. Alhough the SLP is simple and easy to use, it cannot handle nonlinear relations. On the other hand, MLP are complex, but they can be applied to nonlinear relations.

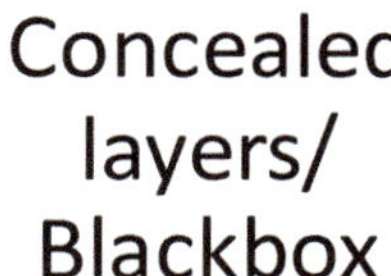

Figure 3. Illustration of ANN.

Mathematically, an MLP operates in following way:
Step 1: The inputs are summed and weighted as

$$s_j = \sum_{i=1}^{n} \omega_{ij} I_i + bj \; j \; = \; 1, 2, 3, \ldots\ldots, h \tag{1}$$

where n = number of total inputs, I_i = current input number, ω_{ij} = weight between the previous layer, and the jth neuron and b are used to define the termination of process.

Step 2: This step includes an activation function. There are various types of activation functions such as sigmoid, ramp, and Gaussian functions. However, this research utilizes sigmoid function which is defined as

$$s_j = \frac{1}{1 + e^{-s_j}} \; j \; = \; 1, 2, 3, \ldots\ldots\ldots, h \tag{2}$$

Step 3: This represents the final outputs. The final outputs depend on the outputs calculated by hidden nodes. The final outcome can be expressed as

$$O_k = \sum_{j=1}^{h} \left(\omega_{jk} \cdot s_j \right) + b'_k, \; k \; = \; 1, 2, \ldots, m \tag{3}$$

$$O_k = sigmoid \, (O_k) = \frac{1}{(1 + e^{-O_k})}, \; k \; = \; 1, 2, \ldots, m \tag{4}$$

In above equation, ω_{jk} = weighted connection between kth output node to jth hidden node. Similarly, b'_k = bias output of kth output node.

In this research, 70 percent of the data points are selected randomly for the training of data, and 30 percent for validation.

3.1.2. ANFIS

ANFIS is a technique that utilizes the combined effect of artificial neural networks and fuzzy logic [45]. Figure 4 shows a brief illustration of the ANFIS technique. An artificial neural network is used to minimize the chances of error in the output data. Thus, the fuzzy logic is implied to demonstrate the expert knowledge [42]. Fuzzy logic rules are applied as if-then while mathematically programming for the desired input and output datasets. An ANFIS model consists of five layers normally. These are (1) fuzzification, (2) set of rules, (3) normalization, (4) defuzzification, (5) aggregation.

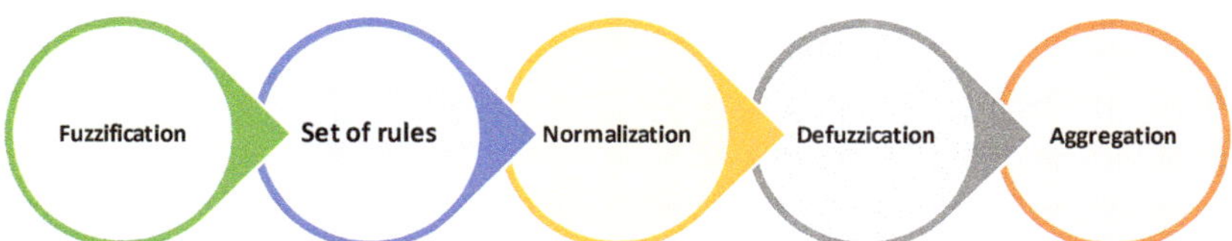

Figure 4. Illustration of adaptive neuro-fuzzy inference system (ANFIS).

Layer 1 is the fuzzification layer. It contains all the function members of the input variables. A Gaussian function is used in this layer to predict the outcome. Mathematically, it can be expressed as

$$\mu_{ui}\left(x\right) = \exp\left[-\frac{(x-a_i)}{2\,\varepsilon_i^{\,2}}\right] \tag{5}$$

where a_i and ε_i are parameters of a function membership.

Layer 2 contains nodes which send the output by multiplying the input by certain weightages. This layer utilizes the fuzzy AND logic by using the equation listed below:

$$w_i = \mu_{ui}\left(x\right) \times \mu_{vi}\left(y\right) \tag{6}$$

Layer 3 has an aim to normalize the data. It normalizes the functions of membership. It calculates the ratios between different firing strength using the following expression:

$$\overline{w} = \frac{w_i}{\sum_i w_i} \tag{7}$$

Layer 4 is known as the defuzzification layer. It contains nodes that conclude the fuzzy logic rules. This layer contains square nodes, which can be expressed by following function:

$$\overline{w}_i f_i = w_i \times \left(m_i x + n_i y + r_i\right) \tag{8}$$

where m_i, n_i, and r_i are linear parameters.

Layer 5 has a function of aggregation. It sums up the previous layers and presents the final output mathematically as follows:

$$\sum_i \overline{w} f_i = \frac{\sum_i w_i f_i}{\sum_i w_i} \tag{9}$$

All the data points are used for the training of data.

Off the shelf functionality of MATLAB is used for ANN and ANFIS techniques in this research.

3.1.3. MNLR

MNLR is a technique which is used to model a random nonlinear relationship between the dependent and independent variables. The following equation represents the MNLR [41]:

$$Y = a + \beta_1 X_i + \beta_2 X_j + \beta_3 X_i^2 + \beta_4 X_j^2 + \ldots \ldots + \beta_k X_i X_j \tag{10}$$

where a = intercept, β = slope or coefficient, K = number of observations. The above equation can make an estimate for the value of Y for each value of X.

3.1.4. LR

LR is a technique in which there is linear relationship between the dependent and independent variables. It can be represented mathematically as follows:

$$Y = a + \beta_1 X_1 + \beta_2 X_2 + \beta_3 X_3 + \ldots \ldots + \beta_i X_i \tag{11}$$

The above equation can be utilized to find values of Y for each input value X.

In above equations of MNLR and LR, "Y" stands for compressive strength of RHA. Similarly, the values of "X" represents inputs which are age, water content, RHA content, SP content, and the percentage of aggregates.

The models are developed in Microsoft Excel by authors for MNLR and LR techniques using the above equations.

4. Results

A total of 192 data points are used for all the models and techniques. A total of 134 data points are used for training, and 58 data points for validation. The results of machine learning techniques and regression models are given in Appendix A.

4.1. ANN

Parametric adjustments are made before running the proposed ANN model. These parameters include number of hidden layers, total number of neurons per hidden layer, training function for neural networks, epochs, and maximum number of iterations. These parameters are determined through the hit and trial method in this research. The details of the parametric adjustment are given in Table 4.

Table 4. Parametric adjustment of the developed model.

Parameters	Description
Total number of hidden layers	2
Maximum number of neurons per hidden layer	10
Training function	Levenberg–Marquardt
Epochs	3
Training completed at epoch	2

MATLAB was used to predict the compressive strength of RBC through ANN. ANN gave the results which were closest to the experimental results. The same can be verified through the statistical parameters of ANN.

It is noteworthy that the correlation factor for ANN predicted CS (R^2 = 0.98) is quite high. The prediction result for ANN is plotted in Figure 5.

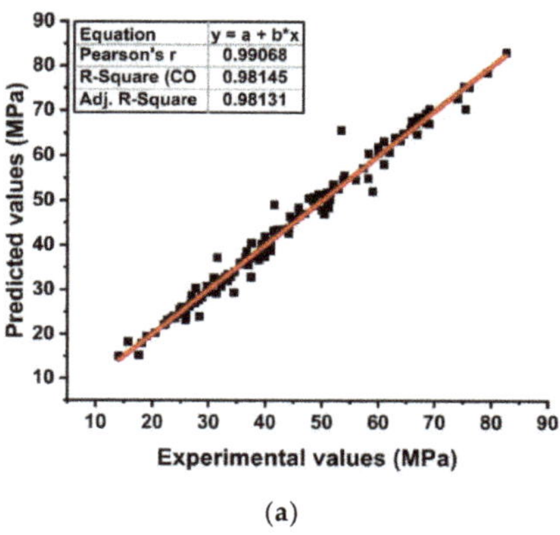

(a)

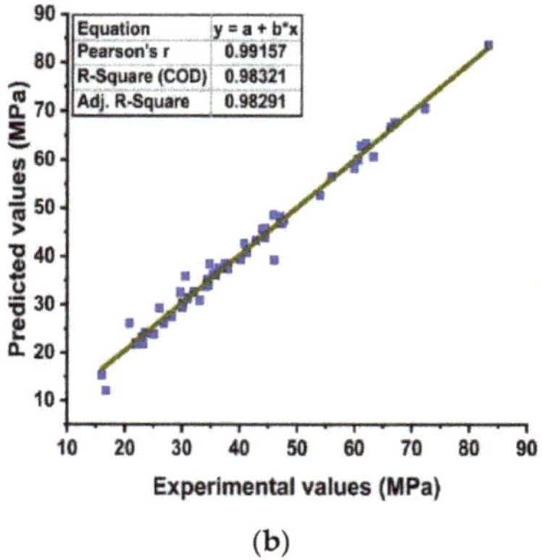

(b)

Figure 5. *Cont.*

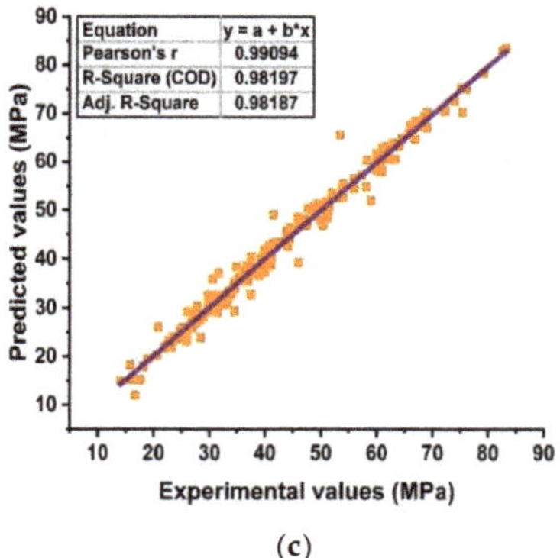

(c)

Figure 5. ANN (**a**) training, (**b**) validation, (**c**) testing.

4.2. ANFIS

Similarly, before training the data on ANFIS, parametric adjustments were made. These included total number of epochs and function used for the activation of ANFIS. The parametric adjustments for ANFIS are given in Table 5.

Table 5. Parametric adjustments for ANFIS.

Parameters	Description
Training function	trimf
Epochs	6
Training completed at epoch	2

MATLAB is used for ANFIS. The correlation factor for ANFIS predicted CS ($R^2 = 0.89$) is also quite high. Figure 6 shows that the predicted results are quite close to the experimental values.

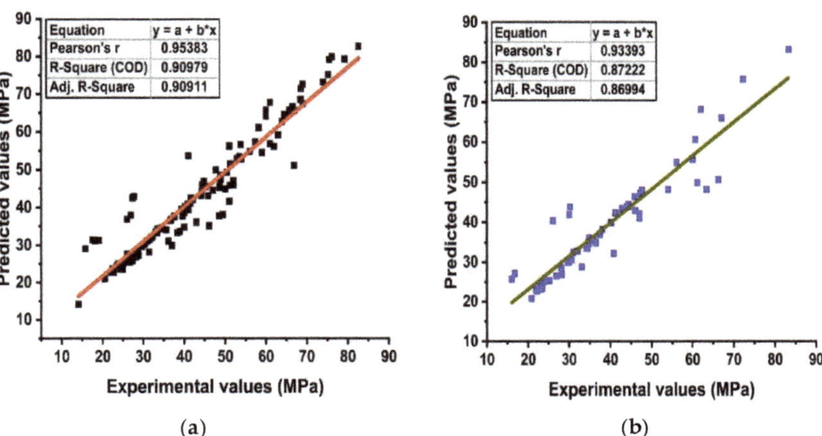

(a)   (b)

Figure 6. *Cont.*

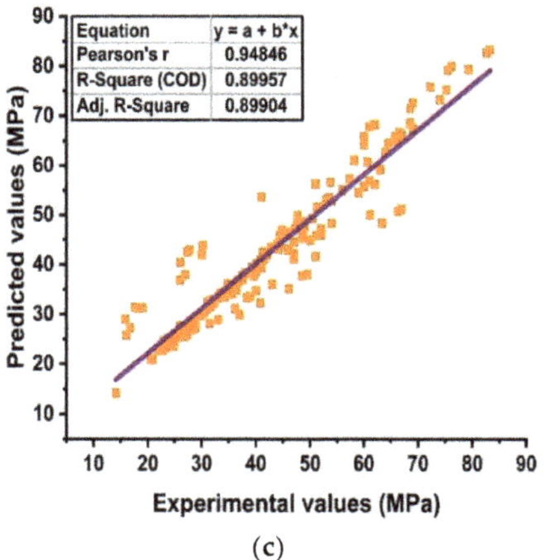

(**c**)

Figure 6. ANFIS (**a**) training, (**b**) validation, (**c**) testing.

Figure 7 shows the rules assigned to ANFIS for the optimum outcome.

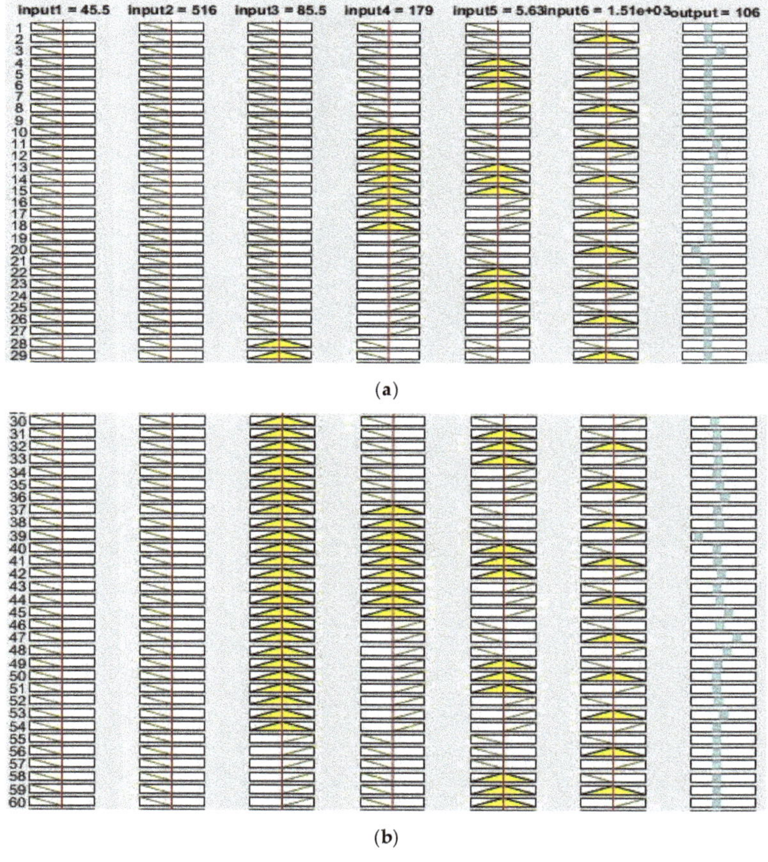

(**a**)

(**b**)

Figure 7. *Cont.*

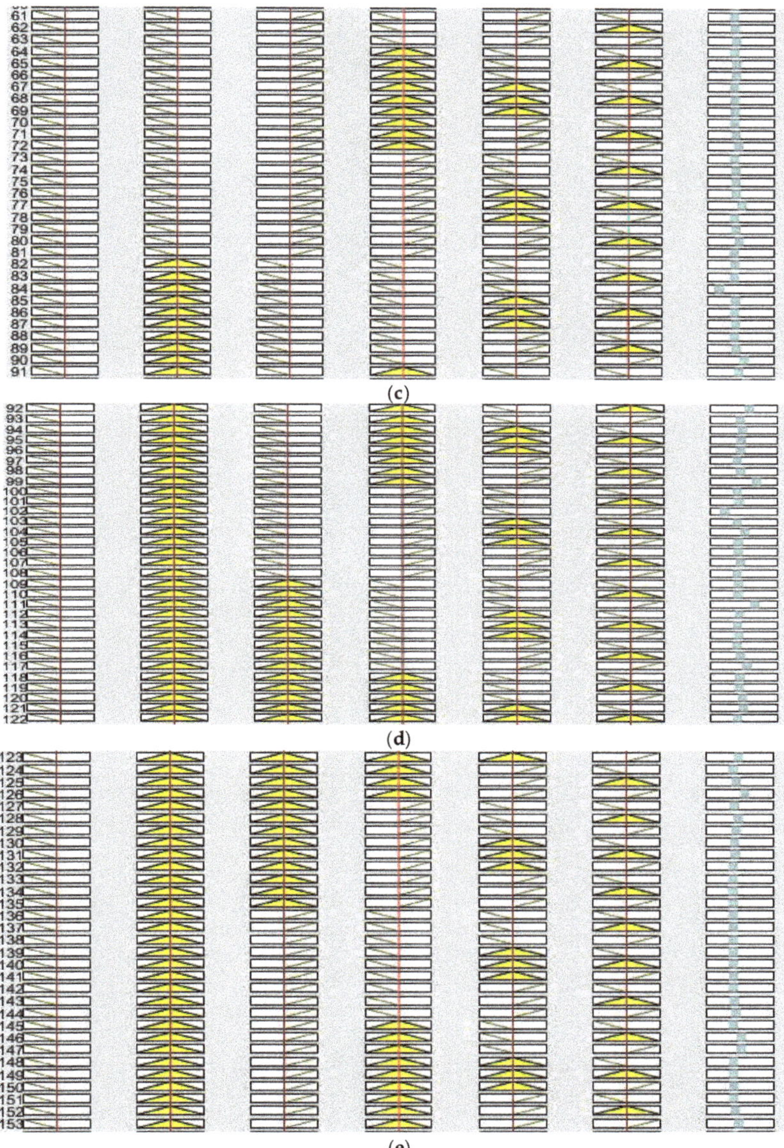

(c)

(d)

(e)

Figure 7. *Cont.*

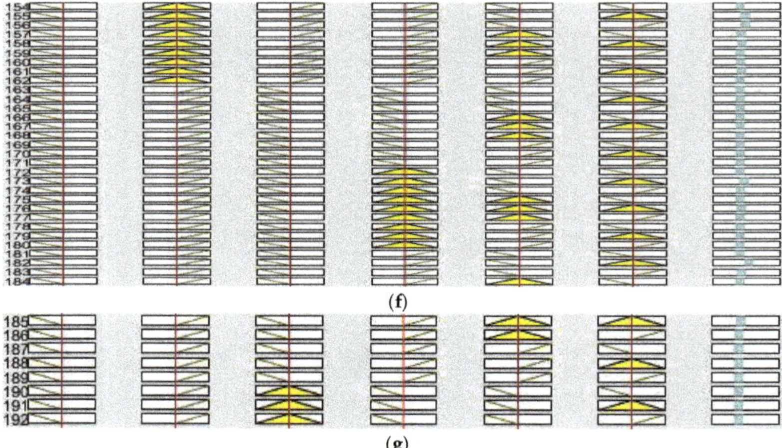

Figure 7. ANFIS modeling rules. (**a**) 1–29, (**b**) 30–60, (**c**) 61–91, (**d**) 92–122, (**e**) 123–153, (**f**) 154–184, (**g**) 185–192.

4.3. MNLR

The results predicted by MNLR were not close to the experimental results. The correlation factor for MNLR predicted CS ($R^2 = 0.70$) confirms the same. The correlation factor for training and validation is 0.75 and 0.69, respectively. The same can be confirmed by the dispersion of points in Figure 8 below.

4.4. LR 40

LR gave the results which were far away from the experimental results. A weak correlation ($R^2 = 0.63$) existed between the experimental and predicted results. The correlation factor for LR training and LR validation is 0.64 and 0.62, respectively. It can be seen in Figure 9 below that points are dispersed.

4.5. Sensitivity and Parametric Analysis

Different variables are used to find the CS of RBC. Sensitivity analysis (SA) is used to determine the relative contribution of these variables to the result. SA is carried out mathematically by using the following Equations:

$$N_i = f_{\max}(x_i) - f_{\min}(x_i) \tag{12}$$

$$SA = \frac{N_i}{\sum_n^{j=1} N_j} \tag{13}$$

where $f_{\max}(x_i)$ is the maximum, and $f_{\min}(x_i)$ is the minimum output of the predictive models, respectively. Thus, i represents the input domain and other input variables that are kept constant. It is obvious from the graphical representation (shown in Figure 10) that the contribution of different input variables on the CS of RBC is same as that in real life.

(a)

(b)

(c)

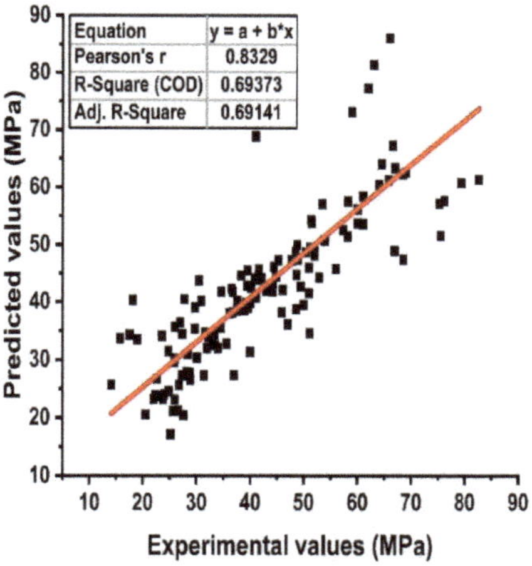

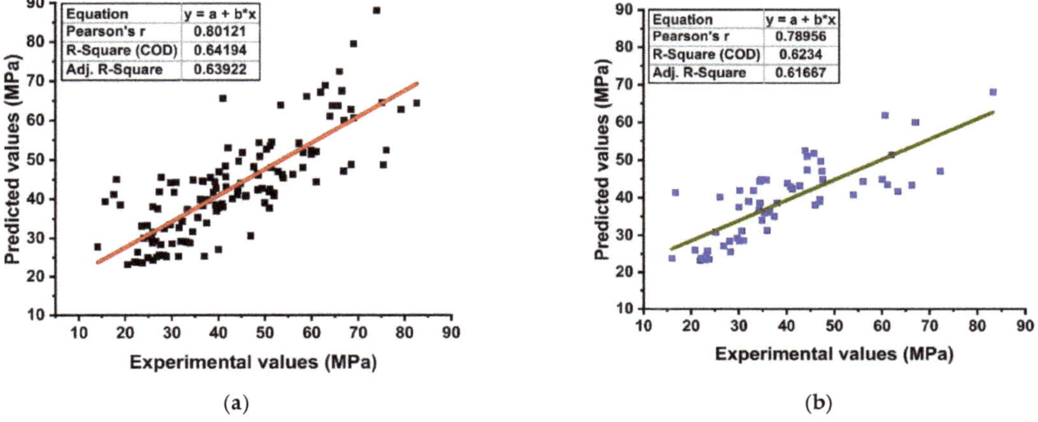

Figure 8. MNLR (**a**) training, (**b**), validation, (**c**) testing.

(a)

(b)

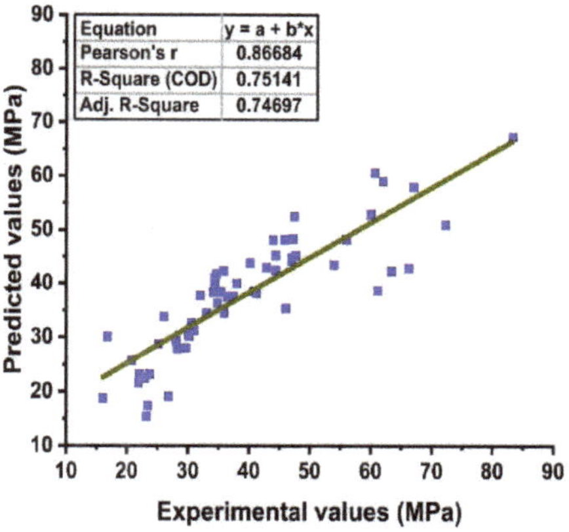

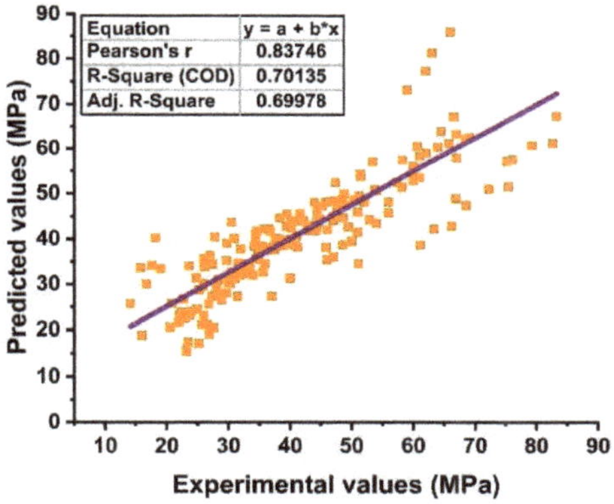

Figure 9. *Cont.*

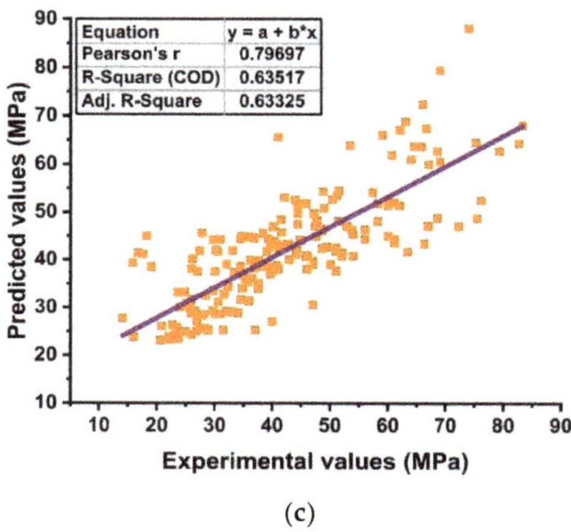

(c)

Figure 9. LR (**a**) training, (**b**) validation, (**c**) testing.

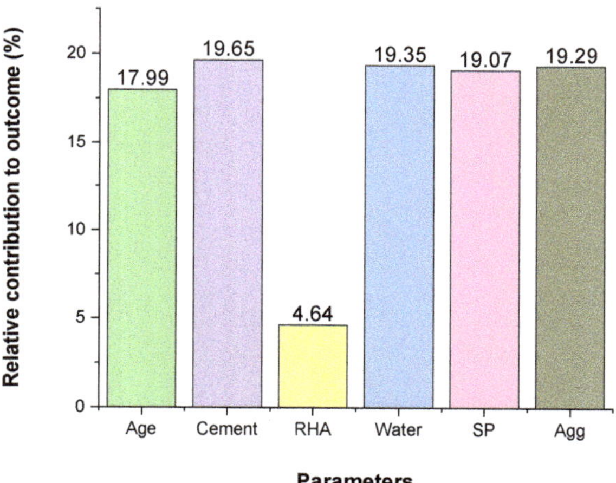

Figure 10. Contribution of inputs to the output.

Along with sensitivity analysis, a parametric analysis (PA) is also carried out. This helps in determining the influence of the input parameters on the output parameter. This shows the trend of CS when all the input variables are kept constant at their mean value except one input. The change in CS is recorded when one input variable is varied from its minimum value to its maximum value. All the results of PA are shown in Figure 11 below.

The sublots in front of each graph in Figure 11 represent the constant parameters of parametric analysis for each input. The literature used for obtaining experimental values includes [12,35–39]. It can be observed from the results that when water is increased from a certain limit, a reduction in CS occurs. This is also obvious from previous studies. De Sensale [39] conducted research in which a water to cement ratio (w/c) of 0.4 gave more CS than w/c of 0.5. RHAP contributes towards the enhancement of strength, but when RHAP is increased by 30 percent, it results in decrease of compressive strength. This is due to the fact that, as discussed in Section 1, RHA contains 90 percent silica. By increasing the RHA percentage, the amount of silica is also increased. This silica remains unreacted by increment of RHA and results in reduced CS of RBC [37].

It can be seen from the above results that the regression models did not show satisfactory results as compared to the machine learning processes. This is due to certain limitations in regression models, such as pre-defined equations that cannot learn the relationship between input variables and the function properly. Whereas, machine learning has efficiently predicted the relationship between input and output variables. The machine learning techniques gave results closer to the experimental values.

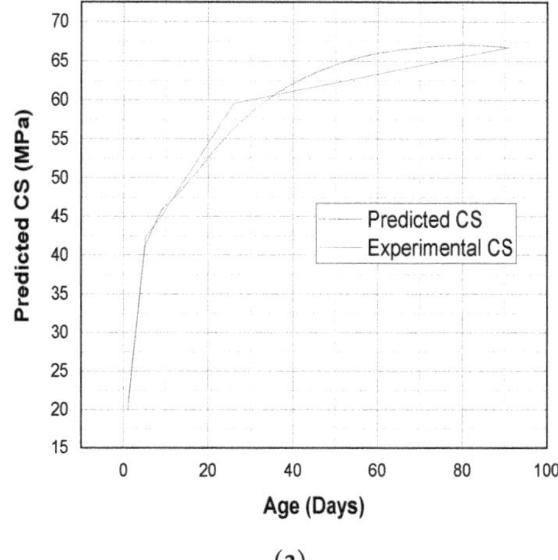

Cement (kg/m³)	RHA (kg/m³)	W (kg/m³)	SP (kg/m³)	A (kg/m³)
409	60	192	3	1700

(a)

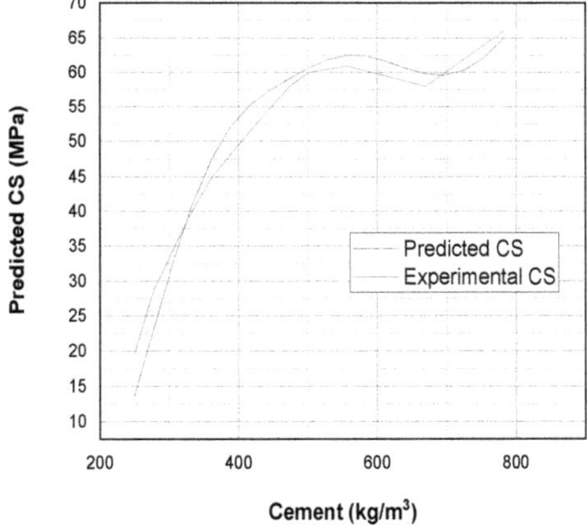

Age (day)	RHA (kg/m³)	W (kg/m³)	SP (kg/m³)	A (kg/m³)
35	24.9	192	3	1700

(b)

Figure 11. *Cont.*

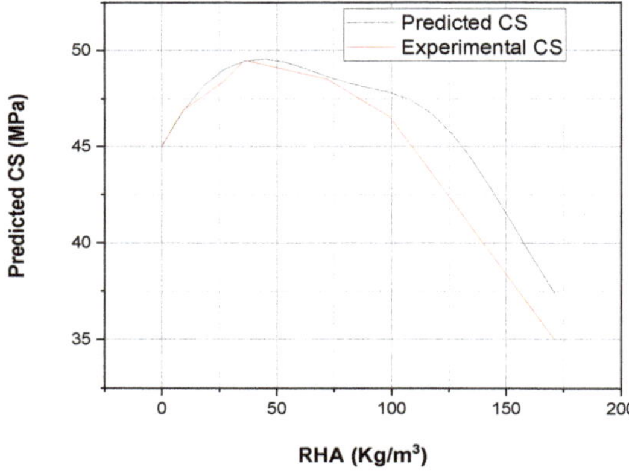

Age (day)	Cement (kg/m³)	W (kg/m³)	SP (kg/m³)	A (kg/m³)
35	409	192	3	1700

(**c**)

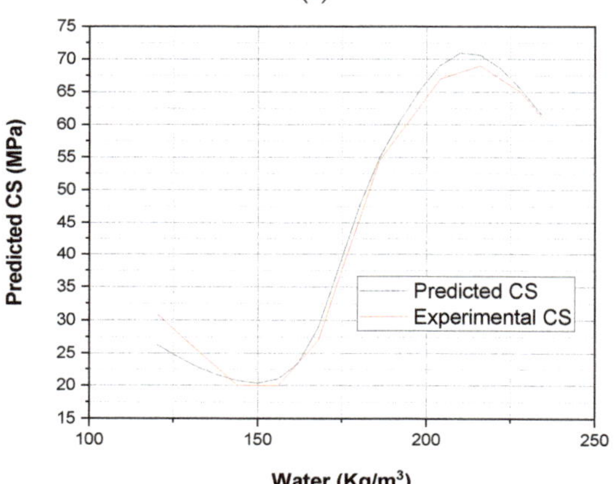

Age (day)	Cement (kg/m³)	RHA (kg/m³)	SP (kg/m³)	A (kg/m³)
35	409	60	3	1700

(**d**)

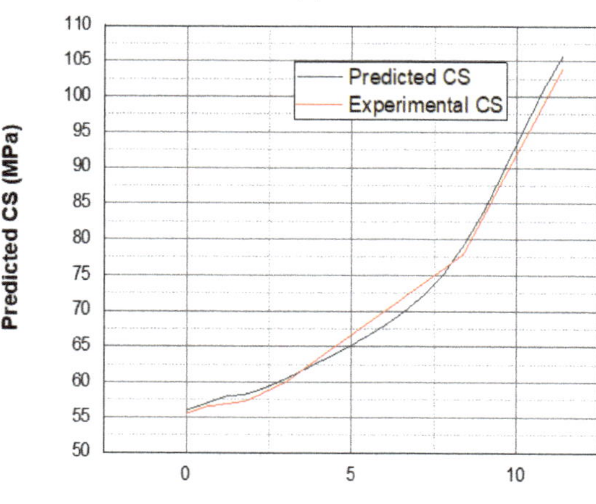

Age (day)	Cement (kg/m³)	RHA (kg/m³)	W (kg/m³)	A (kg/m³)
35	409	60	192	1700

(**e**)

Figure 11. *Cont.*

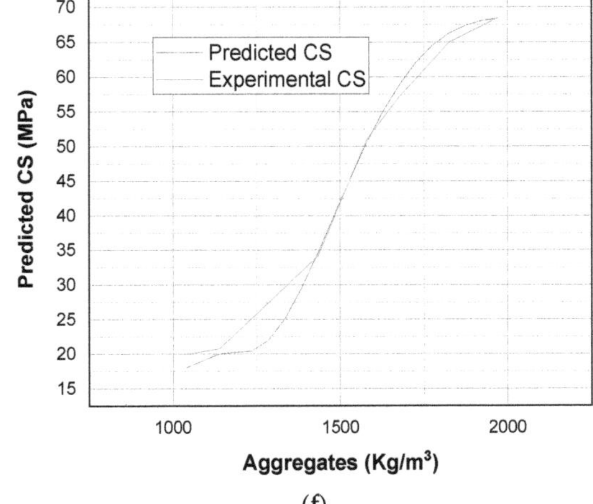

Age (day)	Cement (kg/m³)	RHA (kg/m³)	W (kg/m³)	SP (kg/m³)
35	409	60	192	3

(f)

Figure 11. Parametric analysis of inputs. (**a**) Age, (**b**) cement, (**c**) RHA, (**d**) water content, (**e**) superplasticizer, (**f**) aggregates.

5. Conclusions

Different models for prediction of CS of RBC are developed in this study. The models developed in this study are based on wide range of data which consist of different parameters demonstrated by experimental studies that are available in the literature. The models considered the most influential parameters on CS as inputs. The results obtained in this research are closer to the experimental research. The following conclusions can be drawn from the obtained results:

1. The PA has shown that the input parameters used in this research are effectively utilized by the model to predict the CS. Moreover, the statistical parameter R^2 shows the accuracy of the data used for the training and validation of different models.
2. The R^2 for the predicted strengths of ANN, ANFIS, MNLR, and LR is 0.98, 0.89, 0.70, and 0.63, respectively.
3. It is evident by the comparison of ANN and ANFIS with the regression models that both ANN and ANFIS have a high command on prediction of CS of RBC. Therefore, they are suitable for the predesign of RBC.
4. The proposed models can provide the basis for using RBC in different structures rather than discarding it.

Concrete containing RHA has a great potential to replace OPC concrete. It is recommended that extensive research be carried out by including more parameters. These parameters should include temperature, corrosion, and resistance to chlorine and acid attacks. Other advanced programming techniques such as an M5P tree and gene expression programming can be used to make further predictions.

Author Contributions: A.I.: Conceptualization, data analysis, writing original draft and preparation, N.B.K.: Formal analysis and modelling, S.K.-u.-R.: Supervision, review and editing, M.F.J.: Investigation and review, F.A.: Methodology, review and editing, R.A.: Software, validation check and supervision, H.A.: Review, editing and supervision, A.M.: Validation, review, and supervision. All authors have read and agreed to the published version of the manuscript.

Funding: This research received no external funding.

Data Availability Statement: Not applicable.

Acknowledgments: Amir Mosavi would like to thank Alexander von Humboldt Foundation.

Conflicts of Interest: The authors declare no conflict of interest.

Abbreviations

Nomenclature	Definition
AGG	Amount of aggregates
AI	Artificial intelligence
ANFIS	Adaptive neuro-fuzzy inference system
ANN	Artificial neural network
CA	Curing age
CCA	Corn cob ash
CS	Compressive strength
FA	Fly ash
FFBP	Feed forward back propagation
GGBFS	Ground granulated blast furnace slag
GWP	Global warming potential
LR	Linear regression
MLP	Multi layer perceptron
MNLR	Multiple nonlinear regression
OPC	Ordinary Portland cement
OPCP	Amount of OPC
RBC	RHA blended concrete
RH	Rice Husks
RHA	Rice husk ash
RHAP	Amount of RHA
SCBA	Sugarcane baggase ash
SCM	Secondary cementitious material
SLP	Single layer perceptron
SP	Superplasticizer
W	Water used

Appendix A

Table A1. Compressive strength (CS) (MPa) results obtained through different models.

Experimental	ANN Prediction	ANFIS Prediction	LR Prediction	MNLR
18.16	17.91	31.12	45.00	40.31
16.72	12.06	27.21	41.43	30.10
17.6	15.19	31.43	41.21	34.24
15.76	18.26	29.00	39.39	33.65
27.76	30.28	42.86	45.59	40.46
30.24	30.12	43.81	42.02	30.25
27.36	27.30	42.54	41.88	34.41
26.08	29.29	40.42	40.21	33.86
38.32	37.24	33.33	45.51	44.64
33.04	30.89	28.89	41.93	34.44
38.96	36.67	33.62	41.71	38.58
36.16	35.72	31.05	39.89	37.99
14.08	14.95	14.11	27.75	25.72
48.64	50.77	45.60	46.09	44.79
51.12	49.99	45.85	42.52	34.58
48.56	50.10	45.25	42.38	38.75
45.84	48.30	43.04	40.71	38.20
48.48	49.97	37.76	46.52	48.79
40.8	42.73	32.25	42.95	38.59
49.44	51.43	38.02	42.72	42.73
24	23.64	24.81	30.02	24.01

Table A1. *Cont.*

Experimental	ANN Prediction	ANFIS Prediction	LR Prediction	MNLR
25.2	23.86	25.39	30.85	28.80
26	23.83	27.60	31.67	30.22
28.4	23.95	28.54	32.50	31.01
24.8	24.43	25.31	33.33	31.46
22.08	22.04	22.70	23.90	23.28
23.76	24.07	25.01	23.52	23.27
20.56	20.25	20.97	23.16	20.56
46.08	46.67	35.13	40.90	42.14
66.88	64.77	51.09	47.10	48.94
25.92	23.20	25.88	28.85	29.66
61.12	62.83	49.95	43.53	38.73
66.24	66.69	50.67	43.39	42.90
63.36	60.64	48.28	41.73	42.35
28.24	27.61	26.93	25.55	27.87
28.88	28.21	27.26	25.17	26.58
58.24	60.42	60.99	51.83	57.53
47.68	50.46	49.89	48.25	47.32
58.16	54.92	61.09	48.03	51.46
35.6	36.04	34.11	35.33	32.74
36.4	37.66	34.88	36.16	37.53
39.6	38.16	37.72	36.98	38.95
40	37.39	38.61	37.81	39.75
34.4	35.20	33.55	38.64	40.19
29.68	32.60	29.98	29.33	28.05
33.44	33.46	34.23	28.96	32.49
30.08	29.41	30.17	28.59	30.37
53.76	54.72	56.59	46.21	50.87
76.16	75.21	79.90	52.41	57.68
68.56	67.37	71.43	48.84	47.47
75.44	70.38	79.14	48.70	51.63
32.24	30.73	32.25	34.34	33.92
37.52	32.79	37.51	33.94	38.46
72.24	70.61	75.79	47.04	51.08
41.2	40.88	42.37	42.41	38.23
42.8	43.35	43.43	43.24	43.02
44.8	45.59	46.88	44.06	44.44
47.6	47.49	48.07	44.89	45.24
41.6	49.06	42.34	45.72	45.68
66.56	66.98	66.55	67.50	67.30
53.44	65.68	53.44	63.93	57.09
65.76	66.02	65.76	63.71	61.23
60.64	60.01	60.65	61.89	60.64
34.64	34.06	34.65	44.88	41.78
36.8	36.36	36.85	44.51	41.77
29.76	30.66	29.74	44.14	39.06
44.4	45.77	44.01	51.01	42.51
45.2	45.74	45.49	51.83	47.30
50.4	46.92	49.25	52.66	48.72
51.2	48.40	51.52	53.49	49.52
48.8	49.83	48.52	54.31	49.96
47.2	46.94	47.20	49.75	48.38
83.28	83.60	83.29	68.09	67.45
75.2	75.35	75.20	64.52	57.24
82.64	83.02	82.64	64.38	61.40
79.28	78.48	79.27	62.71	60.85
39.5	39.69	39.50	42.75	45.55
30.5	30.47	30.50	41.75	43.76

Table A1. *Cont.*

Experimental	ANN Prediction	ANFIS Prediction	LR Prediction	MNLR
29.7	29.35	29.70	33.47	35.31
23.6	23.61	23.60	33.19	34.09
22.7	23.21	22.70	26.39	26.79
20.8	26.18	20.80	26.05	25.81
51.4	50.15	51.40	48.06	54.28
47.4	47.08	47.40	47.06	52.50
40.8	40.61	40.80	38.78	44.04
39.4	40.25	39.40	38.50	42.82
34.5	29.33	34.50	31.70	35.53
35.9	36.13	35.90	31.36	34.54
64.5	64.94	64.50	63.74	64.05
68.5	69.42	68.50	62.74	62.27
51.5	51.89	51.50	54.46	53.81
57.3	57.27	57.30	54.17	52.59
44.4	43.98	44.40	47.38	45.30
52.9	52.55	52.90	47.04	44.31
25.2	26.10	25.23	30.56	17.15
25.68	25.92	25.73	30.00	21.15
26.64	26.24	26.88	29.25	21.23
27.6	26.90	27.46	28.31	20.45
26.88	26.18	26.53	27.20	19.16
23.44	24.15	23.45	25.90	17.47
23.2	21.95	23.16	24.42	15.46
33.36	31.93	33.37	37.39	34.36
34.16	33.71	33.61	36.82	38.37
35.36	36.30	35.56	36.07	38.45
37.44	38.46	36.98	35.14	37.67
34.8	38.45	36.07	34.03	36.38
31.6	37.15	31.50	32.73	34.69
30.56	35.91	30.67	31.25	32.68
39.28	37.51	39.44	44.47	39.86
40.16	39.41	39.95	43.90	43.86
41.68	42.31	42.42	43.15	43.94
44.24	43.81	44.16	42.22	43.16
44.16	42.63	43.10	41.10	41.87
37.6	40.41	37.76	39.81	40.18
36.72	38.56	36.63	38.33	38.17
42.08	43.41	41.98	53.07	44.13
43.92	45.62	43.95	52.50	48.14
45.84	48.62	46.40	51.75	48.21
48.96	48.95	47.45	50.82	47.44
44.4	46.41	45.86	49.70	46.15
41.52	43.25	41.14	48.40	44.46
40.16	40.79	40.18	46.92	42.45
41	40.43	53.59	65.61	68.86
30	30.26	41.92	37.55	31.04
27	28.62	37.95	37.54	36.42
26	26.13	36.87	38.13	35.68
19	19.53	31.29	38.54	33.46
16	15.26	25.75	23.81	18.84
59	52.03	54.44	66.11	73.19
46	39.32	42.89	38.05	35.37
41	38.71	39.35	38.05	40.75
38	37.54	38.27	38.64	40.02
32	32.63	32.88	39.04	37.80
26	25.29	27.13	24.32	23.18
62	60.76	56.13	67.12	77.34
50	47.89	44.83	39.06	39.53
47	48.19	42.14	39.06	44.90

Table A1. *Cont.*

Experimental	ANN Prediction	ANFIS Prediction	LR Prediction	MNLR
47	47.53	41.06	39.65	44.17
43	43.50	36.06	40.05	41.95
37	35.53	29.91	25.33	27.33
63	64.01	59.09	68.89	81.37
54	52.69	48.22	40.83	43.56
52	53.58	47.03	40.83	48.93
52	52.95	45.93	41.42	48.20
51	48.99	41.63	41.82	45.98
40	41.96	34.77	27.10	31.36
66	67.67	65.01	72.43	86.08
56	56.63	55.01	44.37	48.26
61	58.11	56.80	44.37	53.64
60	58.28	55.69	44.96	52.90
54	55.56	52.77	45.36	50.68
47	47.06	44.48	30.64	36.06
69	70.31	72.59	79.51	91.57
60	61.89	64.09	51.45	53.75
62	63.30	68.19	51.45	59.13
61	63.18	67.77	52.04	58.39
60	60.58	65.74	52.44	56.17
51	51.74	56.18	37.72	41.55
74	72.72	73.15	88.11	95.85
67	67.64	66.03	60.05	58.03
67	68.51	65.54	60.04	63.41
69	67.15	67.40	60.63	62.67
64	63.34	62.65	61.04	60.45
56	54.59	54.77	46.32	45.83
22.08	22.04	22.70	23.90	23.28
22.4	22.26	23.66	23.78	23.84
23.44	23.49	24.42	23.65	23.75
23.76	24.07	25.01	23.52	23.27
22.96	23.29	24.63	23.40	22.55
21.92	21.81	23.13	23.28	21.64
20.56	20.25	20.97	23.16	20.56
27.36	27.33	25.81	25.67	27.31
28.24	27.61	26.93	25.55	27.87
28.8	28.72	27.69	25.42	27.78
31.44	29.08	28.12	25.29	27.30
28.88	28.21	27.26	25.17	26.58
26.8	26.79	25.47	25.05	25.67
24.88	25.41	23.52	24.93	24.59
32	32.36	32.04	29.21	32.01
33.04	32.86	33.46	29.09	32.58
33.44	33.46	34.23	28.96	32.49
34	32.88	34.33	28.83	32.00
31.04	31.26	32.52	28.71	31.28
30.08	29.41	30.17	28.59	30.37
28.08	27.85	28.61	28.47	29.29
34.64	34.06	34.65	44.88	41.78
35.84	36.36	35.81	44.76	42.35
36.56	37.27	36.54	44.64	42.26
36.8	36.36	36.85	44.51	41.77
34.4	34.58	34.26	44.39	41.05
30.96	32.60	31.06	44.26	40.14
29.76	30.66	29.74	44.14	39.06

References

1. Chen, C.; Habert, G.; Bouzidi, Y.; Jullien, A. Environmental impact of cement production: Detail of the different processes and cement plant variability evaluation. *J. Clean. Prod.* **2010**, *18*, 478–485. [CrossRef]

2. Farooq, F.; Ahmed, W.; Akbar, A.; Aslam, F.; Alyousef, R. Predictive modeling for sustainable high-performance concrete from industrial wastes: A comparison and optimization of models using ensemble learners. *J. Clean. Prod.* **2021**, *292*, 126032. [CrossRef]
3. Mahlia, T.M.I. Emissions from electricity generation in Malaysia. *Renew. Energy* **2002**, *27*, 293–300. [CrossRef]
4. Akbar, A.; Farooq, F.; Shafique, M.; Aslam, F.; Alyousef, R.; Alabduljabbar, H. Sugarcane bagasse ash-based engineered geopolymer mortar incorporating propylene fibers. *J. Build. Eng.* **2021**, *33*, 101492. [CrossRef]
5. Mehta, A.; Siddique, R. An overview of geopolymers derived from industrial by-products. *Constr. Build. Mater.* **2016**, *127*, 183–198. [CrossRef]
6. Tosti, L.; van Zomeren, A.; Pels, J.R.; Damgaard, A.; Comans, R.N.J. Life cycle assessment of the reuse of fly ash from biomass combustion as secondary cementitious material in cement products. *J. Clean. Prod.* **2020**, *245*, 118937. [CrossRef]
7. Zain, M.F.M.; Islam, M.N.; Mahmud, F.; Jamil, M. Production of rice husk ash for use in concrete as a supplementary cementitious material. *Constr. Build. Mater.* **2011**, *25*, 798–805. [CrossRef]
8. Adesanya, D.A.; Raheem, A.A. Development of corn cob ash blended cement. *Constr. Build. Mater.* **2009**, *23*, 347–352. [CrossRef]
9. Crossin, E. The greenhouse gas implications of using ground granulated blast furnace slag as a cement substitute. *J. Clean. Prod.* **2015**, *95*, 101–108. [CrossRef]
10. Khan, K.; Ullah, M.F.; Shahzada, K.; Amin, M.N.; Bibi, T.; Wahab, N.; Aljaafari, A. Effective use of micro-silica extracted from rice husk ash for the production of high-performance and sustainable cement mortar. *Constr. Build. Mater.* **2020**, *258*, 119589. [CrossRef]
11. Parveen, S.; Pham, T.M. Enhanced properties of high-silica rice husk ash-based geopolymer paste by incorporating basalt fibers. *Constr. Build. Mater.* **2020**, *245*, 118422. [CrossRef]
12. Ameri, F.; Shoaei, P.; Bahrami, N.; Vaezi, M.; Ozbakkaloglu, T. Optimum rice husk ash content and bacterial concentration in self-compacting concrete. *Constr. Build. Mater.* **2019**, *222*, 796–813. [CrossRef]
13. Chao-Lung, H.; Anh-Tuan, B.L.; Chun-Tsun, C. Effect of rice husk ash on the strength and durability characteristics of concrete. *Constr. Build. Mater.* **2011**, *25*, 3768–3772. [CrossRef]
14. Chindaprasirt, P.; Kanchanda, P.; Sathonsaowaphak, A.; Cao, H.T. Sulfate resistance of blended cements containing fly ash and rice husk ash. *Constr. Build. Mater.* **2007**, *21*, 1356–1361. [CrossRef]
15. Thomas, B.S. Green concrete partially comprised of rice husk ash as a supplementary cementitious material—A comprehensive review. *Renew. Sustain. Energy Rev.* **2018**, *82*, 3913–3923. [CrossRef]
16. Rattanachu, P.; Toolkasikorn, P.; Tangchirapat, W.; Chindaprasirt, P.; Jaturapitakkul, C. Performance of recycled aggregate concrete with rice husk ash as cement binder. *Cem. Concr. Compos.* **2020**, *108*, 103533. [CrossRef]
17. Gursel, A.P.; Maryman, H.; Ostertag, C. A life-cycle approach to environmental, mechanical, and durability properties of 'green' concrete mixes with rice husk ash. *J. Clean. Prod.* **2016**, *112*, 823–836. [CrossRef]
18. Moraes, C.A.M.; Kieling, A.G.; Caetano, M.O.; Gomes, L.P. Life cycle analysis (LCA) for the incorporation of rice husk ash in mortar coating. *Resour. Conserv. Recycl.* **2010**, *54*, 1170–1176. [CrossRef]
19. Rahman, M.E.; Muntohar, A.S.; Pakrashi, V.; Nagaratnam, B.H.; Sujan, D. Self compacting concrete from uncontrolled burning of rice husk and blended fine aggregate. *Mater. Des.* **2014**, *55*, 410–415. [CrossRef]
20. Prasara, J.-A.; Grant, T. Comparative life cycle assessment of uses of rice husk for energy purposes. *Int. J. Life Cycle Assess.* **2011**, *16*, 493–502. [CrossRef]
21. Saraswathy, V.; Song, H.-W. Corrosion performance of rice husk ash blended concrete. *Constr. Build. Mater.* **2007**, *21*, 1779–1784. [CrossRef]
22. Javed, M.F.; Amin, M.N.; Shah, M.I.; Khan, K.; Iftikhar, B.; Farooq, F.; Aslam, F. Applications of Gene Expression Programming and Regression Techniques for Estimating Compressive Strength of Bagasse Ash based Concrete. *Crystals* **2020**, *10*, 737. [CrossRef]
23. Ahmad, A.; Farooq, F.; Niewiadomski, P.; Ostrowski, K.; Akbar, A.; Aslam, F.; Alyousef, R. Prediction of Compressive Strength of Fly Ash Based Concrete Using Individual and Ensemble Algorithm. *Materials* **2021**, *14*, 794. [CrossRef]
24. Hammoudi, A.; Moussaceb, K.; Belebchouche, C.; Dahmoune, F. Comparison of artificial neural network (ANN) and response surface methodology (RSM) prediction in compressive strength of recycled concrete aggregates. *Constr. Build. Mater.* **2019**, *209*, 425–436. [CrossRef]
25. Jalal, M.; Arabali, P.; Grasley, Z.; Bullard, J.W.; Jalal, H. Behavior assessment, regression analysis and support vector machine (SVM) modeling of waste tire rubberized concrete. *J. Clean. Prod.* **2020**, *273*, 122960. [CrossRef]
26. Nehdi, Y.D.M.; Khan, A. Neural Network Model for Preformed-Foam Cellular Concrete. *ACI Mater. J.* **2001**, *98*, 5. [CrossRef]
27. Atici, U. Prediction of the strength of mineral admixture concrete using multivariable regression analysis and an artificial neural network. *Expert Syst. Appl.* **2011**, *38*, 9609–9618. [CrossRef]
28. Yaseen, Z.M.; Deo, R.C.; Hilal, A.; Abd, A.M.; Bueno, L.C.; Salcedo-Sanz, S.; Nehdi, M.L. Predicting compressive strength of lightweight foamed concrete using extreme learning machine model. *Adv. Eng. Softw.* **2018**, *115*, 112–125. [CrossRef]
29. Deng, F.; He, Y.; Zhou, S.; Yu, Y.; Cheng, H.; Wu, X. Compressive strength prediction of recycled concrete based on deep learning. *Constr. Build. Mater.* **2018**, *175*, 562–569. [CrossRef]
30. Bachir, R.; Mohammed, A.M.S.; Habib, T. Using Artificial Neural Networks Approach to Estimate Compressive Strength for Rubberized Concrete. *Period. Polytech. Civ. Eng.* **2018**, *62*, 858–865. [CrossRef]
31. Altun, F.; Kişi, Ö.; Aydin, K. Predicting the compressive strength of steel fiber added lightweight concrete using neural network. *Comput. Mater. Sci.* **2008**, *42*, 259–265. [CrossRef]

32. Perera, R.; Barchín, M.; Arteaga, A.; Diego, A.D. Prediction of the ultimate strength of reinforced concrete beams FRP-strengthened in shear using neural networks. *Compos. Part B Eng.* **2010**, *41*, 287–298. [CrossRef]

33. Tanarslan, H.M.; Secer, M.; Kumanlioglu, A. An approach for estimating the capacity of RC beams strengthened in shear with FRP reinforcements using artificial neural networks. *Constr. Build. Mater.* **2012**, *30*, 556–568. [CrossRef]

34. Öztaş, A.; Pala, M.; Özbay, E.; Kanca, E.; Caglar, N.; Bhatti, M.A. Predicting the compressive strength and slump of high strength concrete using neural network. *Constr. Build. Mater.* **2006**, *20*, 769–775. [CrossRef]

35. Bui, D.D.; Hu, J.; Stroeven, P. Particle size effect on the strength of rice husk ash blended gap-graded Portland cement concrete. *Cem. Concr. Compos.* **2005**, *27*, 357–366. [CrossRef]

36. Ganesan, K.; Rajagopal, K.; Thangavel, K. Rice husk ash blended cement: Assessment of optimal level of replacement for strength and permeability properties of concrete. *Constr. Build. Mater.* **2008**, *22*, 1675–1683. [CrossRef]

37. Ramezanianpour, A.A.; Mahdikhani, M.; Ahmadibeni, G. The Effect of Rice Husk Ash on Mechanical Properties and Durability of Sustainable Concretes. *Int. J. Civ. Eng.* **2009**, *7*, 83–91. Available online: https://www.sid.ir/en/journal/ViewPaper.aspx?ID=140499 (accessed on 12 September 2020).

38. Sakr, K. Effects of Silica Fume and Rice Husk Ash on the Properties of Heavy Weight Concrete. *J. Mater. Civ. Eng.* **2006**, *18*, 367–376. [CrossRef]

39. De Sensale, G.R. Strength development of concrete with rice-husk ash. *Cem. Concr. Compos.* **2006**, *28*, 158–160. [CrossRef]

40. Golafshani, E.M.; Behnood, A.; Arashpour, M. Predicting the compressive strength of normal and High-Performance Concretes using ANN and ANFIS hybridized with Grey Wolf Optimizer. *Constr. Build. Mater.* **2020**, *232*, 117266. [CrossRef]

41. Golafshani, E.M.; Behnood, A. Automatic regression methods for formulation of elastic modulus of recycled aggregate concrete. *Appl. Soft Comput.* **2018**, *64*, 377–400. [CrossRef]

42. Jaafari, A.; Panahi, M.; Pham, B.T.; Shahabi, H.; Rezaie, F.; Lee, S. Meta optimization of an adaptive neuro-fuzzy inference system with grey wolf optimizer and biogeography-based optimization algorithms for spatial prediction of landslide susceptibility. *CATENA* **2019**, *175*, 430–445. [CrossRef]

43. Gandomi, A.H.; Roke, D.A. Assessment of artificial neural network and genetic programming as predictive tools. *Adv. Eng. Softw.* **2015**, *88*, 63–72. [CrossRef]

44. Alavi, A.H.; Gandomi, A.H. Prediction of principal ground-motion parameters using a hybrid method coupling artificial neural networks and simulated annealing. *Comput. Struct.* **2011**, *89*, 2176–2194. [CrossRef]

45. Çaydaş, U.; Hasçalık, A.; Ekici, S. An adaptive neuro-fuzzy inference system (ANFIS) model for wire-EDM. *Expert Syst. Appl.* **2009**, *36 Pt 2*, 6135–6139. [CrossRef]

Article

Effects of High-Volume Ground Slag Powder on the Properties of High-Strength Concrete under Different Curing Conditions

Yuqi Zhou [1,2], Jianwei Sun [2,3,*] and Zengqi Zhang [4]

1   China Construction First Group Construction and Development Co., Ltd., Beijing 100102, China; zhouyuqi@chinaonebuild.com
2   Department of Civil Engineering, Tsinghua University, Beijing 100084, China
3   School of Civil Engineering, Qingdao University of Technology, Qingdao 266033, China
4   School of Metallurgical and Ecological Engineering, University of Science and Technology Beijing, Beijing 100083, China; zzq4816@163.com
*   Correspondence: jianwei_68@126.com

Abstract: Massive high-strength concrete structures tend to have a high risk of cracking. Ground slag powder (GSP), a sustainable and green industrial waste, is suitable for high-strength concrete. We carried out an experimental study of the effects of GSP with a specific surface area of 659 m^2/kg on the hydration, pore structure, compressive strength and chloride ion penetrability resistance of high-strength concrete. Results show that adding 25% GSP increases the adiabatic temperature rise of high-strength concrete, whereas adding 45% GSP decreases the initial temperature rise. Incorporating GSP refines the pore structure to the greatest extent and improves the compressive strength and chloride ion penetrability resistance of high-strength concrete, which is more obvious under early temperature-matching curing conditions. Increasing curing temperature has a more obvious impact on the pozzolanic reaction of GSP than cement hydration. From a comprehensive perspective, GSP has potential applications in the cleaner production of green high-strength concrete.

Keywords: GSP; high strength; hydration; strength; penetrability

Citation: Zhou, Y.; Sun, J.; Zhang, Z. Effects of High-Volume Ground Slag Powder on the Properties of High-Strength Concrete under Different Curing Conditions. *Crystals* **2021**, *11*, 348. https://doi.org/10.3390/cryst11040348

Academic Editors: Yifeng Ling, Chuanqing Fu, Peng Zhang and Peter Taylor

Received: 7 March 2021
Accepted: 25 March 2021
Published: 29 March 2021

Publisher's Note: MDPI stays neutral with regard to jurisdictional claims in published maps and institutional affiliations.

1. Introduction

Currently, high-rise buildings have become increasingly widespread in China due to their advantages in space, stability and unique design [1,2]. They typically symbolize the landscape architecture and construction of a city, such as the Shanghai Tower (632 m), the Shenzhen Ping'an Finance Centre (599 m) and the China Zun in Beijing (528 m). The foundation structures of high-rise buildings with heavy loads are extremely deep and wide, which is typical for massive high-strength concrete structures [1]. During the hydration process, substantial hydration heat is generated in massive high-strength concrete, resulting in a high internal temperature rise because of its slow heat dissipation [3–6]. After hardening, large tensile stresses are formed due to restrained thermal and autogenous shrinkage deformations, which are the main driving forces of cracking in concrete [7].

Using supplementary cementitious materials (SCMs) to lower early heat and attendant volume changes is the most common preventative method [8–12]. The application of SCMs in concrete also has positive effects on workability, pumpability, strength and permeability resistance to chemical attacks [13–21]. Meanwhile, SCMs, as mineral admixtures to replace cement in high-strength concrete, reduce the carbon footprint in cement and concrete production and are conducive to sustainable development due to the conservation of natural resources [22–24]. Slag, as one of the most suitable SCMs, has been extensively identified and used to directly replace cement, minimizing cracking in massive concrete applications. Slag is a non-metallic residual generated from blast furnaces when iron is extracted from its ore [25,26]. Molten slag, comprising mostly silicates and alumina, is swiftly quenched with abundant water [27]. The rapid cooling method results in amorphous phases of slag (nearly

80% content), which is responsible for its pozzolanic activity [28]. Compared to Portland cement, slag has a lower specific gravity [29]. The colour of slag varies from dark grey to white depending on the moisture inside, its chemical composition and its granulation efficiency [30]. The replacement rate for slag during the production of concrete varies from 30% to 85%, and 50% is typically used in most applications [10,31]. Incorporating as little as 30% slag can reduce the cumulative heat by 25% after the initial 48 h [31,32]. The cementitious activity of slag needs to be further improved for wider application. These inherent attributes are not easy to change, including the chemical composition, amorphous phase content and alkali concentration of the cement system [33,34]. However, the fineness of slag can be further enhanced by drying and subsequent grinding in a rotating ball mill to a finer powder with a specific surface area of 600 m^2/kg – 700 m^2/kg, which is called "ground slag powder (GSP)" in this study.

Many studies have proven that the pozzolanic reaction rate is increased by improving the fineness of slag, which has a great impact on the development of strength and durability. He et al. [35] used two methods to prepare GSP, including the wet-milling method and the dry-separation method, to improve its early reactivity. They verified that the setting time gradually decreased with the dry-separation slag and increased with the wet-milling slag [35]. Moreover, a system containing wet-milling slag had higher electrical resistivity and better mechanical properties [35]. Liu et al. [36] investigated the contribution ratios of GSP to a GSP-cement-steel slag ternary cementitious material system and found that GSP caused an obvious improvement in the hydration and mechanical properties at every stage due to its close-packing effect [36]. Zhang et al. [37] found that the ultimate hydration heat initially increased and then decreased sharply with increasing proportions of GSP [37]. Meanwhile, adding GSP had a slight impact on chemical shrinkage but increased the chloride binding capacity [37]. Mohan et al. [38] conducted experiments on the influence of silica fume and GSP on the properties of self-compacting concrete. They found that incorporating GSP is a better way to decrease free shrinkage due to the diminished water withholding and improved sulphate and acid attack resistance of self-compacting concrete [38]. Pradeep Kumar et al. [39] reported a modification of the corrosion property of concrete with the use of GSP. Using GSP reduced the workability and water absorption of concrete, enhanced the bond strength of the steel rebar, and remarkably reduced rebar corrosion [39].

Based on the literature available, there is still a lack of research on the properties of GSP high-strength concrete. Hence, this study aims to investigate the feasibility of using GSP as a mineral admixture in high-strength concrete. In this paper, high-strength cement concrete with a design strength of C75 was prepared as the reference sample. Two substitution rates of GSP (25% and 45%) and two curing conditions (a standard curing condition and a temperature-matching curing condition) were selected. The adiabatic temperature rise, pore structure, hydration products, compressive strength and chloride ion penetrability resistance of high-strength concrete were measured. Effects of elevated early temperatures on the properties of plain cement concrete and GSP concrete were analyzed. The results of this study can provide considerable theoretical guidance for the use of GSP in massive high-strength concrete applications.

2. Materials and Methods

2.1. Raw Materials

Water-quenched blast-furnace slag powder was produced by Xingye Materials Co., Ltd., Xingtai, Hebei province, China. Portland cement with a strength grade of 42.5 was supplied by Jinyu Cement Co., Ltd., Beijing, China. The slag powder is further ground into GSP in the laboratory. The chemical compositions of raw materials are given in Table 1. The mass coefficient ($K = w(CaO + MgO + Al_2O_3)/w(SiO_2 + MnO + TiO_2)$) of the GSP used was 1.97 according to the Chinese national standard (GB/T 203-2008). The specific surface areas of the cement and GSP were 341 m^2/kg and 659 m^2/kg, respectively. Coarse aggregates of concrete are crushed limestone of 5 to 20 mm in size. Fine aggregates of concrete are river sand with a fineness modulus of 2.9. The workability of fresh concrete was adjusted using a polycarboxylate superplasticizer.

Table 1. Chemical compositions of the cement and ground slag powder (GSP)/%.

	CaO	SiO$_2$	Al$_2$O$_3$	MgO	Fe$_2$O$_3$	SO$_3$	Na$_2$O$_{eq}$ *	f-CaO	LOI
Cement	62.71	22.33	4.75	1.98	2.78	2.37	0.68	0.64	2.03
GSP	39.47	30.14	18.64	8.68	0.75	0.24	0.86	-	1.04

* Na$_2$O$_{eq}$ = Na$_2$O + 0.658K$_2$O; LOI: loss on ignition.

2.2. Mix Proportions

Table 2 shows the mix proportions of the high-strength concrete. The total amount of binder was 550 kg/m^3. A water/binder ratio of 0.28 and a sand ratio of 0.43 were selected. Plain cement concrete was regarded as the reference sample (sample C). Two substitution rates of GSP (25% and 45% by mass) were used, corresponding to sample S25 and sample S45. Cubic concrete samples with side lengths of 100 mm were prepared. Each sample had a total of 36 square concrete specimens. The paste had the same water/binder ratio and substitution rates as the concrete. Fresh pastes were cast into plastic tubes and cured under the same curing conditions as the concrete.

Table 2. Mix proportions of high-strength concrete/kg·m^{-3}.

Sample	Cement	GSP	Fine Aggregate	Coarse Aggregate	Water	Superplasticizer
C	550	-	751	995	154	5.5
S25	412.5	137.5	751	995	154	8.25
S45	302.5	247.5	751	995	154	11

2.3. Curing Conditions and Test Methods

In this study, two curing conditions for the concretes and pastes were set—the standard curing condition (symbol S) and the temperature-matching curing condition (symbol M). Thus, sample SS25 represents S25 concrete cured under the standard curing condition, and sample MS45 represents S45 concrete cured under the temperature-matching curing condition. The standard curing condition required a constant temperature (20 °C ± 2 °C) and relative humidity (>95%). The temperature-matching curing condition needed to be adjusted according to the adiabatic temperature rise curve of the concrete. The adiabatic temperature rise curve of high-strength concrete for the initial 7 d was determined by a temperature measuring instrument (50 L) with an accuracy of ±0.1 °C.

The compressive strength and chloride ion penetrability resistance of concrete for each concrete mixture were obtained from an average of three specimens. The pastes were prepared for the tests of pore structure, Ca(OH)$_2$ (CH) content, and non-evaporable water content. First, the hardened paste was broken into small pieces (less than 5 mm). Then, the pieces were soaked in ethanol (Tongguang fine chemicals company, Beijing, China) for 24 h at test ages. Finally, all pieces were dried in the oven at 110 °C. For tests of CH content and non-evaporable content water content, dried pieces were further ground into ultrafine powder. The pore structure of hardened paste at ages of 28 d and 90 d was measured with a mercury intrusion porometer (MIP, AUTOPORE II 9220 manufactured by Micromeritics, America) with a maximum mercury intrusion pressure of 300 MPa. The CH content was determined by thermogravimetric (TG) and derivative thermogravimetric (DTG) analyses. TG and DTG curves were obtained using Instrument TGA 3+ (METTLER TOLEDO, Switzerland) in an N$_2$ atmosphere from 30 °C to 900 °C at 14 d, 28 d and 90 d. The non-evaporable water content W$_n$ was determined using the following Equation (1):

$$W_n = (m_1 - m_2)/m_1 - (1 - \alpha) \, LOI_C - \alpha LOI_S, \tag{1}$$

where m_1 is the dried mass of hardened paste at 110 °C, m_2 is the mass of hardened paste after heating at 1050 °C, α represents the replacement rate of GSP, and LOI_C and LOI_S represent the loss on ignition of cement and GSP, respectively.

3. Results and Discussion

3.1. Adiabatic Temperature Rise

The adiabatic temperature rise curves of plain cement concrete and GSP concrete for the initial 7 d are depicted in Figure 1. It is obvious that the growth trends of the three temperature curves are similar. The temperature rose sharply before 24 h and remained stable after 150 h. The final temperature rises of samples C, S25 and S45 were 56.56 °C, 60.58 °C and 54.07 °C, respectively. This illustrates that the incorporation of 25% GSP increases the adiabatic temperature rise by 4.02 °C and incorporating 45% GSP decreases the adiabatic temperature rise by 2.49 °C. Note that sample S25 exhibited the maximum temperature rise. This indicates that the promoting effect of GSP on cement hydration exceeds the negative effect due to cement reduction. The promoting effects derived from two main contributing factors [34]. One is the heterogenous nucleation effect [34]. Compared to cement particles, GSP has finer particles, of which the specific surface area is 659 m²/kg. Finer slag particles can serve as heterogeneous nucleation and crystallization sites of C–S–H gel and CH, thus improving the degree of cement hydration. The other is related to the higher reactivity of slag, which participates in pozzolanic reactions at an early stage, thus promoting cement hydration. The drop in the adiabatic temperature rise is attributed to a significant reduction in cement content.

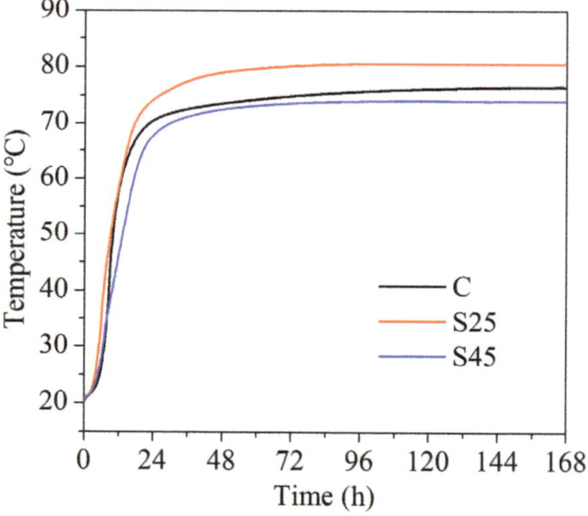

Figure 1. Adiabatic temperature rise curves of concrete.

3.2. Compressive Strength

The compressive strengths of plain cement concrete and GSP concrete under standard curing conditions are shown in Figure 2a. It can be easily observed that the compressive strength of concrete slightly increased with GSP at all ages. The compressive strengths of concrete at ages of 7 d, 28 d, 90 d and 180 d were more than 70 MPa, approximately 80 MPa, 85 MPa and 90 MPa, respectively. Compared to the 7 d compressive strength, the growth rates of the strength at the ages of 28 d, 90 d and 180 d were approximately 12%, 20% and 24%, respectively. Thus, the growth rates of the compressive strengths of plain cement concrete and GSP concrete showed little difference at the same ages under standard curing conditions. Compared to the compressive strengths of plain cement concrete, the growth rates in strength due to the addition of GSP were calculated and are presented in Figure 2b. The growth rate of sample SS25 at all ages was relatively low, at no more than 2%. The growth rate of sample SS45 was higher than that of sample SS25 at all ages. In particular, the 28 d growth rate of sample SS45 reached 6%. The growth rate is related to pore structure and hydration products. Compared to sample SS25, sample SS45 has a higher substitution rate and the presence of GSP with finer particles has a more positive influence on early hydration, resulting in a higher growth rate.

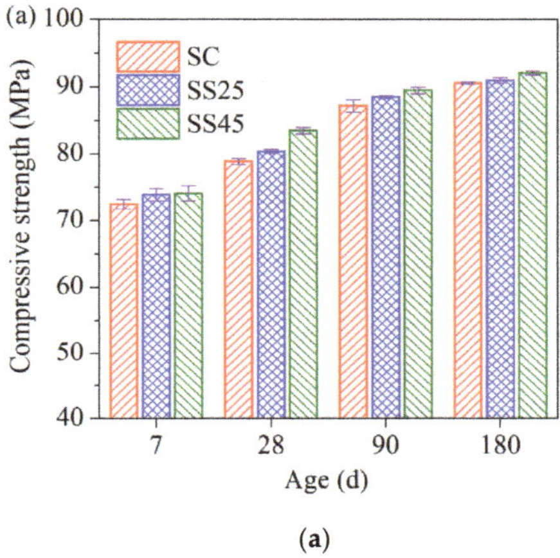
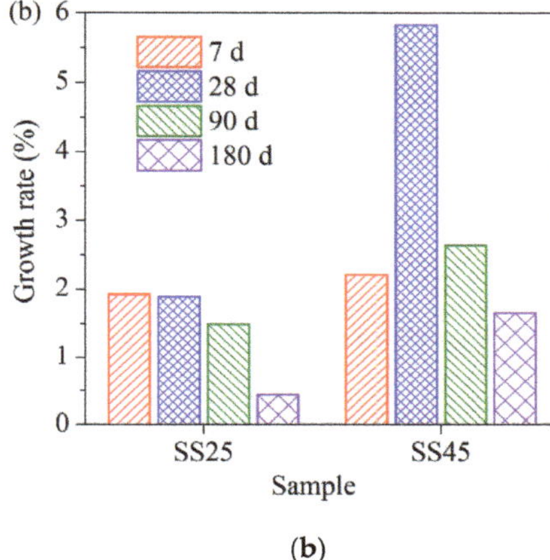

Figure 2. (**a**) Compressive strength of high-strength concrete under standard curing condition; (**b**) growth rate of compressive strength at different ages.

The compressive strength and growth rate of compressive strength under temperature-matching curing conditions are presented in Figure 3a,b, respectively. The compressive strength of high-strength concrete significantly increases with GSP at all ages under temperature-matching curing conditions, which is different from the trend under standard curing conditions. Compared to the 7 d compressive strength, the growth rates of the strength at 28 d and 90 d were approximately 11% and 21%, respectively. However, the growth rates of 180 d compressive strength were approximately 22%, 28% and 30%, respectively. The growth rate of GSP concrete at 180 d was higher than that of plain cement concrete. Meanwhile, compared to the compressive strength of plain cement concrete, the growth rates of strength at different ages due to the addition of GSP under temperature-matching curing conditions (Figure 3b) were higher than those under standard curing temperatures (Figure 2b). In addition, the growth rate increased with GSP. In particular, the 180 d growth rates were the highest. This result indicates that the temperature-matching curing conditions have a more positive effect on the development of the late compressive strength of GSP concrete. Elevated temperatures promote the pozzolanic reaction of GSP. The pozzolanic reaction of GSP consumes CH and forms C–S–H gel, improving the density of the interfacial transition zone between the cement and aggregates [40,41]. Furthermore, C–S–H gel plays a key role in mechanical performance. When GSP is added to the cementitious system, Al^{3+} is released from the slag and finally forms a C–(A)–S–H gel, leading to an increase in the Al/Si molar ratio and a decrease in the Ca/Si molar ratio [42,43]. C–S–H with higher Al/Si and lower Ca/Si ratios has a higher bonding capacity and thus improves the compressive strength [40–44].

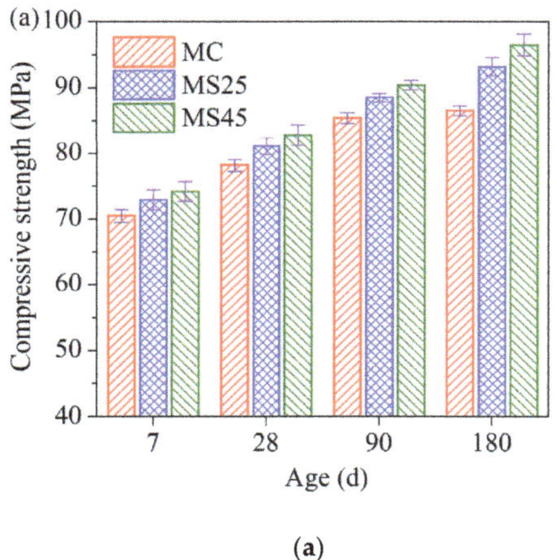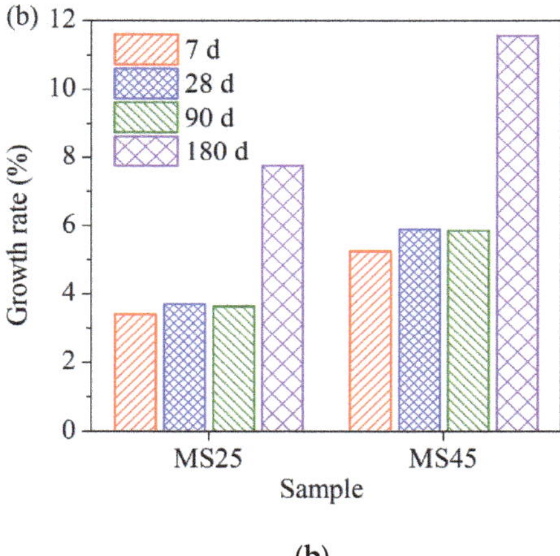

(a)   **(b)**

Figure 3. (**a**) Compressive strength of high-strength concrete under temperature-matching curing conditions; (**b**) growth rate of compressive strength at different ages.

3.3. Chloride Ion Penetrability Resistance

The chloride ion penetrability resistance of concrete at ages of 28 d and 180 d under standard curing conditions are shown in Figure 4. It can be seen in Figure 4 that the chloride ion penetrability grades of sample SC were "moderate" and "low" at 28 d and 180 d, respectively. However, the penetrability grades of samples SS25 and SS45 fell to the "low" level and the "very low" level at the two ages. Therefore, substitution with GSP can improve the chloride ion penetrability resistance of high-strength concrete, and the effect increases with increasing GSP. This is because the filling effect of grinding slag fills the pore structure of concrete, and the pozzolanic reaction consumes CH in the transition zone, resulting in more C–S–H gel, which refines the pore structure.

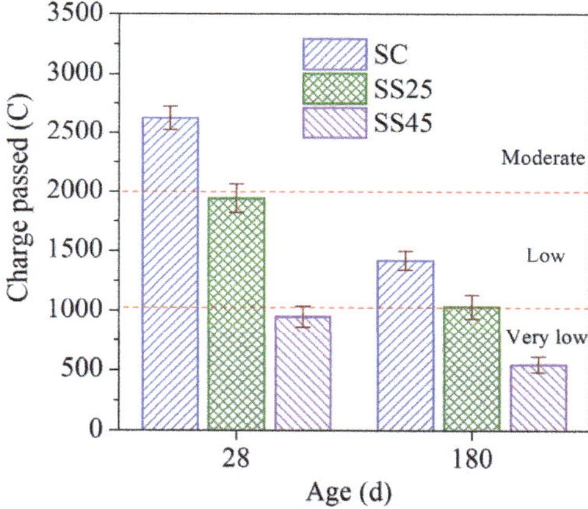

Figure 4. Chloride ion penetrability resistance of high-strength concrete under standard curing conditions.

The chloride ion penetrability resistance of concrete at 28 d and 180 d under temperature-matching curing conditions are presented in Figure 5. Significantly, as the age increases, the chloride ion penetration resistance of concrete did not change. The chloride ion penetrability grades of sample MC were "moderate" at the two ages. The penetrability grades of both samples MS25 and MS45 fell to the "very low" level at the same time. The GSP content has little effect on the chloride ion penetration resistance of concrete with further hydration. This is because the pozzolanic reaction of GSP mainly occurred at an early age and increasing the early curing temperature promoted the reaction of GSP, which had an adverse effect on the late reaction. In terms of the chloride ion penetration resistance, combined with the results under standard curing conditions, increasing the curing temperature has a greater influence on high-strength concrete mixed with 25% GSP.

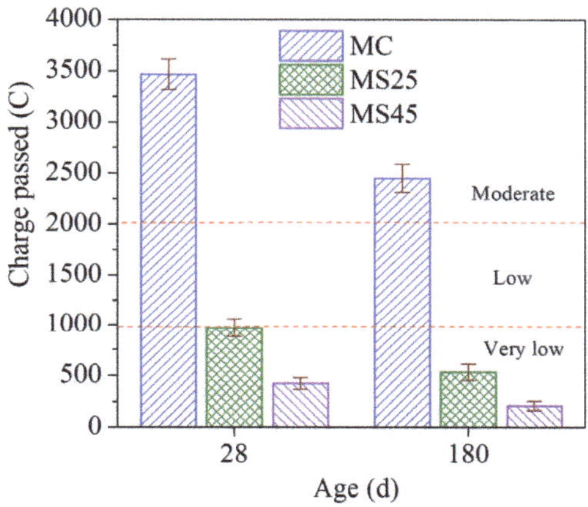

Figure 5. Chloride ion penetrability resistance of high-strength concrete under temperature-matching curing conditions.

3.4. Pore Structure

The differential pore size distributions of different hardened pastes under standard curing conditions are presented in Figure 6. As shown in Figure 6, the most likely pore sizes of samples SC, SS25 and SS45 at 28 d were 62.1 nm, 30.2 nm and 13.94 nm, respectively. With further hydration, the pore structure of the hardened paste becomes dense. The most likely pore sizes of samples SC, SS25 and SS45 at 28 d were 40.7 nm, 5.42 nm and 3.39 nm, respectively. The most likely pore size obviously decreased with GSP due to its positive effects on pore structure. Pores in hardened paste can be divided into four types, depending on their diameters and functions, according to Mehta's opinion: >100 nm, 50–100 nm, 4.5–50 nm and <4.5 nm [45,46]. Pores larger than 100 nm (also called harmful pores) have adverse effects on the development of mechanical strength and the permeability of hardened matrices [36]. Thus, the detailed pore size distribution of the hardened paste is presented in Figure 7. Figure 7a shows that the porosities of samples SC, SS25 and SS45 at 28 d were 21.52%, 20.81% and 17.29%, respectively. The porosities at 90 d, shown in Figure 7b, were 16.77%, 16.79% and 19.09%, respectively. Compared to the plain cement paste (sample SC), the total porosity and the volume of harmful pores in hardened paste at 28 d were significantly reduced due to the addition of GSP. However, the total porosity in hardened paste containing 45% GSP (sample SS45) was the largest at 28 d due to more harmless pores (<4.5 nm).

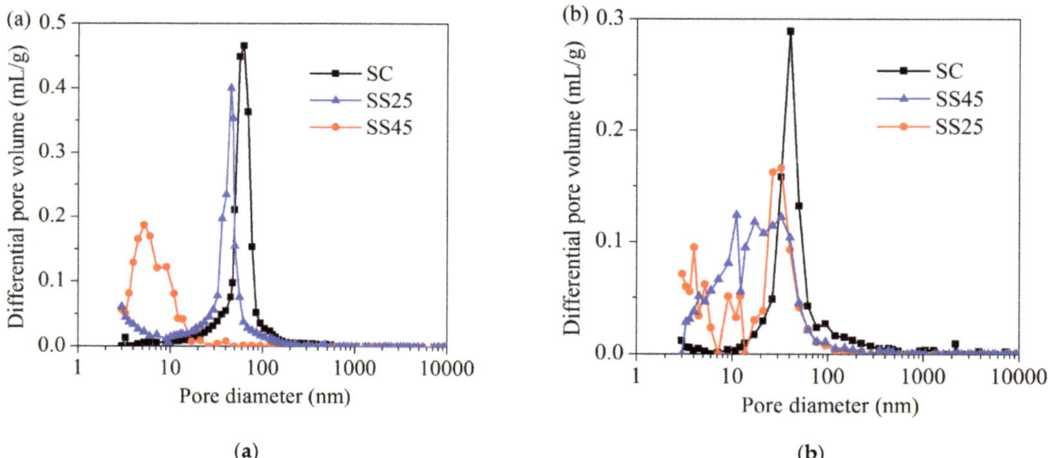

Figure 6. Differential pore volume of hardened pastes under standard curing conditions: (**a**) at 28 d; (**b**) at 90 d.

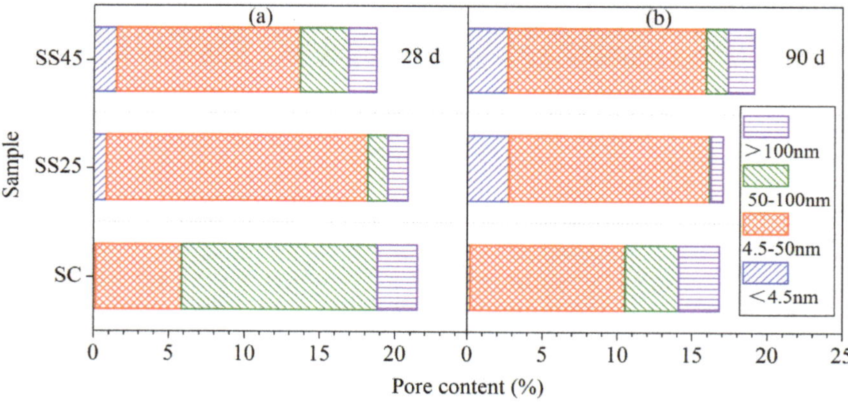

Figure 7. The pore size distribution of hardened pastes under standard curing conditions: (**a**) at 28 d; (**b**) at 90 d.

The differential pore size distributions of different hardened pastes under temperature-matching curing conditions are presented in Figure 8. The most likely pore sizes of samples MC, MS25 and MS45 at 28 d were 24.3 nm, 49.0 nm and 5.4 nm, respectively. With further hydration, the pore structure of the hardened paste becomes denser. The most likely pore sizes of samples MC, MS25 and MS45 at 90 d were 17.1 nm, 5.7 nm and 5.0 nm, respectively. The detailed pore size distribution of the hardened paste is presented in Figure 9. Figure 9a shows that the porosities of samples MC, MS25 and MS45 at 28 d were 24.74%, 17.09% and 16.23%, respectively. The porosities at 90 d, shown in Figure 9b, were 22.20%, 16.35% and 11.31%, respectively. Compared to the plain cement paste (sample MC), the total porosity and the volume of harmful pores in hardened paste containing 25% GSP were significantly reduced. Sample MS25 had lower total porosity than sample MC. However, the volumes of harmful pores of both pastes showed little difference.

products. At an early age, hydration of cement in the cementitious system is dominant. At 28 d, the CH contents in the cement paste after calculation and the corresponding hardened paste containing GSP showed little difference. However, with further hydration, the pozzolanic reaction of GSP consumed more CH, leading to a lower CH content in composite systems at 90 d.

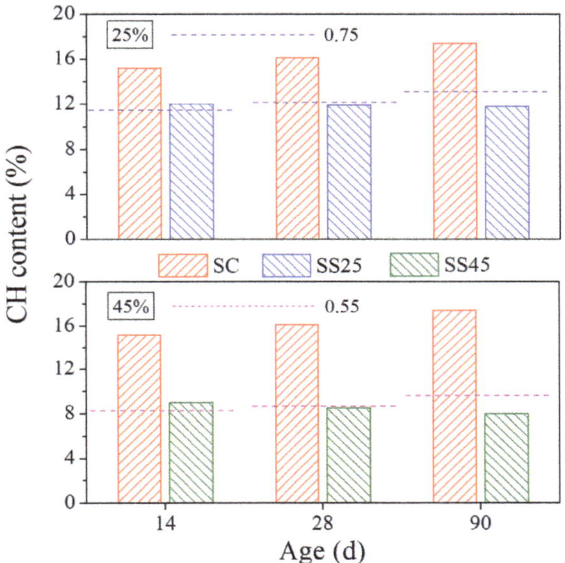

Figure 11. CH content of hardened pastes under standard curing conditions.

The TG and DTG curves of hardened pastes at different ages under temperature-matching curing conditions are presented in Figure 12. As shown in Figure 12, the trend in mass losses was obviously different from that observed under standard curing conditions (Figure 10). The gaps between samples MC, MS25 and MS45 were relatively small at all ages. There are also two endothermic peaks on the DTG curves at approximately 100 °C and 450 °C, corresponding to the sequential decomposition of C–S–H gel and CH crystals. Another absorption peak is located at approximately 650 °C, which represents the decomposition of $CaCO_3$.

The CH contents of different hardened pastes, including measured values and normalization marks, under temperature-matching curing conditions are indicated in Figure 13. At 14 d, the CH contents in the cement paste after calculation and the corresponding hardened paste containing GSP showed little difference. This indicates that the CH content remained balanced between production from cement hydration and consumption due to the pozzolanic reaction of GSP. Currently, the promoting effect of GSP on the cementitious system can compensate for its negative effect on CH content due to the pozzolanic effect. At 28 d, compared to those of the cement after calculation with reduction factors of 0.75 and 0.55, the CH contents of samples SS25 and SS45 were significantly lower. At 90 d, the gap in CH contents between the cement paste and hardened paste containing GSP became wider. This indicates that the pozzolanic effect of GSP proves its importance with further hydration. Compared to that observed under standard curing conditions, the gap in the CH content at 14 d was obviously smaller under temperature-matching curing condition. Elevated curing temperature has a greater impact on the pozzolanic reaction of GSP, consuming more CH.

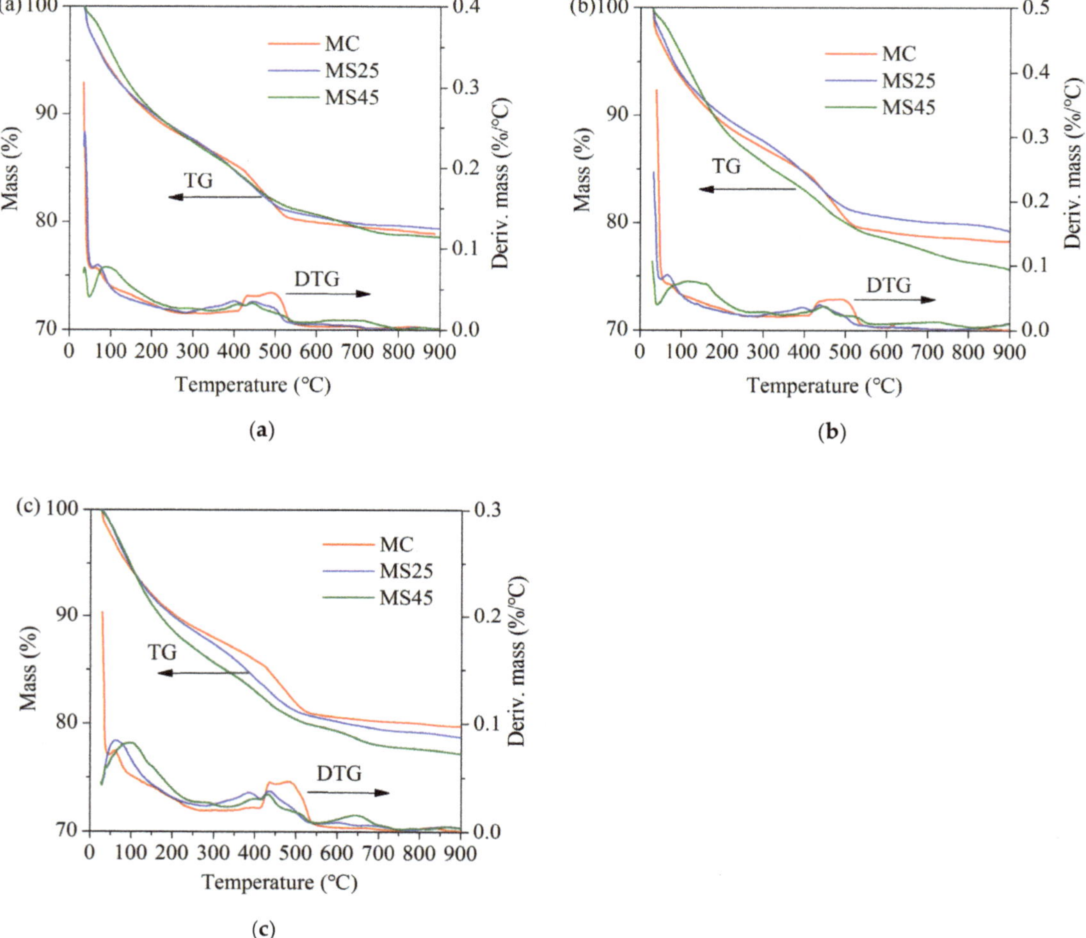

Figure 12. TG and DTG curves of hardened pastes under temperature-matching curing conditions: (**a**) at 14 d; (**b**) at 28 d; (**c**) at 90 d.

3.6. Non-Evaporable Water Content

The non-evaporable water contents of hardened paste under two curing conditions are presented in Figure 14a,b, respectively. The lowest non-evaporable water content of hardened paste exceeded 12% under both curing conditions. Under standard curing condition, the non-evaporable water content of hardened paste containing GSP was slightly higher than that of sample SC. In particular, hardened paste containing 25% GSP showed the highest non-evaporable water content. This indicates that the promoting effects, including the dilution effect and the nucleation effect of GSP on the cement hydration, were obvious at early and late ages. The results agree with the results of the adiabatic temperature rise of concrete (Figure 1). In theory, the hydration of Portland cement per gram produces 0.25 g of non-evaporable water, but the pozzolanic reaction of slag per gram generates 0.30 g of non-evaporable water [48]. Meanwhile, increasing the fineness of active powder has a more positive effect on the pozzolanic reaction [48]. Therefore, the faster reaction of GSP results with a higher temperature rise rate was observed for sample SS25 (Figure 1), along with the higher non-evaporable water content of sample SS25. However, the gap in non-evaporable water contents between samples SC and SS45 was relatively small, resulting from a sharp reduction in cement content. Combined with the results of compressive strength under two curing conditions, sample SS45 had the highest compressive strength. The compressive strength of concrete depends not only on the amount of hydration products, it is also closely related to the pore structure. Sample SS45 had the lowest total porosity and volume of harmful pores, which resulted in the highest compressive strength.

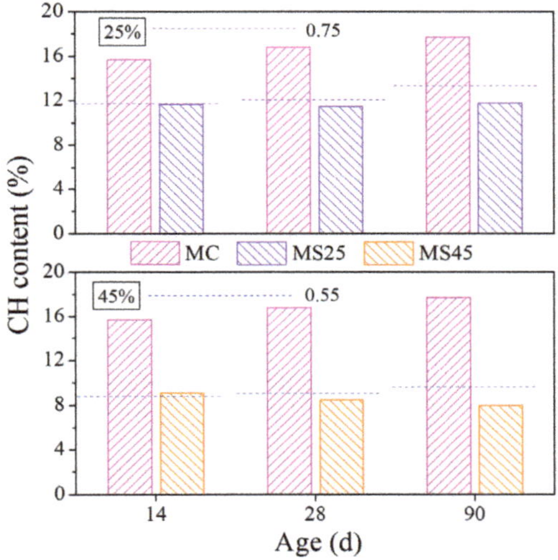

Figure 13. CH content of hardened pastes under temperature-matching curing conditions.

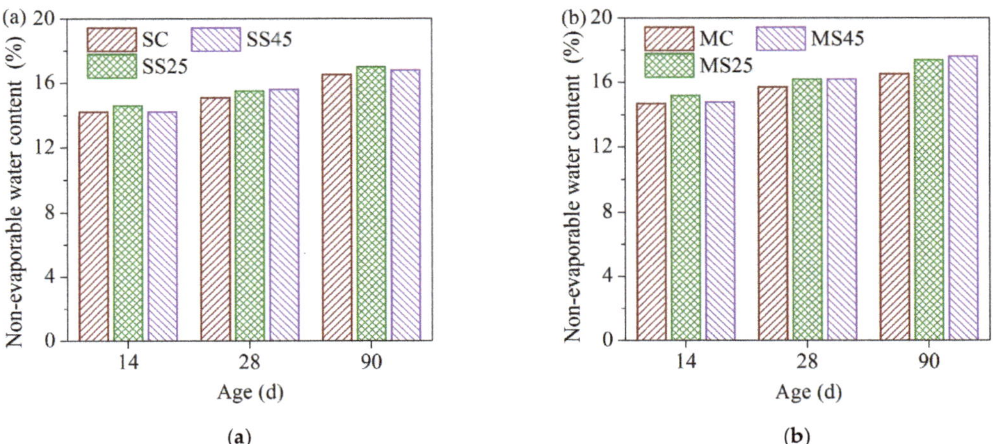

(a)

(b)

Figure 14. Non-evaporable water content of hardened pastes: (**a**) under standard curing conditions; (**b**) under temperature-matching curing conditions.

Under temperature-matching curing conditions, the trend of the non-evaporable water content was similar to that under standard curing conditions (Figure 14a). However, the non-evaporable water content of hardened paste became higher than that under standard curing conditions. Increasing the early curing temperature evidently improves the hydration rate of the plain cement and composite binder. Meanwhile, the highest non-evaporable water content was observed for sample SS45 at 90 d. Replacing cement with GSP markedly increased the late non-evaporable water content under temperature-matching curing conditions. It is worth noting that the non-evaporable water content of hardened paste containing GSP under two curing conditions reached or exceeded the content of cement paste after 14 d of curing. A previous study on ordinary slag showed that the same non-evaporable water content with replacement rates of 50%–70% can be achieved after 60 d of curing [28,49]. This indicates that the fineness of slag plays an important role in cement hydration and the pozzolanic reaction of mineral admixtures.

4. Conclusions

(1) Adding 25% GSP increases the adiabatic temperature rise of high-strength concrete due to the promoting effects on cement hydration, whereas adding 45% GSP decreases the adiabatic temperature rise, which can be attributed to a reduction in cement content.

(2) Compared to the compressive strength of plain cement concrete, the growth rates of strength at different ages due to the addition of GSP under temperature-matching curing conditions are higher than those under standard curing temperature. Temperature-matching curing conditions have a positive effect on the development of the late compressive strength of GSP concrete.

(3) Compared to plain cement systems, cementitious material systems containing GSP tend to have lower total porosity and a lower volume of harmful pores. The dense pore structure of the GSP system leads to better chloride ion penetrability resistance of the concrete, which is more distinct under early temperature-matching curing conditions. Increasing the curing temperature has a greater influence on high-strength concrete mixed with 25% GSP.

(4) The improvement effects of incorporating GSP on the late non-evaporable water content of hardened paste are significant under early temperature matching curing conditions. However, the effect of elevated temperatures on the early non-evaporable water content are limited. Increasing early curing temperatures has a more obvious effect on the pozzolanic reaction of GSP than the hydration of cement.

Author Contributions: Conceptualization, Y.Z.; methodology, Y.Z. and Z.Z.; software, Y.Z. and Z.Z.; validation, J.S. and Z.Z.; writing—original draft preparation, J.S.; writing—review and editing, J.S.; All authors have read and agreed to the published version of the manuscript.

Funding: This research was funded by the National Natural Science Foundation of China, grant number 52008229.

Institutional Review Board Statement: Not applicable.

Informed Consent Statement: Not applicable.

Data Availability Statement: Data is contained within the article.

Conflicts of Interest: The authors declare no conflict of interest.

References

1. Mengxiao, S.; Qiang, W.; Zhikai, Z. Comparison of the properties between high-volume fly ash concrete and high-volume steel slag concrete under temperature matching curing condition. *Constr. Build. Mater.* **2015**, *98*, 649–655. [CrossRef]

2. Alaskar, A.; Albidah, A.; Alqarni, A.S.; Alyousef, R.; Mohammadhosseini, H. Performance evaluation of high-strength concrete reinforced with basalt fibers exposed to elevated temperatures. *J. Build. Eng.* **2021**, *35*, 102108. [CrossRef]

3. Shen, D.; Liu, C.; Jiang, J.; Kang, J.; Li, M. Influence of super absorbent polymers on early-age behavior and tensile creep of internal curing high strength concrete. *Constr. Build. Mater.* **2020**, *258*, 120068. [CrossRef]

4. Shen, D.; Kang, J.; Jiao, Y.; Li, M.; Li, C. Effects of different silica fume dosages on early-age behavior and cracking resistance of high strength concrete under restrained condition. *Constr. Build. Mater.* **2020**, *263*, 120218. [CrossRef]

5. Evram, A.; Akçaoğlu, T.; Ramyar, K.; Çubukçuoğlu, B. Effects of waste electronic plastic and marble dust on hardened properties of high strength concrete. *Constr. Build. Mater.* **2020**, *263*, 120928. [CrossRef]

6. Al-Yousuf, A.; Pokharel, T.; Lee, J.; Gad, E.; Abdouka, K.; Sanjayan, J. Effect of fly ash and slag on properties of normal and high strength concrete including fracture energy by wedge splitting test: Experimental and numerical investigations. *Constr. Build. Mater.* **2021**, *271*, 121553. [CrossRef]

7. Kushnir, A.R.; Heap, M.J.; Griffiths, L.; Wadsworth, F.B.; Langella, A.; Baud, P.; Reuschlé, T.; Kendrick, J.E.; Utley, J.E. The fire resistance of high-strength concrete containing natural zeolites. *Cem. Concr. Compos.* **2021**, *116*, 103897. [CrossRef]

8. Liu, S.; Zhang, T.; Guo, Y.; Wei, J.; Yu, Q. Effects of SCMs particles on the compressive strength of micro-structurally designed cement paste: Inherent characteristic effect, particle size refinement effect, and hydration effect. *Powder Technol.* **2018**, *330*, 1–11. [CrossRef]

9. Aprianti, E. A huge number of artificial waste material can be supplementary cementitious material (SCM) for concrete production—A review part II. *J. Clean. Prod.* **2017**, *142*, 4178–4194. [CrossRef]

10. Samad, S.; Shah, A. Role of binary cement including Supplementary Cementitious Material (SCM), in production of environmentally sustainable concrete: A critical review. *Int. J. Sustain. Built Environ.* **2017**, *6*, 663–674. [CrossRef]

11. Zhuang, S.; Wang, Q. Inhibition mechanisms of steel slag on the early-age hydration of cement. *Cem. Concr. Res.* **2021**, *140*, 106283. [CrossRef]

12. Papayianni, I.; Anastasiou, E. Production of high-strength concrete using high volume of industrial by-products. *Constr. Build. Mater.* **2010**, *24*, 1412–1417. [CrossRef]

13. Wang, D.; Wang, Q.; Huang, Z. Reuse of copper slag as a supplementary cementitious material: Reactivity and safety. *Resour. Conserv. Recycl.* **2020**, *162*, 105037. [CrossRef]

14. Khalil, E.A.B.; Anwar, M. Carbonation of ternary cementitious concrete systems containing fly ash and silica fume. *Water Sci.* **2015**, *29*, 36–44. [CrossRef]

15. Palod, R.; Deo, S.; Ramtekkar, G. Effect on mechanical performance, early age shrinkage and electrical resistivity of ternary blended concrete containing blast furnace slag and steel slag. *Mater. Today Proc.* **2020**, *32*, 917–922. [CrossRef]

16. Cordeiro, G.; Filho, R.T.; Tavares, L.; Fairbairn, E. Experimental characterization of binary and ternary blended-cement concretes containing ultrafine residual rice husk and sugar cane bagasse ashes. *Constr. Build. Mater.* **2012**, *29*, 641–646. [CrossRef]

17. Lemonis, N.; Tsakiridis, P.; Katsiotis, N.; Antiohos, S.; Papageorgiou, D.; Beazi-Katsioti, M. Hydration study of ternary blended cements containing ferronickel slag and natural pozzolan. *Constr. Build. Mater.* **2015**, *81*, 130–139. [CrossRef]

18. Chindaprasirt, P.; Kroehong, W.; Damrongwiriyanupap, N.; Suriyo, W.; Jaturapitakkul, C. Mechanical properties, chloride resistance and microstructure of Portland fly ash cement concrete containing high volume bagasse ash. *J. Build. Eng.* **2020**, *31*, 101415. [CrossRef]

19. Jeong, Y.; Park, H.; Jun, Y.; Jeong, J.-H.; Oh, J.E. Microstructural verification of the strength performance of ternary blended cement systems with high volumes of fly ash and GGBFS. *Constr. Build. Mater.* **2015**, *95*, 96–107. [CrossRef]

20. Cheah, C.B.; Tiong, L.L.; Ng, E.P.; Oo, C.W. The engineering performance of concrete containing high volume of ground granulated blast furnace slag and pulverized fly ash with polycarboxylate-based superplasticizer. *Constr. Build. Mater.* **2019**, *202*, 909–921. [CrossRef]

21. Sun, W.; Yan, H.; Zhan, B. Analysis of mechanism on water-reducing effect of fine ground slag, high-calcium fly ash, and low-calcium fly ash. *Cem. Concr. Res.* **2003**, *33*, 1119–1125. [CrossRef]

22. Wang, Y.; Liu, X.; Zhang, W.; Li, Z.; Zhang, Y.; Li, Y.; Ren, Y. Effects of Si/Al ratio on the efflorescence and properties of fly ash based geopolymer. *J. Clean. Prod.* **2020**, *244*, 118852. [CrossRef]

23. Wang, D.; Wang, Q.; Huang, Z. New insights into the early reaction of NaOH-activated slag in the presence of $CaSO_4$. *Compos. Part B Eng.* **2020**, *198*, 108207. [CrossRef]

24. Juenger, M.C.; Snellings, R.; Bernal, S.A. Supplementary cementitious materials: New sources, characterization, and performance insights. *Cem. Concr. Res.* **2019**, *122*, 257–273. [CrossRef]

25. Piatak, N.M.; Parsons, M.B.; Seal, R.R. Characteristics and environmental aspects of slag: A review. *Appl. Geochem.* **2015**, *57*, 236–266. [CrossRef]

26. Tiwari, M.K.; Bajpai, S.; Dewangan, U.; Tamrakar, R.K. Suitability of leaching test methods for fly ash and slag: A review. *J. Radiat. Res. Appl. Sci.* **2015**, *8*, 523–537. [CrossRef]

27. Jindal, B.B.; Jangra, P.; Garg, A. Effects of ultra fine slag as mineral admixture on the compressive strength, water absorption and permeability of rice husk ash based geopolymer concrete. *Mater. Today Proc.* **2020**, *32*, 871–877. [CrossRef]

28. Hussain, F.; Kaur, I.; Hussain, A. Reviewing the influence of GGBFS on concrete properties. *Mater. Today Proc.* **2020**, *32*, 997–1004. [CrossRef]

29. Samchenko, S.; Kozlova, I.; Zorin, D. The effect of ultrafine fillers on the properties of cement-sand mortars. *Mater. Today Proc.* **2019**, *19*, 2096–2099. [CrossRef]

30. Siddique, R.; Bennacer, R. Use of iron and steel industry by-product (GGBS) in cement paste and mortar. *Resour. Conserv. Recycl.* **2012**, *69*, 29–34. [CrossRef]

31. Özbay, E.; Erdemir, M.; Durmuş, H.I. Utilization and efficiency of ground granulated blast furnace slag on concrete properties—A review. *Constr. Build. Mater.* **2016**, *105*, 423–434. [CrossRef]

32. Gao, D.; Meng, Y.; Yang, L.; Tang, J.; Lv, M. Effect of ground granulated blast furnace slag on the properties of calcium sulfoaluminate cement. *Constr. Build. Mater.* **2019**, *227*, 116665. [CrossRef]

33. Rashad, A.M. An overview on rheology, mechanical properties and durability of high-volume slag used as a cement replacement in paste, mortar and concrete. *Constr. Build. Mater.* **2018**, *187*, 89–117. [CrossRef]

34. Amran, M.; Murali, G.; Khalid, N.H.A.; Fediuk, R.; Ozbakkaloglu, T.; Lee, Y.H.; Haruna, S.; Lee, Y.Y. Slag uses in making an ecofriendly and sustainable concrete: A review. *Constr. Build. Mater.* **2021**, *272*, 121942. [CrossRef]

35. He, X.; Ma, M.; Su, Y.; Lan, M.; Zheng, Z.; Wang, T.; Strnadel, B.; Zeng, S. The effect of ultrahigh volume ultrafine blast furnace slag on the properties of cement pastes. *Constr. Build. Mater.* **2018**, *189*, 438–447. [CrossRef]

36. Liu, Y.; Zhang, Z.; Hou, G.; Yan, P. Preparation of sustainable and green cement-based composite binders with high-volume steel slag powder and ultrafine blast furnace slag powder. *J. Clean. Prod.* **2021**, *289*, 125133. [CrossRef]

37. Zhang, T.; Tian, W.; Guo, Y.; Bogush, A.; Khayrulina, E.; Wei, J.; Yu, Q. The volumetric stability, chloride binding capacity and stability of the Portland cement-GBFS pastes: An approach from the viewpoint of hydration products. *Constr. Build. Mater.* **2019**, *205*, 357–367. [CrossRef]

38. Mohan, A.; Mini, K. Strength and durability studies of SCC incorporating silica fume and ultra fine GGBS. *Constr. Build. Mater.* **2018**, *171*, 919–928. [CrossRef]

39. Kumar, M.P.; Mini, K.; Rangarajan, M. Ultrafine GGBS and calcium nitrate as concrete admixtures for improved mechanical properties and corrosion resistance. *Constr. Build. Mater.* **2018**, *182*, 249–257. [CrossRef]

40. Zhao, H.; Jiang, K.; Yang, R.; Tang, Y.; Liu, J. Experimental and theoretical analysis on coupled effect of hydration, temperature and humidity in early-age cement-based materials. *Int. J. Heat Mass Transf.* **2020**, *146*, 118784. [CrossRef]

41. Hou, D.; Zhang, W.; Sun, M.; Wang, P.; Wang, M.; Zhang, J.; Li, Z. Modified Lucas-Washburn function of capillary transport in the calcium silicate hydrate gel pore: A coarse-grained molecular dynamics study. *Cem. Concr. Res.* **2020**, *136*, 106166. [CrossRef]

42. Tennis, P.D.; Jennings, H.M. A model for two types of calcium silicate hydrate in the microstructure of Portland cement pastes. *Cem. Concr. Res.* **2000**, *30*, 855–863. [CrossRef]

43. Brouwers, H. The work of Powers and Brownyard revisited: Part 1. *Cem. Concr. Res.* **2004**, *34*, 1697–1716. [CrossRef]

44. Jennings, H.M.; Dalgleish, B.J.; Pratt, P.L. Morphological Development of Hydrating Tricalcium Silicate as Examined by Electron Microscopy Techniques. *J. Am. Ceram. Soc.* **1981**, *64*, 567–572. [CrossRef]

45. Mehta, P.K.; Manmohan, D. Pore size distribution and permeability of hardened cement paste. In Proceedings of the 7th International Congress on the Chemistry of Cement, Paris, France, 1–4 July 1980; pp. 1–5.

46. Feng, J.; Liu, S.; Wang, Z. Effects of ultrafine fly ash on the properties of high-strength concrete. *J. Therm. Anal. Calorim.* **2015**, *121*, 1213–1223. [CrossRef]

47. Yan, C.; Zhao, H.; Zhang, J.; Liu, S.; Yang, Z. The cementitious composites using calcium silicate slag as partial cement. *J. Clean. Prod.* **2020**, *256*, 120514. [CrossRef]

48. Han, F.; Luo, A.; Liu, J.; Zhang, Z. Properties of high-volume iron tailing powder concrete under different curing conditions. *Constr. Build. Mater.* **2020**, *241*, 118108. [CrossRef]

49. Attari, A.; McNally, C.; Richardson, M.G. A combined SEM–Calorimetric approach for assessing hydration and porosity development in GGBS concrete. *Cem. Concr. Compos.* **2016**, *68*, 46–56. [CrossRef]

Article

Interpretable Machine Learning Models for Punching Shear Strength Estimation of FRP Reinforced Concrete Slabs

Yuanxie Shen, Junhao Sun and Shixue Liang *

School of Civil Engineering and Architecture, Zhejiang Sci-Tech University, Hangzhou 310018, China; shenyuanxiee@163.com (Y.S.); sunjunh98@163.com (J.S.)
* Correspondence: liangsx@zstu.edu.cn

Abstract: Fiber reinforced polymer (FRP) serves as a prospective alternative to reinforcement in concrete slabs. However, similarly to traditional reinforced concrete slabs, FRP reinforced concrete slabs are susceptible to punching shear failure. Accounts of the insufficient consideration of impact factors, existing empirical models and design provisions for punching strength of FRP reinforced concrete slabs have some problems such as high bias and variance. This study established machine learning-based models to accurately predict the punching shear strength of FRP reinforced concrete slabs. A database of 121 groups of experimental results of FRP reinforced concrete slabs are collected from a literature review. Several machine learning algorithms, such as artificial neural network, support vector machine, decision tree, and adaptive boosting, are selected to build models and compare the performance between them. To demonstrate the predicted accuracy of machine learning, this paper also introduces 6 empirical models and design codes for comparative analysis. The comparative results demonstrate that adaptive boosting has the highest predicted precision, in which the root mean squared error, mean absolute error and coefficient of determination of which are 29.83, 23.00 and 0.99, respectively. GB 50010-2010 (2015) has the best predicted performance among these empirical models and design codes, and ACI 318-19 has the similar result. In addition, among these empirical models, the model proposed by El-Ghandour et al. (1999) has the highest predicted accuracy. According to the results obtained above, SHapley Additive exPlanation (SHAP) is adopted to illustrate the predicted process of AdaBoost. SHAP not only provides global and individual interpretations, but also carries out feature dependency analysis for each input variable. The interpretation results of the model reflect the importance and contribution of the factors that influence the punching shear strength in the machine learning model.

Keywords: machine learning; FRP reinforced concrete slab; punching shear strength; SHAP

Citation: Shen, Y.; Sun, J.; Liang, S. Interpretable Machine Learning Models for Punching Shear Strength Estimation of FRP Reinforced Concrete Slabs. *Crystals* **2022**, *12*, 259. https://doi.org/10.3390/cryst12020259

Academic Editors: Yifeng Ling, Chuanqing Fu, Peng Zhang and Peter Taylor

Received: 26 January 2022
Accepted: 13 February 2022
Published: 14 February 2022

Publisher's Note: MDPI stays neutral with regard to jurisdictional claims in published maps and institutional affiliations.

1. Introduction

Reinforced concrete slabs are one of the common horizontal load-carrying members in civil engineering, and widely applied in bridges, ports and hydro-structures. Since there is no beam in the flat slab under longitudinal load, the punching failure of reinforced concrete slab occurred easily. Many researchers have carried out numerous experiments on the punching shear resistance of reinforced concrete slabs, and obtained successful results [1–3]. However, steel bars are prone to corrosion, which will result in the shortening of the actual service life [4]. In recent years, with the development of technology, the durability of structure has attracted people's attention. For coastal areas and the areas of using chlorides such as deicing salt, the actual service life of structures is often much lower than their design service life, resulting in massive losses [5].

Fiber reinforced polymer (FRP) is a material which has many advantages such as light, high strength and corrosion resistance. In a corrosion environment, to solve the problem of short actual service life of structure, FRP bar can be applied as an alternative to steel bars in concrete structures. This is because FRP bar can be appropriate for service demands of concrete structure in severe environments which contain plenty of chloride ion and sulfate ion. However, FRP bar has a low elastic modulus, which will result in FRP reinforced concrete slabs being more prone to punching failure than reinforced slabs (Figure 1). Many experimental studies show that [6–11] under the same conditions, the punching shear strength and stiffness of the cracked FRP reinforced concrete slabs decrease faster than reinforced concrete slabs. Matthys and Taerwe [6] concluded that the crack development of slabs and brittleness of punching failure were significantly affected by the bond performance of FRP grid reinforcement through experimental results. The experimental results of Ospina et al. [7] indicated that the bond behavior between FRP bar and concrete have a significant impact on punching shear capacity of FRP reinforced concrete slabs. It is shown that the slab thickness and the concrete compressive strength were of considerable influence on the punching shear capacities of FRP reinforced concrete slabs by Bouguerra et al. [9]. Ramzy et al. [12] reported that the punching shear behavior of flat slab was related to the size effect of construction and the Young's modulus of FRP reinforcement.

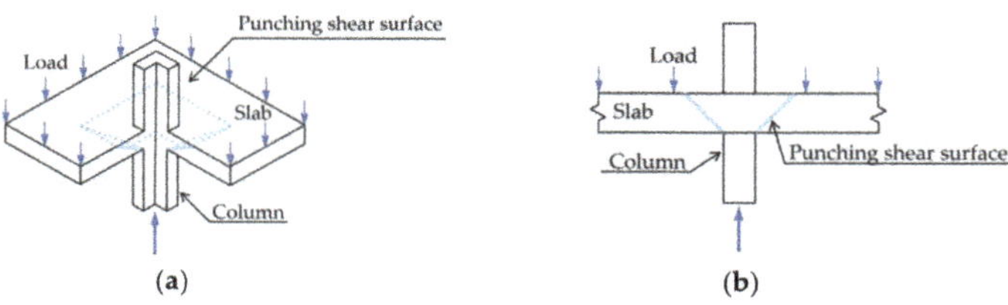

Figure 1. Punching shear failure mode of FRP reinforced concrete slabs. (**a**) Stereogram; (**b**) Profile.

In terms of theoretical models, most of the punching shear strength computational formulas of FRP reinforced concrete slabs were derived from traditional reinforced concrete flat with modifications to account for FRP [13]. Some current design specifications such as the GB 50010-2010 [14] and the ACI 318-14 [15] regard the eccentric shear stress model as theoretical basis. Based on the ACI 318-11, El-Ghandour et al. [16] suggested that one considers the impact of elastic modulus of FRP bars, and come up with an improved equation for punching shear strength. After that, El-Ghandour et al. [17] used the same method to modify the design formula of the BS 8110-97. Matthys and Taerwe [6] considered the effect of the equivalent reinforcement ratio, and proposed a modification of the BS 8110-97. On this basis, Ospina et al. [7] proposed an empirical model for computing the punching shear strength of FRP slabs by modifying the relation between the punching shear capacity and the equivalent reinforcement ratio. On the basis of the probability of exceedance, Ju et al. [18] proposed a new approach for analyzing the punching shear strength of FRP reinforced two-way concrete slabs by using Monte Carlo simulations. However, the aforementioned empirical models adopted some simplifications during theoretical derivations, thus the empirical models were unable to consider all of the influential factors. What's more, the parameters in the aforementioned empirical models were determined by the traditional regression analyses from experimental results. Therefore, the accuracy of the models is highly dependent on the choices of theoretical models and quality of the databases.

In recent years, with the development of artificial intelligence, some algorithms with data at the core have emerged [19]. Among these algorithms, machine learning has received remarkable attention of researchers, and there have been many successful examples [20–24]. In structure engineering, Hoang et al. [25] constructed machine learning based alternatives for estimating the punching shear capacity of steel fiber reinforced concrete (SFRC) flat slabs. Hoang et al. [26] presented the development of an ensemble machine learning model to predict the punching shear resistance of R/C interior slabs. Mangalathu et al. [27] build an explainable machine learning model to predict the punching shear strength of flat slabs without transverse reinforcement. In addition, some researchers [28,29] even use the atomistic simulations as the input parameters of machine learning to predict the performance of materials and structures, which has also seen success. To the best of the authors' knowledge, however, no study examined the interpretable machine learning models in predicting the punching shear strength of FRP reinforced concrete slabs. Therefore, this study intends to fill this research gap. The deep relationship between material properties and punching shear strength will be found if the machine learning research on FRP reinforced concrete slabs is carried out. It is helpful to refine the traditional empirical models and design codes and make them more accurate than before. In this paper, an experimental database for the punching shear strength of FRP reinforced concrete slabs is first compiled, and used for training, validating, and testing machine learning models.

Four machine learning algorithms, namely artificial neural network (ANN), support vector machine (SVM), decision tree (DT) and adaptive boosting (AdaBoost), are employed for punching shear strength prediction. Then, by comparing the performance of machine learning models, their efficiency and accuracy can be determined. These comparisons are valuable for identifying the efficiency and prediction ability of the machine learning models. However, there is problems which need to be solved; for example, the values of hyper-parameters in machine learning are often difficult to determine. Therefore, to assist the improvement of the predicted performance of machine learning models, optimization methods such as particle swarm optimization (PSO) and empirical method will be used. In addition, since previous models based on machine learning found it difficult to explain the predicted mechanism, the predicted results of models were somehow unconvincing [27]. This paper uses SHapley Additive exPlanation (SHAP) [30] to explain the predicted result of model. Distinct from other machine learning papers explained by SHAP, the kernel explainer will be used in this paper, which is appropriate for all the machine learning models. Not only can SHAP carry out analysis of feature importance for input factors, it can also determine whether the impact of input features on predictions is positive or negative. In addition, SHAP can make researchers realize the interrelation between input features, and how each input feature will influence the final predicted value for a single sample. Consequently, the emergence of SHAP renders the predicted results of machine learning more convincing than before.

2. Experimental Database of FRP Reinforced Concrete Slab

To ensure the accuracy of predicted results, a database with adequate samples is required for model training. In this paper, 121 groups of experimental results [6–10,12,31–49] of FRP reinforced concrete slabs under punching shear tests were collected, and made it into a database. In the data set, 80% of the data was used as a training set (10% of this was used as validation set), and remaining 20% were used as test set. In this study, 6 critical parameters are selected to characterize the punching shear strength of FRP slabs such as the types of column section, cross-section area of column (A/cm^2), slab's effective depth (d/mm), compressive strength of concrete (f'_c/Mpa), Young's modulus of FRP reinforcement (E_f/Gpa), and reinforcement ratio ($\rho_f/\%$). According to the findings of Ju et al. [18], these 6 parameters have a great influence on punching shear strength of FRP slabs. Therefore, these 6 parameters are considered as input parameters and the punching shear strength of FRP slabs is considered as an output parameter of machine learning algorithms. To simplify the names of the 6 input parameters and the 1 output parameter, we used x_1 to x_6 and y to represent them, respectively. The column section has 3 types: square ($x_1 = 1$), circle ($x_1 = 2$), rectangle ($x_1 = 3$). The distribution of the dataset is shown as Table 1 and Figure 2, and the detailed information is shown as Appendix A.

Crystals **2022**, 12, 259

Table 1. Parameters in the punching shear strength samples.

Parameter	Unit	Minimum	Maximum	Std. Dev	Mean	Type
x_1: Types of column section	-	1.00	3.00	0.75	1.44	Input
x_2: cross-section area of column	cm^2	50.27	2025.00	515.94	817.72	Input
x_3: slab's effective depth	mm	55.00	284.00	52.71	136.38	Input
x_4: compressive strength of concrete	Mpa	22.16	118.00	14.53	41.17	Input
x_5: Young's modulus of FRP reinforcement	Gpa	28.40	147.00	24.34	60.36	Input
x_6: reinforcement ratio	%	0.15	3.76	0.67	1.03	Input
y: punching shear strength	kN	61.00	1600.00	288.87	402.34	Output

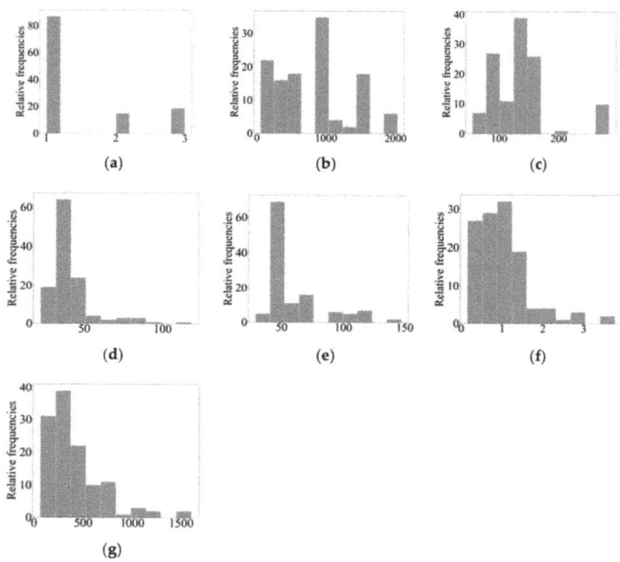

Figure 2. Histograms of input and output variables. (**a**) x_1; (**b**) x_2; (**c**) x_3; (**d**) x_4; (**e**) x_5; (**f**) x_6; (**g**) y.

Before starting the machine learning procedure, the data of input variables should be preprocessed. In this paper, we used deviation standardization as the preprocessing method; this can be written as:

$$x_i^* = \frac{x_i - \min_{1 \le j \le m}\{x_j\}}{\max_{1 \le j \le m}\{x_j\} - \min_{1 \le j \le m}\{x_j\}} \tag{1}$$

where x is the sample before feature scaling; x^* is the sample after feature scaling; m is the total number of samples. Since the sample x contains features of several input parameters, it can also be named the feature vector.

3. Machine Learning Algorithms

Generally, the implementation of machine learning has four stages: (a) divide the database into training set and test set; (b) apply the training set to model training; (c) check whether the accuracy requirements are met; d) output the predicted model for test or adjust the values of hyper-parameters. The flowchart for this procedure is shown as Figure 3. A total of 4 machine learning models, namely, ANN, SVM, DT and Adaboost, are selected in this study. As for ANN, an input layer, hidden layers and an output layer are formed for the predicted results, where several neurons are applied. As for SVM, a prediction equation is formed based on the predicted output in which the error should be smaller than a fixed value. As for DT, the whole database is separated into tree-like decisions which are based on one or several input characteristics, and the output is the tree which falls into the data. As for the AdaBoost, a group of weak learners are collected to form a strong learner which achieves better predicted results. Notably, the predicted performance of model has relations with the value of hyper-parameters. Therefore, in this paper, some methods will be selected to better determine the value of hyper-parameters.

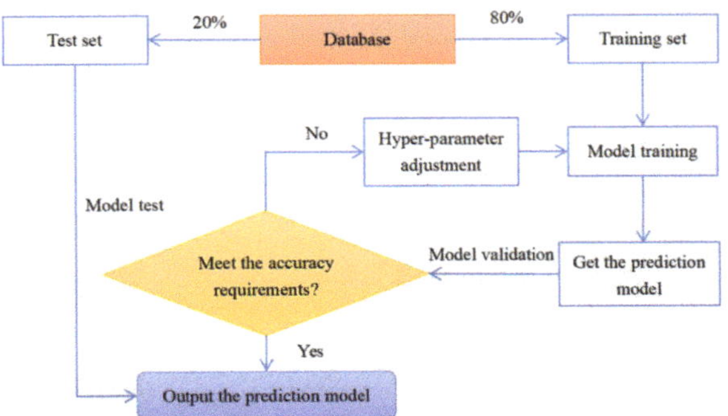

Figure 3. Flowchart for implementation of machine learning algorithm.

3.1. Artificial Neural Network

Artificial neural network (ANN) is a supervised learning algorithm that was originally used to simulate the neural network of human brain [50]. Generally, a simple ANN model consists of an input layer, one or two hidden layers, and an output layer. There are several neurons in each layer of ANN, where the number of neurons in the input layer and output layer depends on the number of input parameters and output parameters, respectively. In this paper, the number of neurons in the input layer and output layer can be seen in Table 1, which equals 6 and 1, respectively.

Except for the output layer, the neurons of each layer needs to be activated by activation function. In ANN, generally, the sigmoid function is used as activation function [51], which expression can be written as:

$$g(z) = \frac{1}{1 + e^{-z}} \tag{2}$$

$$z = w^{\mathrm{T}} x^* + b \tag{3}$$

where b is the bias unit; w denotes the weight vector of sample. According to Rumelhart et al. [52] described, when ANN is used for regression forecasting, the cost function of model as follows:

$$J(w, b) = \frac{1}{m} \sum_{i=1}^{m} \left(y_{\mathrm{pred}}^{(i)} - y^{(i)} \right)^2 + \frac{C}{2m} \|w\|^2 \tag{4}$$

where y_{pred} denotes the predicted value of machine learning; C is the regularization coefficient.

Since the cost function represents the error between the actual value and the predicted value, to improve the accuracy of predicted result, a method similar to gradient descent, in order to decrease the value of cost function, is required. The gradient descent is expressed as:

$$\begin{cases} w_j := w_j - \alpha \frac{\partial}{\partial w_j} J(w, b) \\ b := b - \alpha \frac{\partial}{\partial b} J(w, b) \end{cases} \tag{5}$$

where α is learning rate.

Before starting the training procedure, there are some values of hyper-parameters need to be set. For neuron number of hidden layer, Shen et al. [53] mentioned a method to determine their value range, which method as follows:

$$\begin{cases} \sum\limits_{i=0}^{X} C_H^i > m \\ H = \sqrt{X+O} + a \\ H = \log_2 X \end{cases} \tag{6}$$

where X is the number of neurons in the input layer; H is the number of neurons in the hidden layer; O is the number of neurons in the output layer; a is the constant between 1 to 10. According to the Equation (6), it can be ascertained that the range of neurons in hidden layer ranges from 7 to 13.

In order to better compare the performance of models which have different neuron numbers in the hidden layer, 3 indicators are introduced, which names root mean squared error (RMSE), mean absolute error (MAE) and coefficient of determination (R^2), which are, respectively, defined as:

$$\text{RMSE} = \sqrt{\frac{1}{m} \sum\limits_{i=1}^{m} \left(y_{\text{pred}}^{(i)} - y^{(i)} \right)^2} \tag{7}$$

$$\text{MAE} = \frac{1}{m} \sum\limits_{i=1}^{m} \left| y_{\text{pred}}^{(i)} - y^{(i)} \right| \tag{8}$$

$$R^2 = 1 - \frac{\sum\limits_{i=1}^{m} \left(y_{\text{pred}}^{(i)} - y^{(i)} \right)^2}{\sum\limits_{i=1}^{m} \left(y^{(i)} - \frac{1}{m} \sum\limits_{i=1}^{m} y^{(i)} \right)^2} \tag{9}$$

The performance of ANN which have different neuron numbers in hidden layer is shown as Table 2. According to the comparative results, the number of neurons in the hidden layer H is determined as 8. In addition, the values of other hyper-parameters confirmed by manual adjustment are: the number of hidden layers equal 1; the training times equal 5000; the learning rate α equals 5×10^{-5}; the regularization coefficient C equals 20.

Table 2. Performance and ranking of ANN which have different neuron numbers in hidden layer.

Neuron Number of Hidden Layer	Validation Set			Ranking		
	RMSE	MAE	R^2	RMSE	MAE	R^2
7	100.24	80.92	0.89	5	5	5
8	87.28	71.72	0.91	2	1	2
9	86.91	73.00	0.91	1	3	1
10	87.92	72.60	0.91	3	2	3
11	94.09	79.04	0.90	4	4	4
12	101.66	83.38	0.88	6	6	6
13	109.43	89.78	0.86	7	7	7

3.2. Support Vector Machine

Support vector machine (SVM) was originated from statistical learning theory, and it was firstly proposed by Cortes and Vapnik [54] in 1995. When SVM is used for regression analysis, it can also be called support vector regression (SVR). The predicted method of SVR maps the data to high-dimensional space and use hyperplane for fitting, and these data are generally hard to fit into low-dimensional space [55]. The expression of SVR reflect the relation between input factors and predicted value [56], it can be written as:

$$y_{\text{pred}} = w^{\text{T}}\phi(x) + b \tag{10}$$

where $\phi(x)$ means feature vector after mapping. With the same as ANN, SVR also has the cost function to represents the performance of model. The cost function can be written as:

$$J(w, b, \xi_i, \xi_i^*) = \frac{1}{2}\|w\|^2 + C\sum_{i=1}^{m}(\xi_i + \xi_i^*) \tag{11}$$

where ξ and ξ^* are the slack variables, they represent the errors of the sample exceeding the lower and upper bounds of the hyperplane, respectively.

To minimize the value of cost function, Lagrange multiplier method is used for transforming the original problem to its dual problem. The Lagrange multiplier method is formulated in the following form:

$$\begin{aligned}
L(w, b, \lambda, \lambda^*, \xi, \xi^*, \mu, \mu^*) &= \tfrac{1}{2}\|w\|^2 + C\sum_{i=1}^{m}(\xi_i + \xi_i^*) - \sum_{i=1}^{m}\mu_i\xi_i - \sum_{i=1}^{m}\mu_i^*\xi_i^* \\
&+ \sum_{i=1}^{m}\lambda_i\left(y_{\text{pred}}^{(i)} - y^{(i)} - \varepsilon - \xi_i\right) + \sum_{i=1}^{m}\lambda_i^*\left(y^{(i)} - y_{\text{pred}}^{(i)} - \varepsilon - \xi_i^*\right)
\end{aligned} \tag{12}$$

where μ, μ^*, λ, λ^* denote the Lagrange multiplier; ε is the maximum allowable error. After transforming, the expression of dual problem as follows:

$$\text{Maximise} \sum_{i=1}^{m} y^{(i)}(\lambda_i^* - \lambda_i) - \varepsilon(\lambda_i^* + \lambda_i) - \frac{1}{2}\sum_{i=1}^{m}\sum_{j=1}^{m}(\lambda_i^* - \lambda_i)\left(\lambda_j^* - \lambda_j\right)x_i^{\text{T}}x_j \tag{13}$$

subjected to $\sum_{i=1}^{m}(\lambda_i^* - \lambda_i) = 0, 0 \leq \lambda_i, \lambda_i^* \leq C$.

The Karush-Kuhn-Tucker (KKT) conditions need to be met for Equation (13), which can be written as:

$$\begin{cases}
\lambda_i\left(y_{\text{pred}}^{(i)} - y^{(i)} - \varepsilon - \xi_i\right) = 0, \\
\lambda_i^*\left(y^{(i)} - y_{\text{pred}}^{(i)} - \varepsilon - \xi_i^*\right) = 0, \\
\lambda_i\lambda_i^* = 0, \xi_i\xi_i^* = 0, \\
(C - \lambda_i)\xi_i = 0, (C - \lambda_i^*)\xi_i^* = 0.
\end{cases} \tag{14}$$

According to Equations (13) and (14), the weight vector w and bias unit b can be acquired. Finally, the expression of SVR as follows:

$$y_{\text{pred}} = \sum_{i=1}^{m}(\lambda_i^* - \lambda_i)\kappa(x, x_i) + b \tag{15}$$

$$\kappa(x_i, x_j) = \phi(x_i)^{\text{T}}\phi(x_j) \tag{16}$$

where $\kappa(x_i, x_j)$ denotes kernel function, which can used to express the inner product of eigenvectors in high-dimensional space. The kernel function has many types, and the radial basis function (RBF) kernel function is selected in this paper, which can be expressed as:

$$\kappa(x_i, x_j) = \exp\left(-\gamma\|x_i - x_j\|^2\right) \tag{17}$$

where $\gamma > 0$ is the coefficient of RBF kernel function, which is a hyper-parameter. Except that, in this paper, the values of other hyper-parameters in SVR need to be determined: the maximum allowable error ε, the regularization coefficient C. There are many methods which can be used for numerical adjustment of hyper-parameters, in this paper we have selected particle swarm optimization.

Particle swarm optimization (PSO) is an optimized algorithm inspired by the foraging behavior of bird flock [57,58]. On account of the fact that PSO has a large number of advantages, such as being easily comprehendible and containing, it is usually used for hyper-parameter adjustment in SVM [59,60]. For every particle in PSO, there is two parameters to determine the search result, which are position p and velocity v, respectively. These 2 parameters are defined as [58]:

$$v_{d*}^{t+1} = W v_{d*}^t + r_1 C_1 \left(P_{d*}^t - p_{d*}^t \right) + r_2 C_2 \left(G_{d*}^t - p_{d*}^t \right) \tag{18}$$

$$p_{d*}^{t+1} = p_{d*}^t + v_{d*}^{t+1} \tag{19}$$

where t is the number of iterations; d^* denotes the dimension of search space; W is the inertial coefficient; C_1 and C_2 are the self-learning factor and the global learning factor; r_1 and r_2 are random number between (0, 1); P is the location with the best fitness among all the visited locations of each particle; G is the location with the best fitness among all the visited locations of all the particles.

In this paper, some hyper-parameters in Equations (18) and (19) have been confirmed: the maximum of iteration equals 150; the dimension of search space d^* equals 3; the inertial coefficient W equals 0.8; the total number of particles equals 40; the self-learning factor C_1 and global learning factor C_2 are all equal to 2. The hybrid machine learning approach consists of PSO and SVR can be called PSO-SVR, whose best value of hyper-parameters also can be determined by calculating: the maximum allowable error ε equals 10; the regularization coefficient C equals 4238.58; the coefficient of RBF kernel function γ equals 2.88.

3.3. Decision Tree

Decision tree (DT) served as a supervised learning algorithm that is proposed by Quinlan [61]. In the beginning, DT only could be used for conducting the classification forecasting, and its types less such as iterative dichotomiser 3 (ID3) and classifier 4.5 (C4.5). In this paper, classification and regression tree (CART) [62] is selected to make a study of regression prediction. The dividing evidence of CART is the variance of samples, which can be written as:

$$\text{Minimise} \left[\sum_{x_i \in R_1} \left(y^{(i)} - c_1 \right)^2 + \sum_{x_i \in R_2} \left(y^{(i)} - c_2 \right)^2 \right] \tag{20}$$

where R_1 and R_2 denote the sample set after dividing; c_1 and c_2 are mean of samples in R_1 and R_2, respectively. With the division of sample set, gradually, the purity of leaf nodes will improve.

Pruning is a method to avoid the over-fitting in DT, and it contains two basic methods: pre-pruning and post-pruning [63,64]. In this paper, the specific way of pruning is to adjust the maximum of depth, which equals 5.

3.4. Adaptive Boosting

Adaptive boosting (AdaBoost) is one of the ensemble learning algorithms that can combine multiple weak learners into strong learner [65]. To be distinct from aforementioned machine learning algorithms, each sample in AdaBoost has a subsampling weight which will constantly adjust [66]. At the beginning of the training process, the weights of samples need to be initialized [67], which can be written as:

$$w_{ki}^* = \frac{1}{m} \tag{21}$$

where w^*_{ki} denotes the weight of the i-th sample in the k-th weak learner. The largest error of the k-th weak learner on the training set as follows:

$$E_k = \max\left|y^{(i)} - G_k(x_i)\right| \tag{22}$$

where $G_k(x_i)$ denotes the predicted value of the i-th sample in the k-th weak learner. The relative error e_{ki} of each sample and the error ratio e_k of the k-th weak learner can be written as:

$$e_{ki} = \frac{\left(y^{(i)} - G_k(x_i)\right)^2}{E_k^2} \tag{23}$$

$$e_k = \sum_{i=1}^{m} w_{ki} e_{ki} \tag{24}$$

After that, the weight β_k of the k-th weak learner can be calculated, the value of which will increase as the error ratio e_k decreases. The weight β_k can be defined as:

$$\beta_k = \frac{e_k}{1 - e_k} \tag{25}$$

Afterwards, the error ratio of the next weak learner can be reckoned, and the weights of samples need to be updated:

$$w^*_{k+1,i} = \frac{w^*_{ki}}{Z_k}\beta_k^{1-e_{ki}} \tag{26}$$

$$Z_k = \sum_{i=1}^{m} w_{ki}\beta_k^{1-e_{ki}} \tag{27}$$

where Z_k is the normalization factor. Finally, the expression of the strong learner can be written as:

$$f(x) = \sum_{k=1}^{K}\left(\ln\frac{1}{\beta_k}\right)g(x) \tag{28}$$

where K is the total number of weak learners; $g(x)$ is the median of all the $\beta_k G_k(x)$. The algorithm of the weak learner needs to be confirmed, and PSO-SVR is chosen in this paper.

3.5. Predicted Results

The result comparison between ANN, PSO-SVR, DT and AdaBoost are provided in Table 3. The four machine learning models shows high predicted accuracy reflected in R^2. Especially AdaBoost and PSO-SVR, the predicted accuracy of the former is highest among these four AI models in training set, and the latter is in test set. It is noted that although the forecasting performance of ANN for the test set is better than DT and just slightly lower than PSO-SVR and AdaBoost, the performance for training set is much lower than the other three models.

Table 3. Result comparison between machine learning algorithms.

Machine Learning Model	Training Set			Test Set		
	RMSE	MAE	R^2	RMSE	MAE	R^2
ANN	102.68	60.09	0.88	44.80	38.59	0.97
PSO-SVR	54.85	27.92	0.96	33.46	26.27	0.99
DT	59.52	37.89	0.96	66.57	52.93	0.94
AdaBoost	27.29	20.70	0.99	38.40	32.28	0.98

Figures 4–7 depict the scatter distribution of predicted results for each machine learning model. In Figure 4, it can be found that the fitting condition of ANN is not very

suitable when the true value is larger than 1400 kN. Maybe this situation is related to insufficient data of punching shear strength larger than 1400 kN. Therefore, it is noted that if we want to use this ANN model for predicting the punching shear strength of FRP reinforced concrete slabs, the value range of the predicted object should be restrained under 1400 kN. In addition, from the figure, it can be seen that ANN focuses more on fitting samples which are located in the interval with dense sample distribution. Consequently, a way to improve the overall predicted accuracy of ANN is to ensure that the samples are evenly distributed. It is demonstrated in Figure 5 that PSO-SVR shows the good predicted performance. Although there some predicted values have a relatively larger error between the actual values, the overall predicted accuracy is higher than ANN and DT. Figure 6 depicts the predictions of DT on the dataset; there doubt that the predicted result of DT for the test set is worse than ANN and PSO-SVR. In training set, the predictions of DT are almost no error with the punching shear strength is larger than 1200 kN, it is related to the predicted mechanism of DT [68]. Moreover, the situation also can be seen from the predicted results of DT on punching shear strength less than 1200 kN, the predicted value of which is same when the punching strength is in the identical range. Meanwhile, in test set, DT has a low predicted accuracy for samples with punching shear strength of around 1200 kN, i.e., the DT model proposed in this paper has the risk of over-fitting when the punching shear strength larger than 1200 kN. Obviously, no matter ANN or DT, the predicted accuracy of them is closely related to the number of samples in each value range. Furthermore, in this paper, it can be concluded that neither ANN nor DT has mined the true intrinsic connection between the data. Figure 7 shows the predicted results of AdaBoost for the training set and the test set, it can be seen that all of the dots are closely distributed around the best fitting line. Since PSO-SVR is chosen as the weak learner of AdaBoost, some wildly inaccurate predicted values are improved, and some comparatively accurate predicted values are retained. Moreover, compare Figure 4 to Figure 7, it is no hard to see that the fitting situation of AdaBoost reveals the best predicting performance.

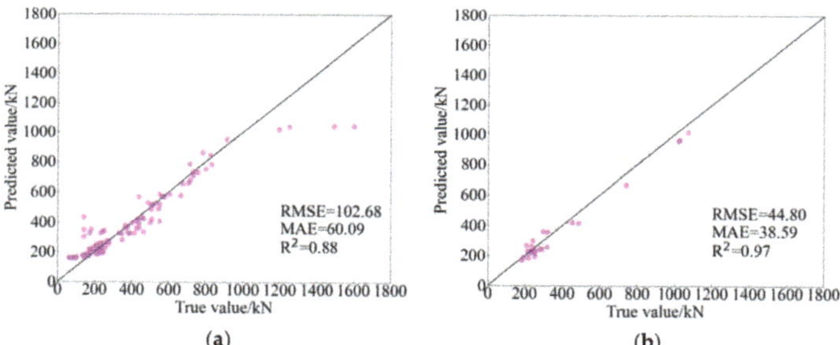

Figure 4. The predictions of ANN for data set. (**a**) Training set; (**b**) Test set.

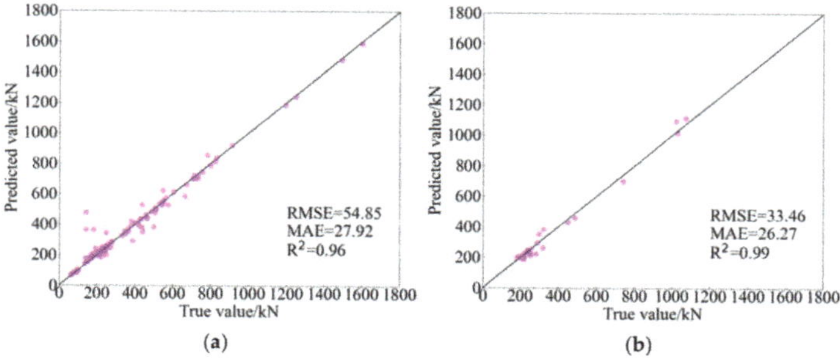

Figure 5. The predictions of PSO-SVR for data set. (**a**) Training set; (**b**) Test set.

3.6. Comparison with Traditional Empirical Models

In order to reflect the accuracy of the machine learning calculation results, this paper introduces design specifications and empirical models proposed by researchers. In this study, 3 punching shear strength models are selected from design codes and reviewed: GB 50010-2010 (2015) [14], ACI 318-19 [69] and BS 8110-97 [70]. Most design codes adopt the same function for both FRP slabs and reinforced concrete slabs. The punching shear strength of slabs as a function of the concrete compressive strength. Some design codes also take the reinforcement ratio, size effect, depth of the slab and so on into consideration. Three design specifications and three empirical models are adopted in comparative analysis between calculations. The formulas of selected traditional empirical model are shown in Table 4.

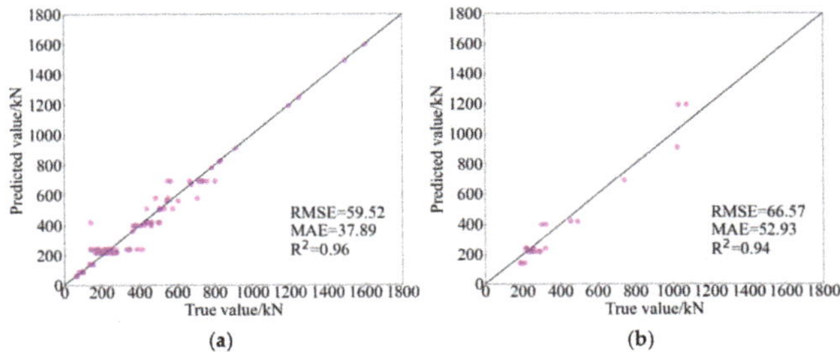

Figure 6. The predictions of DT for data set. (**a**) Training set; (**b**) Test set.

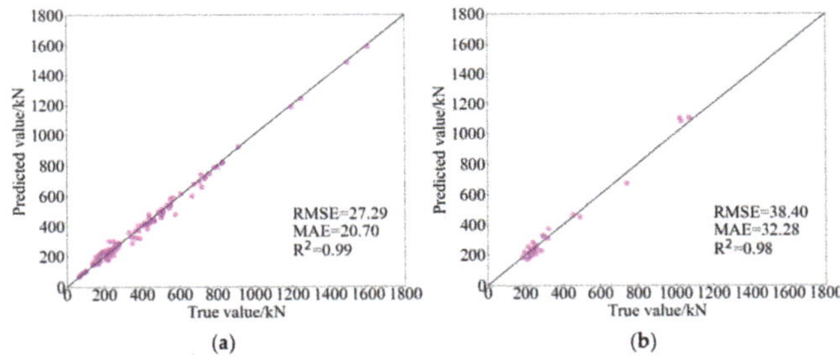

Figure 7. The predictions of AdaBoost for data set. (**a**) Training set; (**b**) Test set.

Table 4. Calculation formulas of traditional empirical model.

Number of Formula	Origin	Expression
Formula (1)	GB 50010-2010 (2015) [14]	$V_1 = 0.7\beta_h f_t \eta b_{0,0.5d} d;\ \eta = \min\begin{cases} \eta_1 = 0.4 + \frac{1.2}{\beta_s} \\ \eta_2 = 0.5 + \frac{\alpha_s d}{4b_{0,0.5d}} \end{cases}$
Formula (2)	ACI 318-19 [62]	$V_2 = \min\left[\frac{1}{3}, \frac{1}{6}\left(1 + \frac{2}{\beta_s}\right), \frac{1}{12}\left(2 + \frac{\alpha_s d}{b_{0,0.5d}}\right)\right]\lambda_s \sqrt{f'_c} b_{0,0.5d} d;\ \lambda_s = \sqrt{\frac{2}{1+0.004d}} \leq 1$
Formula (3)	BS 8110-97 [63]	$V_3 = 0.79(100\rho_f)^{1/3}\left(\frac{400}{d}\right)^{1/4}\left(\frac{f_{cu}}{25}\right)^{1/3} b_{0,1.5d} d$
Formula (4)	El-Ghandour et al. (1999) [16]	$V_4 = 0.33\sqrt{f'_c}\left(\frac{E_f}{E_s}\right)^{1/3} b_{0,0.5d} d$
Formula (5)	El-Ghandour et al. (2000) [17]	$V_5 = 0.79\left(100\rho_f 1.8\left(\frac{E_f}{E_s}\right)\right)^{1/3}\left(\frac{400}{d}\right)^{1/4}\left(\frac{f_{cu}}{25}\right)^{1/3} b_{0,1.5d} d$
Formula (6)	Ospina et al. [7]	$V_6 = 2.77(\rho_f f'_c)^{1/3}\sqrt{\frac{E_f}{E_s}} b_{0,1.5d} d$

β_h is the sectional depth influence coefficient; f_t is the design value of tensile strength; $b_{0,0.5d}$ is the the perimeter of the critical section for slabs and footings at a distance of $d/2$ away from the column face; β_s is the ratio of the long side to the short side of the sectional shape under local load or concentrated reaction force; α_s is the influence coefficient of column type; f_{cu} is the compressive strength of concrete cube; $b_{0,1.5d}$ is the perimeter of the critical section for slabs and footings at a distance of $1.5d/2$ away from the loaded area; E_s is the Young's modulus of steel bar.

Table 5 summaries experimental results of machine learning models and traditional models. Obviously, the predicted performance of AdaBoost is better than other predicted models reflected in RMSE, MAE and R^2. ANN, DT and PSO-SVR, the other three AI models, although their predicted results are not best, they show a better performance compared with traditional empirical models. Among these traditional models, Formula (1) (GB 50010-2010 (2015)) has the highest predicted accuracy, the RMSE, MAE and R^2 of which are 150.41, 98.48 and 0.73, respectively. Notably, the predictions between Formulas (1) and (2) only has a little error; perhaps it is related to the fact that they are both derived from the eccentric shear stress model [14,69]. In addition, the forecasting performance of the model proposed by El-Ghandour et al. is better than the other traditional models proposed by other researchers. In brief, the RMSE and MAE of traditional empirical models are generally greater than 100, and the R^2 of them ranged from 0.6 to 0.8. However, the RMSE and MAE of machine learning models are less than 100, and the R^2 of them are generally greater than 0.9 and even close to 1.0. Consequently, it can be said that the difference in the predicted accuracy between traditional empirical models and machine learning models is obvious.

Table 5. Result comparison between machine learning and traditional empirical models.

Indicator	Formula (1)	Formula (2)	Formula (3)	Formula (4)	Formula (5)	Formula (6)	ANN	PSO-SVR	DT	AdaBoost
RMSE	150.41	151.71	176.15	155.01	178.09	174.47	94.08	51.32	60.99	29.83
MAE	98.48	97.10	127.06	113.41	121.75	117.97	55.82	27.59	40.87	23.00
R^2	0.73	0.72	0.63	0.71	0.62	0.63	0.89	0.97	0.96	0.99

Figure 8 shows the predicted results of traditional empirical models and machine learning models for the whole data set. To facilitate comparison between different types of models in these figures, the red dots are used to represent the predicted result of national codes. Afterwards, the blue and purple dots are used to represent the predicted results of traditional models proposed by researchers and machine learning models, respectively. It can be found that the predicted error of traditional models is relatively large, and the machine learning models reach the opposite conclusion. The forecasting result of Figure 8a is similar to Figure 8b, and this conclusion is also reflected in Table 5. In addition, although the predicted accuracy of Formulas (1) and (2) is better than other traditional empirical models, their predicted results are still unsafe. On the contrary, the predicted results of Formula (3) to Formula (6) tend to be conservative even though they do not have the best predicted performance. From the predicted performance of the machine learning models for the complete database, it can be found that the predicted results of ANN and PSO-SVR are likely to be unsafe when the punching shear strength smaller than 200 kN. On the side, when the punching shear strength larger than 1200 kN, the forecasting result of ANN is conservative. Furthermore, except formula 4, the predicted accuracy of all the empirical models are better than ANN when the punching strength larger than 1400 kN. Consequently, it can be concluded that ANN is very dependent on samples.

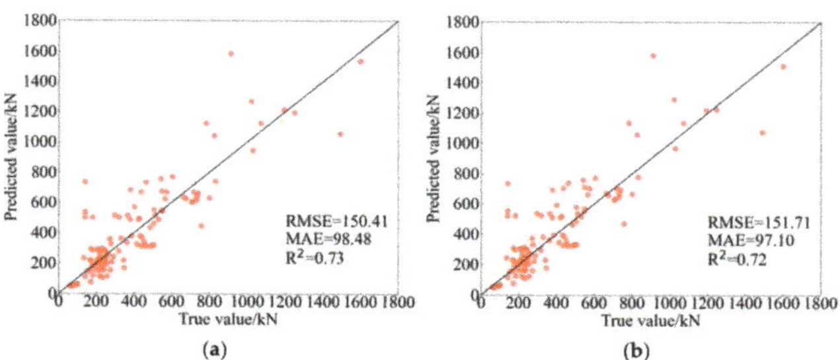

Figure 8. *Cont.*

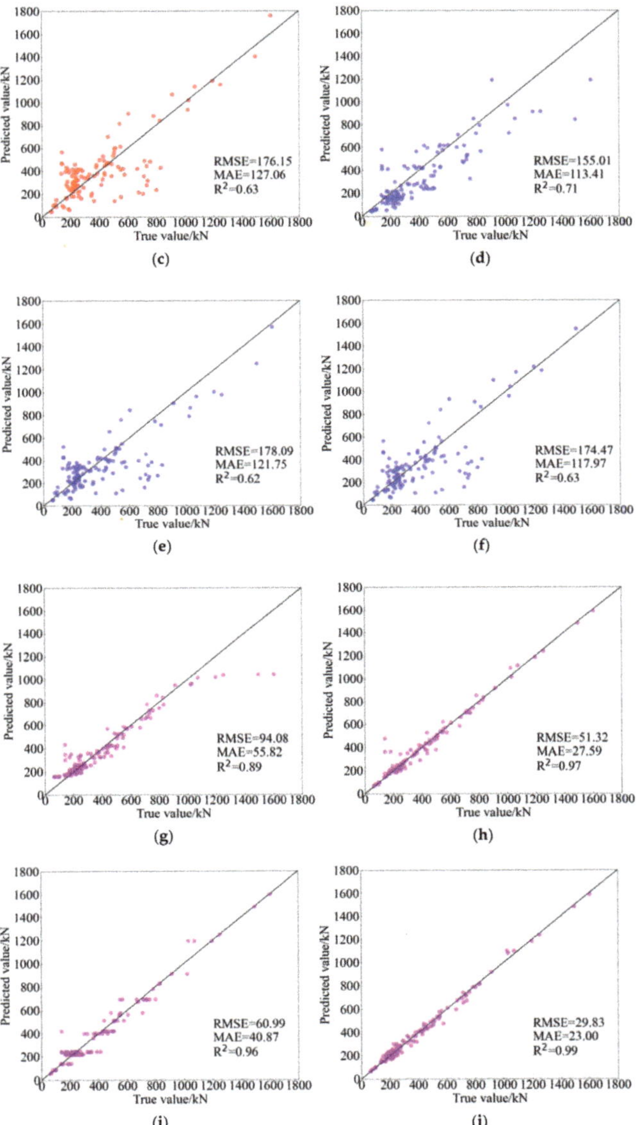

Figure 8. The predictions of models for the whole data set. (**a**) Formula (1); (**b**) Formula (2); (**c**) Formula (3); (**d**) Formula (4); (**e**) Formula (5); (**f**) Formula (6); (**g**) ANN; (**h**) PSO-SVR; (**i**) DT; (**j**) AdaBoost.

4. Model Interpretations

4.1. Shapley Additive Explanation

In this study, SHapley Additive exPlanation (SHAP) is applied to explain the predicting results given by the AdaBoost model. SHAP originates from game theory and is an additive model, i.e., the output of the model is a linear addition of the input variables [71]. The expression of SHAP can be defined as:

$$y_{\text{pred}}^{(i)} = y_{\text{base}} + \sum_{j=1}^{n} f\left(x_{ij}\right) \tag{29}$$

where y_{base} is the baseline value of model; n is the total number of input variables; $f(x_{ij})$ is the SHAP value of the x_{ij}.

According to Equation (29), it can be concluded that the predicted value of each sample depends on the SHAP value of each feature in the sample. In other words, each feature is a contributor for final predicted value. The SHAP value can be positive or negative, depending on the influence trend of each feature to predicted result. SHAP contains several explainers, in current machine learning papers it often uses the tree explainer to illustrate its models, which is only suitable for a part of machine learning algorithms. Consequently, the kernel explainer will be used to illustrate the machine learning model which has the best predicted performance in this paper.

4.2. Global Interpretations

According to the experimental results obtained above, the predicted results of AdaBoost, which has the best predicted performance, was explained. The predicted results of AdaBoost can be interpreted by SHAP in different ways. Firstly, the feature importance analysis of input parameters is shown in Figure 9a. It can be seen that the importance of each input parameter is non-negligible, that is each input parameter will have a different degree of influence on the predicted results. This is consistent with the existing experimental conclusions [4,72]. According to the analysis results, it can be understood that the SHAP value of each parameter are x_1 equals 53.45, x_2 equals 44.50, x_3 equals 126.36, x_4 equals 21.57, x_5 equals 17.23, x_6 equals 52.96, respectively. Obviously, the slab's effective depth (x_3) has the most critical influence on the punching shear capacity, more than twice as much as types of column section (x_1). Moreover, though the Young's modulus of FRP reinforcement (x_5) has the least importance on the predicted results, sometimes ignoring this feature will cause unnecessary errors in the predicted results of model. Consequently, using these 6 features as the input parameters of machine learning algorithm is reasonable.

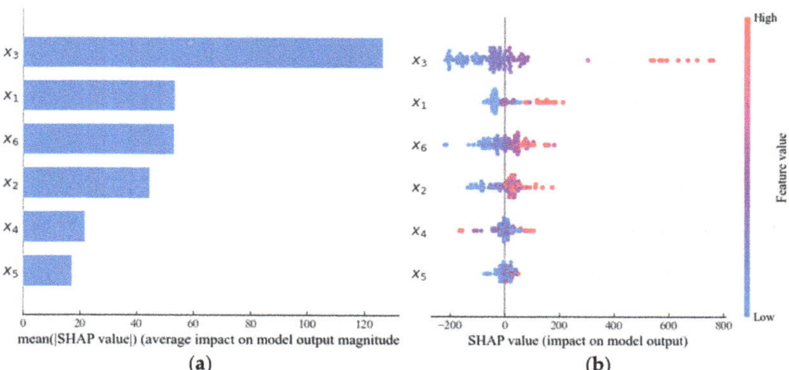

Figure 9. Global interpretations of PSO-SVR model by SHAP values. (**a**) SHAP feature importance; (**b**) SHAP summary plot.

Figure 9b is a SHAP summary plot of the features, which displays the distribution of the SHAP values of each input variable in each sample and demonstrates the overall influence trends. In the figure, the x-axis is the SHAP value of input parameters, the y-axis is the input parameters, ranked by importance. The points in the figure represent the samples in the dataset and their colors represent the feature value, for which color from blue to red represents the value from small to large. From the figure, it can be seen that a few of SHAP values of samples in x_3 are very high, which are around 600. It can be said that in these samples, the value of x_3 has decisive effect on the predicted results. In these input variables, the slab's effective depth (x_3), types of column section (x_1), reinforcement ratio (x_6), cross-section area of column (x_2), Young's modulus of FRP reinforcement (x_5) all have the positive influence on the punching shear strength. However, the influence trend of the compressive strength of concrete (x_4) on the predicted results is hard to definite, which is very complex.

4.3. Individual Interpretations

Except the global interpretations, SHAP also can provides the individual interpretation for each sample in the dataset. In this study, 2 samples are selected, the predicted process of which are illustrated as Figure 10. In the figure, the gray line represents the baseline value, which equals 403.62, and the colorful lines represents the decisive process of each feature. In Figure 10, it can be seen that from the baseline value, every feature will change the value to a different degree, and finally predicted values 419.27 and 501.00, respectively, will be obtained. In the predicted process of the 2 samples, types of column section (x_1), slab's effective depth (x_3), the compressive strength of concrete (x_4) have negative influence on predicted results; whereas, the cross-section area of column (x_2), Young's modulus of FRP reinforcement (x_5), reinforcement ratio (x_6) have positive influence on it. Furthermore, reinforcement ratio (x_6) acts as determinant in forecasting process of sample 1, and the cross-section area of column (x_2) plays an essential role in sample 2. In summary, both the global interpretations and the individual interpretations give a detailed insight into the process by which each input feature affects the final predicted value.

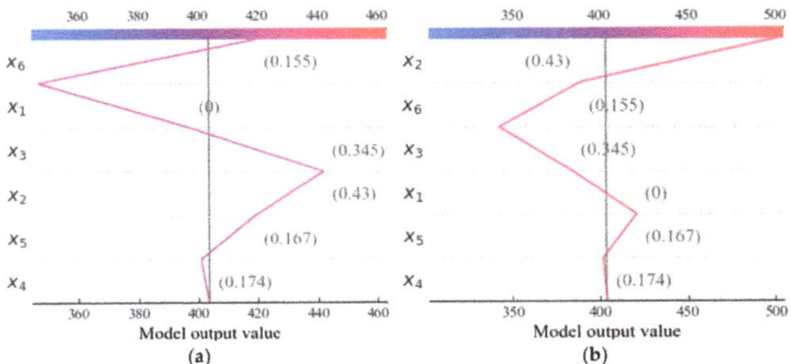

Figure 10. Individual interpretations for selected samples. (**a**) Sample 1: Specimen $G_{(1.6)}30/20$; (**b**) Sample 2: Specimen $G_{(1.6)}45/20$-B.

4.4. Feature Dependency

In addition to explaining the predicted mechanism of model, SHAP can also reveal the interaction between the specific feature and its most closely related feature, which is shown in Figure 11. It can be seen from the Figure that with the values increasing of x_1, x_2, x_3, x_6, their SHAP values will also be increasing, but the relationships between x_4, x_5 and their SHAP values are complicated and nonlinear. As the values of x_4, x_5 increase, the SHAP values of them will increase or decrease; thus, it is difficult to describe the concrete relationships. However, the one that interacts most closely with x_1, x_2, x_5, x_6 is x_3, the features most closely related to x_3 and x_4 are x_2 and x_6, respectively. According to the results of feature dependency, further experiments can be carried out to analyze the relationship between the two variables that interact most closely, and analyze the collaborative impact of these two variables to output. It might be helpful to improve the punching shear resistance of FRP reinforced concrete slab-column connections, and the traditional empirical models cannot accomplish this because the traditional models, relatively, have the worse predicted performance.

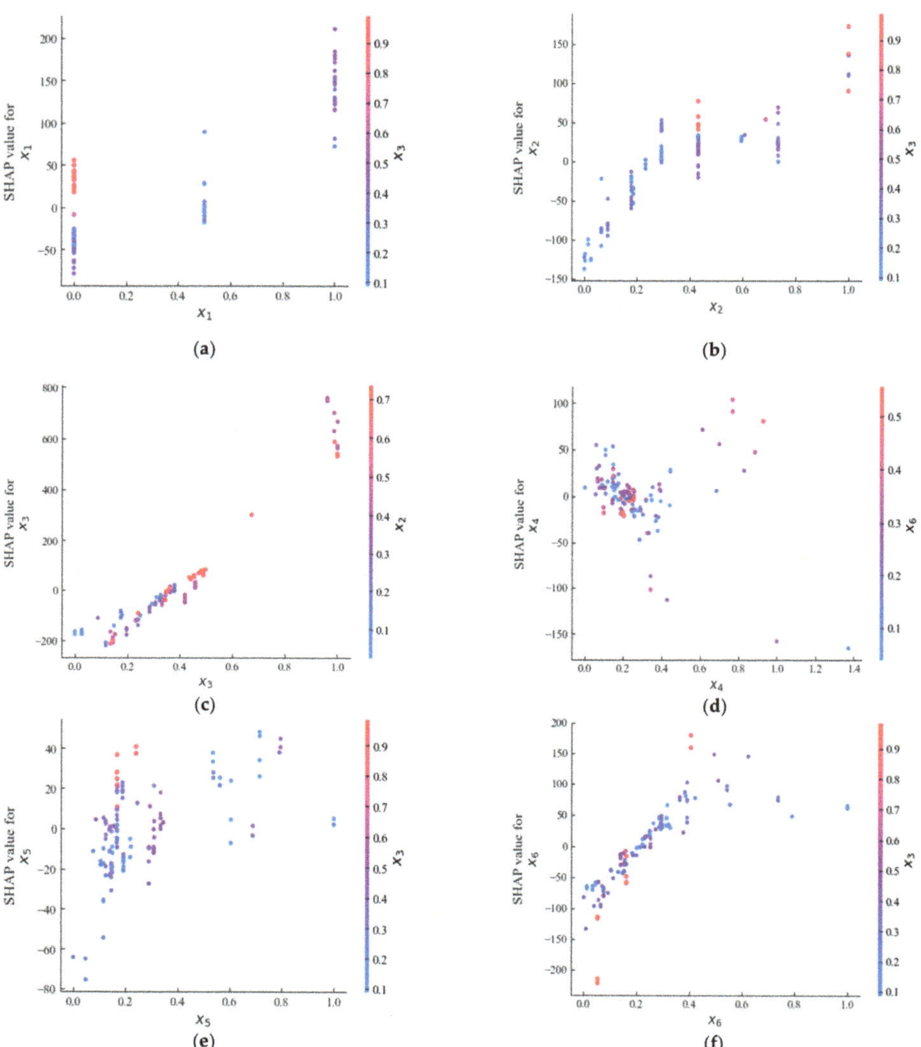

Figure 11. Feature dependence plots. (**a**) x_1; (**b**) x_2; (**c**) x_3; (**d**) x_4; (**e**) x_5; (**f**) x_6.

5. Conclusions

In this paper, several machine learning algorithms, ANN, SVR, DT and AdaBoost, are adopted for punching shear strength of FRP reinforced concrete slabs. A total of 121 groups of experimental tests are collected and the test variables are divided into 6 inputs (types of column section, cross-section area of column, slab's effective depth, compressive strength of concrete, Young's modulus of FRP reinforcement, reinforcement ratio) and 1 output (punching shear strength). Random subsets of 80% and 20% of the data were used for training and testing, respectively, and 10% of the data in training set were used for validating. In addition, 3 indicators (RMSE, MAE, R^2) are introduced to assist the predicted performance. For improving the predicted performance, the empirical method and PSO are utilized for adjusting the values of hyper-parameters in ANN and SVR, respectively. After tuning, both ANN and SVR display the good predicted performance. According to the experimental results, the best performance model is AdaBoost, the RMSE, MAE, R^2 of which are 29.83, 23.00, 0.99, respectively. The predicted performance of ANN, DT and PSO-SVR are worse than AdaBoost, but they are all better than the 6 traditional empirical models introduced in this paper. The predicted result of ANN for training set has a relatively large deviation when the punching shear strength exceeding 1400 kN, but DT has almost no error. Among these traditional models, GB 50010-2010 (2015) has the least deviation reflected in RMSE and R^2, the values of them are 150.41 and 0.73, respectively. Moreover, the model proposed by El-Ghandour et al. (1999) has the best forecasting performance among three traditional models proposed by researchers.

SHAP is used to interpret the predicted process of AdaBoost with the best performance in this study. On the basis of the result of global interpretations, it can be known that every input variable has the different degrees of influence on predicted results, which demonstrates the rationality of input variables selection. Furthermore, the slab's effective depth has the highest impact factor on the predicted results, equaling 126.36, more than twice as much as types of column section, which equals 53.45. In addition, the Young's modulus of FRP reinforcement has the lowest impact factor, equaling 17.23. From the SHAP summary plot, it can be understood that all input variables have positive influence on predicted results except the compressive strength of concrete, it is because its influence trend is relatively fuzzy. According to the individual interpretations, the forecasting process of single sample is shown, which can help researchers to understand the predicted process of model roundly. In addition, SHAP also provides the results of feature dependency, which is helpful to researchers in understanding the relationship between the specific feature and its most closely related feature. According to the analysis results of SHAP, the relationship between material properties and punching shear strength may really be revealed.

Author Contributions: Conceptualization, Y.S. and S.L.; software, Y.S. and J.S.; validation, Y.S. and S.L.; formal analysis, S.L.; writing—original draft preparation, Y.S.; writing—review and editing, S.L.; visualization, Y.S.; supervision, S.L.; project administration, S.L. All authors have read and agreed to the published version of the manuscript.

Funding: This research was funded by Science Foundation of Zhejiang Province of China, grant number LY22E080016; National Science Foundation of China, grant number 51808499.

Institutional Review Board Statement: Not applicable.

Informed Consent Statement: Not applicable.

Data Availability Statement: Not applicable.

Acknowledgments: This study is supported by the Engineering Research Centre of Precast Concrete of Zhejiang Province. The help of all members of the Engineering Research Centre is sincerely appreciated. We would also like to express our sincere appreciation to the anonymous referee for valuable suggestions and corrections.

Conflicts of Interest: The authors declare no conflict of interest.

Appendix A Test Data of FRP Reinforced Concrete Slabs

The sources of the database and the specific material properties are listed in Table A1. The input variables are the types of column section x_1, cross-section area of column x_2 (A/cm^2), slab's effective depth x_3 (d/mm), compressive strength of concrete x_4 (f'_c/MPa), Young's modulus of FRP reinforcement x_5 (E_f/GPa), and reinforcement ratio x_6 ($\rho_f/\%$), respectively. The output variable is the experimental value of punching shear strength y (V/kN). The column section has three types: square ($x_1 = 1$), circle ($x_1 = 2$), rectangle ($x_1 = 3$). In order to use the experimental data for comparative analysis between models, the concrete compressive strength of columns with different cross-section types (cylinder and cube) was unified. Furthermore, to prevent errors caused by different data precision, retain two significant digits to the decimal point for input parameters except for column section type x_1.

Table A1. Database of punching shear strength for FRP reinforced concrete slabs.

Reference	Specimen	x_1	x_2	x_3	x_4	x_5	x_6	y
Ahmad et al. [31] (1994)	SN1	1	56.25	61.00	42.40	113.00	0.95	93.00
	SN2	1	56.25	61.00	39.60	113.00	0.95	78.00
	SN3	1	100.00	61.00	36.00	113.00	0.95	96.00
	SN4	1	100.00	61.00	36.60	113.00	0.95	99.00
Banthia et al. [32] (1995)	I	2	78.54	55.00	41.00	100.00	0.31	65.00
	II	2	78.54	55.00	52.90	100.00	0.31	61.00
Matthys et al. [6] (2000)	C1	2	176.72	96.00	36.70	91.80	0.27	181.00
	C1$'$	2	415.48	96.00	37.30	91.80	0.27	189.00
	C2	2	176.72	95.00	35.70	95.00	1.05	255.00
	C2$'$	2	415.48	95.00	36.30	95.00	1.05	273.00
	C3	2	176.72	126.00	33.80	92.00	0.52	347.00
	C3$'$	2	415.48	126.00	34.30	92.00	0.52	343.00
	CS	2	176.72	95.00	32.60	147.00	0.19	142.00
	CS$'$	2	415.48	95.00	33.20	147.00	0.19	150.00
	H1	2	176.72	95.00	118.00	37.30	0.62	207.00
	H2	2	176.72	89.00	35.80	40.70	3.76	231.00
	H2$'$	2	50.27	89.00	35.90	40.70	3.76	171.00
	H3	2	176.72	122.00	32.10	44.80	1.22	237.00
	H3$'$	2	50.27	122.00	32.10	44.80	1.22	217.00
Khanna et al. [33] (2000)	1	3	1250.00	138.00	35.00	42.00	2.40	756.00
El-Ghandour et al. [34] (2003)	SG1	1	400.00	142.00	32.00	45.00	0.18	170.00
	SC1	1	400.00	142.00	32.80	110.00	0.15	229.00
	SG2	1	400.00	142.00	46.40	45.00	0.38	271.00
	SG3	1	400.00	142.00	30.40	45.00	0.38	237.00
	SC2	1	400.00	142.00	29.60	110.00	0.35	317.00
Ospina et al. [7] (2003)	GFR-1	1	625.00	120.00	29.50	34.00	0.73	217.00
	GFR-2	1	625.00	120.00	28.90	34.00	1.46	260.00
	NEF-1	1	625.00	120.00	37.50	28.40	0.87	206.00
Hussein et al. [35] (2004)	G-S1	1	625.00	100.00	40.00	42.00	1.18	249.00
	G-S3	1	625.00	100.00	29.00	42.00	1.67	240.00
	G-S4	1	625.00	100.00	26.00	42.00	0.95	210.00
El-Gamal et al. [36] (2005)	G-S1	3	1500.00	163.00	49.60	44.60	1.00	740.00
	G-S2	3	1500.00	159.00	44.30	38.50	1.99	712.00
	G-S3	3	1500.00	159.00	49.20	46.50	1.21	732.00
	C-S1	3	1500.00	156.00	49.60	122.50	0.35	674.00
	C-S2	3	1500.00	165.00	44.30	122.50	0.69	799.00
Zhang et al. [37] (2005)	GS2	1	625.00	100.00	35.00	42.00	1.05	218.00
	GSHS	1	625.00	100.00	71.00	42.00	1.18	275.00
Zaghloul [38] (2007)	ZJF5	1	625.00	75.00	44.80	100.00	1.33	234.00
Ramzy et al. [12] (2008)	F1	1	400.00	82.00	37.40	46.00	1.10	165.00
	F2	1	400.00	112.00	33.00	46.00	0.81	170.00
	F3	1	400.00	82.00	38.20	46.00	1.29	210.00
	F4	1	400.00	82.00	39.70	46.00	1.54	230.00
Lee et al. [8] (2009)	GFU1	1	506.25	110.00	36.30	48.20	1.18	222.00
	GFB2	1	506.25	110.00	36.30	48.20	2.15	246.00
	GFB3	1	506.25	110.00	36.30	48.20	3.00	248.00

Table A1. *Cont.*

Reference	Specimen	x_1	x_2	x_3	x_4	x_5	x_6	y
Xiao [39] (2010)	A	1	225.00	130.00	22.16	45.60	0.42	176.40
	B-2	1	225.00	130.00	32.48	45.60	0.42	209.40
	B-3	1	225.00	130.00	32.40	45.60	0.55	245.30
	B-4	1	225.00	130.00	32.80	45.60	0.29	166.60
	B-6	1	225.00	130.00	33.20	45.60	0.42	217.20
	B-7	1	225.00	130.00	28.32	45.60	0.42	221.50
	C	1	225.00	130.00	46.05	45.60	0.42	252.50
Bouguerra et al. [9] (2011)	G-200-N	3	1500.00	155.00	49.10	43.00	1.20	732.00
	G-175-N	3	1500.00	135.00	35.20	43.00	1.20	484.00
	G-150-N	3	1500.00	110.00	35.20	43.00	1.20	362.00
	G-175-H	3	1500.00	135.00	64.80	43.00	1.20	704.00
	G-175-N-0.7	3	1500.00	135.00	53.10	43.00	0.70	549.00
	G-175-N-0.35	3	1500.00	137.00	53.10	43.00	0.35	506.00
	C-175-N	3	1500.00	140.00	40.30	122.00	0.40	530.00
Hassan et al. [40] (2013)	$G_{(0.7)}30/20$	1	900.00	134.00	34.30	48.20	0.71	329.00
	$G_{(1.6)}30/20$	1	900.00	131.50	38.60	48.10	1.56	431.00
	$G_{(0.7)}45/20$	1	2025.00	134.00	45.40	48.20	0.71	400.00
	$G_{(1.6)}45/20$	1	2025.00	131.50	32.40	48.10	1.56	504.00
	$G_{(0.3)}30/35$	1	900.00	284.00	34.30	48.20	0.34	825.00
	$G_{(0.7)}30/35$	1	900.00	284.00	39.40	48.10	0.73	1071.00
	$G_{(0.3)}45/35$	1	2025.00	284.00	48.60	48.20	0.34	911.00
	$G_{(0.7)}45/35$	1	2025.00	281.50	29.60	48.10	0.73	1248.00
	$G_{(1.6)}30/20$-H	1	900.00	131.00	75.80	57.40	1.56	547.00
	$G_{(1.2)}30/20$	1	900.00	131.00	37.50	64.90	1.21	438.00
	$G_{(1.6)}30/35$	1	900.00	275.00	38.20	56.70	1.61	1492.00
	$G_{(1.6)}30/35$-H	1	900.00	275.00	75.80	56.70	1.61	1600.00
	$G_{(0.7)}30/20$-B	1	900.00	131.00	38.60	48.20	0.73	386.00
	$G_{(1.6)}45/20$-B	1	2025.00	131.00	39.40	48.10	1.56	511.00
	$G_{(0.3)}30/35$-B	1	900.00	284.00	39.40	48.20	0.34	781.00
	$G_{(1.6)}30/20$-B	1	900.00	131.00	32.40	48.10	1.56	451.00
	$G_{(0.3)}45/35$-B	1	2025.00	284.00	32.40	48.20	0.34	1020.00
	$G_{(0.7)}30/35$-B-1	1	900.00	281.00	29.60	48.10	0.73	1027.00
	$G_{(0.7)}30/35$-B-2	1	900.00	281.00	46.70	48.10	0.73	1195.00
	$G_{(0.7)}37.5/27.5$-B-2	1	1406.25	209.00	32.30	48.20	0.72	830.00
Nguyen-Minh et al. [10] (2013)	GSL-0.4	1	400.00	129.00	39.00	48.00	0.48	180.00
	GSL-0.6	1	400.00	129.00	39.00	48.00	0.68	212.00
	GSL-0.8	1	400.00	129.00	39.00	48.00	0.92	244.00
Elgabbas et al. [41] (2016)	S2-B	3	1500.00	167.00	48.81	64.80	0.70	548.30
	S3-B	3	1500.00	169.00	42.20	69.30	0.69	664.60
	S4-B	3	1500.00	167.00	42.20	64.80	0.70	565.90
	S5-B	3	1500.00	167.00	47.90	64.80	0.99	716.40
	S6-B	3	1500.00	167.00	47.90	64.80	0.42	575.80
	S7-B	3	1500.00	167.00	47.90	64.80	0.42	436.40
Gouda et al. [42,43] (2016)	GN-0.65	1	900.00	160.00	42.00	68.00	0.65	363.00
	GN-0.98	1	900.00	160.00	38.00	68.00	0.98	378.00
	GN-1.30	1	900.00	160.00	39.00	68.00	1.13	425.00
	GH-0.65	1	900.00	160.00	70.00	68.00	0.65	380.00
	G-00-XX	1	900.00	160.00	38.00	68.00	0.65	421.00
	G-30-XX	1	900.00	160.00	42.00	68.00	0.65	296.00
	R-15-XX	1	900.00	160.00	40.00	63.10	0.65	320.00
Hussein et al. [44] (2018)	H-1.0-XX	1	900.00	160.00	80.00	65.00	0.98	461.00
	H-1.5-XX	1	900.00	160.00	84.00	65.00	1.46	541.00

Table A1. *Cont.*

Reference	Specimen	x_1	x_2	x_3	x_4	x_5	x_6	y
Eladawy et al. [45] (2019)	G1$_{(1.06)}$	1	900.00	151.00	52.00	62.60	1.06	140.00
	G2$_{(1.51)}$	1	900.00	151.00	46.00	62.60	1.51	140.00
	G3$_{(1.06)}$-SL	1	900.00	151.00	46.00	62.60	1.06	180.00
Gu [46] (2020)	A30-1	1	900.00	88.00	27.40	51.10	1.28	191.00
	A30-2	1	900.00	108.00	27.30	51.10	1.05	289.00
	A30-3	1	900.00	138.00	26.20	51.10	0.82	413.00
	A30-4	1	1225.00	86.00	26.80	51.10	1.31	209.00
	A40-1	1	1225.00	88.00	28.20	51.10	1.28	232.00
	A40-2	1	1225.00	88.00	26.40	54.10	0.89	221.00
	A40-3	1	900.00	88.00	28.60	51.10	1.28	236.00
	A50-1	1	900.00	88.00	29.20	51.10	1.28	253.00
	A50-2	1	900.00	90.00	32.20	54.10	0.87	237.00
	A50-3	1	1225.00	88.00	26.70	51.10	1.28	280.00
Zhou [47] (2020)	S40-1	1	900.00	88.00	32.30	51.10	0.98	187.00
	S50-1	1	900.00	86.00	43.20	54.40	0.70	134.00
Eladawy et al. [48] (2020)	G4$_{(1.06)}$-H	1	900.00	151.00	92.00	62.60	1.06	140.00
Mohammed et al. [49] (2021)	0F-60S	1	625.00	125.00	38.20	50.60	2.81	463.00
	0F-80F	1	625.00	125.00	38.20	50.60	2.11	486.00
	0F-110S	1	625.00	125.00	38.20	50.60	1.53	436.00
	1.25F-60S	1	625.00	125.00	39.80	50.60	2.81	455.00
	1.25F-80S	1	625.00	125.00	39.80	50.60	2.11	506.00
	1.25F-110S	1	625.00	125.00	39.80	50.60	1.53	498.00

References

1. Hawkins, N.M.; Criswell, M.E.; Roll, F. Shear Strength of Slabs without Shear Reinforcement. *ACI. Struct. J.* **1974**, *42*, 677–720.
2. Muttoni, A. Punching shear strength of reinforced concrete slabs without transverse reinforcement. *ACI. Struct. J.* **2008**, *105*, 440–450.
3. Shu, J.; Belletti, B.; Muttoni, A.; Scolari, M.; PLoS, M. Internal force distribution in RC slabs subjected to punching shear. *Eng. Struct.* **2017**, *153*, 766–781. [CrossRef]
4. Truong, G.T.; Choi, K.K.; Kim, C.S. Punching shear strength of interior concrete slab-column connections reinforced with FRP flexural and shear reinforcement. *J. Build. Eng.* **2022**, *46*, 103692. [CrossRef]
5. Zhang, D.; Jin, W.; Gao, J. Investigation and detection on corrosion of concrete structure in marine environment. *Proc. Spie* **2009**, *15*, 317–324.
6. Matthys, S.; Taerwe, L. Concrete Slabs Reinforced with FRP Grids. II: Punching Resistance. *J. Compos. Constr.* **2000**, *4*, 154–161. [CrossRef]
7. Ospina, C.E.; Alexander, S.D.B.; Cheng, J.J.R. Punching of Two-Way Concrete Slabs with Fiber-Reinforced Polymer Reinforcing Bars or Grids. *ACI. Struct. J.* **2003**, *100*, 589–598.
8. Lee, J.H.; Yoon, Y.S.; Cook, W.D.; Mitchell, D. Improving Punching Shear Behavior of Glass Fiber-Reinforced Polymer Reinforced Slabs. *ACI. Struct. J.* **2009**, *106*, 427–434.
9. Bouguerra, K.; Ahmed, E.A.; El-Gamal, S.; Benmokrane, B. Testing of full-scale concrete bridge deck slabs reinforced with fiber-reinforced polymer (FRP) bars. *Constr. Build. Mater.* **2011**, *25*, 3956–3965. [CrossRef]
10. Nguyen-Minh, L.; Rovnak, M. Punching Shear Resistance of Interior GFRP Reinforced Slab-Column Connections. *J. Compos. Constr.* **2013**, *17*, 2–13. [CrossRef]
11. El-Gamal, S.E. Finite Element Analysis of Concrete Bridge Decks Reinforced with Fiber Reinforced Polymer Bars. *J. Eng. Res-Kuwait.* **2014**, *11*, 50.
12. Mahmoud, Z.; Salma, T. Punching behavior and strength of slab-column connection reinforced with glass fiber rebars. In Proceedings of the 7th International Conference on FRP Composites in Civil Engineering, Vancouver, BC, Canada, 20–22 August 2014.
13. Vu, D.T.; Hoang, N.D. Punching shear capacity estimation of FRP-reinforced concrete slabs using a hybrid machine learning approach. *Struct. Infrastruct. Eng.* **2015**, *12*, 1153–1161. [CrossRef]
14. MOHURD (Ministry of Housing and Urban-Rural Development of the People's Republic of China). *GB 50010—2010 Code for Design of Concrete Structures*; China Building Industry Press: Beijing, China, 2015; pp. 232–236.

15. ACI (American Concrete Institute). *Building Code Requirements for Structure Concrete (ACI 318-14) and Commentary*; American Concrete Institute: Farmington Hills, MI, USA, 2014; Volume 69, p. 5.
16. El-Ghandour, A.W.; Pilakoutas, K.; Waldron, P. New approach for punching shear capacity prediction of fiber reinforced polymer reinforced concrete flat slabs. *ACI. Struct. J.* **1999**, *188*, 135–144.
17. El-Ghandour, A.W.; Pilakoutas, K.; Waldron, P. Punching shear behavior and design of FRP RC flat slabs. In Proceedings of the International Workshop on Punching Shear Capacity of RC Slabs, Stockholm, Sweden, 1 January 2000.
18. Ju, M.; Ju, J.; Sim, J. A new formula of punching shear strength for fiber reinforced polymer (FRP) or steel reinforced two-way concrete slabs. *Compos. Struct.* **2021**, *259*, 113471. [CrossRef]
19. Geetha, N.K.; Bridjesh, P. Overview of machine learning and its adaptability in mechanical engineering. *Mater. Today* **2020**, *4*. [CrossRef]
20. Mangalathu, S.; Karthikeyan, K.; Feng, D.C.; Jeon, J.S. Machine-learning interpretability techniques for seismic performance assessment of infrastructure systems. *Eng. Struct.* **2022**, *250*, 112883. [CrossRef]
21. Chen, S.Z.; Feng, D.C.; Han, W.S.; Wu, G. Development of data-driven prediction model for CFRP-steel bond strength by implementing ensemble learning algorithms. *Constr. Build. Mater.* **2021**, *303*, 124470. [CrossRef]
22. Feng, D.C.; Wang, W.J.; Mangalathu, S.; Hu, G.; Wu, T. Implementing ensemble learning methods to predict the shear strength of RC deep beams with/without web reinforcements. *Eng. Struct.* **2021**, *235*, 111979. [CrossRef]
23. Fu, B.; Chen, S.Z.; Liu, X.R.; Feng, D.C. A probabilistic bond strength model for corroded reinforced concrete based on weighted averaging of non-fine-tuned machine learning models. *Constr. Build. Mater.* **2022**, *318*, 125767. [CrossRef]
24. Nasiri, S.; Khosravani, M.R. Machine learning in predicting mechanical behavior of additively manufactured parts. *J. Mater. Res. Technol.* **2021**, *14*, 1137–1153. [CrossRef]
25. Hoang, N.D. Estimating punching shear capacity of steel fibre reinforced concrete slabs using sequential piecewise multiple linear regression and artificial neural network. *Measurement* **2019**, *137*, 58–70. [CrossRef]
26. Nguyen, H.D.; Truong, G.T.; Shin, M. Development of extreme gradient boosting model for prediction of punching shear resistance of r/c interior slabs. *Eng. Struct.* **2021**, *235*, 112067. [CrossRef]
27. Mangalathu, S.; Shin, H.; Choi, E.; Jeon, J.S. Explainable machine learning models for punching shear strength estimation of flat slabs without transverse reinforcement. *J. Build. Eng.* **2021**, *39*, 102300. [CrossRef]
28. Rahman, A.; Deshpande, P.; Radue, M.S.; Odegard, G.M.; Gowtham, S.; Ghosh, S.; Spear, A.D. A machine learning framework for predicting the shear strength of carbon nanotube-polymer interfaces based on molecular dynamics simulation data. *Compos. Sci. Technol.* **2021**, *207*, 108627. [CrossRef]
29. Ilawe, N.V.; Zimmerman, J.A.; Wong, B.M. Breaking Badly: DFT-D2 Gives Sizeable Errors for Tensile Strengths in Palladium-Hydride Solids. *J. Chem. Theory Comput.* **2015**, *11*, 5426–5435. [CrossRef]
30. Lundberg, S.M.; Lee, S.I. A Unified Approach to Interpreting Model Predictions. In Proceedings of the 31st Conference on Neural Information Processing Systems (NIPS 2017), Long Beach, CA, USA, 25 November 2017.
31. Ahmad, S.H.; Zia, P.; Yu, T.J.; Xie, Y. Punching shear tests of slabs reinforced with 3 dimensional carbon fiber fabric. *Concr. Int.* **1994**, *16*, 36–41.
32. Banthia, N.; Al-Asaly, M.; Ma, S. Behavior of Concrete Slabs Reinforced with Fiber-Reinforced Plastic Grid. *J. Mater. Civil. Eng.* **1995**, *7*, 252–257. [CrossRef]
33. Khanna, O.S.; Mufti, A.A.; Bakht, B. Experimental investigation of the role of reinforcement in the strength of concrete deck slabs. *Can. J. Civil. Eng.* **2000**, *27*, 475–480. [CrossRef]
34. El-Ghandour, A.W.; Pilakoutas, K.; Waldron, P. Punching Shear Behavior of Fiber Reinforced Polymers Reinforced Concrete Flat Slabs: Experimental Study. *J. Compos. Constr.* **2003**, *7*, 258–265. [CrossRef]
35. Hussein, A.; Rashid, I. Two-way concrete slabs reinforced with GFRP bars. In Proceeding of the 4th International Conference on Advanced Composite Materials in Bridges and Structures, Calgary, AB, Canada, 10–12 July 2004.
36. El-Gamal, S.; El-Salakawy, E.; Benmokrane, B. Behavior of Concrete Bridge Deck Slabs Reinforced with Fiber-Reinforced Polymer Bars Under Concentrated Loads. *ACI Struct. J.* **2005**, *102*, 727–735.
37. Zhang, Q.; Marzouk, H.; Hussein, A. A preliminary study of high-strength concrete two-way slabs reinforced with GFRP bars. In Proceedings of the 33rd CSCE Annual Conference: General Conference and International History Symposium, Toronto, ON, Canada, 1 June 2005.
38. Zaghloul, A. Punching Shear Strength of Interior and Edge Column-Slab Connections in CFRP Reinforced Flat Plate Structures Transferring Shear and Moment. Ph.D. Thesis, Carleton University, Ottawa, ON, Canada, 2007.
39. Xiao, Z. Expremental Study on Two-Way Concrete Slab Subjected to Punching Shear. Master's Thesis, Zhengzhou University, Zhengzhou, China, 2010.
40. Hassan, M.A.W. Punching Shear Behaviour of Concrete Two-Way Slabs Reinforced with Glass Fiber-Reinforced Polymer (GFRP) Bars. Ph.D. Thesis, Université de Sherbrooke, Sherbrooke, QC, Canada, 2013.
41. Elgabbas, F.; Ahmed, E.A.; Benmokrane, B. Experimental Testing of Concrete Bridge-Deck Slabs Reinforced with Basalt-FRP Reinforcing Bars under Concentrated Loads. *J. Bridge. Eng.* **2016**, *21*, 04016029. [CrossRef]
42. Gouda, A.; El-Salakawy, E. Behavior of GFRP-RC Interior Slab-Column Connections with Shear Studs and High-Moment Transfer. *J. Compos. Constr.* **2016**, *20*, 04016005. [CrossRef]

43. Gouda, A.; El-Salakawy, E. Punching Shear Strength of GFRP-RC Interior Slab–Column Connections Subjected to Moment Transfer. *J. Compos. Constr.* **2016**, *20*, 04015037. [CrossRef]
44. Hussein, A.H.; El-Salakawy, E.F. Punching Shear Behavior of Glass Fiber-Reinforced Polymer–Reinforced Concrete Slab-Column Interior Connections. *ACI Struct. J.* **2018**, *115*, 1075–1088. [CrossRef]
45. Eladawy, B.M.; Hassan, M.; Benmokrane, B. Experimental Study of Interior Glass Fiber-Reinforced Polymer-Reinforced Concrete Slab-Column Connections under Lateral Cyclic Load. *ACI Struct. J.* **2019**, *116*, 165–180. [CrossRef]
46. Gu, S. Study on The Punching Shear Behavior of FRP Reinforced Concrete Slabs Subjected to Concentric Loading. Master's Thesis, Zhejiang University of Technology, Zhejiang, China, 2020.
47. Zhou, X. Expremental Study on the Punching Shear Behavior of Square GFRP Reinforced Concrete Slabs. Master's Thesis, Zhejiang University of Technology, Zhejiang, China, 2020.
48. Eladawy, M.; Hassan, M.; Benmokrane, B.; Ferrier, E. Lateral cyclic behavior of interior two-way concrete slab–column connections reinforced with GFRP bars. *Eng. Struct.* **2020**, *209*, 109978. [CrossRef]
49. AlHamaydeh, M.; Anwar Orabi, M. Punching Shear Behavior of Synthetic Fiber–Reinforced Self-Consolidating Concrete Flat Slabs with GFRP Bars. *J. Compos. Constr.* **2021**, *25*, 04021029. [CrossRef]
50. Kim, J.; Kim, D.K.; Feng, M.Q.; Yazdani, F. Application of Neural Networks for Estimation of Concrete Strength. *J. Mater. Civil. Eng.* **2004**, *16*, 257–264. [CrossRef]
51. Yonaba, H.; Anctil, F.; Fortin, V. Comparing Sigmoid Transfer Functions for Neural Network Multistep Ahead Streamflow Forecasting. *J. Hydrol. Eng.* **2010**, *15*, 275–283. [CrossRef]
52. Rumelhart, D.E.; Hinton, G.E.; Williams, R.J. Learning representations by back-propagating errors. *Nature* **1986**, *323*, 533–536. [CrossRef]
53. Shen, H.; Wang, Z.; Gao, C.; Qin, J.; Yao, F.; Xu, W. Determining the number of BP neural network hidden layer units. *J. Tianjin Univ. Technol.* **2008**, *24*, 13–15.
54. Cortes, C.; Vapnik, V. Support-Vector Networks. *Mach. Learn.* **1995**, *20*, 273–297. [CrossRef]
55. Yu, Y.; Li, W.; Li, J.; Nguyen, T.N. A novel optimised self-learning method for compressive strength prediction of high performance concrete. *Constr. Build. Mater.* **2018**, *184*, 229–247. [CrossRef]
56. Smola, A.J.; Schölkopf, B. A tutorial on support vector regression. *Stat. Comput.* **2004**, *14*, 199–222. [CrossRef]
57. Lin, J.; Wang, G.; Xu, R. Particle Swarm Optimization–Based Finite-Element Analyses and Designs of Shear Connector Distributions for Partial-Interaction Composite Beams. *J. Bridge. Eng.* **2019**, *24*, 04019017. [CrossRef]
58. Kennedy, J.; Eberhart, R. Particle swarm optimization. In Proceedings of the International Conference on Neural Networks, Perth, Australia, 27 November–1 December 1995.
59. Huang, W.; Liu, H.; Zhang, Y.; Mi, R.; Tong, C.; Xiao, W.; Shuai, B. Railway dangerous goods transportation system risk identification: Comparisons among SVM, PSO-SVM, GA-SVM and GS-SVM. *Appl. Soft. Comput.* **2021**, *109*, 107541. [CrossRef]
60. Jiang, W.; Xie, Y.; Li, W.; Wu, J.; Long, G. Prediction of the splitting tensile strength of the bonding interface by combining the support vector machine with the particle swarm optimization algorithm. *Eng. Struct.* **2021**, *230*, 111696. [CrossRef]
61. Quinlan, J.R. Induction on decision tree. *Mach. Learn.* **1986**, *1*, 81–106. [CrossRef]
62. Rutkowski, L.; Jaworski, M.; Pietruczuk, L.; Duda, P. The CART decision tree for mining data streams. *Inform. Sci.* **2014**, *266*, 1–15. [CrossRef]
63. Almuallim, H. An efficient algorithm for optimal pruning of decision trees. *Artif. Intell.* **1996**, *83*, 347–362. [CrossRef]
64. Qin, Z.; Lawry, J. ROC Analysis of a Linguistic Decision Tree Merging Algorithm. *Comput. Intell.* **2004**, 33–42.
65. Feng, D.; Cetiner, B.; Azadi Kakavand, M.; Taciroglu, E. Data-Driven Approach to Predict the Plastic Hinge Length of Reinforced Concrete Columns and Its Application. *J. Struct. Eng.* **2021**, *147*, 04020332. [CrossRef]
66. Feng, D.; Liu, Z.; Wang, X.; Chen, Y.; Chang, J.; Wei, D.; Jiang, Z. Machine learning-based compressive strength prediction for concrete: An adaptive boosting approach. *Constr. Build. Mater.* **2020**, *230*, 117000. [CrossRef]
67. Freund, Y.; Schapire, R.E. A Decision-Theoretic Generalization of On-Line Learning and an Application to Boosting. *J. Comput. Syst. Sci.* **1997**, *55*, 119–139. [CrossRef]
68. Tiryaki, B. Predicting intact rock strength for mechanical excavation using multivariate statistics, artificial neural networks, and regression trees. *Eng. Geol.* **2008**, *99*, 51–60. [CrossRef]
69. ACI (American Concrete Institute). *Building Code Requirements for Structure Concrete (ACI 318-19) and Commentary*; American Concrete Institute: Farmington Hills, MI, USA, 2019.
70. BSI (British Standards Institution). *Structure Use of Concrete BS 8110: Part 1: Code of Practice for Design and Construction: BS 8110: 1997*; BSI: London, UK, 1997.
71. Feng, D.; Wang, W.; Mangalathu, S.; Taciroglu, E. Interpretable XGBoost-SHAP Machine-Learning Model for Shear Strength Prediction of Squat RC Walls. *J. Struct. Eng.* **2021**, *147*, 04021173. [CrossRef]
72. Deifalla, A. Punching shear strength and deformation for FRP-reinforced concrete slabs without shear reinforcements. *Case Stud. Constr. Mat.* **2022**, *16*, e00925. [CrossRef]

Article

FE Modelling and Analysis of Beam Column Joint Using Reactive Powder Concrete

Afnan Nafees [1], Muhammad Faisal Javed [1,*], Muhammad Ali Musarat [2], Mujahid Ali [2], Fahid Aslam [3] and Nikolai Ivanovich Vatin [4]

[1]   Department of Civil Engineering, COMSATS University Islamabad, Abbottabad Campus, Abbottabad 22060, Pakistan; afnan@cuiatd.edu.pk

[2]   Department of Civil and Environmental Engineering, Universiti Teknologi PETRONAS, Bandar Seri Iskandar 32610, Perak, Malaysia; muhammad_19000316@utp.edu.my (M.A.M.); mujahid_19001704@utp.edu.my (M.A.)

[3]   Department of Civil Engineering, College of Engineering in Al-Kharj, Prince Sattam Bin Abdulaziz University, Al-Kharj 16273, Saudi Arabia; f.aslam@psau.edu.sa

[4]   Institute of Civil Engineering, Peter the Great St. Petersburg Polytechnic University, 195251 St. Petersburg, Russia; vatin@mail.ru

*   Correspondence: arbab_faisal@yahoo.com

Abstract: Reactive powder concrete (RPC) is used in the beam-column joint region in two out of four frames. Finite element modeling of all specimens is developed by using ABAQUS software. Displacement controlled analysis is used rather than load control analysis to obtain the actual response of the structure. The prepared models were verified by using experimental results. The results showed that using RPC in the joint region increased the overall strength of the structure by more than 10%. Moreover, it also helped in controlling the crack width. Furthermore, using RPC in the joint region increased the ductility of the structures. Comparisons were made by varying the size of the mesh and viscosity parameter values. It was found that by increasing the mesh size and viscosity parameter value, analysis time and the number of steps during analysis were reduced. This study provides a new modeling approach using RPC beam-column joint to predict the behavior and response of structures and to improve the shear strength deformation against different structural loading.

Keywords: reactive powder concrete; beam-column joint; FE modeling; crack; concrete

Citation: Nafees, A.; Javed, M.F.; Musarat, M.A.; Ali, M.; Aslam, F.; Vatin, N.I. FE Modelling and Analysis of Beam Column Joint Using Reactive Powder Concrete. *Crystals* **2021**, *11*, 1372. https://doi.org/10.3390/cryst11111372

Academic Editor: José L. García

Received: 20 September 2021
Accepted: 8 November 2021
Published: 11 November 2021

Publisher's Note: MDPI stays neutral with regard to jurisdictional claims in published maps and institutional affiliations.

1. Introduction

The beam-column joint is a sensitive and crucial part of a structure, where a failure in it can cause the sudden collapse of a building [1]. It is the most seismically vulnerable component in a structure that is typically designed for gravity loads [2]. Recently, extensive research has been on the behavior of reinforced concrete (RC) beam-column joints brought under monotonic loading [3–5]. It was found that many beam-column joints designed with the concept of strong column weak beam concept undergo severe shear force during a seismic event causing joint failure [6].

Shear failures are brittle and more vulnerable causing the catastrophic collapse of structures. To achieve ductile design, ductile material or appropriate reinforcement should be used to improve shear capacity. The latter technique is mostly done by providing stirrups and ties in beams and columns, respectively, with appropriate spacing for good bonding between concrete and reinforcement [7]. Shear capacity can also be enhanced by following various techniques. De Corte and Boel [8] examined the use of rectangular spiral reinforcement (RSR) by testing RC beams under continuous four-point test and results showed increased shear capacity. Yang, Kim [9] explored the effectiveness of Spiral type wire rope as a shear reinforcement by testing three two-span reinforced concrete T-beams in the four-point test under static loading conditions, and results demonstrated increased ductility and controlled crack width. Similarly, Al-Nasra and Asha [10] utilized swimmer bars as transverse reinforcement with three types of connection (weld, bolt, and U-link bolt), and results depicted that it is a convenient method for improving shear strength, ductility, and controlling crack width. Another research by Ghobarah and Said [11] suggested different retrofitting and reinforcing techniques for improving shear resistance of beam-column joints by using concrete jacketing, bolted steel plates, and corrugated steel sheets, etc. [12]. Moreover, Gencoglu and Mobasher [13] concentrated on the use of external steel plates on each side of the column face by bolting it through epoxy bonding and steel angles welded to the plates and the joint region inflamed with concrete fillet in a two-way beam column slab system to provide additional strength to structure against different loadings. As the columns were prefabricated, these approaches were extremely beneficial in terms of construction time. All these methods were very effective in enhancing the shear strength of a beam-column joint, but such techniques are neither cost-effective nor time-efficient.

Reactive powder concrete (RPC) exhibiting strain hardening processes can be utilized to improve beam-column joint strength. During the 1990s, ultra high strength performance mortar known as reactive powder concrete was developed having the compressive strength of 200 MPa [14]. The RPC concept was first developed in 1990 by P. Richard and M. Cheyrezy [15]. It was first utilized in 1997 for the construction of the Sherbrooke bridge in Canada [16]. RPC provides many advanced and high strength and ductility properties in comparison to conventional concrete [17]. RPC constituents include cement, sand, silica fumes, quartz powder, superplasticizers, and steel fibers (optional) [18]. The compressive strength of the RPC used in high prestressed bridge girders is more than 200 MPa while its flexural strength is 50 MPa with high workability. Moreover, it possesses strong ductility and energy absorption characteristics [19]. These properties of RPC make it a significant material. Therefore, RPC is widely used in the construction industry for the construction of different structures like prestressed girders, sewer pipes, blast resistance structures, and high-pressure pipes [20]. Experimental investigation on RPC showed significant improvement in the strength, ductility, strain capacity, and energy dissipation of structures. Furthermore, during the uniaxial compression test, RPC sustained a significant amount of load after initial cracking [21]. The presence of silica fumes and fine particles in the material provides pozzolanic characteristics, agitating the hydration reaction and increasing strength [14]. RPC sometimes shows brittle behavior due to its ultra-high strength. This can be mitigated by adding steel fibers. RPC is gaining momentum and recently has been used in a number of construction fields including bridge erection, mining engineering and high-rise buildings [22,23].

RPC can be used for retrofitting structures. Al-Jubory [24] evaluated the bond strength and durability of RPC using as a repairing material. The addition of silica fume and quartz powder to RPC improved temperature resistance and rendered the structure impermeable. Furthermore, employing RPC as a retrofitting material increased the structure's compressive and flexural strength by more than 12%. It was observed that the abrasion coefficient of RPC was 7.58% more than ordinary concrete. Results indicated no drastic declination for RPC which proved it to be more durable than reinforced concrete.

The experimental study was employed on reinforced RPC (having 1% and 2%) with and without steel fibers. On both of these samples, several strength tests were performed, including compressive strength, tensile strength, and flexural strength. It was discovered that the inclusion of steel fibers increases compressive strength, flexural strength, and split tensile strength by more than 10%. Compressive strength for samples without and with reinforcement was 50–67 MPa and 74.5 MPa, respectively. Low values indicated the presence of higher calcium aluminate content. Experimental results showed that RPC has 250 times greater durability and 200% more compressive strength and 150% more flexural strength than conventional and high strength mortar (HSM). Furthermore, RPC has an abrasion coefficient that is eight times that of normal and four times that of HSM. Freeze and thaw cycles have less effect on RPC which makes it more durable. All these factors lead RPC to be one of the best retrofitting materials [25].

RPC has improved material usage in the concrete industry by providing economic benefits and builds considerably strong, efficient, and durable structures. Experimental research on RPC is conducted by many researchers, however, there is little study on modeling of RPC beam-column joints. This research focused on the numerical modeling of RPC beam-column joint besides experimental work. Numerical modeling provided complete diagnoses about the cause and extent of damage to the structures. Moreover, it is an efficient technique, and it is gaining momentum as it is not only cost-effective but also time efficient. The numerical modeling of the beam-column joint was done using ABAQUS software which is capable of simulating the nonlinear behavior and gives more realistic results in comparison to other software. The experimental results obtained were validated against the numerical results.

2. Experimental Investigation

Four triangular frames as shown in Figure 1 were cast and tested under simple monotonic loading for the determination of tensile strength of beam-column joints. Two out of four frames consisted of conventional concrete. RPC was used in the beam-column joint in the remaining two frames.

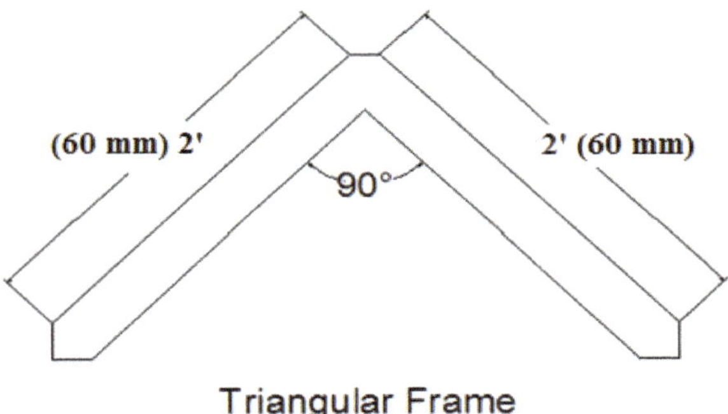

Figure 1. Triangular frames of 2′ × 2′.

The cross-section and long section details of specimens are shown in Figures 1 and 2. Column and beam dimensions are 4″ × 4″ × 24″ and 4″ × 6″ × 24″, respectively, with a cover of 0.5″ from all sides. All the frames were brought under a monotonic loading machine for testing. During the application of load, roller support was provided to the beam and the column was kept fixed. Sensors were installed both at the joints and the beam ends.

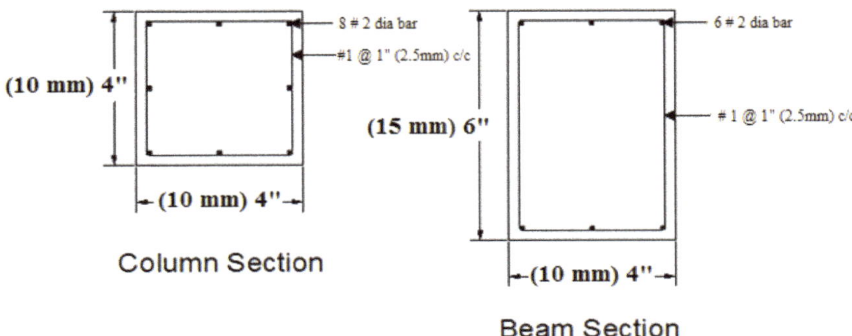

Figure 2. Cross-section and reinforcement detailing of beam and column.

Constituents of RPC are shown in Table 1. RPC mix requires a higher cement quantity as compared to conventional concrete. The quality of cement is also of immense significance in this case [26]. Previous studies have employed high-performance cement with a low sodium oxide and low calcium aluminate content [27–30]. Reinforcement of grade 60 (60 ksi) was used in the specimens. RPC specimens in this study were made with low C3A Portland cement Type V complying with ASTM C150-2. Silica fume was utilized as an auxiliary binder. This was done as RPC requires a pozzolanic substance containing microparticles to reduce small voids in the paste. It also contributed to enhancing the strength and durability properties of the mix as a result of improved dense packing. According to ASTM C 494, a superplasticizer was used to recompense for the decreased water/cement ratio [31]. In the end, quartz mineral was employed to produce high-performance RPC. As the attributes of RPCs not only depend mainly on the order in which the components are inserted into the combination, but also on the speed and length of the process of mixing [32–34]. Approximately 7 min of gradual mixing of dry materials made out of silica fume, Portland cement, and quartz. The superplasticizer was added to water and the whole combination from the superplasticizer with water was added to the components immediately. The blend was then mixed up at around 10 min of progressively escalating speed. Beam column joint for two out of the four frames were left (4 inches for beam and 6 inches for column) for RPC concrete as shown in Figure 3. Joints were cast using RPC concrete monolithically with the conventional concrete as shown in Figure 4. The burlap curing method was adopted. In this method, the triangular specimens were kept under a burlap that was kept wet. Both controlled conventional concrete and RPC specimens were brought under a monotonic loading machine having a capacity of 200 tons as depicted in Figures 5 and 6. In monotonic load testing, the load is steadily escalated at a constant rate, with no reversals from test start to ultimate fracture. Casting and testing of RPC frames are shown in Figures 3–6.

Table 1. Mix design of RPC.

Ordinary Portland Cement	Silica Fume	Quartz	Fine Aggregate	W/C Ratio	Steel Fibers	Superplasticizers
1	0.25	0.4	1.1	0.17	0.03	0.015

Figure 3. Joint left for RPC.

Figure 4. Joint filled with RPC concrete.

Figure 5. Load setup for specimens.

Figure 6. RPC specimens before application of load.

As seen in Figures 7–10, shear cracking was the primary cause of failure in all the specimens. The distribution of cracks in the RPC sample was distributed uniformly due to the presence of steel fibers. As no coarse aggregates were involved in the case of RPC specimens, beam-column joint resulted in decreased stiffness as discussed in the results section of the article. RPC resulted in an increase of 10–15 percent of the tensile strength (the ability of a material to stretch when pulled apart) as compared to controlled conventional concrete samples.

Figure 7. Failure of the conventional concrete specimen (CC_S1) after application of load.

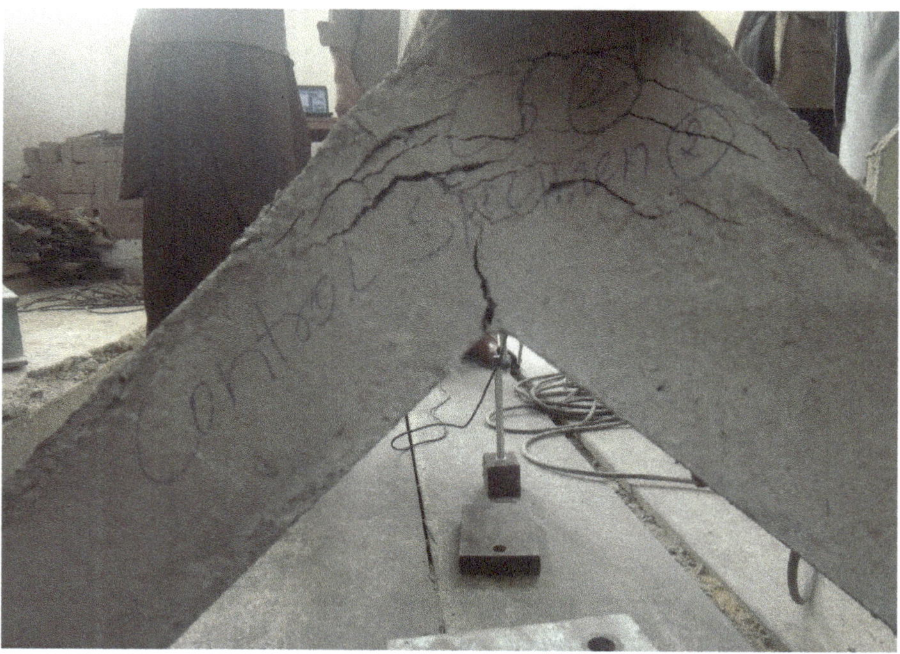

Figure 8. Failure of the conventional concrete specimen (CC_S2) after application of load.

Figure 9. Failure of RPC specimen (RPC_S1) after application of load.

Figure 10. Failure of RPC specimen (RPC_S2) after application of load.

Table 2 shows the values of experimental results. Strength (fc'), Elastic modulus (Ec), maximum load, and displacement for all the specimens were studied. RPC specimens reached ultimate strength at a later stage and have shown higher Ec. Moreover, the load taken by RPC specimens was greater in comparison to controlled concrete specimens.

Table 2. Experimental results.

Specimens	fc' (MPa)	Ec (MPa)	Max Load in Experimental (N)	Max Displacement Experimental (mm)
CC_S1	21.03	27,497.88	15,500	35.98
CC_S2	21.03	27,497.88	13,330	46.28
RPC_S1	45.24	37,433.67	18,000	40.15
RPC_S2	45.24	37,433.67	16,020	43.33

3. Modelling

Recently, numerical modeling has been increasingly adopted to simulate the damaging effect of structures. These numerical models can predict the failure events by analysis of nonlinear behavior such as buckling, large displacements, cracking, and inter-surface contacts. Finite element analysis (FEA) model-based software ABAQUS was used to model and to simulate and determine the response of RPC in improving the shear strength deformation of vulnerable beam-column joint. Different parameters for linear and nonlinear analysis were taken from experimental work of shear strength-deformation improvement of vulnerable beam-column connection using RPC.

3.1. Finite Element Modeling of Nonlinear Behavior of Beam-Column Joint

3.1.1. Finite Element Method

The finite element method (FEM) is the most widely used in numerical simulation of structures [35]. Finite element models have the potential to solve a wide range of complex problems from elastic linear models for linear elements to highly plastic models for nonlinear and solid elements. FEM is one of the leading methods to simulate all types of structures (timber, steel, concrete, masonry) [36].

3.1.2. Abaqus Software

To perform numerical simulation of beam-column joint using RPC a FEM-based software ABAQUS/CAE was selected, which is general-purpose analysis software having the capability of solving the elastic and inelastic problems of the static and dynamic response of components [37].

ABAQUS/CAE 6.14-1 VERSION was used for modeling and analysis of beam-column joint using RPC.

3.1.3. Concrete Damage Plasticity Model

The concrete damage plasticity (CDP) model was selected as it has the capability and potential for modeling reinforced concrete and other quasi-brittle material for different types of structure. CDP model can define the nonlinear behavior of the RPC beam-column joint. Additionally, it takes into account the isotropic damage elasticity concepts with isotropic tensile and compressive plasticity. It also considers the degradation of elastic stiffness produced by plastic straining both in compression and tension [38]. The CDP model can show damage characteristics of a material. The main failure mechanism that this model assumes is the tensile cracking and the compressive crushing [39].

Different parameters required in the CDP model were studied and selected based on available literature both for conventional as well RPC specimens. The dilation angle for the model was taken as 36°. It is the angle obtained due to a change in volumetric strain produced due to plastic shearing. It depends on the angle of internal friction. Dilation angle controls the amount of plastic volumetric strain produced due to plastic shearing. Normally dilation angle is taken between 30° and 40° for concrete to avoid large variation between experimental work and numerical modeling. For the seismic design of reinforced concrete, the value of dilation angle is normally between 35° to 38° [40]. Moreover, eccentricity is the deviation from the center. The default value for eccentricity was taken, i.e., 0.1. If the value is increased by 0.1 the curvature of flow potential is increased. If the value is decreased from the default value, the convergence problem may occur if confinement pressure is not high enough. Furthermore, the ratio of biaxial loading (f_b) to uniaxial loading (f_{c0}) is normally taken as 1 or greater than 1. In this case default value was taken i.e., $f_b/f_{c0} = 1.16$. K is the shape factor and default value for K = 0.667 [41]. The viscosity parameter shows the amount of flow potential in a material. A lower viscosity parameter value is better as higher values result in a high force of reaction. Therefore, the viscosity parameter, in this case, was taken as 0.001 [42].

3.2. Compressive and Tensile Behavior Determination by Using Eurocode

Compressive behavior and tensile behavior of both normal concrete and RPC were determined by using EN 1992 Eurocode 2: Design of concrete structures part 1–1 [43]. It describes different principles and requirements for the safety, serviceability, and durability of concrete structures with specific provisions of buildings. Eurocode 2 applies to the design of civil engineering works such as buildings, roads, bridges, etc. It is applied to plain, reinforced, and prestressed concretes. It complies with the specifications and requirements given in EN 1992-1-1 about safety, serviceability of the structures, the basis of their design, and verification of structures given in EN 1990; basis of structural design [38]. Compressive and tensile stress-strain curves are shown in Figures 11 and 12, respectively. The limitation of the Eurocode 2 for concrete structures is that it is concerned only with the requirements for resistance, safety, serviceability, durability, and fire resistance of the structures. Moreover, it does not consider the other requirements like thermal or sound insulation, etc. [38].

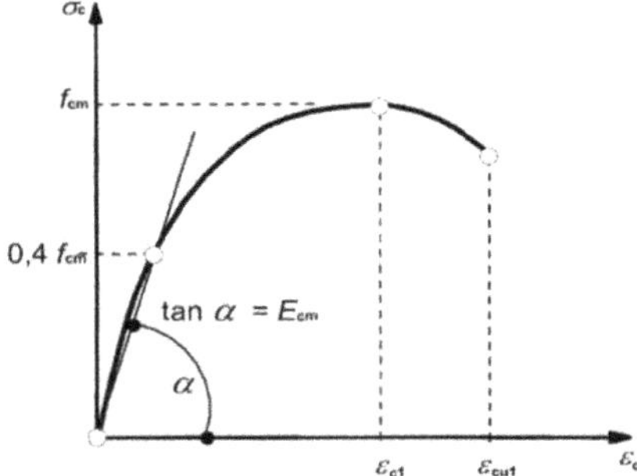

Figure 11. Stress-strain diagram for analysis of concrete using Eurocode 2 [43].

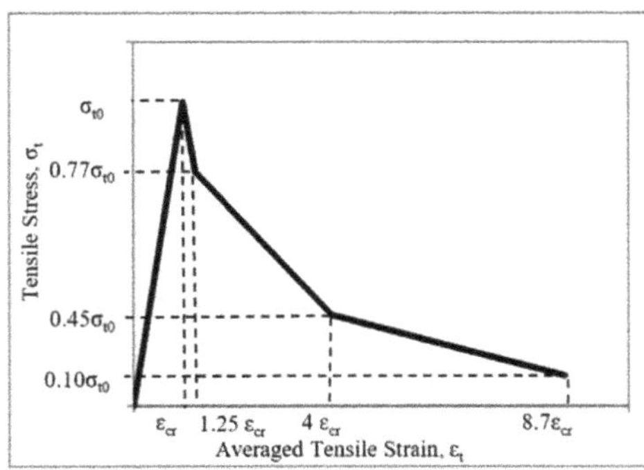

Figure 12. Tensile stress-strain curve of concrete from Eurocode [43].

3.2.1. Compressive Behavior

The compressive behavior of concrete was calculated by using the relations of Eurocode [43] given in Equation (1).

$$Ecm = 22(0.1fcm)^{0.3} \tag{1}$$

where:

f_{cm} (MPa) is the compressive strength
E_{cm} (GPa) is the modulus of elasticity
Other values showing the position of characteristics points are strain ε_{c1} at average compressive strength and ultimate strain ε_{cu} at 0.

$$\varepsilon c1 = 0.7(fcm)^{0.31} \tag{2}$$

$$\varepsilon cu = 0.35\% \tag{3}$$

where:

ε_{c1} is the strain at peak stress
ε_{cu} is the ultimate strain at which concrete fails
Equations (2) and (3) are only pertinent to concrete having a cylindrical compressive strength of 50 MPa and cube compressive strength of 60 MPa at the most. On the basis of a list of the experimental results, Kmiecik and Kamiński [44] proposed the quite accurate approximating Equations (4) and (5):

$$\varepsilon c1 = 0.0014 \, [2 - \exp(-0.024 \, fcm) - \exp(0.140 \, fcm) \tag{4}$$

$$\varepsilon cu = 0.004 - 0.0011[1 - \exp(-0.0215 \, fcm)] \tag{5}$$

Knowing the values of the output in Equations (4) and (5) one can determine the points at which the graphs intersect. Compressive stress values can be determined at any point using these relations [43].
According to Eurocode EN 1992-1-1

$$\sigma c = fcm(k\eta - \eta^2) / (1 + (k - 2)\eta \tag{6}$$

where:

$$k = 1.05 * Ecm\left(\frac{\varepsilon c1}{fcm}\right) \tag{7}$$

and:

$$\eta = \frac{\varepsilon c}{\varepsilon c1} \tag{8}$$

3.2.2. Tensile Behaviors

The tensile behavior of concrete was calculated by using the Equations (9)–(12).
If $\varepsilon t \leq \varepsilon_{cr}$

$$\sigma t = Ec * \varepsilon t \tag{9}$$

and if $\varepsilon t > \varepsilon_{cr}$

$$\sigma t = fcm(\frac{\varepsilon cr}{\varepsilon t})^{0.4} \tag{10}$$

$$ft = 0.33 * fc^{0.5} \tag{11}$$

$$ftr = 0.30 fck^{2/3} \tag{12}$$

For the determination of the complete stress-strain curve for compressive behavior and the tensile behavior of normal concrete and RPC, Eurocode has been used which is capable of determining the actual response of structures closer to the experimental setup. As RPC is a composite material and there is no official code for RPC developed yet, therefore small modifications based on literature have been made in normal concrete formulas for the determination of stress-strain curves of RPC.

Simple modifications were incorporated into the Equations (1), (3), and (12). These equations were utilized in assigning material properties during numerical modeling in ABAQUS to obtain more realistic results of RPC concrete. Modified equations are shown in Equations (13)–(15).

$$Ecm = 22(0.13fcm)^{0.3} \tag{13}$$

$$\varepsilon cu = 0.40\% \tag{14}$$

$$ft = 0.40 * fc^{0.5} \tag{15}$$

3.3. Steps and Boundary Conditions

After the completion of assembly, a step was formed. In steps, a time period was provided for which the load is applied to the assembly. The load was then applied to the designated location according to the magnitude of the sample and boundary conditions were applied according to experimental work in which two specimens (CC_S1 and RPC_S1) have hinge boundary condition, i.e., (U1 = U2 = UR3 = 0) while the other two specimens have fixed boundary condition (U1 = U2 = U3 = UR1 = UR2 = UR3 = 0) at column end while roller support (U1 = UR2 = UR3 = 0) at the beam end in all specimens. "U" refers to translatory motion while "UR" refers to rotation of the support. Both the boundary conditions for the column were studied and their effect on the strength and load values were observed.

3.4. Meshing

Meshing is the process of dividing the whole finite element model into a smaller number of chunks by the formation of different nodes at different points. Meshing is an important process as it allows us to apply load and find displacement or any other desired result at any point in the model. The greater the size of the mesh, the smaller will be the number of iterations taken to analyze the whole model and vice versa. In a greater size mesh, a lesser number of nodes are formed, hence the number of iterations and time of analysis is reduced. In our case, the size of the mesh taken was 25 mm, 40 mm, and 50 mm. Independent types of meshing for concrete and steel are selected in Figures 13–15 [42].

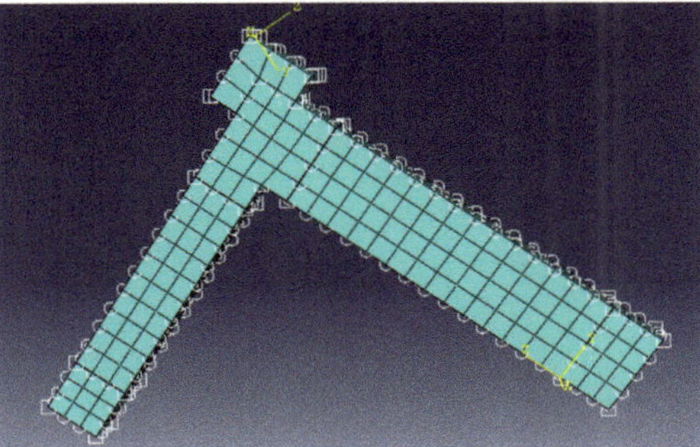

Figure 13. Meshing of 25 mm size for concrete.

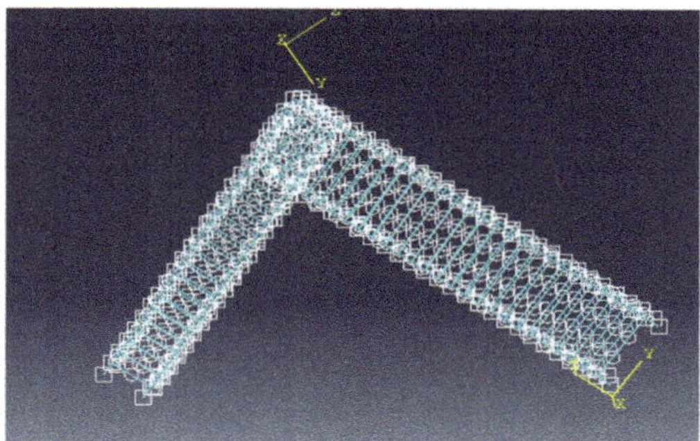

Figure 14. Meshing of steel embedded in concrete.

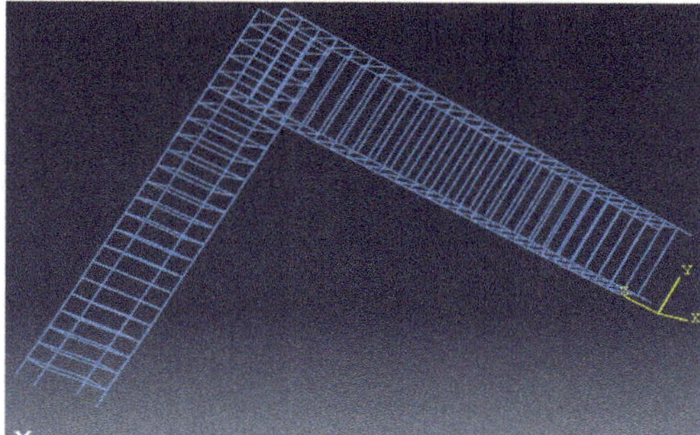

Figure 15. Typical view of Abaqus model of reinforcement.

3.5. Load

A series of analysis was performed on conventional concrete controlled specimens and RPC specimens to simulate and predict the actual response of linear and nonlinear behavior on beam-column joint and to show the behavior of RPC in improving the shear strength deformation against different structural loading. A monotonic load of 0 to 20 kN was applied at the top of the exterior joint for all specimens till the specimens reached the ultimate value. After the application of load, step was created for static analysis. Time period and increment values were given to all specimens. Figures 16 and 17 show the analysis of RPC samples with fixed and hinge boundary conditions, respectively. As seen in Figure 16, a small amount of buckling was observed when the column boundary condition was kept fixed, whereas no buckling was observed in the case of hinge column conditions as seen in Figure 17.

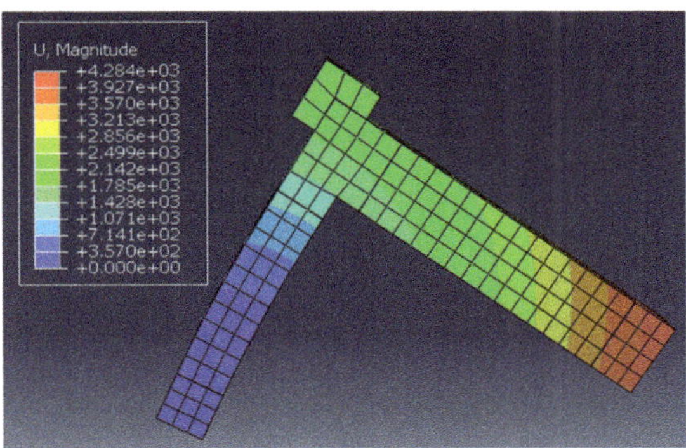

Figure 16. Analysis of RPC with fixed column boundary condition.

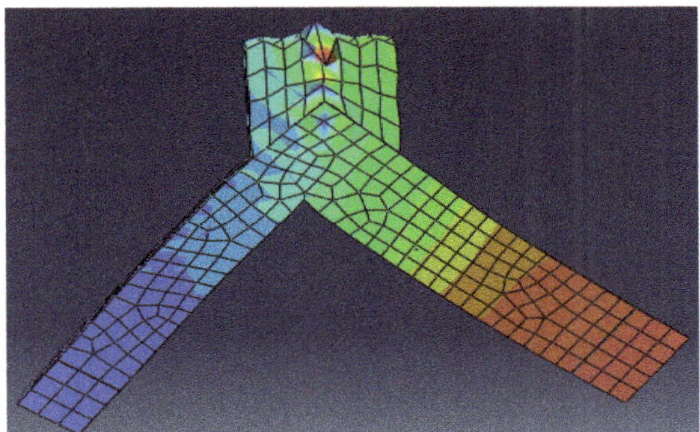

Figure 17. Analysis of concrete with hinge column boundary condition.

4. Results and Discussion

Experimental results of shear strength-deformation improvement for vulnerable beam-column connection using RPC were used to validate the developed FEM approach. Different parameters from experimental work were used in numerical modeling. This approach provided a more realistic response simulation of the actual beam-column joint. The numerical results were compared with experimental results for the verification of the model as shown in Table 3. There was a negligible deviation of numerical results from experimental results for controlled concrete samples in case of maximum loads. It was 1.13% and 0.63% for CC_S1 and CC_S2, respectively. For RPC samples, the divergence was comparatively higher. It was 6.05% for RPC_S1 and 6.73% for RPC_S2. Maximum displacement variation was 7.06%, 4.18%, 3.12%, and 6.54% for CC_S1, CC_S2, RPC_S1, and RPC_S2, respectively. The maximum variation observed was 7.06% for CC_S1 for displacement. This shows that numerical results were in strong agreement with the experimental results.

Table 3. Comparison of load and displacement between experimental and modeling.

Specimens	f_c' (MPa)	E_c (MPa)	Max Load in Experimental (N)	Max Load in Modeling (N)	Max Displacement Experimental (mm)	Max Displacement Modeling (mm)	Difference b/w Modeling and Experimental Displacement Max Values (mm)
CC_S1	21.03	27,497.88	15,500	15,675.21	35.98	33.44	2.54
CC_S2	21.03	27,497.88	13,330	13,413.50	46.28	48.21	1.93
RPC_S1	45.24	37,433.67	18,000	19,090.02	40.15	38.89	1.25
RPC_S2	45.24	37,433.67	16,020	17,097.80	43.33	40.50	2.83

4.1. Load Displacement Curve

4.1.1. Conventional Concrete Controlled Specimens

The comparison of the load-displacement curve obtained from experimental and ABAQUS simulations are shown in Figures 18–21. The shape of the ABAQUS simulations curves is quite close to the experimental curves. The maximum average discrepancy between modeling and experimental results of conventional concrete was 3–7%. Almost linear behavior was obtained using ABAQUS modeling for CC_S1 whereas in experimental work the pattern of the graph showed nonlinearity which might be due to non-uniform increment of load in the experimental setup.

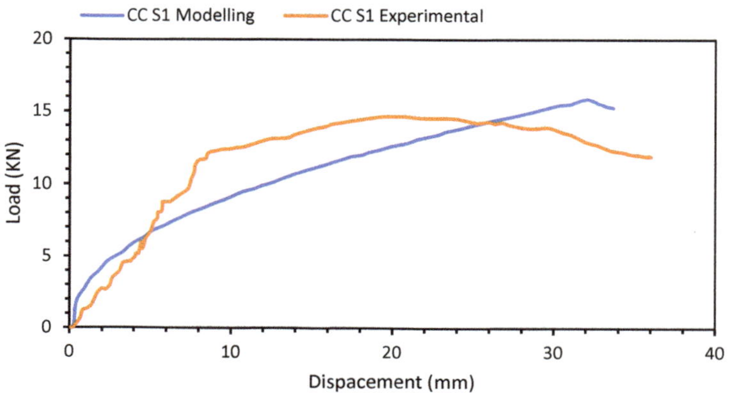

Figure 18. Displacement at beam end (roller support) with a hinge boundary condition at column end for CC_S1 (mesh size 25 mm).

4.1.2. RPC Specimens

The maximum discrepancy between modeling and experimental results of RPC in the case of RPC_S1 was 6.05% while that of RPC_S2 was 6.7%. The deviation of experimental results from modeling in RPC_S1 was due to non-uniform increment of load and time period in the experimental setup while RPC_S2 showed quite accurate results. Mesh size effect was studied for RPC specimens and compared with the experimental results Figures 22 and 23. Mesh size 25 was considered for RPC_S1 and mesh size 40 for RPC_S2 for comparison with the experimental values.

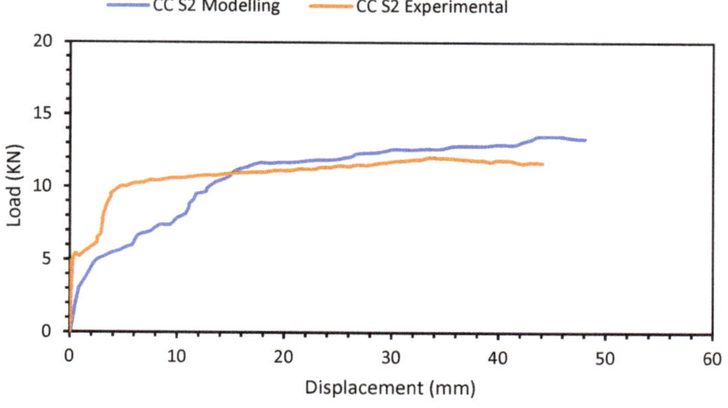

Figure 19. Displacement at beam end (roller support) with a fixed boundary condition at column end for CC_S2 (mesh size 25 mm).

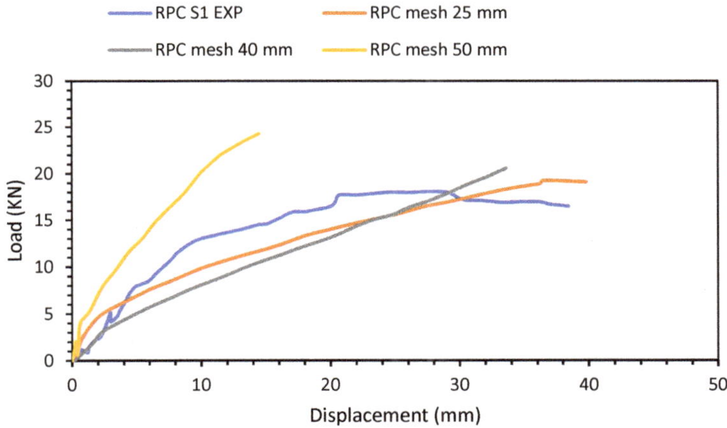

Figure 20. Displacement at beam end (roller support) with hinge boundary condition at column end for RPC_S1.

4.2. Comparison between Conventional Concrete and RPC Specimens

Comparison between conventional concrete and RPC specimen is shown in Figures 22 and 23. RPC specimens took 10–15% more load as compared to conventional concrete-controlled specimens. Delayed peaks were obtained for RPC specimens which shows delayed damaging effect in the samples.

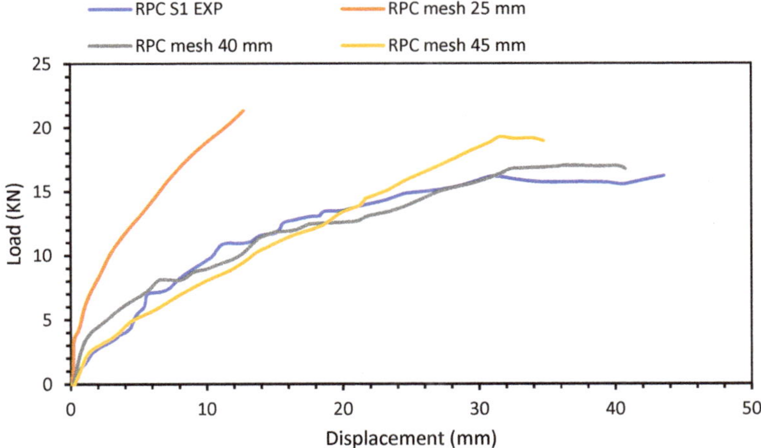

Figure 21. Displacement at beam end (roller support) with fixed boundary condition at column end for RPC_S2.

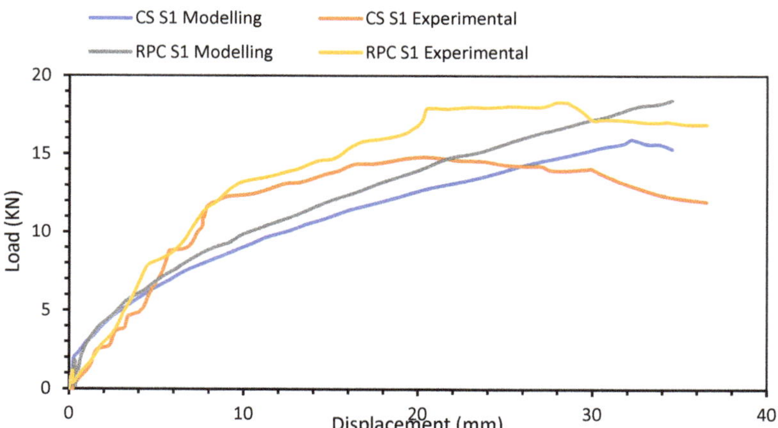

Figure 22. Comparison between concrete and RPC specimens for S1 (column hinge condition).

4.3. Stiffness of Concrete and RPC Specimens

For the validation of the model, the stiffness of all specimens was calculated. It can be seen from Tables 4 and 5 that the initial stiffness in the RPC specimens is low in comparison to conventional concrete. Moreover, it can be observed that as the load increased the structure lost its rigidity and stiffness (the ability of a structure to resist deformation when subjected to the applied force). However, after 20% of loading RPC was still taking more load in comparison to conventional concrete as shown in Tables 4 and 5. It can also be observed from Figures 24–26 that the initial stiffness of RPC specimens is low compared to controlled concrete specimens. Figures 25 and 26 depict that as the load was increased to 25% and 50% of the ultimate load, RPC showed to have high stiffness comparatively.

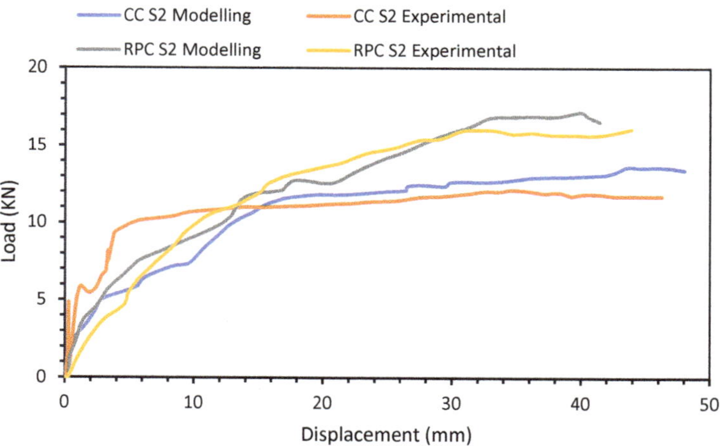

Figure 23. Comparison between concrete and RPC specimens for S2 (column fixed condition).

Table 4. Stiffness of all specimens from experimental work.

(% of Ultimate Load)	Specimens			
	CC_S1	CC_S2	RPC_S1	RPC_S2
	Stiffness (kN/mm)	Stiffness (kN/mm)	Stiffness (kN/mm)	Stiffness (kN/mm)
5%	21.36	19.18	14.66	14.19
25%	2.24	3.23	4.59	4.38
50%	1.48	1.96	2.53	2.43

Table 5. Stiffness of all specimens at different loading rates (numerically).

(% of Ultimate Load)	Specimens			
	CC_S1	CC_S2	RPC_S1	RPC_S2
	Stiffness (kN/mm)	Stiffness (kN/mm)	Stiffness (kN/mm)	Stiffness (kN/mm)
5%	18.561	20.82	15.76	15.98
10%	12.67	17.49	12.23	12.76
15%	5.42	7.18	8.88	8.48
20%	3.45	4.31	5.44	5.73
25%	2.50	3.06	4.11	4.03
30%	1.99	2.38	3.10	2.88
40%	1.42	1.34	1.98	1.56
50%	1.09	0.97	1.03	1.29
60%	0.88	0.76	0.85	0.81

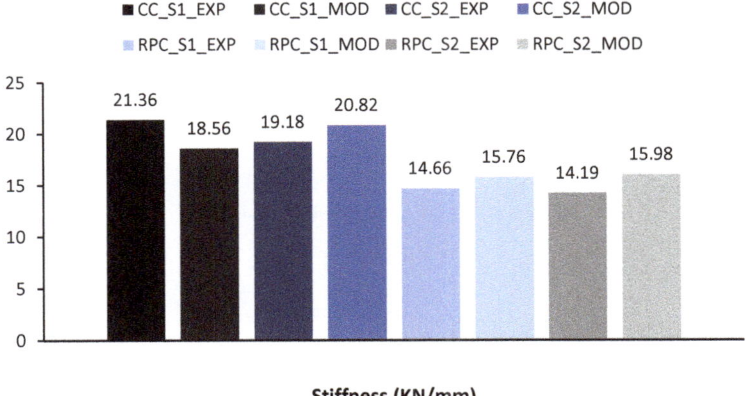

Figure 24. Initial stiffness of all specimens at 5% of ultimate loading.

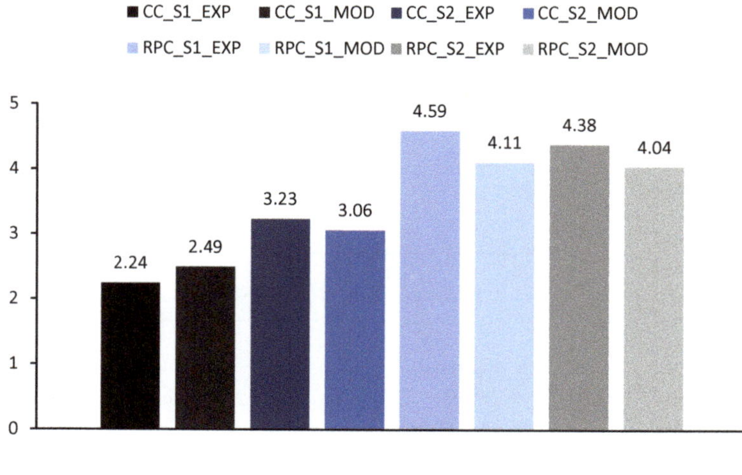

Figure 25. Stiffness of all specimens at 25% of ultimate loading.

4.4. Ductility of Concrete and RPC Specimens

The ductility displacement factor (R), as depicted in Figure 27, according to the Committee Euro International Du Beton, 1996, is defined as the ratio between failure displacement and yield displacement. The yield displacement is the lateral displacement at 80% of the ultimate load at ascending part of the curve while the failure displacement is the lateral displacement at 80% of the ultimate load at descending part of the curve [26]. Ductility factor R and ductility displacement (DD) can be obtained from Equations (16) and (17), respectively.

$$R = \frac{\Delta f}{\Delta y} \tag{16}$$

where Δf = failure displacement Δy = yield displacement

$$DD = \frac{\Delta i}{\Delta y} \tag{17}$$

where Δi = maximum displacement in any cycle I Δy = yield displacement

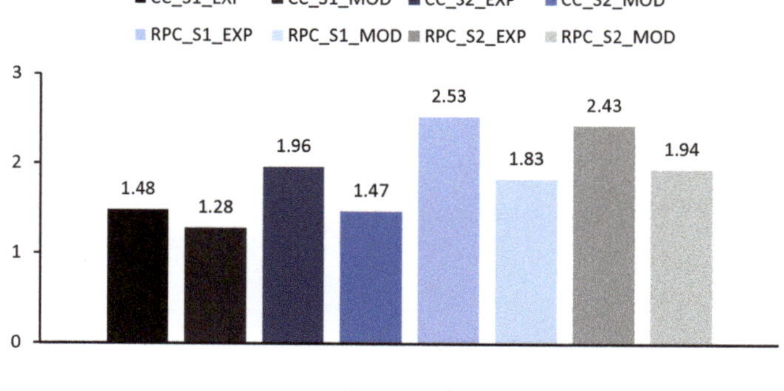

Figure 26. Stiffness of all specimens at 50% of ultimate loading.

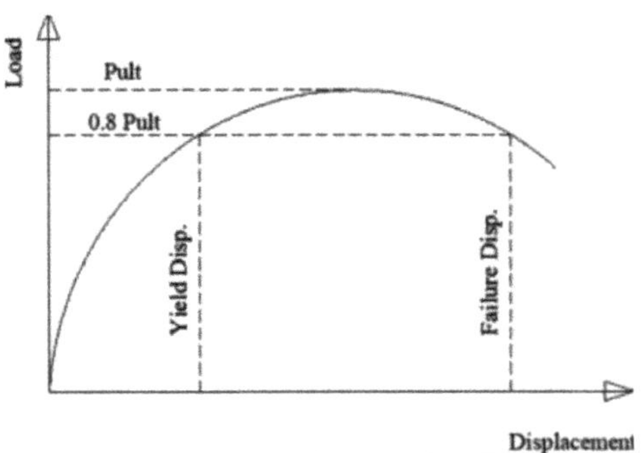

Figure 27. Ductility displacement.

Table 6 displays ductility factor R and displacement ductility (DD) for all the specimens both for experimental and numerical modeling. Experimental results showed an increase of 27% and 11% in R for S1 and S2, respectively. Similarly, DD for RPC was enhanced by 29% and 12% for S1 and S2, respectively. The same trend was shown by numerical modeling results.

Table 6. Ductility factor and ductility displacement.

Sample	Experimental		Modeling	
	Ductility Factor (R)	Displacement Ductility (DD)	Ductility Factor (R)	Displacement Ductility (DD)
CC_S1	4.91	4.78	4.671	4.699
CC_S2	5.56	5.71	5.18	5.33
RPC_S1	6.257	6.19	6.03	6.153
RPC_S2	6.2	6.39	5.98	6.13

5. Conclusions

A series of analysis were performed on conventional concrete and RPC beam-column joint specimens. Following conclusions were made based on experimental and numerical testing.

1. The use of RPC only in the joint region increased the overall strength of the structure by 10–15% and also delayed the crack propagations.
2. The maximum average discrepancy between modeling and experimental results of conventional concrete and RPC was 3–7%. This discrepancy was due to the non-uniform increment of load and time period in the experimental setup.
3. It was observed that with an increment in the mesh size, a reduction in the number of analysis increments occurred. This caused variation of modeling results from experimental results. Therefore, finer mesh size is recommended.
4. Increasing the value of viscosity reduced the analysis time but produced more errors in results. The lower value of the viscosity parameter is better as higher values cause a high peak of reaction force. Therefore, smaller values are preferable i.e., 0.001, 0.002, 0.003, or 0.005 etc.
5. Fixed column end conditions caused an increase in column stresses which resulted in buckling of column. No buckling was observed for hinged column conditions. Maximum deformation was observed at the beam end irrespective of the column end conditions.

6. To obtain actual results, displacement control analysis should be used rather than load control analysis. With displacement control analysis it is easier to obtain the converged solutions in ABAQUS in case of highly nonlinear problems.

6. Recommendations

Based on this research following recommendations can be used for future research work.

In the case of RPC, a decrease in the initial stiffness of the specimen was observed in the joint region as coarse aggregates were not used. Therefore, the use of suitable size of coarse aggregate will not only increase the initial stiffness but also increase the strength of the structure.

Steel fibers of a longer length should be used so that they can help in controlling the crack from widening. In the case of RPC, shear cracking was observed in the joint region. Combining the RPC technique with some other technique will convert the failure mechanism from the joint to the beam through their combined effect.

Besides the CDP model smeared crack modeling and brittle concrete modeling of the RPC can be used to determine the complex behavior of RPC in structure.

As there is no official Eurocode for RPC. Therefore, the development of Eurocode for RPC with and without steel fibers will enable us to clearly understand the complicated behavior of the material.

Author Contributions: Supervision, review, and editing, A.N.; data curation and methodology, M.A.; investigation and review, M.F.J.; conceptualization, data analysis, writing original draft preparation, M.F.J.; formal analysis and modeling, F.A.; validation, proofreading, review, M.A.M. and N.I.V. All authors have read and agreed to the published version of the manuscript.

Funding: The research is partially funded by the Ministry of Science and Higher Education of the Russian Federation under the strategic academic leadership program 'Priority 2030' (Agreement 075-15-2021-1333 dated 30 September 2021).

Data Availability Statement: The data used in this research was collected from published literature.

References

1. Al-Salloum, Y.A.; Alrubaidi, M.A.; Elsanadedy, H.M.; Almusallam, T.H.; Iqbal, R.A. Strengthening of precast RC beam-column connections for progressive collapse mitigation using bolted steel plates. *Eng. Struct.* **2018**, *161*, 146–160. [CrossRef]
2. Shannag, M.J.; Abu-Dyya, N.; Abu-Farsakh, G. Lateral load response of high performance fiber reinforced concrete beam–column joints. *Constr. Build. Mater.* **2005**, *19*, 500–508. [CrossRef]
3. Chiu, C.-K.; Lays, D.P.; Ricky; Krasna, W.A. Anchorage strength development of headed bars in HSRC external beam-column joints considering side-face blowout failure under monotonic loading. *Eng. Struct.* **2021**, *239*, 112218. [CrossRef]
4. Yuan, H.; Hu, S.; Du, X.; Yang, L.; Cheng, X.; Theofanous, M. Experimental behaviour of stainless steel bolted T-stub connections under monotonic loading. *J. Constr. Steel Res.* **2019**, *152*, 213–224. [CrossRef]
5. Allam, S.M.; Elbakry, H.M.; Arab, I.S. Exterior reinforced concrete beam column joint subjected to monotonic loading. *Alex. Eng. J.* **2018**, *57*, 4133–4144. [CrossRef]
6. Ganesan, N.; Indira, P.; Abraham, R. Steel fibre reinforced high performance concrete beam-column joints subjected to cyclic loading. *ISET J. Earthq. Technol.* **2007**, *44*, 445–456.
7. Uma, S.; Jain, S.K. Seismic design of beam-column joints in RC moment resisting frames—Review of codes. *Struct. Eng. Mech.* **2006**, *23*, 579–597. [CrossRef]
8. De Corte, W.; Boel, V. Effectiveness of spirally shaped stirrups in reinforced concrete beams. *Eng. Struct.* **2013**, *52*, 667–675. [CrossRef]
9. Yang, K.-H.; Kim, G.-H.; Yang, H.-S. Shear behavior of continuous reinforced concrete T-beams using wire rope as internal shear reinforcement. *Constr. Build. Mater.* **2011**, *25*, 911–918. [CrossRef]
10. Al-Nasra, M.; Asha, N. Shear reinforcements in the reinforced concrete beams. *Am. J. Eng. Res. (AJER)* **2013**, *2*, 191–199.
11. Ghobarah, A.; Said, A. Shear strengthening of beam-column joints. *Eng. Struct.* **2002**, *24*, 881–888. [CrossRef]
12. Engindeniz, M.; Kahn, L.F.; Abdul-Hamid, Z. Repair and strengthening of reinforced concrete beam-column joints: State of the art. *ACI Struct. J.* **2005**, *102*, 1.

13. Gencoglu, M.; Mobasher, B. The strengthening of the deficient RC exterior beam-column joints using CFRP for seismic excitation. In Proceedings of the 3rd International Conference on Structural Engineering, Mechanics and Computation, Cape Town, South Africa, 10–12 September 2007.

14. Richard, P.; Cheyrezy, M. Reactive powder concrete. *Cem. Concr. Res.* **1995**, *25*, 1501–1511. [CrossRef]

15. Sadrekarimi, A. Development of a Light Weight Reactive Powder Concrete. *J. Adv. Concr. Technol.* **2004**, *2*, 409–417. [CrossRef]

16. Aïtcin, P.-C.; Lachemi, M.; Adeline, R.; Richard, P. The Sherbrooke Reactive Powder Concrete Footbridge. *Struct. Eng. Int.* **1998**, *8*, 140–144. [CrossRef]

17. Yi, N.-H.; Kim, J.-H.J.; Han, T.-S.; Cho, Y.-G.; Lee, J.H. Blast-resistant characteristics of ultra-high strength concrete and reactive powder concrete. *Constr. Build. Mater.* **2012**, *28*, 694–707. [CrossRef]

18. Shi, C.; Long, M.; Cao, C.; Long, G.; Lei, M. Mechanical property test and analytical method for Reactive Powder Concrete columns under eccentric compression. *KSCE J. Civ. Eng.* **2017**, *21*, 1307–1318. [CrossRef]

19. Schmidt, M.; Fehling, E. Ultra-high-performance concrete: Research, development and application in Europe. *ACI Spec. Publ.* **2005**, *228*, 51–78.

20. Blais, P.Y.; Couture, M. PRECAST, PRESTRESSED PEDESTRIAN BRIDGE-WORLD'S FIRST REACTIVE POWDER CONCRETE BRIDGE. *PCI J.* **1999**, *44*. [CrossRef]

21. Jungwirth, J.; Muttoni, A. Underspanned bridge structures in reactive powder concrete (RPC). In Proceedings of the 4th International PhD Symposium in Civil Engineering, Munich, Germany, 19–21 September 2002.

22. Shao, X.; Pan, R.; Zhan, H.; Fan, W.; Yang, Z.; Lei, W. Experimental Verification of the Feasibility of a Novel Prestressed Reactive Powder Concrete Box-Girder Bridge Structure. *J. Bridg. Eng.* **2017**, *22*, 04017015. [CrossRef]

23. Abdulraheem, M.S.; Kadhum, M.M. Experimental investigation of fire effects on ductility and stiffness of reinforced reactive powder concrete columns under axial compression. *J. Build. Eng.* **2018**, *20*, 750–761. [CrossRef]

24. Al-Jubory, N.H. Mechanical Properties of Reactive Powder Concrete (RPC) with Mineral Admixture-ENG. *AL Rafdain Eng. J. (AREJ)* **2013**, *21*, 92–101. [CrossRef]

25. Lee, M.-G.; Wang, Y.-C.; Chiu, C.-T. A preliminary study of reactive powder concrete as a new repair material. *Constr. Build. Mater.* **2007**, *21*, 182–189. [CrossRef]

26. Cwirzen, A.; Penttala, V.; Vornanen, C. Reactive powder based concretes: Mechanical properties, durability and hybrid use with OPC. *Cem. Concr. Res.* **2008**, *38*, 1217–1226. [CrossRef]

27. Tai, Y.-S.; Pan, H.-H.; Kung, Y.-N. Mechanical properties of steel fiber reinforced reactive powder concrete following exposure to high temperature reaching 800 °C. *Nucl. Eng. Des.* **2011**, *241*, 2416–2424. [CrossRef]

28. Tai, Y. Flat ended projectile penetrating ultra-high strength concrete plate target. *Theor. Appl. Fract. Mech.* **2009**, *51*, 117–128. [CrossRef]

29. Zdeb, T. Influence of the Physicochemical Properties of Portland Cement on the Strength of Reactive Powder Concrete. *Procedia Eng.* **2015**, *108*, 419–427. [CrossRef]

30. Singh, S.; Munjal, P.; Thammishetti, N. Role of water/cement ratio on strength development of cement mortar. *J. Build. Eng.* **2015**, *4*, 94–100. [CrossRef]

31. Mailvaganam, N.P.; Rixom, M.R.; Manson, D.P.; Gonzales, C. *Chemical Admixtures for Concrete*; CRC Press: Abingdon, UK, 1999.

32. Ipek, M.; Yilmaz, K.; Sümer, M.; Saribiyik, M. Effect of pre-setting pressure applied to mechanical behaviours of reactive powder concrete during setting phase. *Constr. Build. Mater.* **2011**, *25*, 61–68. [CrossRef]

33. Tam, C.M.; Tam, V.W.; Ng, K. Assessing drying shrinkage and water permeability of reactive powder concrete produced in Hong Kong. *Constr. Build. Mater.* **2012**, *26*, 79–89. [CrossRef]

34. Yunsheng, Z.; Wei, S.; Sifeng, L.; Chujie, J.; Jianzhong, L. Preparation of C200 green reactive powder concrete and its static–dynamic behaviors. *Cem. Concr. Compos.* **2008**, *30*, 831–838. [CrossRef]

35. Adam, J.M. *Special Issue on Analysis of Structural Failures Using Numerical Modeling*; American Society of Civil Engineers: Reston, VA, USA, 2013.

36. LLourenço, P.; Krakowiak, K.; Fernandes, F.; Ramos, L. Failure analysis of Monastery of Jerónimos, Lisbon: How to learn from sophisticated numerical models. *Eng. Fail. Anal.* **2007**, *14*, 280–300. [CrossRef]

37. Gardner, J.D.; Vijayaraghavan, A.; Dornfeld, D.A. *Comparative Study of Finite Element Simulation Software*; Laboratory for Manufacturing and Sustainability: Lafayette, CA, USA, 2005.

38. Narayanan, R.; Beeby, A. *Designers' Guide to EN 1992-1-1 and EN 1992-1-2. Eurocode 2: Design of Concrete Structures: General Rules and Rules for Buildings and Structural Fire Design*; Thomas Telford: Telford, UK, 2005; Volume 17.

39. Lubliner, J.; Oliver, J.; Oller, S.; Onate, E. A plastic-damage model for concrete. *Int. J. Solids Struct.* **1989**, *25*, 299–326. [CrossRef]

40. Abaqus, F. *Analysis User's Manual 6.14*; Dassault Systemes Simulia Corp.: Providence, RI, USA, 2011.

41. Elchalakani, M.; Karrech, A.; Dong, M.; Ali, M.M.; Yang, B. Experiments and Finite Element Analysis of GFRP Reinforced Geopolymer Concrete Rectangular Columns Subjected to Concentric and Eccentric Axial Loading. *Structures* **2018**, *14*, 273–289. [CrossRef]

42. Chaudhari, S.; Mukane, K.; Chakrabarti, M. Comparative study on exterior RCC beam column joint subjected to monotonic loading. *Int. J. Comput. Appl.* **2014**, *102*, 35–40.

43. Institution, B.S. *Eurocode 2: Design of Concrete Structures*; BSI: London, UK, 1992.
44. Kmiecik, P.; Kamiński, M. Modelling of reinforced concrete structures and composite structures with concrete strength degradation taken into consideration. *Arch. Civil Mech. Eng.* **2011**, *11*, 623–636. [CrossRef]

Article

Study on Energy Evolution and Damage Constitutive Model of Siltstone

Ruihe Zhou *, **Longhui Guo and Rongbao Hong**

School of Civil Engineering and Architecture, Anhui University of Science and Technology, Huainan 232001, China; guolonghui7864@163.com (L.G.); cherishrb2020@163.com (R.H.)
* Correspondence: zrhaust@163.com; Tel.: +86-187-5602-3671

Abstract: In order to study the energy evolution characteristics and damage constitutive relationship of siltstone, the conventional triaxial compression tests of siltstone under different confining pressures are performed, and the evolution laws of input energy, elastic strain energy and dissipative energy of siltstone with axial strain and confining pressure are analyzed. According to the test results, the judgment criterion of the rock damage threshold is improved, and an improved three-shear energy yield criterion is proposed., The damage constitutive equation of siltstone is established based on the damage mechanics theory through the principle of minimum energy consumption and by considering the residual strength of rock, and lastly, the rationality of the model is verified by experimental data. The results reveal that (1) both the input energy and dissipative energy gradually increase with the increase of axial strain, and the elastic strain energy first increases and then decreases with the increase of axial strain, and reaches its maximum at the peak. (2) The input energy and dissipation energy increase exponentially with the increase of the confining pressure, and the elastic strain energy increases linearly with the increase of confining pressure. (3) According to the linear relationship between the sum of shear strain energy and hydrostatic pressure, an improved three-shear energy yield criterion is established. (4) The model curve can better describe the strain softening stage and the residual strength characteristics of siltstone. The relative standard deviation between the model results and the test results is only 4.35%, which verifies the rationality and feasibility of the statistical damage constitutive model that is established in this paper.

Keywords: energy evolution; minimum energy dissipation principle; three-shear energy yield criterion; damage variable; constitutive model

Citation: Zhou, R.; Guo, L.; Hong, R. Study on Energy Evolution and Damage Constitutive Model of Siltstone. *Crystals* **2021**, *11*, 1271. https://doi.org/10.3390/cryst11111271

Academic Editors: Peter Taylor, Per-Lennart Larsson, Yifeng Ling, Chuanqing Fu and Peng Zhang

Received: 14 September 2021
Accepted: 14 October 2021
Published: 20 October 2021

Publisher's Note: MDPI stays neutral with regard to jurisdictional claims in published maps and institutional affiliations.

1. Introduction

During the excavation of coal mine shafts and roadways, the stress of the surrounding rock is redistributed, and the surrounding rock is repeatedly disturbed by construction. Therefore, there are a large number of micro defects such as cracks and cavities in the rock mass, resulting in the nonlinear mechanical characteristics of the rock mass under various stress states [1,2]. The mechanical properties of rocks are particularly important for the stability analysis of surrounding rocks [3,4], so it is necessary to analyze the mechanical properties of rock under different stress states. Establishing the constitutive model of rock, using statistical damage theory, has become one of the important methods to study the nonlinear mechanical properties of rock [5–9].

Experts at home and abroad have produced much research on the statistical damage constitutive model of rock materials. The key to statistical damage theory is the reasonable measurement of rock micro element strength [10]. Tang Chunan [11] proposed to measure the rock micro-element strength through axial strain, which has achieved satisfactory results, but has ignored the influence of the stress state of the rock micro-element on its strength, therefore, this method has some shortcomings. Cao Wengui et al. [12] first proposed a new rock micro element strength measurement method based on Mohr-Coulomb criterion. Mohr-Coulomb criterion can accurately reflect the bearing capacity of rock, but it ignores the influence of intermediate principal stress, so the constitutive model established also has some limitations. Xu Weiya et al. [13] assumed that rock micro-element failure conforms to the Drucker–Plager criterion and established a statistical damage constitutive model of rock based on Weibull distribution. However, the Drucker–Plager criterion is relatively conservative, which limits the rationality of the rock micro-element strength that is determined based on this criterion. In view of the shortcomings of the above yield criterion, the Hoek–Brown criterion proposed by Hoek et al. [14,15] can better reflect the nonlinear failure characteristics of rock and the confining pressure effect of rock strength; Cao Ruilang et al. [16] established a three-dimensional statistical damage constitutive model under the condition that the micro-element failure of rock conforms to the Hoek–Brown criterion, which can reflect the post peak softening characteristics of rock by using the post-peak residual strength of rock to modify the damage variable. Most of the above studies use the mechanical parameters of rock as the basis for establishing the statistical damage constitutive equation, whereas only a few studies introduce the energy evolution characteristics of rock into the constitutive relationship [17]. Gao Wei et al. [18] established the damage constitutive model of granite under uniaxial compression by using the principle of minimum energy consumption. Sun Mengcheng et al. [19] established a new damage constitutive model by introducing the rock unified energy yield criterion as the energy consumption constraint condition, which is based on the principle of minimum energy consumption and continuous damage theory, but did not consider the influence of the post-peak softening stage on the damage constitutive model. It can be seen that, when studying the nonlinear constitutive relationship of rock under complex stress conditions, the minimum energy consumption principle is rarely applied to the establishment of rock damage evolution equation, and the influence of post-peak strain softening stage on the constitutive model is rarely considered in the existing damage constitutive model based on the energy principle.

In summary, much progress has been made in the research on rock damage mechanics, but there are still some problems, specifically the following: (1) there are weaknesses in the definition of the damage variable. Some scholars establish damage variables according to the statistical distribution theory, but the statistical distribution function cannot accurately describe the crack development and deformation in rock materials. Therefore, the rock damage constitutive model based on the statistical distribution theory is not in accordance with the actual situation of rock damage and deformation. (2) There are weaknesses in the establishment of the damage model. Hooke's law can only describe the elastic stress–strain relationship of rock in the pre-peak stage, but cannot describe the stress–strain relationship in the post-peak strain softening stage. Some studies do not consider the strain softening stage, and therefore there are weaknesses in the damage constitutive model.

In order to avoid the above problems, a constitutive model that can accurately describe the characteristics of rock damage deformation and residual strength is constructed. Based on the principle of minimum energy consumption, this paper introduces the improved three-shear energy yield criterion as the energy consumption constraint, considers the influence of the post-peak strain softening stage and deduces the damage evolution equation of rock. According to the principle of effective stress, the damage constitutive model of rock under a conventional triaxial compression is established. In addition, the conventional triaxial compression tests of siltstone under different confining pressures are carried out, and the model parameters are identified according to the test data. The rationality and feasibility of the model in this paper are verified by comparing the model results with the test results.

2. Conventional Triaxial Compression Test of Siltstone

2.1. Test Process

The test material used is Permian upper Shihezi Formation siltstone, which is taken from matoumen—at a 425 m depth of an east air shaft in Yuandian No. 2 mine, Huaibei

City, Anhui Province. According to the standards for test methods of engineering rock mass (GBT50266-2013), the sample size is a cylinder with a diameter of 50 mm, a length of 100 mm and a length diameter ratio of 2. 2S-200 using a vertical coring machine, DQ-4 rock cutter and SHM-200 double face grinder to drill, cut and grind the siltstone rock sample. The disparity between the sample is less than 0.05 mm and the diameter error of the sample is no more than 0.3 mm.

The conventional triaxial compression test of siltstone samples is performed using a ZTCR-2000 low temperature rock triaxial system (Figure 1). During the test, the siltstone samples are preloaded to 0.5 MPa, and then the confining pressure is applied to the predetermined value at a loading speed of 50 N/s according to the load control mode. The test confining pressure is set to 0, 5, 10, 15 and 20 MPa. After the confining pressure reaches the predetermined value, it is stabilized for 30 s, and finally the press applies an axial pressure at the loading speed of 0.06 mm/min in the way of displacement control until the siltstone sample is damaged.

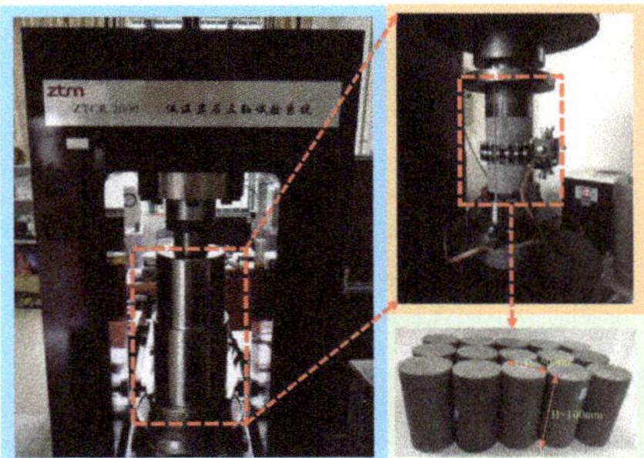

Figure 1. Rock mechanics test system.

2.2. Analysis of Test Results

The siltstone samples were divided into three groups with five samples in each group. The tests were repeated three times under the same loading type, and the representative test data were selected for analysis. Figure 2a shows the stress–strain curve of siltstone samples under different confining pressures. It can be seen from Figure 2a that when the confining pressure is at 0 MPa in the pre-peak stage, due to the lack of confining pressure constraints when the loading exceeds the elastic limit of the sample the siltstone sample quickly reaches the failure state, and the pre-peak curve of the sample has no obvious yield stage. After reaching the peak strength, the stress decreases rapidly, and the post-peak curve is steep, displaying an obvious strain softening. When the confining pressure is at 20 MPa, because the confining pressure effectively limits the propagation speed of microcracks in the sample and slows down the damage degree of the sample, the yield stage is obvious in the pre-peak curve. The post-peak curve and stress drop trend tend to be gentle, and the strain softening phenomenon decreases in the post-peak stage. With the increase of the confining pressure, the peak stress and peak strain increase gradually, indicating that the bearing capacity of the siltstone sample increases gradually, and it becomes more difficult for the sample to enter the failure state. The failure characteristics of siltstone samples under different confining pressures are shown in Figure 2b.

The basic mechanical parameters of siltstone are obtained according to the total stress–strain curve. See Table 1.

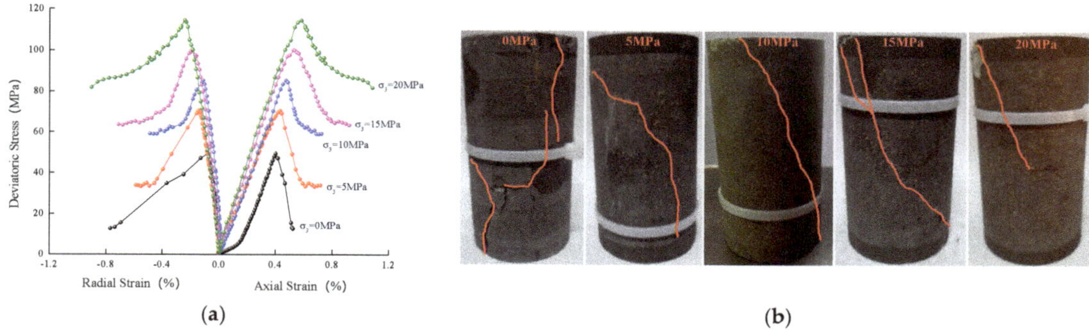

(**a**) (**b**)

Figure 2. Conventional triaxial test results of siltstone samples: (**a**) Full stress–strain curves of siltstone under different confining pressures; (**b**) Failure characteristics of siltstone samples.

Table 1. Mechanical parameters of siltstone under conventional triaxial test.

σ_3/MPa	σ_{sc}/MPa	ε_{sc}/10^{-3}	σ_1^r/MPa	ε_r/10^{-3}	E/GPa	μ
0	49.40	4.01	12.69	5.24	17.4	0.27
5	75.57	4.32	34.86	5.93	17.6	0.26
10	95.45	4.72	61.29	6.37	20.0	0.26
15	115.09	5.26	64.54	8.17	21.0	0.25
20	134.38	5.81	86.40	9.80	21.5	0.25

Notes: σ_3 is the confining pressure, σ_{sc} is the peak stress, ε_{sc} is the peak strain, σ_1^r is the residual stress, ε_r is the strain corresponding to the residual stress, E is the elastic modulus and μ is the Poisson's ratio.

3. Energy Analysis of Triaxial Compression Process of Siltstone

3.1. Theoretical Analysis of Energy Evolution

Assuming that heat exchange does not occur between the rock system and the external environment, according to the first law of thermodynamics, during the process of rock deformation and failure, the input energy W_F is equal to the elastic strain energy W_E plus the dissipative energy W_D [20,21].

The above-mentioned energy relationship is shown in Formula (1):

$$W_F = W_E + W_D \tag{1}$$

The work performed by the axial force and confining pressure during the test is [22]:

$$W_F = \frac{\pi}{4}D^2H\left(\int_0^{\varepsilon_1}\sigma_1 d\varepsilon_1 + 2\int_0^{\varepsilon_3}\sigma_3 d\varepsilon_3\right) = VU_F \tag{2}$$

where σ_1, σ_3 are the axial pressure and confining pressure, respectively, ε_1, ε_3 are the axial and radial strain, respectively, D, H are the diameter and height of the siltstone specimen, respectively, V is the volume of the siltstone sample and U_F is the input energy density.

In the same way, the elastic strain energy and dissipative energy are as follows:

$$\begin{cases} W_E = \frac{\pi}{4}D^2HU_E = VU_E \\ W_D = \frac{\pi}{4}D^2HU_D = VU_D \end{cases} \tag{3}$$

where U_E, U_D are the elastic strain energy density and dissipative energy density, respectively. According to the elastic theory [22], the elastic strain energy density is:

$$U_E = \frac{1}{2}(\sigma_1\varepsilon_1^e + \sigma_2\varepsilon_2^e + \sigma_3\varepsilon_3^e) \tag{4}$$

The three-dimensional constitutive relationship of siltstone is:

$$\varepsilon_{ij}^e = \frac{1+\mu}{E_{ij}}\sigma_{ij} - \frac{\mu}{E_{ij}}\sigma_{kk}\delta_{ij} \tag{5}$$

where $\varepsilon_{ij}^e (i,j = 1,2,3)$ is the elastic strain in the direction of the main stress, $\sigma_{ij}(i,j = 1,2,3)$ is the main stress, $\sigma_{kk} = \sigma_1 + \sigma_2 + \sigma_3$, δ_{ij} is the Kronecker tensor, $E_{ij}(i,j = 1,2,3)$ is replaced by the initial elastic modulus and E [22], μ is the Poisson's ratio.

Equations (4) and (5) can be substituted into Equation (3) to obtain the elastic strain energy W_E:

$$W_E = \frac{1}{2E}\left(\sigma_1^2 + \sigma_2^2 + \sigma_3^2 - 2\mu\sigma_1\sigma_2 - 2\mu\sigma_1\sigma_3 - 2\mu\sigma_2\sigma_3\right)V \tag{6}$$

In the conventional triaxial compression test where $\sigma_2 = \sigma_3$, Formula (6) can be simplified to:

$$W_E = \frac{1}{2E}\left[\sigma_1^2 + 2(1-\mu)\sigma_3^2 - 4\mu\sigma_1\sigma_3\right]V \tag{7}$$

Substituting Formula (7) into Formula (1) and combining this with Formula (2) to obtain the dissipative energy results in the following:

$$W_D = \left\{ \int_0^{\varepsilon_1}\sigma_1 d\varepsilon_1 + 2\int_0^{\varepsilon_3}\sigma_3 d\varepsilon_3 - \frac{1}{2E}\left[\sigma_1^2 + 2(1-\mu)\sigma_3^2 - 4\mu\sigma_1\sigma_3\right]\right\}V \tag{8}$$

3.2. Principle of Minimum Energy Consumption

For dissipative materials such as geotechnical materials, the stress comes from internal variables such as strain and temperature. Therefore, the internal variables reflecting the internal changes of materials need to be considered to truly establish the constitutive relationship of dissipative materials [23].

According to the internal variable theory, for any infinitesimal dissipative micro element, the volume dissipation rate is:

$$\varphi\rho_0 = \sigma_i : \dot{\varepsilon}^N + Y : D + \sum_{i=1}^{k} R_i\dot{\gamma}_i - q\frac{g}{T} \tag{9}$$

where, ρ_0 is the unit weight of rock, φ is the energy consumption rate of rock, D is the damage variable of material, Y is the dual variable corresponding to the damage variable, R_i is the corresponding dual variable of γ_i.

Without considering heat dissipation, the above formula is simplified as:

$$\varphi\rho_0 = \sigma_i : \dot{\varepsilon}^N + Y : D + \sum_{i=1}^{k} R_i\dot{\gamma}_i \tag{10}$$

It is generally assumed that dissipative materials meet the following energy dissipation constraints in the process of energy dissipation:

$$\begin{cases} F_1(\sigma, Y, R_1, \ldots R_k) = 0 \\ F_m(\sigma, Y, R_1, \ldots R_k) = 0 \end{cases} \tag{11}$$

According to the principle of minimum energy consumption, Formula (10) takes the stationary value under the condition of Formula (11) and introduces the Lagrange multiplier λ to obtain:

$$\frac{\partial(\varphi + \lambda_i F_i)}{\partial \xi_i} = 0 \ (i = 1, \ldots, m) \tag{12}$$

where, ξ_i takes $\sigma, Y, R_1, \ldots R_k$ respectively.

The Lagrange multiplier $\lambda_i (i = 1, \ldots, m)$ is introduced into Equation (10) to obtain:

$$
\begin{cases}
\dot{\varepsilon}^N + \sigma : \frac{\partial \dot{\varepsilon}^N}{\partial \sigma} + \sum\limits_{i=1}^{m} \lambda_i \frac{\partial F_i}{\partial \sigma} = 0 \\[2mm]
\dot{D} + Y : \frac{\partial \dot{D}}{\partial Y} + \sum\limits_{i=1}^{m} \lambda_i \frac{\partial F_i}{\partial Y} = 0 \\[2mm]
\dot{\gamma}_1 + R_1 : \frac{\partial \dot{\varepsilon}^N}{\partial R_1} + \sum\limits_{i=1}^{m} \lambda_i \frac{\partial F_i}{\partial R_1} = 0 \\[2mm]
\qquad\qquad \cdot \\
\qquad\qquad \cdot \\
\qquad\qquad \cdot \\
\dot{\gamma}_k + R_k : \frac{\partial \dot{\varepsilon}^N}{\partial R_k} + \sum\limits_{i=1}^{m} \lambda_i \frac{\partial F_i}{\partial R_k} = 0
\end{cases}
\tag{13}
$$

Assuming $\rho_0 \varphi$ is a potential function, we get:

$$
\begin{cases}
\dot{\varepsilon}^N + \sum\limits_{i=1}^{m} \lambda_i \frac{\partial F_i}{\partial \sigma} = 0 \\[2mm]
\dot{D} + \sum\limits_{i=1}^{m} \lambda_i \frac{\partial F_i}{\partial Y} = 0 \\[2mm]
\dot{\gamma}_1 + \sum\limits_{i=1}^{m} \lambda_i \frac{\partial F_i}{\partial R_1} = 0 \\[2mm]
\qquad \cdot \\
\qquad \cdot \\
\qquad \cdot \\
\dot{\gamma}_k + \sum\limits_{i=1}^{m} \lambda_i \frac{\partial F_i}{\partial R_k} = 0
\end{cases}
\tag{14}
$$

The above formula is the internal variable evolution equation of energy dissipation of dissipative materials derived based on the principle of minimum energy dissipation.

3.3. Relationship between Energy Evolution and Axial Strain

The triaxial compression test results are substituted into Equations (2), (7) and (8) to obtain the evolution curves of the input energy, elastic energy and dissipative energy of siltstone with an axial strain under different confining pressures, as shown in Figure 3.

It can be seen from Figure 3 that under different confining pressures, the input energy and dissipative energy increase with the increase of axial strain, and the elastic strain energy increases at first and then decreases. At the initial stage of sample loading, the initial pores in the rock are gradually closed as a result of the action of the external load, most of the work performed by the external load is transformed into elastic energy and stored in the sample and the elastic strain energy gradually increases with axial strain, the dissipative energy at this stage is very small, and the evolution curves of input energy, elastic energy and dissipative energy concave upward. Alongside the increase of the external load, the siltstone sample enters the linear elastic deformation stage, and the external work is essentially transformed into elastic strain energy, and the slope of the three energy evolution curves reaches the maximum. This stage is the main stage of energy storage in the overall process of rock failure. When the external load reaches the yield limit of rock, new cracks will appear in the sample, and part of the energy is required to be dissipated for its initiation and diffusion. Therefore, the slope of the elastic strain energy curve decreases gradually at this stage. When the external load reaches the peak strength, the elastic strain energy reaches the maximum value, and as a result the siltstone sample is damaged, and the elastic strain energy stored in the pre peak stage is released rapidly. Therefore, the elastic strain energy after the peak decreases gradually with the axial strain, and most of the input energy is dissipated in the process of the mutual penetration of cracks to form a macro-fracture surface. When the sample reaches its peak strength, the elastic strain energy decreases gradually until the sample failure reaches the minimum value, and the dissipative energy increases gradually until the sample failure reaches the maximum value.

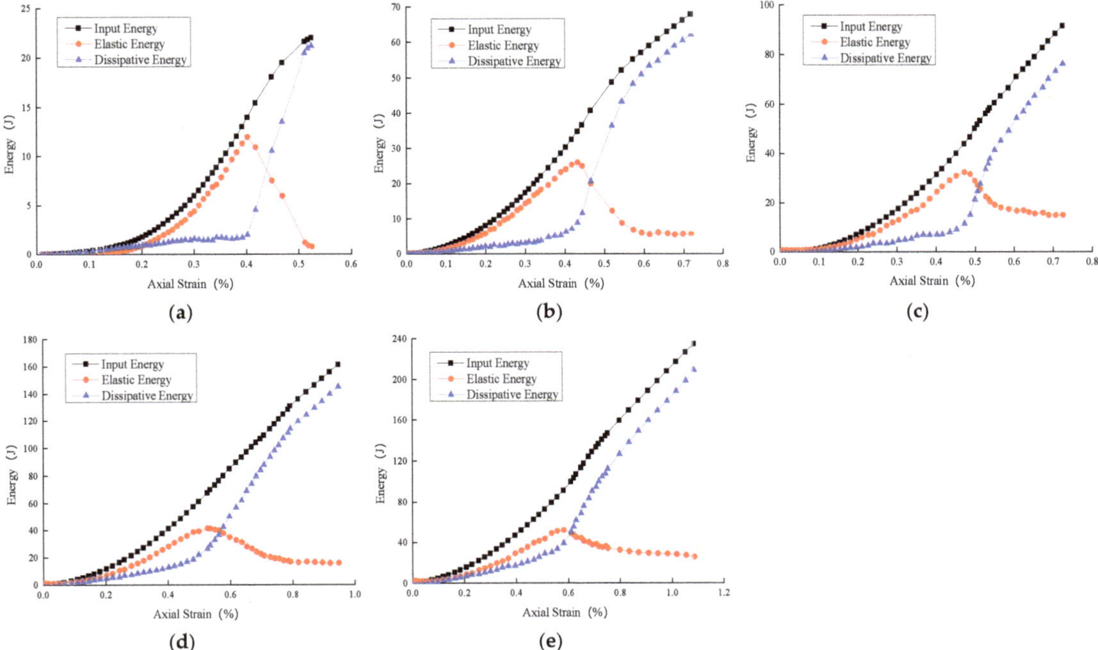

Figure 3. Energy evolution curve of siltstone under different confining pressures: (**a**) 0 MPa; (**b**) 5 MPa; (**c**) 10 MPa; (**d**) 15 MPa; (**e**) 20 MPa.

3.4. Relationship between Energy Evolution and Confining Pressure

Based on the triaxial compression test data of siltstone under different confining pressures, the characteristic energy results of siltstone are calculated by substituting Equations (2), (7) and (8), as shown in Table 2. It can be seen from Table 2 that when the confining pressure is at 0 MPa, the input energy $W_{F(pre)}$, elastic energy $W_{E(pre)}$ and dissipative energy $W_{D(pre)}$, corresponding to the peak stress of the siltstone sample, are 13.97 J, 11.97 J and 2.00 J, respectively. Most of the input energy before the peak of the sample is transformed into storable elastic energy, and only a small amount of the energy is dissipated in the process of damage deformation and of crack propagation of the sample, indicating that the energy storage capacity of the sample is strong, Therefore, the elastic property is the source power of the specimen failure. When the confining pressure is at 20 MPa, the input energy $W_{F(pre)}$, elastic energy $W_{E(pre)}$ and dissipative energy $W_{D(pre)}$, corresponding to the peak stress of the siltstone sample, are 91.46 J, 52.01 J and 39.45 J respectively. The elastic energy that was stored before the peak accounts for 56.8% of the input energy. The energy dissipated via the damage and deformation of the sample accounts for 43.2% of the input energy, indicating that the energy storage capacity of the sample is weakened at this time. The energy required for specimen failure is involved in the work provided by the external testing machine, and the self-sustaining fracture ability of the specimen is weak.

Table 2. Calculation results of characteristic energy of siltstone sample at peak stress.

σ_3/MPa	$W_{F(pre)}$	$W_{E(pre)}$	$W_{D(pre)}$
0	13.97	11.97	2.00
5	34.87	26.00	8.87
10	43.95	32.37	11.58
15	67.82	41.38	26.44
20	91.46	52.01	39.45

The evolution law of input energy, elastic strain energy and dissipative energy of the siltstone sample with confining pressure is shown in Figure 4. It can be seen from Figure 4, that with the increase of confining pressure, the three energies increase at different rates, the input energy and dissipative energy are in an exponential function proportional relationship with the confining pressure, the elastic energy is in a linear function proportional relationship with the confining pressure, and the storage rate of elastic strain energy gradually increases, indicating that the confining pressure has an obvious restrictive effect on crack propagation, thus limiting the siltstone sample to the failure state.

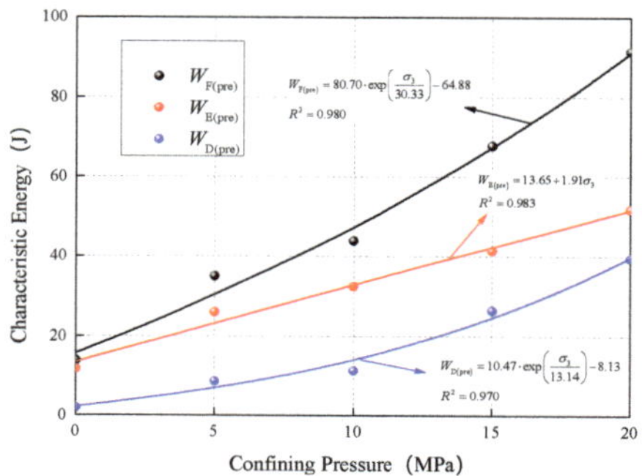

Figure 4. Characteristic energy of siltstone under different confining pressures.

4. Criterion for Determining Rock Damage Threshold

The yield criterion of rock materials generally included the following three categories: the yield criterion established from the angle of stress, strain and energy [24]. There are abundant yield criterion that are established from the angle of stress, mainly the Mises criterion, the Drucker-Plager criterion, the Mohr–Coulomb criterion, the Hoek–Brown criterion and the double shear strength criterion. Among them, the Mohr–Coulomb criterion is the most widely used, is applicable to both plastic and brittle rocks, and can reflect the characteristic that the tensile strength of rock is less than the compressive strength. However, the criterion does not consider the influence of the intermediate principal stress on rock yield, and the envelope line of the criterion on the meridional plane is a straight line, indicating that the internal friction angle does not change with hydrostatic pressure; this is inconsistent with the test results of rock mechanics [25,26]. The Drucker–Plager criterion considers the influence of intermediate principal stress on rock yield, but it cannot distinguish the difference between the tensile meridian and compressive meridian of rock, which is inconsistent with the triaxial test results of rock [27]. Double shear strength criterion is only applicable to materials for which the shear, tensile and compressive strength meet a certain relationship [28]. The accuracy of establishing yield criterion from the point of view of stress depends on the description of yield stress, and the expression does not include material parameters, so its application range is limited. The establishment of a yield criterion from the perspective of energy can accurately express the yield state of rock with the assistance of the description of energy evolution. Therefore, the establishment of an energy yield criterion across a wide range of applications is a fundamental method to analyze the yield deformation of rock.

The yield criterion is established from the perspective of energy, mainly by determining the functional relationship between the strain energy during material yield [29]. The mises energy criterion defines the maximum shape change of specific energy as a constant, but this criterion cannot reflect the characteristics of the different tensile and compressive strengths of rock materials. Gao Hong et al. [30] proposed the three-shear energy yield criterion of rock materials based on the maximum strain specific energy, and introduced the expression of shear strain energy by considering the internal friction angle. The criterion is simple and is a generalization of many common yield criteria. However, the three-shear energy yield criterion assumes that the sum of the shear strain energy of a composite sliding surface at the time of the material yield is a constant, and does not consider the influence of hydrostatic pressure on the material yield, which leads to a considerable difference between and the test results of rock materials.

In conclusion, it is highly necessary to establish a yield criterion that can reflect the internal friction characteristics and hydrostatic pressure effects of rock materials. Based on the test results, this paper explores the functional relationship between the sum of the shear strain energy of the three composite sliding surfaces and the hydrostatic pressure during rock yield, and establishes an improved three-shear energy yield criterion.

4.1. Three-Shear Energy Yield Criterion

The three-shear energy yield criterion assumes that the sum of the shear strain energy w_s of the three composite sliding surfaces is constant when the rock material yields. For the rock material with internal friction characteristics, w_s plays a vital role in the rock yield. The following is a simple derivation of w_s. It agrees that the compressive stress is positive in the derivation process.

The limit value of the shear strain energy of the three yield sliding surfaces of rock mass is:

$$w_{12} = \frac{1}{2G}\left[\frac{\sigma_1 - \sigma_2}{2\cos\varphi_{12}} - \frac{\sigma_1 + \sigma_2}{2}\tan\varphi_{12}\right]^2 \tag{15}$$

$$w_{23} = \frac{1}{2G}\left[\frac{\sigma_2 - \sigma_3}{2\cos\varphi_{23}} - \frac{\sigma_2 + \sigma_3}{2}\tan\varphi_{23}\right]^2 \tag{16}$$

$$w_{13} = \frac{1}{2G}\left[\frac{\sigma_1 - \sigma_3}{2\cos\varphi_{13}} - \frac{\sigma_1 + \sigma_3}{2}\tan\varphi_{13}\right]^2 \tag{17}$$

where, G is the shear modulus.

The three-shear energy yield criterion considers that the rock material begins to yield when the sum of the shear strain energy of the three maximum friction angle action surfaces reaches a certain value, as follows:

$$w_s = w_{12} + w_{23} + w_{13} = k_0 \tag{18}$$

In the case of uniaxial compression, $\sigma_2 = \sigma_3 = 0$, the value of k_0 is obtained:

$$k_0 = \frac{c_0^2}{G} \tag{19}$$

where, c_0 is the cohesion of rock at yield.

4.2. Improved Three-Shear Energy Yield Criterion

The research results of Hao Tiesheng et al. [31] reveals that the sum of shear strain energy of three composite sliding surfaces during rock yield is not constant, and rock yield has an obvious hydrostatic pressure effect. Therefore, the functional relationship between the sum of the shear strain energy during rock yield and the hydrostatic pressure is established as follows:

$$F(w_s, p) = 0 \tag{20}$$

where, p is the hydrostatic pressure and W_s is the shear strain energy of rock.

In order to establish the functional relationship between the sum of shear strain energy and hydrostatic pressure, by considering the sum of shear strain energy W_s as the ordinate and hydrostatic pressure p as the abscissa, the variation law of the sum of shear strain energy of the three composite sliding surfaces with hydrostatic pressure during rock yield can be obtained. The two approximate accordance with the first-order functional relationship, so the energy yield criterion of siltstone is as follows [31]:

$$F(\sigma_1, \sigma_3) = \frac{1}{4G}\left[\frac{\sigma_1 - \sigma_3}{\cos\varphi_s} - (\sigma_1 + \sigma_3)\tan\varphi_s\right]^2 + a(\sigma_1 + 2\sigma_3) + b \tag{21}$$

where, a and b are material parameters, representing the numerical relationship between shear strain energy and hydrostatic pressure and the rock yield, φ_s is the internal friction angle when rock yield and G is the shear modulus of rock.

5. Establishment of Damage Constitutive Model of Siltstone

5.1. Damage Constitutive Relationship

Assuming that the rock is an isotropic material, according to J. Lemaitre's [32] strain equivalence hypothesis, the constitutive relationship of the rock is established as follows:

$$\sigma_i = \sigma_i'(1 - D) + \sigma_i''D \quad (i = 1, 2, 3) \tag{22}$$

where, σ_i is the nominal stress of the rock micro-element, σ_i' is the effective stress of the rock micro-element, σ_i'' is the stress on the damaged part of the rock micro-element, and D is the damage variable.

Since the damaged part of the rock is closely associated with the undamaged part of the rock, according to the deformation coordination principle, $\varepsilon_i = \varepsilon_i' = \varepsilon_i''$, the undamaged part of the rock obeys Hooke's law, and its stress is:

$$\sigma_i' = E\varepsilon_i + \mu(\sigma_j' + \sigma_k') \tag{23}$$

Equation (23) can be written as:

$$\begin{cases} \sigma_1' = E\varepsilon_1 + \mu(\sigma_2' + \sigma_3') \\ \sigma_2' = E\varepsilon_2 + \mu(\sigma_1' + \sigma_3') \\ \sigma_3' = E\varepsilon_3 + \mu(\sigma_1' + \sigma_2') \end{cases} \tag{24}$$

The damage of the rock micro-element can be defined as the result of the reduction of stiffness caused by the change of physical properties of undamaged micro-element. The stress of damaged micro-element and undamaged the micro-element under the external load is related to the stiffness, and γ is defined as the damage correction coefficient [33]. Based on the rock damage mechanism, the damage models of rock micro-elements in different states can be established.

(1) When the rock is not damaged:

$$\sigma_i' = \sigma \ , \quad \sigma_i'' = 0 \tag{25}$$

(2) When random damage occurs to rock:

$$\sigma = \sigma_i'(1 - D) + \sigma_i''D \tag{26}$$

From the stress distribution relationship between damaged micro-elements and un-damaged micro-elements, it can be concluded that:

$$\sigma_i'' = \gamma \cdot \sigma_i' \tag{27}$$

Substituting to formula (26) to obtain:

$$\sigma_i = \sigma_i'(1 - D + \gamma D) \tag{28}$$

(3) When the rock is completely damaged:

$$\sigma_i'' = \sigma, \ \sigma_i' = 0 \tag{29}$$

Substituting Equation (23) into Equation (28) yields:

$$\sigma_1 = E\varepsilon_1(1 - D + \gamma D) + \mu(\sigma_2 + \sigma_3) \tag{30}$$

Equation (30) is the damage constitutive equation of rock. To establish the damage constitutive model of rock, the damage variable must first be determined.

5.2. Damage Evolution Equation of Siltstone

In each stage of rock damage and failure, the thermodynamic change of the system is regarded as a process in which the equilibrium is broken and then reconstructed. When the rock is damaged and deformed, the equilibrium state of the system is broken. When the energy dissipation tends to be stable, the system reaches a new equilibrium state. The energy dissipation process of rock damage and deformation conforms to the minimum energy dissipation principle. According to the minimum energy dissipation principle, all energy dissipation processes occur along the minimum energy dissipation path under corresponding constraints. It illustrates that the instantaneous energy dissipation rate is at the minimum at any time in the energy dissipation process. In the elastic damage model, it is assumed that the irreversible strain caused by damage is the only energy dissipation mechanism in the rock failure process [23]. The energy consumption rate of rock is defined as:

$$\varphi = \sigma_i \dot{\varepsilon}_i \tag{31}$$

where, φ is the rock energy consumption rate, σ_i is the nominal stress of rock micro element, and $\dot{\varepsilon}_i$ is the irreversible strain rate caused by damage.

The three-dimensional constitutive relationship of rock can be obtained from Equation (30) as follows:

$$\begin{cases} \varepsilon_1 = \dfrac{\sigma_1 - \mu(\sigma_2 + \sigma_3)}{[1 - D(t) + \gamma D(t)]E} \\[2mm] \varepsilon_2 = \dfrac{\sigma_2 - \mu(\sigma_1 + \sigma_3)}{[1 - D(t) + \gamma D(t)]E} \\[2mm] \varepsilon_3 = \dfrac{\sigma_3 - \mu(\sigma_1 + \sigma_2)}{[1 - D(t) + \gamma D(t)]E} \end{cases} \tag{32}$$

where, $D(t)$ is the damage variable at t time.

The irreversible strain rate caused by the damage variable is:

$$\begin{cases} \dot{\varepsilon}_1 = \dfrac{(1-\gamma)D'(t)[\sigma_1 - \mu(\sigma_2 + \sigma_3)]}{[1 - D(t) + \gamma D(t)]^2 E} \\[2mm] \dot{\varepsilon}_2 = \dfrac{(1-\gamma)D'(t)[\sigma_2 - \mu(\sigma_1 + \sigma_3)]}{[1 - D(t) + \gamma D(t)]^2 E} \\[2mm] \dot{\varepsilon}_3 = \dfrac{(1-\gamma)D'(t)[\sigma_3 - \mu(\sigma_1 + \sigma_2)]}{[1 - D(t) + \gamma D(t)]^2 E} \end{cases} \tag{33}$$

By substituting Equation (33) into Equation (31), the energy consumption rate of rock is found as follows:

$$\varphi = \frac{(1 - \gamma)D'(t)}{[1 - D(t) + \gamma D(t)]^2 E}\left[\sigma_1^2 + \sigma_2^2 + \sigma_3^2 - 2\mu(\sigma_1\sigma_2 + \sigma_2\sigma_3 + \sigma_1\sigma_3)\right] \tag{34}$$

For a conventional triaxial compression test, $\sigma_1 > \sigma_2 = \sigma_3$, Equation (34) can be written as:

$$\varphi = \frac{(1-\gamma)D'(t)}{[1-D(t)+\gamma D(t)]^2 E}\left[\sigma_1^2 + 2(1-\mu)\sigma_3^2 - 4\mu\sigma_1\sigma_3\right] \tag{35}$$

5.3. Establishment of Damage Constitutive Model

Substitute Equation (21) into Equation (12) to obtain [23]:

$$\frac{\partial(\varphi + \lambda F(\sigma_1,\sigma_3))}{\partial\sigma_i} = 0 \quad (i = 1,3) \tag{36}$$

$$\begin{cases} \frac{\partial(\varphi+\lambda F(\sigma_1,\sigma_3))}{\partial\sigma_1} = 0 \\ \frac{\partial(\varphi+\lambda F(\sigma_1,\sigma_3))}{\partial\sigma_3} = 0 \end{cases} \tag{37}$$

where, $\frac{\partial\varphi}{\partial\sigma_1} = \frac{(1-\gamma)D'}{1-D+\gamma D}\cdot 2\varepsilon_1$, $\frac{\partial\varphi}{\partial\sigma_3} = \frac{(1-\gamma)D'}{1-D+\gamma D}\cdot 4\varepsilon_3$, $\frac{\partial F}{\partial\sigma_1} = \frac{(1-\sin\varphi)^2}{2G\cos^2\varphi}\sigma_1 + \left(-\frac{1}{2G}\right)\sigma_3 + a$, $\frac{\partial F}{\partial\sigma_3} = \left(-\frac{1}{2G}\right)\sigma_1 + \frac{(1+\sin\varphi)^2}{2G\cos^2\varphi}\sigma_3 + 2a$.

The damage evolution equation is derived as follows:

$$D(t) = \frac{1}{1-\gamma} - \frac{1}{1-\gamma}\exp\left(\frac{\lambda\cdot H}{R}t + C_0\right) \tag{38}$$

where, $A = \frac{(1-\sin\varphi)^2}{2G\cos^2\varphi}$, $B = -\frac{1}{2G}$, $C = \frac{(1+\sin\varphi)^2}{2G\cos^2\varphi}$, $H = (B^2 - AC)\sigma_3 + (B - 2A)a$, $R = 2B\varepsilon_1 - 4A\varepsilon_3$, λ and C_0 are parameters related to material properties.

Under a constant loading rate and loading path, the axial strain of the rock sample is directly proportional to time. Assuming that $\lambda\cdot t = \lambda^*\cdot\varepsilon_1$, the damage evolution equation is simplified as:

$$D(t) = \frac{1}{1-\gamma} - \frac{1}{1-\gamma}\exp\left(\frac{\lambda^*\cdot H}{R}\varepsilon_1 + C_0\right) \tag{39}$$

5.4. Parameter Identification of Constitutive Model

According to the triaxial compressive stress–strain curve of rock, when the rock is completely damaged ($D = 1$), Formula (30) is written as:

$$\sigma_1^r = E\varepsilon_1^r\gamma + \mu(\sigma_2 + \sigma_3) \tag{40}$$

$$\gamma = \frac{\sigma_1^r - \mu(\sigma_2 + \sigma_3)}{E\varepsilon_1^r} \tag{41}$$

Two boundary conditions can be determined from the extreme value characteristics of the rock total stress–strain curve. The peak stress of the rock is σ_{sc} and the corresponding strain is ε_{sc}. The two boundary conditions are as follows:

where $\varepsilon_1 = \varepsilon_{sc}$, $\sigma_1 = \sigma_{sc}$, $d\sigma_1/d\varepsilon_1 = 0$.

Substitute conditions into Formula (30) to obtain:

$$\sigma_{sc} = E\varepsilon_{sc}\exp\left(\frac{\lambda^*\cdot H}{R}\varepsilon_{sc} + C_0\right) + 2\mu\sigma_3 \tag{42}$$

In the triaxial compression test, the peak stress σ_{sc} and peak strain ε_{sc} of rock are not the peak strength σ_{sc}' and corresponding strain ε_{sc}' of the rock deviatoric stress–strain curve, because the starting point of the test curve provides that the rock is subjected to deviatoric stress, ignoring the initial strain of the rock under hydrostatic pressure, therefore, the relationship between the theoretical peak value and the test peak value of rock stress–strain curve is as follows:

$$\sigma_{sc} = \sigma_{sc}' + \sigma_3 \tag{43}$$

$$\varepsilon_{sc} = \varepsilon'_{sc} + \varepsilon_c \tag{44}$$

The initial strain can be obtained from the first equation in formula (24):

$$\varepsilon_c = \frac{\sigma_3(1 - 2\mu)}{E} \tag{45}$$

The values of λ^* and C_0 are as follows:

$$\lambda^* = \frac{\left(2B\varepsilon_{1(sc)} - 4A\varepsilon_{3(sc)}\right)^2}{4AH\varepsilon_{1(sc)}\varepsilon_{3(sc)}} = \frac{R^2}{4AH\varepsilon_{1(sc)}\varepsilon_{3(sc)}} \tag{46}$$

$$C_0 = \ln\left[\frac{\sigma_{1(sc)} - 2\mu\sigma_3}{E\varepsilon_{1(sc)}}\right] - \frac{B\varepsilon_{1(sc)} - 2A\varepsilon_{3(sc)}}{2A\varepsilon_{3(sc)}} = \ln\left[\frac{\sigma_{1(sc)} - 2\mu\sigma_3}{E\varepsilon_{1(sc)}}\right] - \frac{R}{4A\varepsilon_{3(sc)}} \tag{47}$$

5.5. Verification of Damage Constitutive Model

In order to verify the applicability and rationality of the damage constitutive model that considers the rock residual strength proposed in this paper, the conventional triaxial compression tests of siltstone under different confining pressure conditions (5 MPa, 10 MPa, 15 MPa and 20 MPa) are carried out. The mechanical parameters of the rock samples are as follows: average uniaxial compressive strength $\sigma_c = 49.4$ MPa, cohesion $c = 12.94$ MPa, internal friction angle $\varphi = 22°$ and Poisson's ratio $\mu = 0.25$. The parameters of the improved three-shear energy criterion are obtained by linear regression, $a = 1.42 \times 10^{-4}$, $b = -0.035$. The calculated parameters of the constitutive model are shown in Table 3. The parameters of the damage statistical model are substituted into the formula, and the theoretical curve is created, which is compared with the test curve of the conventional triaxial compression under four confining pressures, as showed in Figure 5.

Table 3. Parameter values of constitutive model under different confining pressures.

σ_3/MPa	A	B	C	H	λ	C_0
5	3.28×10^{-5}	-7.10×10^{-5}	1.54×10^{-4}	-1.94×10^{-8}	11,250	-2.28
10	2.89×10^{-5}	-6.25×10^{-5}	1.35×10^{-4}	-1.71×10^{-8}	20,533	-3.65
15	2.75×10^{-5}	-5.95×10^{-5}	1.29×10^{-4}	-1.63×10^{-8}	10,435	-2.19
20	2.68×10^{-5}	-5.81×10^{-5}	1.26×10^{-4}	-1.59×10^{-8}	7784	-1.76

Figure 5 presents the comparison results of the theoretical curves and the test curves under different confining pressures. It can be seen from Figure 5 that the damage constitutive equation of siltstone proposed in this paper can better reflect the actual mechanical behavior of rock, and its initial deformation modulus and peak strength are approximately the same as the test results. The consistency between the constitutive model curve and the test curve is high, which overcomes the defect in some constitutive models that cannot describe the residual strength in the post peak stage, improves the accuracy of the model, and illustrates that the constitutive model based on the energy principle is more reasonable than the traditional constitutive model.

In order to further verify the rationality of the damage constitutive model of siltstone that has been established in this paper, the deviation between the conventional triaxial stress–strain curve and the model curve of siltstone under four different confining pressures is analyzed, and the calculation formula is shown in Formula (48).

$$\begin{cases} \eta = \sqrt{\dfrac{\sum\limits_{i=1}^{n}(\sigma_s - \sigma_l)^2}{n-1}} \\ f = \dfrac{\eta}{\sigma_0} \end{cases} \tag{48}$$

where, η is the standard deviation, f is the relative standard deviation, σ_s, σ_l are the test value and theoretical value respectively, σ_0 is the mean value of the test value and n is the number of samples.

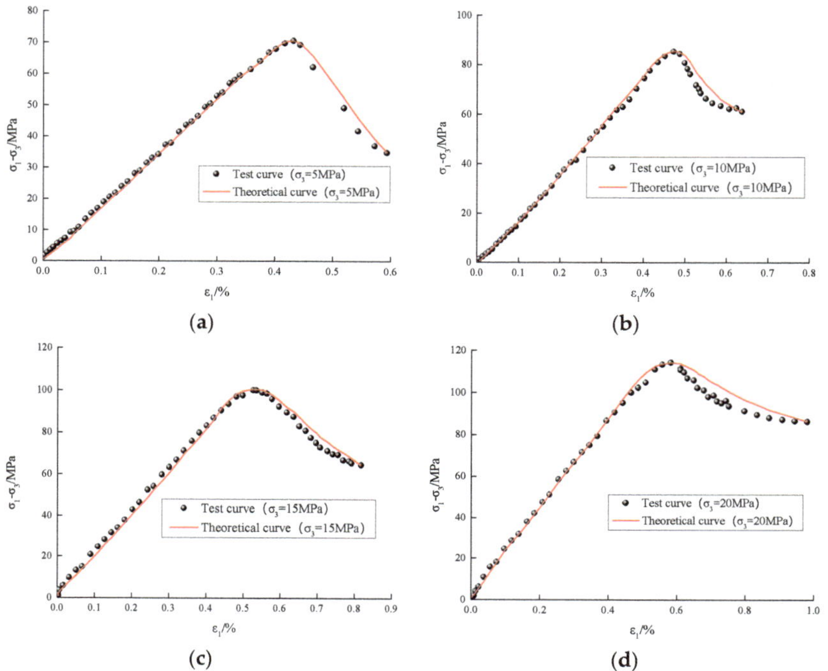

Figure 5. Comparison between model curve and test curve: (**a**) 5 MPa; (**b**) 10 MPa; (**c**) 15 MPa; (**d**) 20 MPa.

The calculation results show that when the confining pressure is at 5, 10, 15 and 20 MPa, the relative standard deviations between the constitutive model results and the test results are 3.75%, 4.77%, 4.05% and 4.83% respectively, and the average relative standard deviation is 4.35%, indicating that the difference between the constitutive model results and the test results is minute, which further proves the rationality of the constitutive model that has been established in this paper.

The change curve of the damage variable with axial strain is obtained by substituting the model parameters into Equation (39), as shown in Figure 6. It can be seen from Figure 6 that the damage evolution curve is approximately an "S" curve. There is no obvious damage accumulation in the initial stage of the curve, then there is a rapid rise stage, indicating that alongside the gradual axial loading, the internal microcracks rub and squeeze each other, and the damage accumulates and converges continuously. Finally, the curve gradually flattens, indicating that the internal structure of the rock is completely destroyed. Along with the increase of the confining pressure, the development trend of cumulative damage slows, because the confining pressure inhibits the development of damage and improves the stress state of rock.

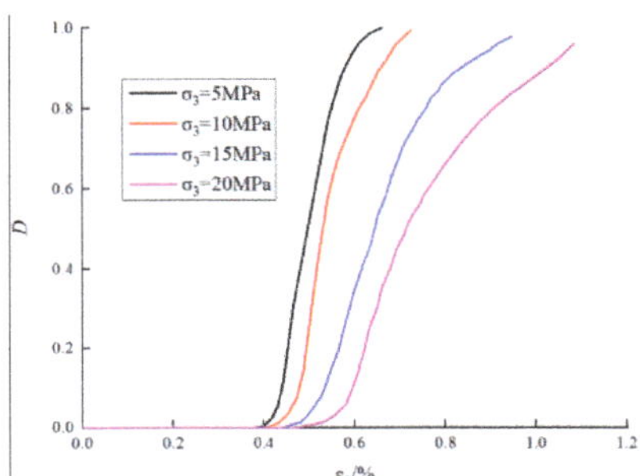

Figure 6. Variation curve of damage variable with axial strain.

6. Conclusions

According to the stress–strain curves of siltstone under different confining pressures, the energy evolution characteristics of siltstone samples under conventional triaxial loading are analyzed, the damage constitutive model of siltstone is established, and the rationality of the model established in this paper is verified by using the conventional triaxial test data of siltstone under different confining pressures. The following conclusions are drawn:

Under different confining pressures, the input energy and dissipative energy of siltstone samples increase with the increase of axial strain, and the elastic strain energy increases at first and then decreases. When the specimen reaches the peak strength, the elastic strain energy gradually decreases and the dissipative energy gradually increases until the specimen is damaged and reaches the maximum and minimum values respectively.

Considering the internal friction characteristics and hydrostatic pressure effect of rock materials, based on the test results of siltstone samples, the three-shear energy yield criterion is improved, the functional relationship between the sum of shear strain energy and hydrostatic pressure is established, and the improved three-shear energy yield criterion is obtained.

Based on the continuous damage theory, the damage evolution equation of rock is derived using the minimum energy consumption principle and the improved three-shear energy yield criterion, and the damage constitutive model of siltstone under complex stress state is established. The model overcomes the defect of some existing damage constitutive models that cannot simulate the residual strength. By comparing the model curve with the test curve, it has been found that the margin of error is small, and the relative standard deviation is 4.35%, which verifies the rationality of the model established in this paper.

Author Contributions: Conceptualization, R.Z.; methodology, R.Z.; validation, R.Z. and L.G.; data curation, R.H.; writing—original draft preparation, R.Z.; writing—review and editing, R.Z. All authors have read and agreed to the published version of the manuscript.

Funding: This research received no external funding.

Data Availability Statement: The data used to support the findings of this study are available upon request from the corresponding author.

Conflicts of Interest: The authors declare no conflict of interest.

References

1. Cai, M.F. *Rock Mechanics and Engineering*; Science Press: Beijing, China, 2002; pp. 149–156. (In Chinese)
2. Wu, Z.J.; Ji, X.K.; Liu, Q.S.; Fan, L.F. Study of microstructure effect on the nonlinear mechanical behavior and failure process of rock using an image-based-FDEM model. *Comput. Geotech.* **2020**, *121*, 103480. [CrossRef]
3. Xie, Z.Z.; Zhang, N.; Feng, X.W.; Liang, D.X.; Wei, Q.; Weng, M.Y. Investigation on the evolution and control of surrounding rock fracture under different supporting conditions in deep roadway during excavation period. *Int. J. Rock Mech. Min. Sci.* **2019**, *123*, 104122. [CrossRef]

4. Zhao, C.X.; Li, Y.M.; Liu, G.; Meng, X.R. Mechanism analysis and control technology of surrounding rock failure in deep soft rock roadway. *Eng. Fail. Anal.* **2020**, *115*, 104611. [CrossRef]
5. Li, X.; Cao, W.G.; Su, Y.H. A statistical damage constitutive model for softening behavior of rocks. *Eng. Geol.* **2012**, *143–144*, 1–17. [CrossRef]
6. Li, H.; Liao, H.; Xiong, G.; Han, B.; Zhao, G. A three-dimensional statistical damage constitutive model for geomaterials. *J. Mech. Sci. Technol.* **2015**, *29*, 71–77. [CrossRef]
7. Zhao, H.; Shi, C.; Zhao, M.; Li, X. Statistical damage constitutive model for rocks considering residual strength. *Int. J. Geomech.* **2016**, *17*, 04016033. [CrossRef]
8. Zhao, H.; Zhang, C.; Cao, W.G.; Zhao, M.H. Statistical meso-damage model for quasi-brittle rocks to account for damage tolerance principle. *Environ. Earth Sci.* **2016**, *75*, 862. [CrossRef]
9. Zhang, L.L.; Cheng, H.; Wang, X.J.; Liu, J.M.; Guo, L.H. Statistical damage constitutive model for high-strength concrete based on dissipation energy density. *Crystals* **2021**, *11*, 800. [CrossRef]
10. Da Rabi, M.K.; Al-Rub, R.; Little, D.N. A continuum damage mechanics framework for modeling micro-damage healing. *Int. J. Solids Struct.* **2012**, *49*, 492–513. [CrossRef]
11. Tang, C.A. *Catastrophe in the Process of Rock Fracture*; Coal Industry Press: Beijing, China, 1993; pp. 10–30. (In Chinese)
12. Cao, W.G.; Xiang, L.I.; Zhao, H. Damage constitutive model for strain-softening rock based on normal distribution and its parameter determination. *J. Cent. South Univ. Technol.* **2007**, *14*, 719–724. [CrossRef]
13. Xu, W.Y.; Wei, L.D. Study on statistical constitutive model of rock damage. *Chin. J. Rock Mech. Eng.* **2002**, *21*, 787–791. (In Chinese)
14. Hoek, E. Hoek-Brown failure criterion—2002 edition. *Proc. N. Am. Rock Mech. Symp.* **2002**, *1*, 267–273.
15. Li, H.Z.; Guo, T.; Nan, Y.L.; Han, B. A simplified three-dimensional extension of Hoek-Brown strength criterion. *J. Rock Mech. Geotech. Eng.* **2021**, *13*, 568–578. [CrossRef]
16. Cao, R.L.; He, S.H.; Wei, J.; Wang, F. Study on statistical constitutive model of rock damage softening based on residual strength correction. *Rock Soil Mech.* **2013**, *34*, 1652–1660.
17. Liu, X.S.; Ning, J.G.; Tan, Y.L.; Gu, Q.H. Damage constitutive model based on energy dissipation for intact rock subjected to cyclic loading. *Int. J. Rock Mech. Min. Sci.* **2016**, *85*, 27–32. [CrossRef]
18. Gao, W.; Wang, L.; Yang, D.Y. Study on energy method of rock damage evolution. *Chin. J. Rock Mech. Eng.* **2011**, *30*, 4087–4092. (In Chinese)
19. Sun, M.C.; Xu, W.Y.; Wang, S.S.; Wang, R.B.; Wang, W. Study on rock damage constitutive model based on the principle of minimum energy consumption. *J. Cent. South Univ.* **2018**, *49*, 2067–2075. (In Chinese)
20. Xie, H.P.; Li, L.Y.; Peng, R.D.; Ju, Y. Energy analysis and criteria for structural failure of rocks. *J. Rock Mech. Geotech. Eng.* **2009**, *1*, 11–20. [CrossRef]
21. Zhou, R.H.; Cheng, H.; Li, M.J.; Zhang, L.L.; Hong, R.B. Energy evolution analysis and brittleness evaluation of high-strength concrete considering the whole failure process. *Crystals* **2020**, *10*, 1099. [CrossRef]
22. Wu, J. *Elasticity*; Higher Education Press: Beijing, China, 2011. (In Chinese)
23. Zhou, Z.B. *Principle of Minimum Energy Consumption and Its Application*; Science Press: Beijing, China, 2001. (In Chinese)
24. Li, Q.M. Strain energy density failure criterion. *Int. J. Solids Struct.* **2001**, *38*, 6997–7013. [CrossRef]
25. Shen, B.T.; Shi, J.Y.; Nick, B. An approximate nonlinear modified Mohr-Coulomb shear strength criterion with critical state for intact rocks. *J. Rock Mech. Geotech. Eng.* **2018**, *10*, 37–44. [CrossRef]
26. Gong, B.; Tang, C.A.; Wang, S.Y.; Bai, H.M.; Li, Y.C. Simulation of the nonlinear mechanical behaviors of jointed rock masses based on the improved discontinuous deformation and displacement method. *Int. J. Rock Mech. Min.* **2019**, *122*, 104076. [CrossRef]
27. Zienkiewicz, O.C. Some useful forms of isotropic yield surfaces for soil and rock mechanics. In *Finite Element in Geomechanics*; John Wiley: London, UK, 1977; pp. 179–190.
28. Yu, M.H.; Zan, Y.W.; Zhao, J.; Yoshimine, M. A unified strength criterion for rock material. *Int. J. Rock Mech. Min. Sci.* **2002**, *39*, 975–989. [CrossRef]
29. Xie, H.P.; Li, L.Y.; Ju, Y.; Peng, R.D.; Yang, Y.M. Energy analysis for damage and catastrophic failure of rocks. *Sci. China Technol. Sci.* **2011**, *54*, 199–209. [CrossRef]
30. Gao, H.; Zheng, Y.R.; Feng, X.T. Study on energy yield criterion of geotechnical materials. *Chin. J. Rock Mech. Eng.* **2007**, *26*, 2437–2443. (In Chinese)
31. Hao, T.S.; Liang, W.G.; Zhang, C.T. Stability analysis of underground horizontal salt rock reservoir cavity wall based on three shear energy yield criterion. *Chin. J. Rock Mech. Eng.* **2014**, *33*, 1997–2006. (In Chinese)
32. Lemaitre, J. How to use damage mechanics. *Nucl. Eng. Des.* **1984**, *80*, 233–245. [CrossRef]
33. Li, T.; Lyu, Y.X.; Zhang, S.L.; Sun, J.C. Development and application of a statistical constitutive model of damaged rock affected by the load-bearing capacity of damaged elements. *J. Zhejiang Univ. Sci. A (Appl. Phys. Eng.)* **2015**, *16*, 644–655. [CrossRef]

Article

Stress–Strain Behavior of FRC in Uniaxial Tension Based on Mesoscopic Damage Model

Weifeng Bai, Xiaofeng Lu, Junfeng Guan *, Shuang Huang *, Chenyang Yuan and Cundong Xu

School of Water Conservancy, North China University of Water Resources and Electric Power, Zhengzhou 450046, China; baiweifeng@ncwu.edu.cn (W.B.); z20201010121@stu.ncwu.edu.cn (X.L.); yuanchenyang@ncwu.edu.cn (C.Y.); xucundong@ncwu.edu.cn (C.X.)
* Correspondence: junfengguan@ncwu.edu.cn (J.G.); z201710103065@stu.ncwu.edu.cn (S.H.)

Abstract: Fiber-reinforced concrete (FRC) is widely used in the field of civil engineering. However, the research on the damage mechanism of FRC under uniaxial tension is still insufficient, and most of the constitutive relations are macroscopic phenomenological. The aim is to provide a new method for the investigation of mesoscopic damage mechanism of FRC under uniaxial tension. Based on statistical damage theory, the damage constitutive model for FRC under uniaxial tension is established. Two kinds of mesoscopic damage mechanisms, fracture and yield, are considered, which ultimately determines the macroscopic nonlinear stress–strain behavior of concrete. The yield damage mode reflects the potential bearing capacity of materials and plays a key role in the whole process. Evolutionary factor is introduced to reflect the degree of optimization and adjustment of the stressed skeleton in microstructure. The whole deformation-to-failure is divided into uniform damage phase and local failure phase. It is assumed that the two kinds of damage evolution follow the independent triangular probability distributions, which could be represented by four characteristic parameters. The validity of the proposed model is verified by two sets of test data of steel fiber-reinforced concrete. Through the analysis of the variation law of the above parameters, the influence of fiber content on the initiation and propagation of micro-cracks and the damage evolution of concrete could be evaluated. The relations among physical mechanism, mesoscopic damage mechanism, and macroscopic nonlinear mechanical behavior of FRC are discussed.

Keywords: fiber-reinforced concrete; damage mechanism; uniaxial tension

Citation: Bai, W.; Lu, X.; Guan, J.; Huang, S.; Yuan, C.; Xu, C. Stress–Strain Behavior of FRC in Uniaxial Tension Based on Mesoscopic Damage Model. *Crystals* **2021**, *11*, 689. https://doi.org/10.3390/cryst11060689

Academic Editors: Yifeng Ling, Chuanqing Fu, Peng Zhang, Peter Taylor and Nima Farzadnia

Received: 23 May 2021
Accepted: 12 June 2021
Published: 16 June 2021

Publisher's Note: MDPI stays neutral with regard to jurisdictional claims in published maps and institutional affiliations.

1. Introduction

Fiber-reinforced concrete (FRC) has been widely used in the field of civil engineering for its excellent physical and mechanical properties. FRC is a kind of cement-based composite material composed of metal fiber, inorganic non-metallic fiber, synthetic fiber, or natural organic fiber as a reinforcing material. The most widely used are steel fiber-reinforced concrete (SFRC) and polypropylene fiber-reinforced concrete (PFRC). A large number of experimental studies have shown that these randomly distributed fibers can effectively hinder the expansion of micro-cracks in the concrete and the formation of macro-cracks, significantly improving the compressive, tensile, bending, impact resistance, and fatigue resistance of concrete [1–12]. Fiber materials are further used in the research of new types of concrete, such as coral concrete, geopolymer concrete, self-compacting concrete, etc. Liu et al. [7] carried out uniaxial compression tests of carbon fiber-reinforced coral concrete and established empirical piecewise constitutive model. At present, the representative theories on the mechanism of fiber reinforcement mainly include the fiber spacing theory proposed by Romualdi and Batson [13], and the reinforcement rules for composite materials proposed by Swamy [14].

The constitutive relation of concrete material is one of the key problems in the nonlinear analysis of concrete structure. Much effort has been devoted in the last decade to model the FRC by considering it as a composite material composed of concrete matrix and fibers [15–19]. Uniaxial stress–strain behavior is the most fundamental constitutive relationship of the concrete material. Due to the limitation of experimental technology and the deficiency of relevant theories, the research achievements on uniaxial compression are abundant, while the research on tension is relatively few. The research on the constitutive relation of FRC is based primarily on the relevant theoretical results of ordinary concrete, mainly focusing on the influence of fiber characteristic parameters on the constitutive relation. In recent years, based on the direct tensile test carried out on the universal test machine, the uniaxial tensile stress–strain curves for different types of FRC were obtained, and the corresponding macroscopic phenomenological constitutive models were proposed [3,8,20–22].

As a typical quasi-brittle material, the nonlinear macroscopic stress–strain behavior of concrete are closely related to the heterogeneity in microstructure. The essential nature lies not only in the initiation and propagation of individual micro-cracks, but also in the interaction and coalescence of crack populations [23,24]. Most of the existing concrete constitutive relations are macroscopic phenomenological models established by polynomial and other mathematical formulas, focusing on the fitting of test data, but lacking the clear physical meaning for the parameters. Therefore, it is difficult to reveal the meso-damage evolution law of concrete in essence by those phenomenological models.

Damage Mechanics (DM) [25–27] is a relatively new field studying the response and reliability of materials with countless randomly distributed irregular microcracks. The fundamental aspect of damage mechanics is the selection of damage parameters. The essential feature of the original Kachanov's model [28] resides in the introduction of a special internal variable defined by the state of local damage and its accumulation. Many macroscopic continuum damage models and micromechanics damage models [25,29] were proposed after Kachanov's work. Li et al. [21] suggested a continuum damage mechanics-based model for FRC in tension, in which the quasi-brittleness of the matrix and the fiber–matrix interfacial properties were taken into consideration. Considering the stochastic characteristics of the distribution of disfigurements in the microstructure, a physically motivated damage model (i.e., the bundle parallel bar system (PBS)) was proposed by Krajcinovic and Silva [30], based on the continuous damage theory and statistical strength theory. After this work, Statistical Damage Mechanics (SDM) was gradually formed, constitutive relations of quasi-brittle material (rock and concrete) have been extensively studied, and some inspiring results were obtained [31–34]. This kind of model abstracts quasi-brittle material into a complex system composed of mesoscopic physical elements. The heterogeneity in the microstructure is introduced by assuming that the characteristic parameters of the mesoscopic unit follow Weibull or other statistical distribution forms. The progressive damage accumulation process of concrete is described by means of probability and statistics. It ignores the complicated physical details of the damage process and avoids the complicated calculation of statistical mechanics. The relationship between meso-damage mechanism and macroscopic mechanical behavior of material is effectively established. Considering there are two fundamental damage modes (fracture and yield) in the microstructure of concrete, the statistical damage models of concrete under uniaxial and multiaxial loading were proposed by Chen et al. [35] and Bai et al. [36–38], which could be used to predict the constitutive behavior of concrete under a complex loading environment and explore the damage mechanism of the concrete material.

In this paper, a statistical damage constitutive model for the FRC is presented based on the macroscopic test phenomenon and the statistical damage theory. Firstly, a deep analysis of the mesoscopic damage mechanism for concrete material under uniaxial tension is elaborated. It indicated that the macroscopic nonlinear stress–strain behavior is determined by the evolution of fracture and yield damage on a meso scale. The yield damage mode, reflecting load redistribution and adjustment of stress skeleton in the microstructure, is emphasized as the key role in the whole deformation and failure process. Evolutionary factor is introduced to reflect the development of the potential mechanical capacity of materials. Subsequently, the triangular probability distributions with four parameters are used to simulate the cumulative evolution of fracture and yield damage. By defining the above parameters as a function of fiber content, the impact of fiber content on the nucleation and growth of microcracks and the mesoscopic damage evolution of the FRC is reflected. The determination method for the model's characteristic parameters is proposed. The validity of the proposed model is demonstrated by comparing the theoretical and experimental results. The influence rule of fiber content on the mesoscopic damage evolution of the FRC is analyzed.

2. Deformation and Failure of Concrete under Uniaxial Tension

Since the failure of concrete is essentially caused by the nucleation and growth of microcracks produced by local tensile strain, the uniaxial tension could be regarded as the most fundamental failure form for concrete.

According to the macroscopic experimental phenomena, the deformation and failure of the concrete specimen under uniaxial tension seems to be divided into two stages, the uniform damage stage (US) and the local fracture stage (LS), as shown in Figure 1a. In the uniform damage stage, the whole specimen remains uniformly loaded and deformed, with the damage evolving mainly by the nucleation and growth of micro-defects, which randomly distribute in the whole specimen. In the local fracture stage, a macroscopic main-crack perpendicular to the tensile direction forms in the fracture process zone (FPZ). The macroscopic response of the specimen strongly depends on the size of the largest crack with a preferential orientation in the FPZ. Local further damage and fracture occurs in the FPZ, while the rest of the region of the specimen shows the unloading behavior.

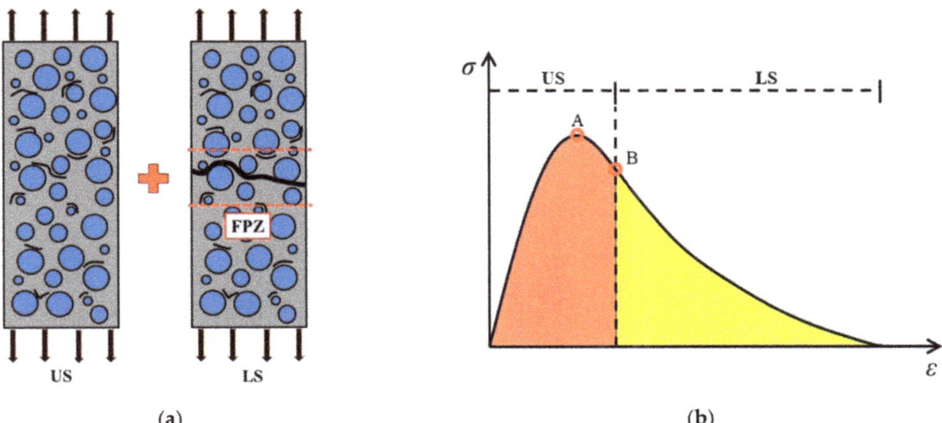

Figure 1. Two-stage feature of macroscopic mechanical behavior of concrete under uniaxial tension: (**a**) deformation-failure process; (**b**) typical nominal stress–strain curve.

As shown in Figure 1b, a typical master stress–strain curve of concrete specimen under uniaxial tension is presented. The signs A and B denote, respectively, the two characteristic states, i.e., (1) the peak nominal stress state and (2) the critical state at which macro-crack starts to develop in the FPZ and the damage process switches into the local failure stage. From the shape of the curve, the stress–strain curve seems to be divided into a stress-strengthening phase and a strain-softening phase bounded by state A. Before A, the tensile stress increases gradually with the growth of tensile strain. After A, the tensile stress decreases gradually with the increase of tensile strain, until to zero. The stress value corresponding to state A, known as the so-called tensile strength, is considered to be one of the most important mechanical indexes for concrete. The descending segment of the tensile stress–strain curve for concrete was firstly measured by Rusch and Hilsdorf [39]. Hughes and Chapman [40] confirmed that the concrete did not break at the maximum load, but the softening phenomenon occurred.

From the deformation and failure features, and also the process of microcracks propagation, the stress–strain behavior seems more suitable to be divided into an even damage phase and a local fracture phase bounded by state B, corresponding to the two-stage features in Figure 1a. During the whole process, the nucleation and growth of the random distributed microcracks typically ranges from 0.01 to 1.0 mm in width, leading to a concentration of these microcracks into a narrow zone and producing a visible macroscopic fissure wider than 1.0 mm [41,42]. Before B, the material response could be considered as statistical homogeneous. The density of micro-defects is modest, retaining a dilute degree. After this threshold, the macroscopic response of the specimen strongly depends on the macroscopic main-crack in the FPZ. Local softening and fracture (decrease of the nominal stress with increase of the strain) occurs in the FPZ, while the rest of the region of the specimen enters the unloading process. This portion of the stress–strain curve becomes heavily dependent on the measure gauge length, which could not be treated as a pure material mechanical behavior. Based on the catastrophe theory, the process from damage to fracture for quasi-brittle materials is divided into two phases, globally stable (GS) mode (relevant to the distributed damage accumulation) and evolution induced catastrophic (EIC) mode [43]. The critical state transforming from GS to EIC plays a key role during the whole process, and exhibits the critical sensitivity, with many physics showing abnormal behavior. Experiments on rocks have also shown that there exists a critical value for the fracture of the quasi-brittle material [24,44].

For the location of the critical state B, many experiments [45,46] showed that it lies in the softening region behind the peak state A in a tensile stress–strain curve. However, the accurate location of B is still uncertain, and almost all the theoretical analysis in literatures ignored the identification of the critical state, treating A and B simply as the same one.

3. Materials and Methods

3.1. Basis of Statistical Damage Theory

It is well known that statistic physics is a ligament that communicates continuum mechanics, damage mechanics, and material mechanics. Statistical damage theory ignores the microscopic details of damage, and the representative volume element (RVE) is abstracted as a complex system composed of N ($N\rightarrow\infty$) mesoscopic units (micro-spring, micro-bar, etc., with the same sectional area dA and stiffness dk). The heterogeneity of the material is introduced by endowing each unit with different mechanical parameters (strength, characteristic strain, etc.). The macroscopic mechanical properties of concrete depend on the statistical mechanical properties of individual mesoscopic elements. The macroscopic mechanical properties could be described by using a phenomenological model with statistical method.

According to the number of mesoscopic mechanical parameters adopted, statistical damage models could be classified into two types, i.e., the single-parameter model and the double-parameter model, which are represented by the bundle parallel bar system (PBS) [30] and the improved parallel bar system (IPBS) [35,36], respectively.

3.1.1. The Series-Parallel Spring Stochastic Damage Model

Based on the PBS, Li and Zhang [47] presented the series-parallel spring stochastic damage model for concrete under uniaxial tensile test. The core idea is to extend the single-layer parallel element model to multiple layers, which can better simulate the mechanical phenomenon such as localized failure and stress drop. The tensile specimen is abstracted as a complex system composed of n typical units (a unit is composed of numerous micro-springs in parallel with equal spacing) in series. F is the tensile force. $H = nh_0$ is the height of the model, where h_0 is the material characteristic height and three times the maximum particle size of aggregate presented by Bažant and Oh [48]. In this model, each micro-spring has the same elastic modulus and cross-sectional area. ε_{Ri} is the fracture strain for micro-spring i, as a random variable obeying the same probability distribution for each layer unit, where Weibull distribution and Lognormal distribution are often used.

Since the prominent feature of concrete failure is local failure, the failure of the series-parallel spring model is caused by the fracture of a unit body, called the main crack unit body (corresponding to the FPZ), and the others as the non-main crack unit bodies. Define a characteristic strain, namely, the critical strain. Before the critical strain, all unit bodies have the same deformation. After the critical strain, the deformation of the main crack unit body continues to increase, while in the non-main crack unit bodies will occur the unloading behavior with the tensile strain no longer increased. Therefore, the average stress–strain curve after the critical state for the concrete tensile specimen shows the obvious size effect, and stress drop will occur when the specimen size is long enough.

3.1.2. The Improved Parallel Bar System (IPBS)

Based on the PBS, the IPBS was proposed by Chen et al. [35] to simulate the damage evolution in the FPZ of quasi-brittle material. The fundamental assumptions for the damage mechanism are as follows: (1) the essence of damage and fracture for a quasi-brittle material is the nucleation, growth, and coalescence of micro-cracks, the stress redistribution and adjustment of the force skeleton in the microstructure. (2) The influence of these irreversible micro-changes on macroscopic mechanical properties of material could be addressed by two dominant aspects: decrease of the effective cross-section area and degradation of the elastic modulus corresponding to the effective bearing position. (3) The two damage modes can be simulated, respectively, by rupture and yield of the micro-bar; and the essence of failure can be interpreted as a continuum accumulation and evolution of the two damage modes.

In this model, each micro-bar is composed of spring, cementation bar, and sliding block. The micro-bar i has two feature strains, i.e., the fracture strain ε_{Ri} and the yield strain ε_{yi}, and it may have two kinds of failure modes (elastic-fracture and yield-fracture) according to the values between ε_{Ri} and ε_{yi} (as shown in Figure 2a). By introducing some simplifying assumptions, the two kinds of damage evolution process in the meso-scale are decoupled (further detailed discussion can be referred to from Chen et al. [35]). As shown in Figure 2b, $q(\varepsilon)$ and $p(\varepsilon)$ are the independent probability density functions to ε_{Ri} and ε_{yi}, respectively; where $\varepsilon_{ymin} = \varepsilon_{Rmin}$ denote the minimums of ε_{yi} and ε_{Ri}; $\varepsilon_{ymax} < \varepsilon_{Rmax}$ are the corresponding maximums. In the course of damage evolution, the cross-sectional area A_0 in the initial state will gradually transform into three parts, i.e., A_1, A_2, A_3, denoting the cross-sectional areas corresponding to the fractured bars, the yielded bars, and the bars in elastic state, respectively. A_E is the effective cross-section area bearing load. It satisfies $A_0 = A_1 + A_2 + A_3$ and $A_E = A_2 + A_3$.

The macroscopic stress–strain curves described by IPBS are drawn in Figure 2c. σ_N is the nominal stress corresponding to A_0; σ_E is the effective stress corresponding to A_E; $\bar{\sigma}$ is the elastic stress corresponding to A_3. $\varepsilon_{cr} = \varepsilon_{ymax}$; $\varepsilon_u = \varepsilon_{Rmax}$.

By ε_{ymax}, the whole process predicted by the IPBS can be divided into two phases, i.e., partial yield phase ($0 \leq \varepsilon < \varepsilon_{ymax}$) and full yield phase ($\varepsilon_{ymax} \leq \varepsilon \leq \varepsilon_{Rmax}$). When $\varepsilon = \varepsilon_{ymax}$, all the micro-bars in IPBS will yield, and σ_E will reach its maximum. After this state, σ_E will keep the maximum constant with the further increase of ε. Therefore, we could use the IPBS to simulate the two-stage deformation and failure characteristics of quasi-brittle materials in the FPZ, by assuming that ε_{ymax} corresponds to the critical state B in Figure 1b.

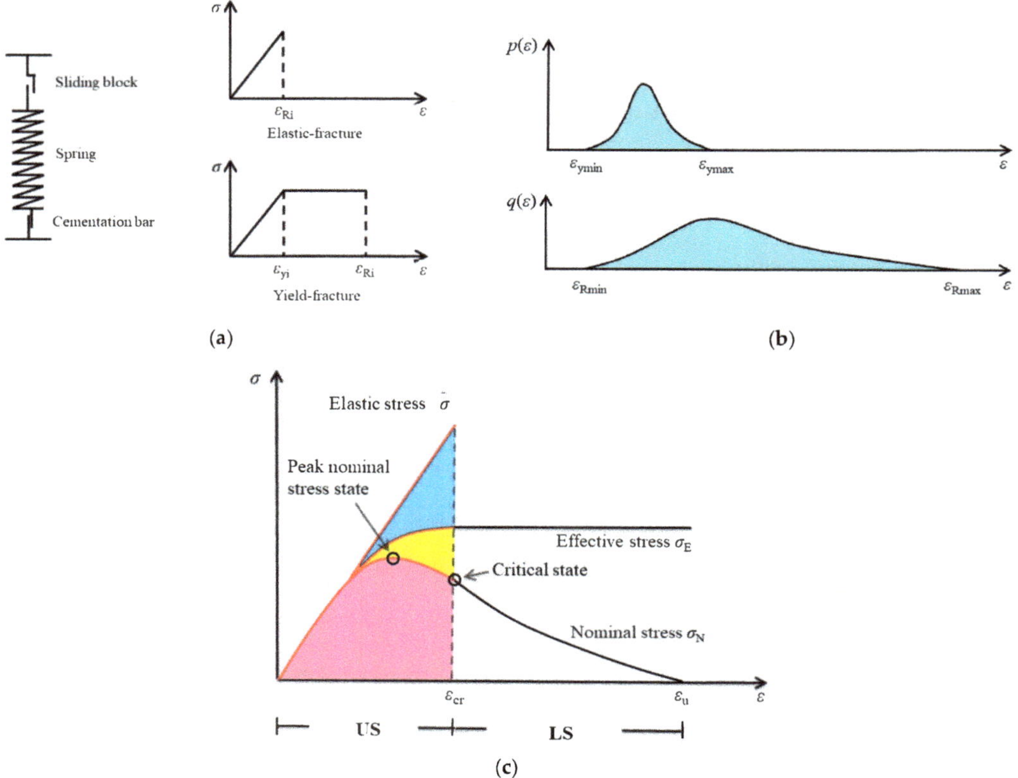

Figure 2. Mechanical behavior of concrete under uniaxial tension described by the Improved Parallel Bar System (IPBS): (**a**) micro bar and failure mode; (**b**) damage evolution process on meso-scale; (**c**) stress–strain curves on macro-scale.

The constitutive relation can be expressed as follows:

(1) Partial yield phase ($0 < \varepsilon < \varepsilon_{ymax}$)

$$\sigma_N = E_0(1 - D_y)(1 - D_R)\varepsilon \tag{1}$$

$$\sigma_E = E_0(1 - D_y)\varepsilon \tag{2}$$

$$D_y = \int_0^\varepsilon p(\varepsilon)\mathrm{d}\varepsilon - \frac{\int_0^\varepsilon p(\varepsilon)\varepsilon\mathrm{d}\varepsilon}{\varepsilon} \tag{3}$$

$$D_R = \int_0^\varepsilon q(\varepsilon)\mathrm{d}\varepsilon \tag{4}$$

where D_y and D_R denote the accumulated damage variables of elastic modulus of IPBS due to the yield and fracture of the micro-bars; D_R also represents the cumulative distribution function of $q(\varepsilon)$.

(2) Full yield phase ($\varepsilon_{ymax} \leq \varepsilon < \varepsilon_{Rmax}$)

$$\sigma_N = E_0(1 - D_{ymax})(1 - D_R)\varepsilon_{ymax} \tag{5}$$

$$\sigma_E = E_0(1 - D_{ymax})\varepsilon_{ymax} \tag{6}$$

$$D_{ymax} = 1 - \frac{\int_0^{\varepsilon_{ymax}} p(\varepsilon)\varepsilon d\varepsilon}{\varepsilon_{ymax}} = \text{constant} \tag{7}$$

$$D_R = \int_0^{\varepsilon_{ymax}} q(\varepsilon)d\varepsilon + \int_{\varepsilon_{ymax}}^{\varepsilon} q(\varepsilon)d\varepsilon \tag{8}$$

where D_{ymax} is the value of D_y corresponding to ε_{ymax}.

IPBS also could simulate the unloading process of the rest of the region of the specimen in the local fracture phase. σ_N and σ_E can be expressed by Equations (1) and (2). D_y and D_R can be expressed by Equations (9) and (10).

$$D_y = \frac{\int_0^{\frac{\varepsilon_{ymax}-\varepsilon}{2}} p(\varepsilon)\varepsilon d\varepsilon}{\varepsilon} + \int_0^{\frac{\varepsilon_{ymax}-\varepsilon}{2}} p(\varepsilon)d\varepsilon + \frac{\varepsilon_{ymax}\int_{\frac{\varepsilon_{ymax}-\varepsilon}{2}}^{\varepsilon_{ymax}} p(\varepsilon)d\varepsilon}{\varepsilon} - \frac{\int_{\frac{\varepsilon_{ymax}-\varepsilon}{2}}^{\varepsilon_{ymax}} p(\varepsilon)\varepsilon d\varepsilon}{\varepsilon} \tag{9}$$

$$D_R = D_R(\varepsilon_{ymax}) = \int_0^{\varepsilon_{ymax}} q(\varepsilon)d\varepsilon = \text{constant} \tag{10}$$

If the length of specimen (H) and the length of the FPZ(h_0) are given, the average stress–strain curve of the specimen during the local fracture phase could be determined by Equations (5) to (10), which would show the obvious size effect [35].

3.1.3. Description of the Damage Evolution Process by IPBS

A typical nominal stress–strain curve and a predicted effective stress–strain curve under uniaxial tension are shown in Figure 3a,b. Signs D and E denote the peak nominal stress state and the critical state, respectively. A denotes the limit elastic state, B and C denote two states in strengthen segment, F and F$'$ denote two states in soften segment. The uniaxial tensile damage evolution process of concrete is simulated by using the microscopic spring beam model and the cylinder model, respectively.

In Figure 3c, the micro-bar has two kinds of failure modes, brittle-fracture and yield-fracture. A corresponds to the limit elastic state, where all the micro-bars remain in the elastic state. B, C, D correspond to the two states in strengthen segment and the peak nominal stress state, where some micro-bars are out of work due to fracture, some are yield, and the others still in the elastic state. E corresponds to the critical state, where all the residual bars are yield. F corresponds to the local breach phase in the FPZ, where the yielded bars continue to fracture, exhibiting the softening phenomenon. F$'$ corresponds to the local breach phase in the other region, exhibiting the unloading phenomenon.

In Figure 3d, the tensile specimen is divided into three regions, the elastic region, the yield-damage region, and the rupture-damage region, during the whole damage process. With the growth of damage, the elastic region will gradually transform into the other two kinds of regions. When it reaches the critical state E, the elastic region will disappear, the tensile traction will be fully borne by the yield-damage region, and then the damage of specimen will change into the local breach phase.

In Figure 2c, it shows the stress–strain curves predicted by the IPBS, where $\bar{\sigma}$ is the ideal elastic stress corresponding to the elastic region, σ_E is the effective stress corresponding to A_E (the cross-sectional area relevant to the elastic region and the yield-damage region), σ_N is the nominal stress corresponding to A_N (the initial total cross-sectional area).

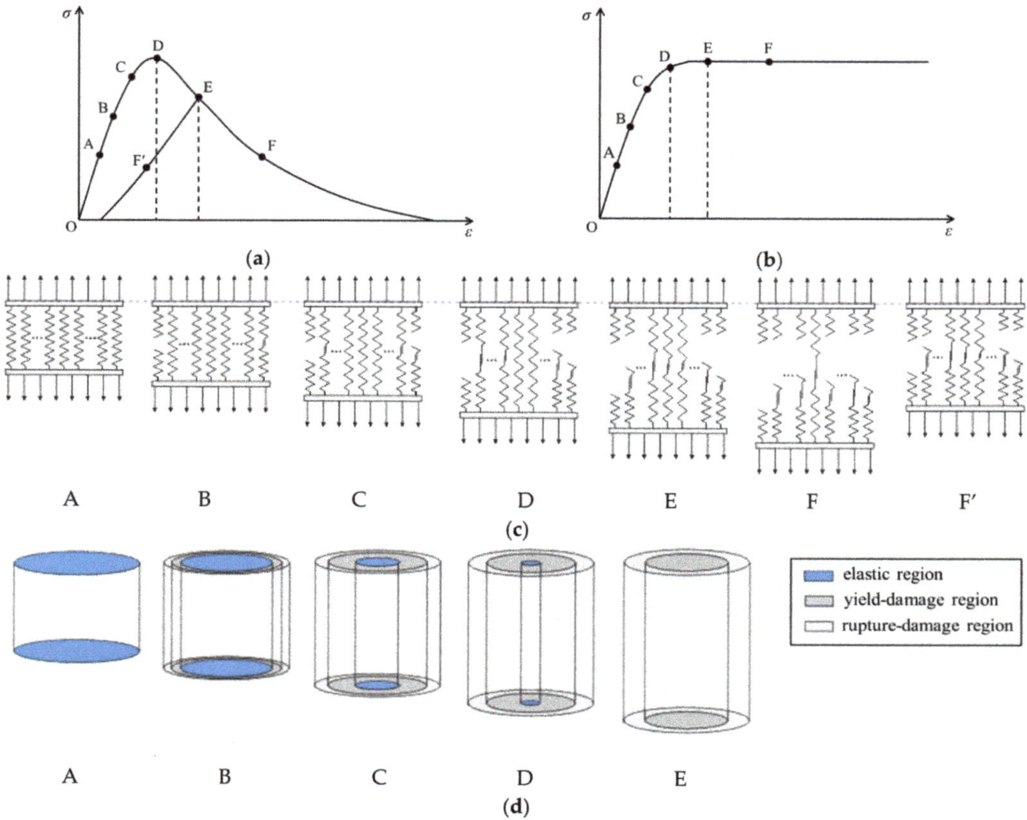

Figure 3. Uniaxial tensile damage evolution process: (**a**) nominal stress–strain curve; (**b**) effective stress–strain curve; (**c**) microscopic spring beam model; (**d**) cylinder model.

3.1.4. Dialectical Unification between Degeneration and Evolution

Natural dialectics [49] holds that, the process of evolution in nature is the unity of opposites between evolution and degeneration, which exist and occur simultaneously. Evolution-oriented processes often inherently involve degradation, and vice versa. Evolution and degradation are two opposite trends in nature. They are closely combined and inseparable. Each side is the condition for the other side to occur. The combination of the two forms a circular spiral propulsion mode for the evolution of nature, which makes the evolution process of nature appear periodic.

A novel fundamental assumption of mesoscopic damage for concrete material was proposed by Chen et al. [35]. As shown in Figure 4a, it indicated that the damage evolution of concrete materials could be summarized as two kinds of mechanisms on a mesoscopic scale. On the one hand, with the increase of deformation, the initiation, propagation, and coalescence of micro-cracks and micro-defects, as well as the acoustic emission phenomenon, will occur randomly in the microstructure, which is the so-called degradation effect. On the other hand, due to the initiation and propagation of microcracks, stress redistribution and adjustment of force skeleton will take place in the microstructure. We may be able to understand the second mechanism as the active adjustment of the material system itself. In this way, the force skeleton of the microstructure is further adjusted and optimized, and the potential mechanical ability of the material is further developed to bear more external loads (effective stress). Therefore, the second type of mechanism is called the strengthening effect here. Berthier et al. [24] indicated the quasi-brittle failure emerges from the interaction between the elements constituting the material. They also highlighted the central role played by the mechanism of load redistribution to control the failure behavior of quasi-brittle solids.

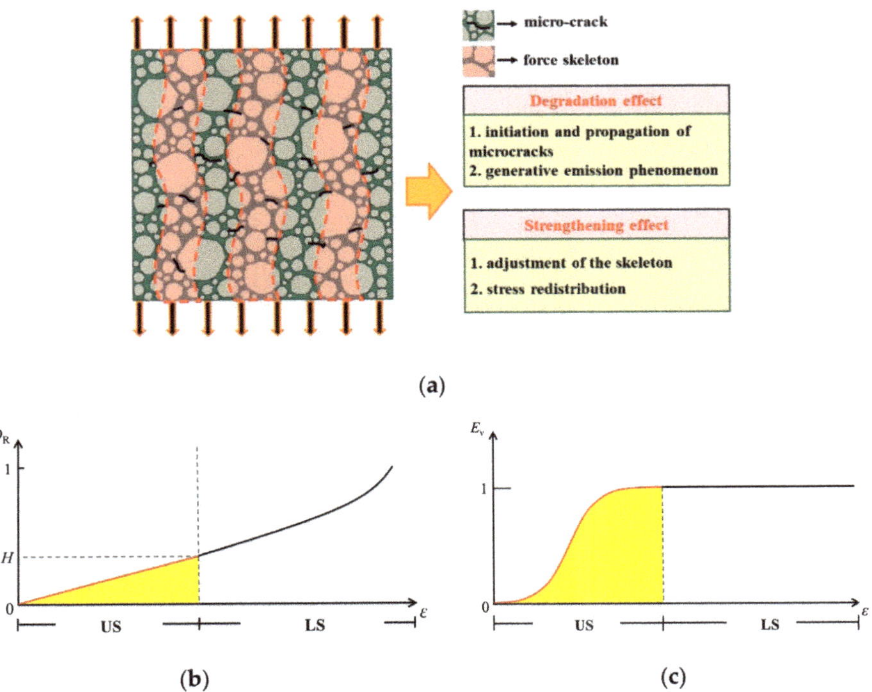

Figure 4. Two types of mesoscopic mechanism of quasi-brittle material: (**a**) damage evolution mechanism on a mesoscopic scale; (**b**) degradation factor; (**c**) evolutionary factor.

According to the assumption, the deformation and failure of concrete material is not only the process of "deterioration" of macroscopic and microscopic mechanical properties, but also the process of "evolution" in which the material goes through their own active-adjustment to adapt to the external load environment. When the inner adjustment capabilities of the material exert to its limit, the damage evolution of the material will switch from the statistical uniform damage to the local breach. This local failure process generally takes place in the weakest region (such as the fracture process zone in concrete).

The above viewpoints are consistent with the catastrophe theory [43], which holds that the failure process of solid materials could be divided into two phases, GS and EIC. The critical state transforming from GS to EIC plays a key role during the whole process, and exhibits the critical sensitivity. The two-stages embody the process from quantitative change to qualitative change.

Figure 4b,c shows the two mechanisms of the deformation and failure of concrete on a mesoscopic scale, i.e., the "degradation" and "strengthening", respectively.

As shown in Figure 4c, E_v is introduced as the evolutionary factor to describe the "strengthening" process, corresponding to the yield damage mode. The expression is as follows:

$$E_v = \int_0^\varepsilon p(\varepsilon)d\varepsilon \quad (0 \leq \varepsilon \leq \varepsilon_{ymax}) \tag{11}$$

E_v could be used to assess the extent to which the potential mechanical capabilities (adjustment capabilities of force skeleton in microstructure) of materials are developed, ranging from 0 to 1. When $E_v = 0$, it corresponds to the initial undamaged state. When $E_v = 1$, it corresponds to the critical state, at which the potential adjustment capabilities of materials reach their limits, σ_E reaches its maximum value, and then the materials enter into the local catastrophic stage. The whole process embodies the characteristics of "quantum" to "qualitative", in which the yield damage mode plays a key role.

As shown in Figure 4b, D_R is as the fracture damage variable, and could be used to describe the "degradation" processes. It has the similar physical meanings with D defined by Dougill [50], which characterizes the initiation and propagation of microcracks and leads to the reduction of the effective stress area. At the uniform damage stage, it ranges from 0 to H (a certain smaller value), and it seems not to be playing a key role in the whole process.

3.2. Statistical Damage Model for FRC in Uniaxial Tension

3.2.1. Influence Mechanism of Fiber

A large number of experimental studies [1–9,12] have shown that, the influencing factors mainly include: the strength of the matrix concrete, the type of fiber, the length diameter ratio and the volume ratio of fiber, the bonding strength between fiber and matrix, and the distribution and orientation of the fiber in the matrix. The existing theories mainly focus on fiber and matrix, and study the interaction between single fiber and matrix.

(1)   Fiber spacing theory

Romualdi and Batson [13] analyzed the limitation and restraint mechanism of steel fiber to concrete crack, and put forward the theory of fiber spacing to explain the mechanism of fracture enhancement of the SFRC. This theory holds that the existence of fiber could significantly reduce the size and quantity of micro-cracks, reduce the stress concentration degree of the crack tip, and restrain the occurrence and expansion of micro-cracks. The key to fiber reinforcement is the average spacing of the fibers. With the increase of average spacing, the fiber's ability to restrain the crack initiation and expansion will be greater, and the strength of the SFRC will be higher. The size of the average spacing of the fibers depends on the number of active fibers in the volume of the unit matrix.

(2)   Reinforcement rules for composite materials

Swamy [14] proposed the reinforcement rules for composite materials based on the mechanics principle of composite materials. The SFRC is simplified as fiber and concrete two-phase composite materials, and the properties of the composite material are cumulative for each phase. Due to the non-homogeneity of ordinary concrete structure, irregular stress concentration would occur within the matrix when the structure is pulled. When the ultimate tensile strength of ordinary concrete is less than the tensile stress of the stress concentration point, the stress concentration point will create cracks. Since the tensile strength of the steel fibers is much higher than the tensile strength of the concrete matrix, the incorporation of steel fiber can effectively inhibit and delay the initiation and expansion of micro-cracks in ordinary concrete.

In other words, these randomly distributed fibers can effectively prevent the expansion of micro-cracks in concrete and delay the formation of macro-cracks. Hence, it can significantly improve the macroscopic mechanical properties of concrete, such as tensile, bending, impact, fatigue resistance, and so on. The main factors that influence fiber behavior include: the species, geometric features, and content of the fibers, the bonding properties of the fibers to the concrete matrix, distribution and orientation of the fibers in matrix. Most of the existing constitutive models for FRC are adopted from the macroscopic phenomenological mathematical expressions (as shown in Table 1), of which parameters lack a clear physical meaning. Therefore, it is difficult for them to reflect the influence of fiber content on the mesoscopic damage evolution of the concrete.

Table 1. Theoretical stress–strain models of SFRC under tension in literature.

Origin of Data	Tensile Strength/MPa	Fiber Types	Main Formula
Gao [20]	2~3	Melt-extracted	$\begin{cases} y = A + (3 - 2A)x + (A - 2)x^2 & (x \le 1) \\ y = \frac{x}{a(x-1)^{1.7}+x} & (x > 1) \end{cases}$
Han et al. [22]	3~5	Large steel fiber	$\begin{cases} y = 1.2x - 0.2x^6 & (x \le 1) \\ y = \frac{x}{a_{ft}(x-1)^{1.7}+x} & (x > 1) \end{cases}$

3.2.2. Practical Expressions of the IPBS

According to the statistical damage theory mentioned above, the macroscopic mechanical behavior (nominal/effective stress–strain curve) of concrete under uniaxial tension, is determined by the cumulative evolution process of the fracture and yield damage modes in a meso-scale, as shown in Figure 5. The whole process includes the homogeneous damage accumulation stage and local failure stage. Two characteristic states are distinguished, namely peak nominal stress state and critical state. ε_{cr} and ε_u denote the strains of critical state and ultimate state in the nominal stress–strain curve.

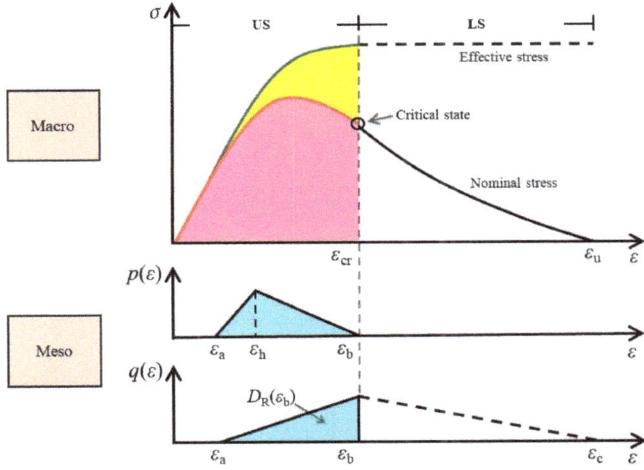

Figure 5. Relationship between constitutive behavior on macro-scale and damage evolution process on meso-scale.

As the probability density functions corresponding to fracture and yield damage in the meso-scale respectively, $q(\varepsilon)$ and $p(\varepsilon)$, may be subject to complex statistical distribution laws in the true case, such as Weibull, Normal, and other distributions. Considering the complexity of the problem, $q(\varepsilon)$ and $p(\varepsilon)$ are both assumed to obey the independent triangle distribution form in a specific calculation, as shown in Figure 5. Analyses [35–38] show that the true stress–strain test curves could be well fitted and the evolution mechanism of non-homogeneous damage on the meso-scale could be well revealed, when $q(\varepsilon)$ and $p(\varepsilon)$ are adopted by the simplified triangular distributions. ε_a is the initial damage strain. ε_h is the strain corresponding to the peak value of $p(\varepsilon)$. ε_b is the strain corresponding to the maximum yield damage state, and also to the peak value of $q(\varepsilon)$. ε_c is the strain corresponding to the maximum fracture damage state. It satisfies $\varepsilon_b = \varepsilon_{cr}$ and $\varepsilon_c = \varepsilon_u$ in uniaxial tension.

Due to the softening segment of the nominal stress–strain curve corresponding to the local failure stage having obvious size effect, the critical state is suggested as the final failure point of the constitutive model in this paper. Hence, the following content in the paper only discusses the constitutive behavior of concrete in the homogeneous damage stage.

At the homogeneous damage stage, σ_N and σ_E can be obtained by Equations (1)–(4), where $q(\varepsilon)$ and $p(\varepsilon)$ can be expressed as the following:

$$q(\varepsilon) = \begin{cases} 0 & (\varepsilon \leq \varepsilon_a) \\ \dfrac{2H(\varepsilon-\varepsilon_a)}{(\varepsilon_b-\varepsilon_a)^2} & (\varepsilon_a < \varepsilon \leq \varepsilon_b) \end{cases} \tag{12}$$

$$p(\varepsilon) = \begin{cases} 0 & (\varepsilon \leq \varepsilon_a) \\ \dfrac{2(\varepsilon-\varepsilon_a)}{(\varepsilon_h-\varepsilon_a)(\varepsilon_b-\varepsilon_a)} & (\varepsilon_a < \varepsilon \leq \varepsilon_h) \\ \dfrac{2(\varepsilon_b-\varepsilon)}{(\varepsilon_b-\varepsilon_h)(\varepsilon_b-\varepsilon_a)} & (\varepsilon_h < \varepsilon \leq \varepsilon_b) \end{cases} \tag{13}$$

where $H = D_R(\varepsilon_b)$ is the fracture damage value relevant to ε_b.

Define S as the energy absorbing capability, which represents the energy absorbed by concrete in the process of stress and deformation. The expression is as follows:

$$S = \int_0^\varepsilon \sigma_N d\varepsilon \tag{14}$$

$$S_p = \int_0^{\varepsilon_p} \sigma_N d\varepsilon \tag{15}$$

$$S_{cr} = \int_0^{\varepsilon_{cr}} \sigma_N d\varepsilon \tag{16}$$

where S_p and S_{cr} are the energy absorption capacity corresponding to peak nominal stress state and critical state, respectively.

3.2.3. Influence of the Fibers on Mesoscopic Damage Mechanism

Statistical damage theory suggests that the change on the macroscopic nonlinear stress–strain behavior of concrete under a complex environment is essentially caused by the microscopic damage mechanism. The fiber reinforcement effect changes the composition and characteristics of the concrete microstructures, the nucleation and growth of microcracks, and the cumulative evolution process of damage, which finally lead to the change on the macro-nonlinear stress–strain behavior of concrete. The above effects can be summarized into two aspects: (1) changes on the composition and mechanical properties of microstructure, measured by E_0; (2) changes on the pattern and law of the initiation and propagation of microcracks; in other words, changes on the cumulative evolution process of the two meso-damage modes (yield and fracture), measured by ε_a, ε_h, ε_b, and H, which determine the shape of the triangle probability distribution of $q(\varepsilon)$ and $p(\varepsilon)$.

As shown in Figure 6, it is assumed that the microstructure characteristics and meso-damage evolution process of concrete under different fiber contents obey a certain regularity, and the above changes at the meso-level lead to the strengthening of the macroscopic constitutive behavior of concrete. Make E_0, ε_a, ε_h, ε_b, and H as a function of fiber volume fraction ρ (%) (here, does not consider the influence of other factors, such as fiber type and geometric features), the expression is as follows:

$$\begin{cases} E_0 = f_1\,(\rho) \\ \varepsilon_a = f_2\,(\rho) \\ \varepsilon_h = f_3\,(\rho) \\ \varepsilon_b = f_4\,(\rho) \\ H = f_5\,(\rho) \end{cases} \tag{17}$$

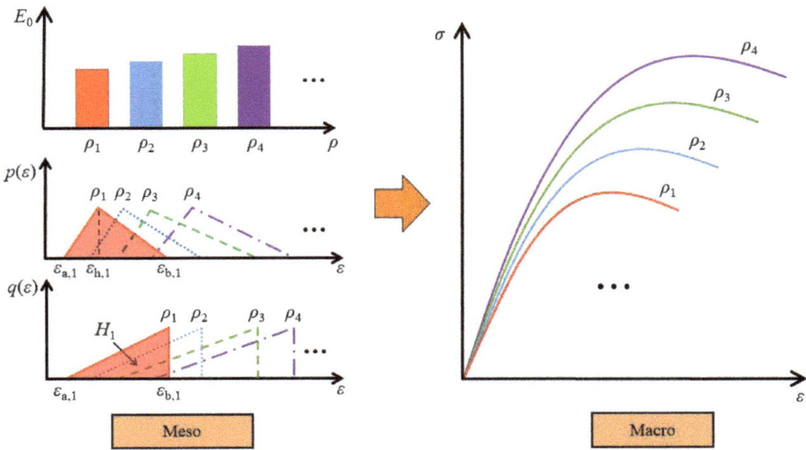

Figure 6. The influence of the fiber volume fraction on the damage mechanism of concrete.

As mentioned above, according to these five parameters, we can determine the uniaxial tensile stress–strain behaviors of concrete with different fiber content, and also the laws of the evolution of the mesoscopic damage mechanism.

3.2.4. Determination of Model Parameters

Five parameters need to be determined for each nominal stress–strain curve under uniaxial tension, they are E_0, ε_a, ε_h, ε_b, and H. E_0 can be obtained directly from the test curve, taken as the secant modulus from 0.2–0.4 times of peak stress point in the ascending part of the curve to the original point. ε_a, ε_h, ε_b, and H can be obtained by the multivariate regression analysis of the genetic algorithm module in the Matlab toolbox. The solution procedure is summarized as follows:

(1) Create a fitness function including ε_a, ε_h, ε_b, and H, and take the sum of the squares of the difference between predicted value and measured value of nominal stress as the optimization criterion.
(2) Initially, set the search interval for the values of the four parameters.
(3) To perform the genetic algorithm, and obtain the optimal solution of the 4 parameters calculated by this iteration. Adjust or narrow the search interval of parameters according to the results.
(4) Repeat step (3), until the optimal solution is obtained.

4. Results and Discussion

The presented model is validated by using two sets of uniaxial tensile tests for steel fiber-reinforced concrete with different fiber contents reported by Han et al. [22] and Gao [20]. The relevant experimental information are summarized in Table 2. The rationality and applicability of the presented model are verified, and the damage evolution mechanism of steel fiber concrete in uniaxial tension is discussed.

Table 2. Summary of uniaxial tensile test information.

		Types	Length to diameter ratio	Volume fraction /%	Equivalent diameter /mm	Average length /mm
Gao [20]	Fiber	Melt-drawn	50	0.5, 1.0, 1.5, 2.0	0.5	25
	Mixture/kg/m^3	Cement	Water	Sand	Stone	Water reducer
		450	225	887.5	887.5	0
Han et al. [22]	Fiber	Types	Length to diameter ratio	Volume fraction /%	Equivalent diameter /mm	Average length /mm
		Large-end	44.34	0, 0.5, 1.0, 1.5, 2.0, 2.5	0.698	30.96
	Mixture/kg/m^3	Cement	Water	Sand	Stone	Water reducer
		450	158	737	1105	4.5

4.1. Comparison with the Test by Han et al., 2006

Han et al. [22] conducted a uniaxial tensile test for a steel fiber-reinforced concrete specimen. The specimen is of dumbbell shape with a total length of 450 mm. The length of the middle tensile region is 170 mm, with a cross section of 100 mm × 100 mm. A large-end steel fiber with a length-to-diameter ratio of 44.34 is adopted. The volume fractions of the steel fiber ρ are 0%, 0.5%, 1.0%, 1.5%, 2.0%, and 2.5%. The theoretical nominal stress–strain curves of steel fiber concrete with different fiber contents calculated by the presented model, corresponding to the homogeneous damage phase, are shown in Figure 7a (the curve with a fiber volume fraction of 1.5% is removed due to the dispersion of the experimental data.). The curves include the ascending and partial descending segments of the stress–strain behavior of the fiber-reinforced concrete. They show a good fitting effect compared to the test data. The predicted effective stress–strain curves are shown in Figure 7b. The relevant calculation parameters are listed in Table 3, where R^2 is the correlation coefficient. For this proposed model, the entire process of deformation and failure of concrete under uniaxial tension is understood from the viewpoint of effective stress. In the uniform damage stage, the nominal stress σ_N first increases and then decreases, involving with a peak nominal stress state (so-called strength state). The effective stress σ_E increases monotonously and reaches its maximum at the critical state. After the critical state, the specimen enters the local failure stage characterized by macroscopic crack propagation. The 3D envelopes of σ_N-ε and σ_E-ε curves predicted by the presented model are shown in Figure 7c,d. They clearly show the variation trend of the curves of concrete with different fiber contents, and the shape of the curves show a good similarity rule, especially the connection part between the ascending and descending segments. With the increase of steel fiber volume ratio ρ from 0% to 2.5%, the values of σ_N and ε corresponding to the peak nominal stress state and the values of σ_E and ε corresponding to the critical state increase significantly.

The relationship curves of ρ-ε_a, ε_h, ε_b are shown in Figure 8a, which represent the evolution law of the yield damage mode on a mesoscale. The curve of H-ρ is shown in Figure 8b, which depicts the evolution law of the fracture damage mode on a mesoscale. The relationship curve of E_0-ρ is shown in Figure 8c. The aforementioned five parameters show an obvious regularity with the increase in steel fiber content, in which ε_h, ε_b, and H increase linearly, whereas E_0 decreases linearly. The fitting curves and fitting expressions are also shown in the figures.

Figure 9a shows the curves of the evolution process of fracture damage variable D_R with different steel fiber contents, respectively. D_R is closely related to microcrack density, and its evolution process under uniaxial tension is significantly delayed with the increase in fiber content from 0% to 2.5%, which is consistent with the physical background in the microstructure.

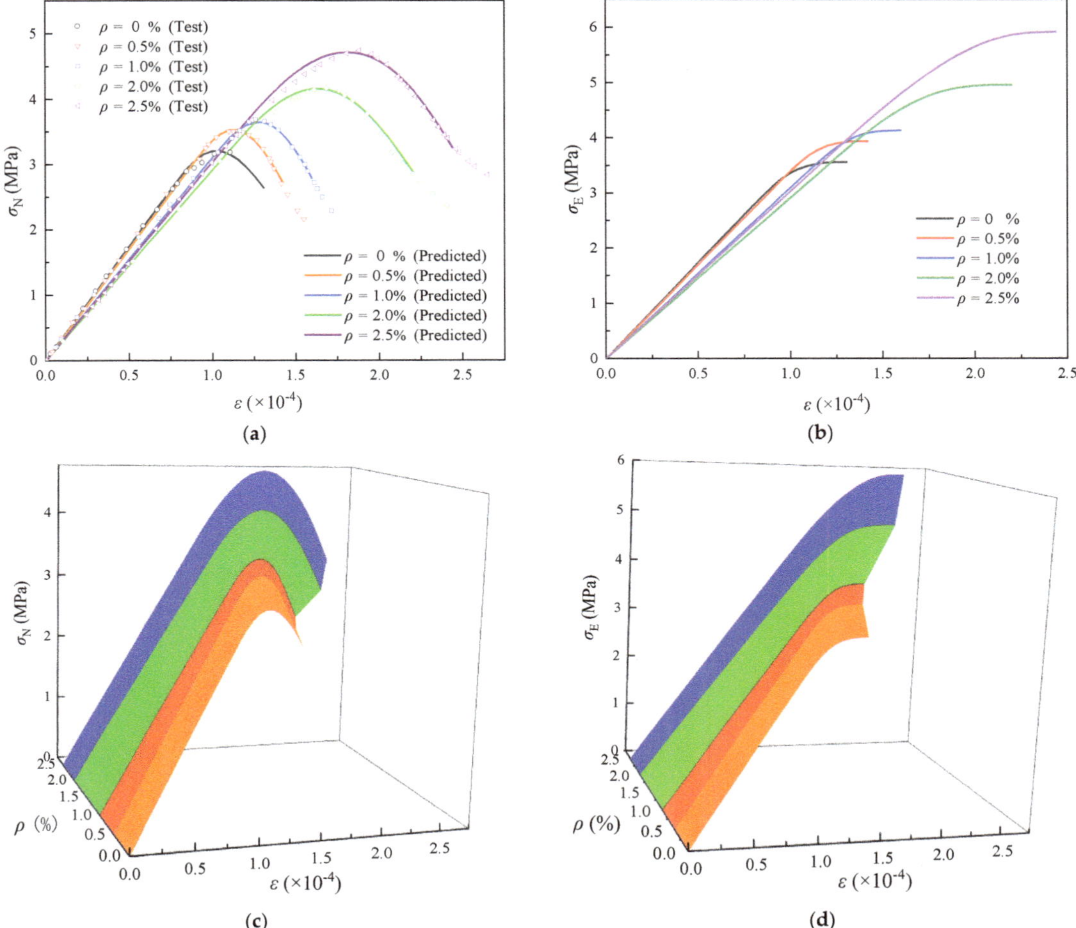

Figure 7. Stress–strain curves under uniaxial tension: (**a**) nominal stress–strain curve (Plan view); (**b**) effective stress–strain curve (Plan view); (**c**) envelope of the nominal stress–strain curve (3D view); (**d**) envelope of the effective stress–strain curve (3D view).

Table 3. Results for calculation parameter.

ρ (%)	$E_0/\times 10$ GPa	$\varepsilon_a/\times 10^{-4}$	$\varepsilon_h/\times 10^{-4}$	$\varepsilon_b/\times 10^{-4}$	H	R^2
0	3.481	0.771	0.993	1.311	0.260	0.9995
0.5	3.413	0.909	1.134	1.427	0.307	0.9997
1.0	3.107	1.032	1.362	1.598	0.327	0.9993
2.0	2.921	1.127	1.771	2.210	0.419	0.9991
2.5	3.046	1.181	2.211	2.441	0.450	0.9994

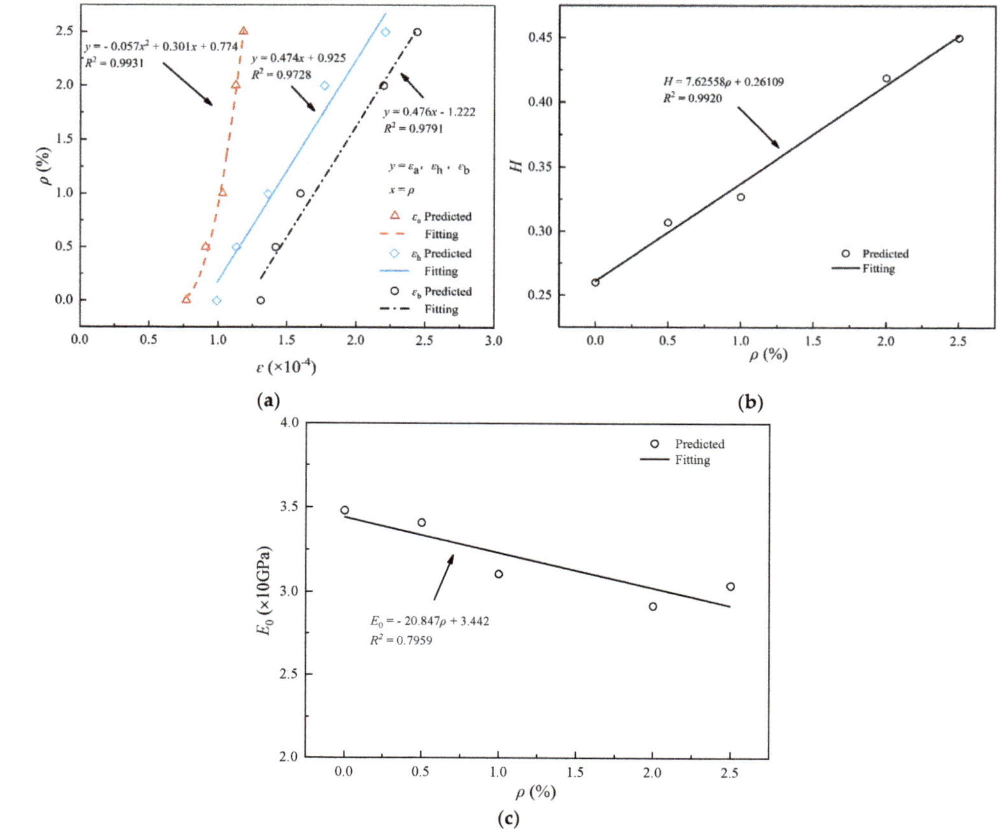

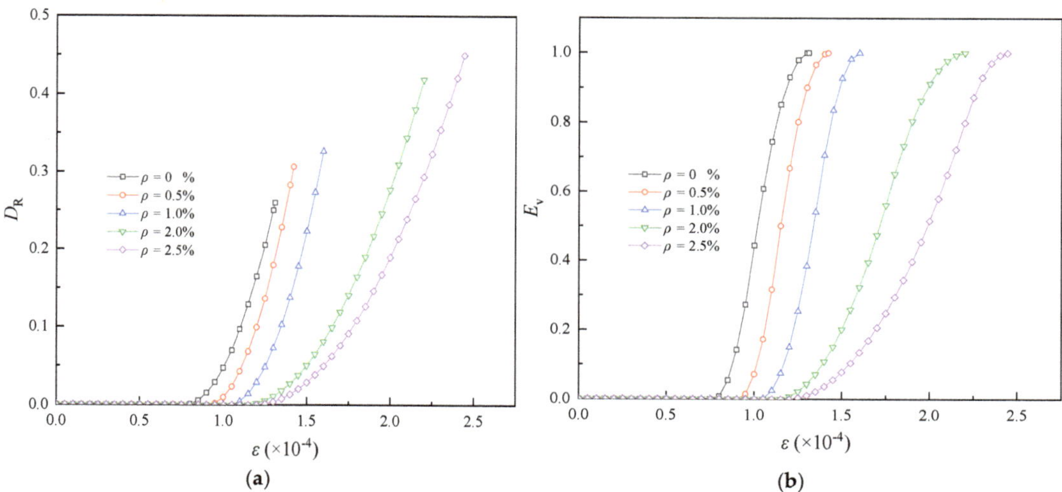

Figure 8. Influence of fiber volume fraction on the characteristic parameters: (**a**) ρ-ε_a, ε_h, ε_b curves; (**b**) H-ρ curve; (**c**) E_0-ρ curve.

Figure 9. Evolution curves of damage variable and evolution factor: (**a**) D_R-ε curves; (**b**) E_v-ε curves.

In Figure 9b, evolutionary factor E_v represents the exerting degree of the underlying mechanical capacity of materials, and its evolution process is delayed with the increase in fiber content. When the critical state, $E_v = 1$, is reached, all the microbars in IPBS will yield and cannot endure much effective stress, then the concrete specimen enters the local failure stage characterized by macroscopic crack growth.

The physical meanings of the above-mentioned characteristic parameters are clear. The prediction results can be used to explore the internal relations among the physical background, the mesodamage evolution mechanism, and the macro-nonlinear mechanical behavior of concrete for adding steel fibers with different contents.

(1)   Damage mechanism on a mesoscale

Local stress concentration will occur in the concrete matrix under tensile load due to the heterogeneity in the microstructure, which will lead to the initiation and expansion of microcracks. After adding steel fiber to the concrete matrix, the composition and mechanical properties of the microstructure will change, and the initiation and penetration of microcracks in the concrete matrix will be significantly inhibited and delayed. The evolution and accumulation process of the mesodamage modes will consequently change. For the yield damage mode, with the increase in ρ from 0% to 2.5%, ε_a increases from 0.771×10^{-4} to 1.181×10^{-4}, ε_h increases linearly from 0.993×10^{-4} to 2.211×10^{-4}, and ε_b increases linearly from 1.311×10^{-4} to 2.441×10^{-4}. For the fracture damage mode, H increases linearly from 0.260×10^{-4} at 0% to 0.450×10^{-4} at 2.5%. The above-mentioned four parameters can be used to determine the shape of triangular probability distribution corresponding to the mesoscopic yield and fracture damage evolution process, which can demonstrate intuitionistic physical pictures to people to understand the entire process on a mesoscale.

(2)   Mechanical behavior in macroscale

The macroscopic nonlinear stress–strain behavior of concrete is determined by the mechanical properties of the microstructure and the evolution of mesodamage. After adding steel fiber to the concrete matrix, with ρ varying from 0% to 2.5%, the composition and mechanical properties of the microstructure will change, which will lead the initial elastic modulus E_0 to decrease from 3.481×10 GPa to 3.046×10 GPa. The nominal stress–strain curves generally show "strengthened" features due to the change in the mesodamage evolution process characterized by the four parameters (ε_a, ε_h, ε_b, and H) with the increase in fiber content. The nominal stresses corresponding to the peak nominal stress state and the critical state, $\sigma_{N,p}$ and $\sigma_{N,cr}$, increase from 3.134 and 2.612 MPa to 4.713 and 3.251 MPa, respectively. The tensile strains corresponding to the peak nominal stress state and the critical state, $\varepsilon_{N,p}$ and $\varepsilon_{N,cr}$, increase from 1.051×10^{-4} and 1.303×10^{-4} to 1.801×10^{-4} and 2.442×10^{-4}, respectively. No significant change is observed in the slope of the descending branch of the nominal stress–strain curve, and the ductility performance is significantly improved.

Figure 10 shows the change curves of energy absorption capacity S with fiber volume fraction, where S_p and S_{cr} represent the energy absorption capacities corresponding to the peak nominal stress state and the critical state, respectively. With the increase in fiber content ρ from 0% to 2.5%, S_p increases from 0.172 to 0.479 kPa, S_{cr} increases from 0.266 to 0.749 kPa, and the difference between S_{cr} and S_p increases from 0.094 to 0.270 kPa. In previous studies, S_p was commonly used to characterize the energy absorption capacity of concrete before failure. In the present study, S_{cr} is suggested to characterize the energy absorption capacity of concrete before failure. This approach allows full consideration of the potential mechanical properties of the material.

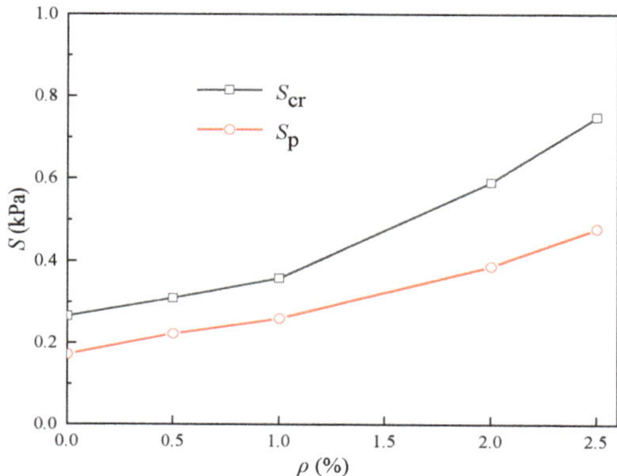

Figure 10. Influence of fiber volume fraction on the energy absorption capacity.

4.2. Comparison with the Test by Gao, 1991

Gao [20] also conducted a uniaxial tensile test for steel fiber-reinforced concrete specimen. The rectangular block with size of 100 mm × 100 mm × 500 mm is used in the experiment. Deformed steel bars with a diameter of 20 mm are inserted into both ends of the specimen. Fiber type is the melt-drawn steel fiber with length-to-diameter ratio of 50. The volume fractions of the steel fiber ρ are 0.5%, 1.0%, 1.5%, and 2.0%.

As shown in Figure 11a, the predicted nominal stress–strain curves relevant to the homogeneous damage phase agree well with the test curves. The predicted effective stress–strain curves are shown in Figure 11b. The calculation parameters are listed in Table 4. The 3-D envelopes of σ_N-ε and σ_E-ε curves predicted are shown in Figure 11c,d, in order to better show the variation trend of the curves. It clearly shows that the shape of the curves has obvious similarity law with the increase in the volume ratio of steel fiber. With the increase of ρ from 0.5% to 2.0%, the stresses and strains relevant to the peak nominal stress state and the critical state increase significantly. The change of the slope of the descending section of the nominal stress–strain curves is not obvious, meanwhile the ductility is further improved.

Figure 12a shows the relationship curves of ρ-ε_a, ε_h, ε_b. With ρ ranging from 0.5% to 2.0%, ε_a decreases linearly from 0.394×10^{-4} to 0.194×10^{-4}, ε_h decreases linearly from 0.164×10^{-4} to 0.030×10^{-4}, meanwhile ε_b increases linearly from 1.601×10^{-4} to 2.411×10^{-4}. Figure 12b shows the relationship curves of H-ρ, H decreases linearly from 0.202 at 0.5% to 0.162 at 2.0%. Figure 12c shows the change curve of E_0 with ρ, which has no obvious changes. The fitting curves and fitting functions of the above five parameters are also shown in the figures.

Figure 13a shows the evolution curves of D_R, which characterize the cumulative evolution process of the fracture damage mode of concrete on a mesoscale, respectively. With the increase of ρ from 0.5% to 2.0%, the evolution process of D_R has been significantly delayed.

Figure 13b shows the evolution curves of E_v, which characterizes the release process of potential mechanical capacity of concrete. Its evolution process has accelerated in early and then obviously delayed in later, with the increase of steel fiber content. When it reaches the critical state, $E_v = 1$. It indicates that the potential mechanical capacity of the material to be fully released, and then the concrete specimen enters into the local failure stage. The whole process embodies the conversion from quantitative change to qualitative change.

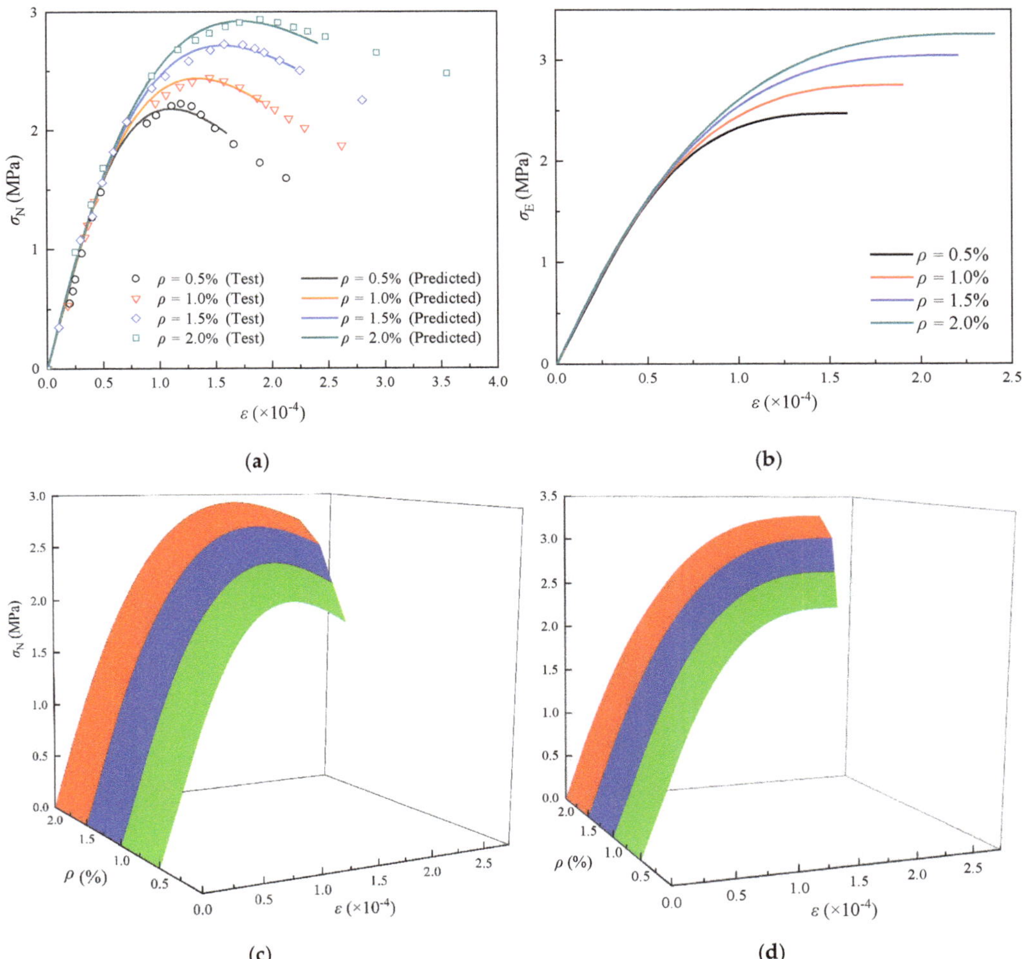

Figure 11. Stress–strain curves under uniaxial tension: (**a**) nominal stress–strain curve (Plan view); (**b**) effective stress–strain curve (Plan view); (**c**) envelope of the nominal stress–strain curve (3D view); (**d**) envelope of the effective stress–strain curve (3D view).

Table 4. Results for calculation parameter.

ρ (%)	$E_0/\times 10$ GPa	$\varepsilon_a/\times 10^{-4}$	$\varepsilon_h/\times 10^{-4}$	$\varepsilon_b/\times 10^{-4}$	H	R^2
0.5	3.427	0.164	0.394	1.601	0.202	0.9773
1.0	3.512	0.088	0.357	1.905	0.186	0.9692
1.5	3.649	0.033	0.261	2.209	0.172	0.9846
2.0	3.712	0.030	0.194	2.411	0.162	0.9721

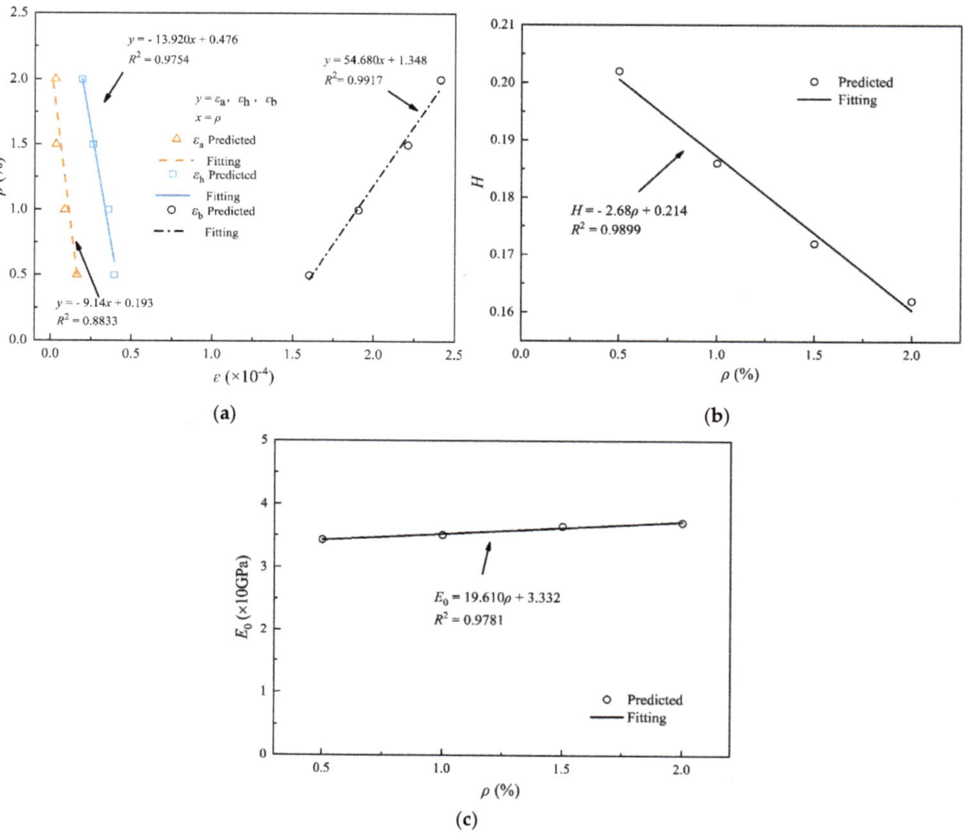

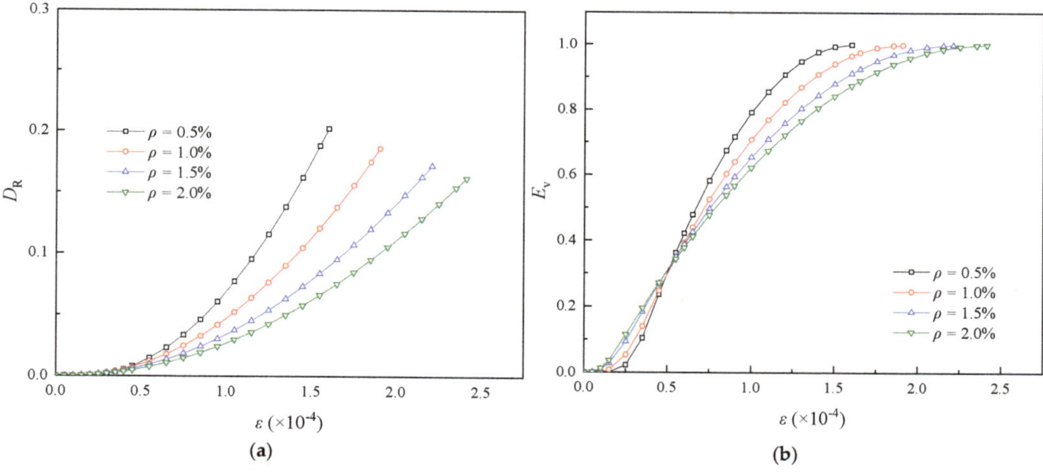

Figure 12. Influence of fiber volume fraction on the characteristic parameters: (**a**) ρ-ε_a, ε_h, ε_b curves; (**b**) H-ρ curve; (**c**) E_0-ρ curve.

Figure 13. Evolution curves of damage variable and evolution factor: (**a**) D_R-ε curves; (**b**) E_v-ε curves.

Figure 14 shows the change curves of S_p and S_{cr} with ρ. With the increase in the fiber content ρ from 0.5% to 2.5%, S_p increases from 0.161 to 0.345 kPa, S_{cr} increases from 0.266 to 0.548 kPa; and the difference between S_{cr} and S_p increases from 0.105 to 0.203 kPa. The results show that the potential mechanical properties of materials could be fully considered if the energy absorption capacity corresponding to the critical state is adopted.

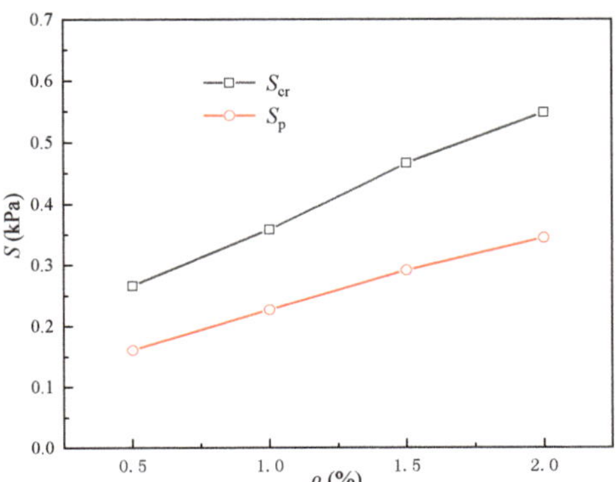

Figure 14. Influence of fiber volume fraction on the energy absorption capacity.

The nonlinear stress–strain behavior of fiber concrete is determined by many factors, which include water/cement contents, fiber type and content, aggregate source, additive type, specimen type, loading method and so on. Therefore, although the aforementioned two sets of tests are both for steel fiber concrete specimens, there are great differences both in macroscopic stress–strain behavior and in mesoscopic damage evolution. The experimental results and theoretical analysis show that, in the same group of tests, when only the content of steel fiber is changed, the macroscopic constitutive behavior and mesoscopic damage evolution process of fiber concrete show good regularity with the change of fiber content.

5. Conclusions

1. The macroscopic stress–strain behavior (including hardening and softening curves) of concrete under uniaxial tension is a continuous process with deformation and damage evolution. For the traditional segmented constitutive models, two independent expressions are used to describe the pre-peak ascending phase and the post-peak descending phase (taking the peak nominal stress state as the boundary), respectively. Therefore, the link of the mesoscopic damage evolution between the two stages has been isolated artificially. This paper discusses the mesoscopic damage evolution mechanism reflected by the IPBS in detail. The fracture and yield damage modes on meso-scale are considered, and the peak nominal stress state and critical state are distinguished. The uniaxial tensile process is divided into uniform damage phase and local failure phase by the critical state. The uniform damage phase, including the pre-peak ascending segment and a portion of the post-peak descending segment, is the main stage for deformation and damage accumulation and reflects the process from quantitative change to qualitative change. The yield damage mode reflects the development of potential mechanical properties of materials and plays a key role during the whole deformation-to-failure process. Due to the size effect on the local failure phase, the critical state is regarded as the ultimate failure point in the suggested constitutive model.

2. A statistical damage model of fiber concrete under uniaxial tension is established, which considers the fiber enhancement effect. In essence, the addition of fiber changes the composition of the microstructure, restricts the initiation and expansion of microcracks, and also changes the evolution and accumulation process of two damage

modes on a meso-scale. This model contains two kinds of feature parameters (E and ε_a, ε_h, ε_b, H) with clear physical meanings, and has the ability to effectively reflect the above changes on meso-scale. Calculations were conducted to simulate the two sets of steel fiber concrete tensile tests in the literature. The experimental and theoretical analysis results show that, when only the fiber content is changed, the shape of the macroscopic nominal stress–strain curve will show a good law of similarity. With the increase of the fiber content, the values of stress and strain corresponding to the peak nominal stress state and the critical state linearly increase, and the curvature of the connecting part of the ascending and descending branch of the nominal stess–strain curve has the changing trend of gradual and orderly. Meanwhile, the characteristic parameters ε_a, ε_h, ε_b, H, representing the two types of damage evolution of yield and fracture on a meso-scale, have obvious linear variation law with the change of fiber content. Through this model, the link among the physical mechanism, the mesoscopic damage mechanism and the macroscopic nonlinear constitutive behavior are effectively established.

3. The macroscopic constitutive behavior of FRC is a complex process of multiple factors. The influence factors include water/cement contents, source of aggregate, fiber type and content, type of additive, specimen size, loading mode, etc. Due to the limitation of the length of articles and test data, only two groups of steel fiber concrete test data are adopted in the validation analysis. Whether this constitutive model could be applicable to the analysis of the influence of other factors on the macroscopic mechanical behavior of fiber concrete, remains to be further researched later.

Author Contributions: Methodology, W.B.; data curation, validation and writing—original draft preparation, J.G. and S.H.; writing—review and editing, C.Y. and X.L.; funding acquisition, C.X. All authors have read and agreed to the published version of the manuscript.

Funding: This research was funded by National Natural Science Foundation of China, grant number "51679092, 51779095"; National Key R&D Program of China, grant number "2018YFC0406803" and the Science Technology Innovation Talents in Universities of Henan Province, China, grant number "20HASTIT013".

Institutional Review Board Statement: Not applicable.

Informed Consent Statement: Not applicable.

Data Availability Statement: The data presented in this study are contained in this article.

Acknowledgments: The author thanks Yao Xianhua and Li Lilie of the School of Civil Engineering and Communication of North China University of Water Resources and Electric Power for their technical support.

Conflicts of Interest: The authors declare no conflict of interest.

References

1. Iqbal, S.; Ali, A.; Holschemacher, K.; Bier, T.A. Mechanical properties of steel fiber reinforced high strength lightweight self-compacting concrete (shlscc). *Constr. Build. Mater.* **2015**, *98*, 325–333. [CrossRef]
2. Wang, Z.L.; Liu, Y.S.; Shen, R.F. Stress–strain relationship of steel fiber-reinforced concrete under dynamic compression. *Constr. Build. Mater.* **2008**, *22*, 811–819. [CrossRef]
3. Zainal, S.M.; Hejazi, F.; Abd Aziz, F.N.; Jaafar, M.S. Constitutive Modeling of New Synthetic Hybrid Fibers Reinforced Concrete from Experimental Testing in Uniaxial Compression and Tension. *Crystals* **2020**, *10*, 885. [CrossRef]
4. Peng, Z.; Li, Q.F. Effect of polypropylene fiber on durability of concrete composite containing fly ash and silica fume. *Compos. Part B Eng.* **2013**, *45*, 1587–1594.
5. Lam, L.; Teng, J.G. Ultimate Condition of Fiber Reinforced Polymer-Confined Concrete. *J. Compos. Constr.* **2004**, *8*, 539–548. [CrossRef]
6. Longbang, Q.; Nie, Y.; Ru, M.U. Influence of steel fibres on the resistance to crack initiation of cementitious composites. *Acta Mater. Compos. Sin.* **2017**, *34*, 1862–1869.
7. Liu, B.; Zhou, J.; Wen, X.; Hu, X.; Deng, Z. Mechanical properties and constitutive model of carbon fiber reinforced coral concrete under uniaxial compression. *Constr. Build. Mater.* **2020**, *263*, 120649. [CrossRef]

8. Shi, X.J.; Park, P.; Rew, Y.; Huang, K.J.; Sim, C. Constitutive behaviors of steel fiber reinforced concrete under uniaxial compression and tension. *Constr. Build. Mater.* **2020**, *233*, 117316. [CrossRef]

9. Ding, X.X.; Li, C.Y.; Zhao, M.L.; Li, J.; Geng, H.B.; Lian, L. Tensile behavior of self-compacting steel fiber reinforced concrete evaluated by different test methods. *Crystals* **2021**, *11*, 251. [CrossRef]

10. Akcay, B.; Tasdemir, M.A. Mechanical behaviour and fibre dispersion of hybrid steel fibre reinforced self-compacting concrete. *Constr. Build. Mater.* **2012**, *28*, 287–293. [CrossRef]

11. Tassew, S.T.; Lubell, A.S. Mechanical properties of glass fiber reinforced ceramic concrete. *Constr. Build. Mater.* **2014**, *51*, 215–224. [CrossRef]

12. Song, P.S.; Hwang, S. Mechanical properties of high-strength steel fiber-reinforced concrete. *Constr. Build. Mater.* **2004**, *18*, 669–673. [CrossRef]

13. Romualdi, J.P.; Batson, G.B. Mechanics of crack arrest in concrete. *J. Eng. Mech. Div.* **1963**, *89*, 147–168. [CrossRef]

14. Swamy, R.N. Influence of slow crack growth on the fracture resistance of fibre cement composites. *Int. J. Cem. Compos.* **1980**, *2*, 43–53.

15. Luccioni, B.; Ruano, G.; Isla, F.; Zerbino, R.; Giaccio, G. A simple approach to model sfrc. *Constr. Build. Mater.* **2012**, *37*, 111–124. [CrossRef]

16. Hameed, R.; Sellier, A.; Turatsinze, A.; Duprat, F. Metallic fiber-reinforced concrete behaviour: Experiments and constitutive law for finite element modeling. *Eng. Fract. Mech.* **2013**, *103*, 124–131. [CrossRef]

17. Chi, Y.; Xu, L.; Yu, H.S. Constitutive modeling of steel-polypropylene hybrid fiber reinforced concrete using a non-associated plasticity and its numerical implementation. *Compos. Struct.* **2014**, *111*, 497–509. [CrossRef]

18. Yaghoobi, A.; Mi, G.C. Meshless modeling framework for fiber reinforced concrete structures. *Comput. Struct.* **2015**, *161*, 43–54. [CrossRef]

19. Mihai, I.C.; Jefferson, A.D.; Lyons, P. A plastic-damage constitutive model for the finite element analysis of fibre reinforced concrete. *Eng. Fract. Mech.* **2016**, *159*, 35–62. [CrossRef]

20. Gao, D.Y. Experimental study on tensile stress-strain relationship of steel fiber reinforced concrete. *J. Hydroelectr. Eng.* **1991**, *11*, 54–58. (In Chinese)

21. Li, F.; Li, Z. Continuum damage mechanics based modeling of fiber reinforced concrete in tension. *Int. J. Solids Struct.* **2001**, *38*, 777–793. [CrossRef]

22. Han, R.; Zhao, S.B.; Qu, F.L. Experimental study on the tensile performance of steel flber reinforced concrete. *China Civ. Eng. J.* **2006**, *39*, 63–67. (In Chinese)

23. Blair, S.C.; Cook, N.G.W. Analysis of compressive fracture in rock using statistical techniques: Part II. Effect of microscale heterogeneity on macroscopic deformation. *Int. J. Rock Mech. Min. Sci.* **1998**, *35*, 849–861. [CrossRef]

24. Berthier, E.; Démery, V.; Ponson, L. Damage spreading in quasi-brittle disordered solids: I. localization and failure. *J. Mech. Phys. Solids* **2017**, *102*, 101–124. [CrossRef]

25. Krajcinovic, D. Damage mechanics: Accomplishments, trends and needs. *Int. J. Solids Struct.* **2000**, *37*, 267–277. [CrossRef]

26. Krajcinovic, D.; Rinaldi, A. Thermodynamics and statistical physics of damage processes in quasi-ductile solids. *Mech. Mater.* **2005**, *37*, 299–315. [CrossRef]

27. Armero, F.; Oller, S. A general framework for continuum damage models. II. Integration algorithm, with application to the numerical simulation of porous metals. *Int. J. Solids Struct.* **2000**, *37*, 7437–7464. [CrossRef]

28. Kachanov, L.M. Rupture time under creep conditions. *Int. J. Fract.* **1999**, *97*, 11–18. [CrossRef]

29. Jia, L. A micromechanics-based strain gradient damage model for fracture prediction of brittle materials–part i: Homogenization methodology and constitutive relations. *Int. J. Solids Struct.* **2011**, *48*, 3336–3345.

30. Krajcinovic, D.; Silva, M.A.G. Statistical aspects of the continuous damage theory. *Int. J. Solids Struct.* **1982**, *18*, 551–562. [CrossRef]

31. Kandarpa, S.; Kirkner, D.J.; Spencer, B.F., Jr. Stochastic Damage Model for Brittle Materials Subjected to Monotonic Loading. *J. Eng. Mech.* **1996**, *122*, 788–795. [CrossRef]

32. Jie, L.; Ren, X. Stochastic damage model for concrete based on energy equivalent strain. *Int. J. Solids Struct.* **2009**, *46*, 2407–2419.

33. Cao, W.G.; Zhao, H.; Xiang, L.; Zhang, Y.J. Statistical damage model with strain softening and hardening for rocks under the influence of voids and volume changes. *Can. Geotech. J.* **2010**, *47*, 857–871. [CrossRef]

34. Li, G.; Tang, C.A. A statistical meso-damage mechanical method for modeling trans-scale progressive failure process of rock. *Int. J. Rock Mech. Min. Sci.* **2015**, *74*, 133–150. [CrossRef]

35. Chen, J.Y.; Bai, W.F.; Fan, S.L.; Lin, G. Statistical damage model for quasi-brittle materials under uniaxial tension. *J. Cent. South Univ. Technol.* **2009**, *16*, 669–676. [CrossRef]

36. Bai, W.F.; Chen, J.Y.; Fan, S.L.; Lin, G. The statistical damage constitutive model for concrete materials under uniaxial compression. *J. Harbin Inst. Technol.* **2010**, *17*, 338–344.

37. Bai, W.F.; Li, W.H.; Guan, J.F.; Wang, J.Y.; Yuan, C.Y. Research on the Mechanical Properties of Recycled Aggregate Concrete under Uniaxial Compression Based on the Statistical Damage Model. *Materials* **2020**, *13*, 3765. [CrossRef] [PubMed]

38. Bai, W.F.; Zhang, S.J.; Guan, J.F.; Chen, J.Y. Orthotropic statistical damage constitutive model for concrete. *J. Hydraul. Eng.* **2014**, *45*, 607–618. (In Chinese)

39. Evans, R.H.; Marathe, M.S. Microcracking and stress-strain curves for concrete in tension. *Matériaux Constr.* **1968**, *1*, 61–64. [CrossRef]

40. Hughes, B.P.; Chapman, B.P. The complete stress-strain curve for concrete in direct tension. *Matls Struct. Res. Test. Fr* **1966**, *30*, 95–97.
41. Chaboche, J.L. Continuum damage mechanics: Present state and future trends. *Nucl. Eng. Des.* **1987**, *105*, 19–33. [CrossRef]
42. Cerrolaza, M.; Garcia, R. Boundary elements and damage mechanics to analyze excavations in rock mass. *Eng. Anal. Bound. Elem.* **1997**, *20*, 1–16. [CrossRef]
43. Bai, Y.L.; Wang, H.Y.; Xia, M.F.; Ke, F.J. Statistical mesomechanics of solid, linking coupled multiple space and time scales. *Adv. Mech.* **2006**, *58*, 286–305. [CrossRef]
44. Yu, S.W.; Feng, X.Q. A micromechanics-based damage model for microcrack-weakened brittle solids. *Mech. Mater.* **1995**, *20*, 59–76. [CrossRef]
45. Gopalaratnam, V.S.; Shah, S.P. Softening response of plain concrete in direct tension. *ACI Mater. J.* **1985**, *82*, 310–323.
46. Guo, Z.H.; Zhang, X.Q. Investigation of complete stress-deformation curves for concrete in tension. *Mater. J.* **1987**, *84*, 278–285.
47. Li, J.; Zhang, Q.Y. Study of stochastic damage constitutive relationship for concrete material. *J. Tongji Univ.* **2001**, *29*, 1135–1141.
48. Bažant, Z.P.; Oh, B.H. Crack band theory for fracture of concrete. *Matériaux Constr.* **1983**, *16*, 155–177. [CrossRef]
49. Foster, J.B. The dialectics of nature and marxist ecology. *Dialectics New Century* **2008**, 50–82. [CrossRef]
50. Dougill, J.W. On stable progressively fracturing solids. *Z. Angew. Math. Phys. ZAMP* **1976**, *27*, 423–437. [CrossRef]

Article

Experimental and Numerical Study of Lattice Girder Composite Slabs with Monolithic Joint

Xuefeng Zhang [1], Huiming Li [1], Shixue Liang [2,*] and Hao Zhang [1]

[1] College of Civil Engineering, Zhejiang University of Technology, Hangzhou 310014, China; xuefeng_zhang@126.com (X.Z.); zjslhm1996@126.com (H.L.); zhanghao@zjut.edu.cn (H.Z.)
[2] School of Civil Engineering and Architecture, Zhejiang Sci-Tech University, Hangzhou 310018, China
* Correspondence: liangsx@zstu.edu.cn

Abstract: This paper studies the behavior of lattice girder composite slabs with monolithic joint under bending. A full-scale experiment is performed to investigate the overall bending resistance, deflection and the final crack distribution of latticed girder composite slab under uniformly distributed load. A finite element model is given for the analysis of the latticed girder composite slabs. The effectiveness and correctness of the numerical simulations are verified against experimental results. The experimental and numerical studies conclude that the lattice girder composite slabs conform to the requirement of existing design codes. A parametric study is provided to investigate the effects of lattice girder with following conclusions: (a) the lattice girder significantly increases the stiffness of the slab when comparing with the precast slab without reinforcement crossing the interface; (b) the additional reinforcement near the joint slightly increases the stiffness and resistance, while it prevents damage near the joint.

Keywords: precast concrete structure; lattice girder semi-precast slabs; bending resistance; FE modelling; concrete damage

Citation: Zhang, X.; Li, H.; Liang, S.; Zhang, H. Experimental and Numerical Study of Lattice Girder Composite Slabs with Monolithic Joint. *Crystals* **2021**, *11*, 219. https://doi.org/10.3390/cryst11020219

Academic Editor: Yifeng Ling

Received: 3 February 2021
Accepted: 18 February 2021
Published: 23 February 2021

Publisher's Note: MDPI stays neutral with regard to jurisdictional claims in published maps and institutional affiliations.

1. Introduction

Among various types of components in the precast concrete structures, slabs are particularly suitable and achievable for precasting since the loads on slabs are usually uniform and their shapes are regular. As a result, precast and composite slabs have been used for concrete structures for more than 200 million m^2 in China, 2018. It is also predicted that in the coming 5 years, the usage of precast slabs will expand to 2 billion m^2 in China, which will bring about CNY 600 billion to the construction market. According to the construction demands and economic considerations, it is of great interest in studying the experiments and numerical simulations of the precast concrete slabs.

One of the most widely used precast floors is the latticed girder composite slab (LGCS) system. As depicted in Figure 1, it consists of steel lattice girders which are cast with the precast plank, and cast-in-place concrete is poured after the installation of precast plank. It combines the cast-in-place and precast slab to utilize the advantages of both forms which increases construction efficiency, quality control and construction savings. In addition, since the precast and cast-in-place planks are fully continuous and tied together without the need for shuttering on site, they show better connecting behaviors when comparing with total precast concrete slabs.

In order to investigate the mechanical behavior of composite slabs, a number of experimental studies [1–5] have been provided for the flexural bearing capacity, shear strength, bonding assessment and interface behaviors of precast planks without reinforcement crossing the interface. Meanwhile, the existence of latticed girder resists the interface shear force and avoids slippage, delamination and debonding at the interface. Some experimental studies have been conducted on the mechanical behavior of LGCS. Du et al. [6] provided an experimental work on the flexural bearing capacity of lattice girder composite slabs.

Newell and Goggins [7] carried out experiments of the lattice girder composite slabs at construction stage and examined the key parameters which influence their behavior at both serviceability and ultimate bearing limit states.

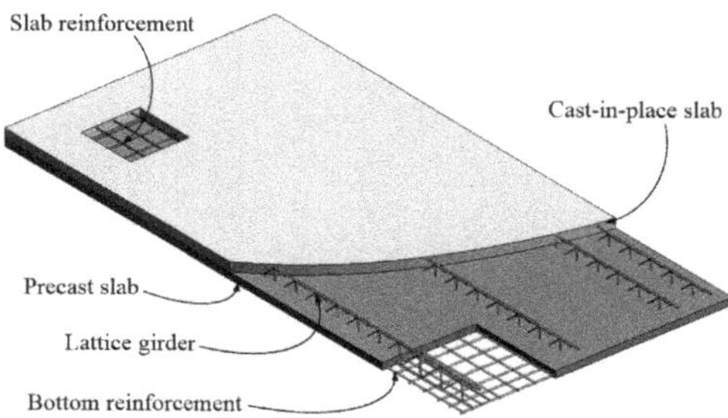

Figure 1. Latticed girder composite slabs (LGCS).

As for the numerical modeling of the precast concrete structures, two kinds of numerical models are developed, namely the macro level element models and three-dimensional (3D) solid finite element (FE) models. In the first kind models, the fiber elements are adopted to simulate the beams and columns [8–10], and layered shell are utilized to simulate the precast slabs [11,12]. It is obvious that the macro level element models is simple and computationally efficient. Nevertheless, the macro level element models cannot precisely simulate the latticed girder and the contact between precast and cast-in-place concrete. Models of the second kind are 3D solid finite element models, which can provide simulations of the local region and detailed set up of RC structures [13–15]. Thanks to the development of computational speed, 3D finite element modelling provides the promising results in the numerical analysis of precast concrete structures, due to its ability to describe the complex connection behavior in an elaborate manner.

Based on the aforementioned background, the aim of this paper is to experimentally and numerically investigate mechanical properties of LGCS under bending. Full-scale experiment of LGCS with monolithic joint under uniformly distributed load is carried out to obtain the load–deflection curve, strain distribution of concrete and reinforcement and the crack distribution of the bottom surface of the slab. It is shown in the experimental results, the mechanical properties of latticed girder composite slab completely satisfy the design criterion under bearing capacity limit state. Furthermore, the safety of the slab can also be achieved under the load of 2.5 times of bearing capacity limit states. In order to thoroughly analysis the mechanical behavior of latticed girder composite slabs, a 3D FE model by ABAQUS is provided, with a particular emphasis on the bending behavior and damage of the slab. In the FE model, the nonlinear material behavior of the steel and concrete all both considered, and all components in contact with the concrete are properly modeled. The proposed FE model is validated by the experimental results of several indexes, such as load–deflection curve, reinforcement strain, final crack width, etc. Then, a parametric study is performed to quantify the influence of the lattice girder and additional reinforcement. It is concluded that: (a) the lattice girder provides stiffness and the bonding between precast plank and cast-in-place concrete; (b) the additional reinforcement has no evident influence on the stiffness of the slab, but it helps preventing the damage of concrete near the joint.

2. Experimental Study

2.1. Test Specimen

A full-scale experiment is undertaken on LGSC with monolithic joint in the Engineering Research Center of Precast Concrete of Zhejiang Province, Hanzhou, China. According to the layout of the frame structure in the engineering applications, the whole slab is supported by reinforced beams on 4 sides (Figure 2). Four columns are settled at the corner of the beams and the bottom of the columns are rigidly connected to the ground. In the experiment, x direction is used to represent the direction along the slab, namely parallel to the monolithic joint direction, and y direction is used to represent the direction vertical to the monolithic joint direction. In this experiment, the size of specimen is 5.0 m × 5.0 m (measured from the axis of beams). Many precast factories require supports at a maximum of 2.4 m to limit deflection during transportation. The whole slab is composed by 2 precast planks with the size of 2.31 m × 4.62 m and this is chosen so that the limitation of supports can be fulfilled. As depicted in Figure 2, the precast bottom plank of the whole slab is composed by 2 precast concrete lattice girder planks. Therefore, one monolithic joint (Figure 2) is settled between precast concrete lattice girder planks.

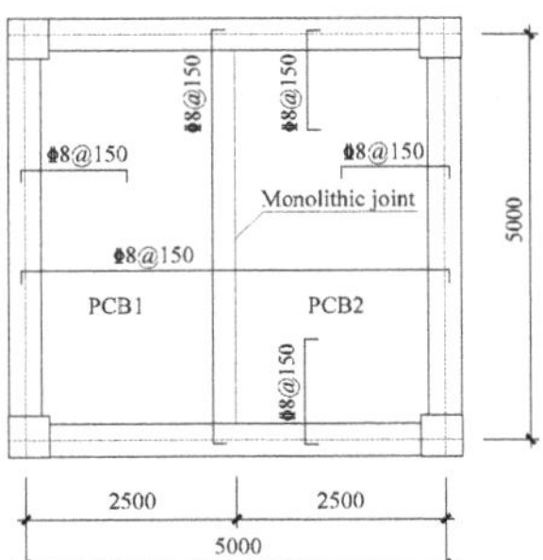

Figure 2. Layout plan of LGCS.

The thickness of precast bottom plank is chosen as 60 mm, and the thickness of the cast-in-place top plank is 80 mm [16]. The beams and columns are cast-in-place, where the strength of precast bottom plank and cast-in-place top plank are both C30 and the steel bars are all made of HRB400. The reinforcement in these two precast planks is C 8 at 150 mm (C 8@150) spacing in both the longitudinal and transverse directions and the concrete cover to the reinforcement at the plank bottom surface is 25 mm. Reinforcement is settled near the support as C 8 at 150 mm (C 8@150). It is shown in Figure 3, four latticed girders are settled in one precast bottom plank (PCB1). The first lattice girder is placed 140 mm from the joint, which is equal to thickness of the slab. The second lattice girder placed 280 mm from the first lattice girder. These two lattice girders are arranged to strengthen the joint. A lattice girder is settled in the middle region of the precast plank at intervals of 850 mm from the second lattice girder. The last lattice girder is placed 200 mm from the joint to the beam to enhance the beam–slab connection.

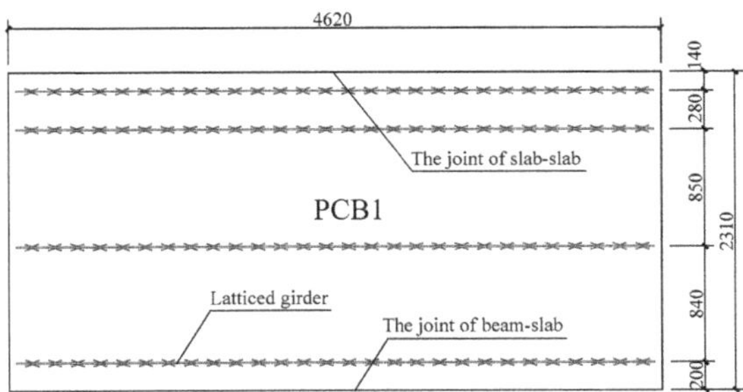

Figure 3. Latticed girder of precast bottom plank.

The configuration of latticed girder is illustrated in Figure 4. The detailed latticed girder and monolithic joint are shown in Figure 5. In order to further enhance the resistance of the slab near monolithic joint, additional reinforcement with C 10@130 is arranged along the joint. The length of the additional reinforcement is 970 mm (Figure 5). As shown in Figure 6, reinforcement near the support is installed to strengthen the connection between slab and beam. The length of the support reinforcement is 120 mm. It is emphasized that the design parameters, such as size, thickness and reinforcement of the slab, are strictly followed "Code for Design of Concrete Structures (GB50010-2010) [17]" and "Technical Specification for Application of Lattice Girder Slab (T/CECS 715-2020) [16]".

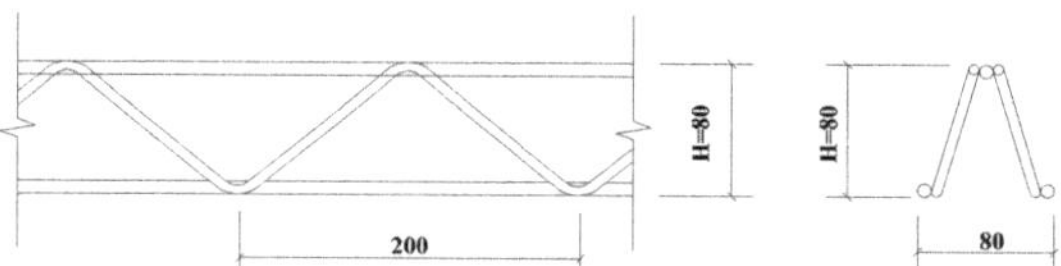

Figure 4. Configuration of latticed girder (elevational and sectional drawing).

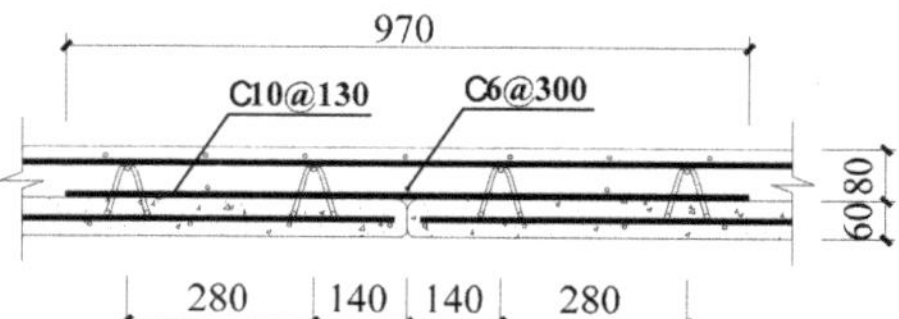

Figure 5. Monolithic joint.

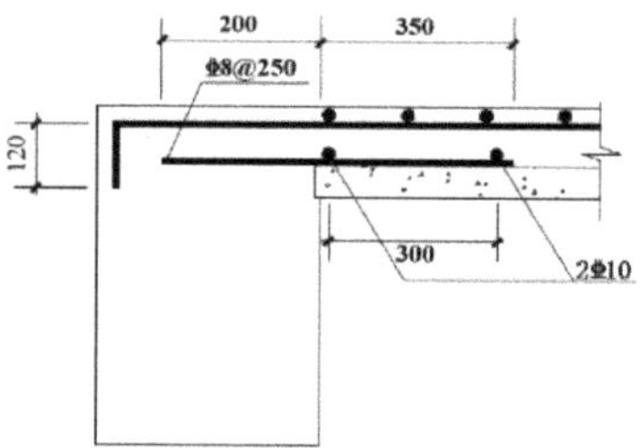

Figure 6. Beam–slab connection.

2.2. Testing Method

In order to investigate the strain condition of different components in the LGCS, electrical resistance (ER) strain gauges are bonded to the longitudinal reinforcement and the concrete surface of the slab. The arrangement of strain measuring point of concrete is illustrated in Figure 7a. It can be seen in Figure 7a that C1a and C1b are symmetrically arranged. In the experiment, the strain is calculated as the mean value of C1a and C1b, the remaining strain measuring points are the same. Since lattice girder and longitudinal reinforcement are both set in the slab, the strain distribution is much complex than that of cast-in-place slab. We apply 13 strain measuring point of bottom slab longitudinal reinforcement depicted in Figure 7b, and the final strain is also calculated as the mean value of S1a and S1b. With regard to the adopted instrumentation, five linear variable differential transformers in Figure 7c are applied to measure the vertical displacement at midspan, in the middle of each precast plank and at edge of the bottom slab. With regard to the adopted instrumentation, five linear variable differential transformers in Figure 7c are applied to measure the vertical displacement at midspan, in the middle of each precast plank and at edge of the bottom slab.

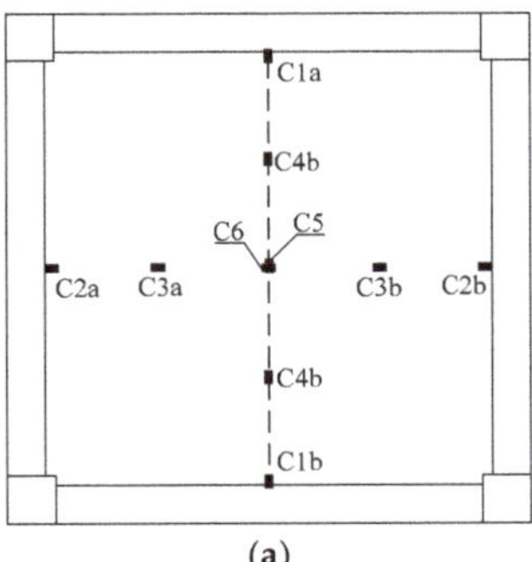

(a)

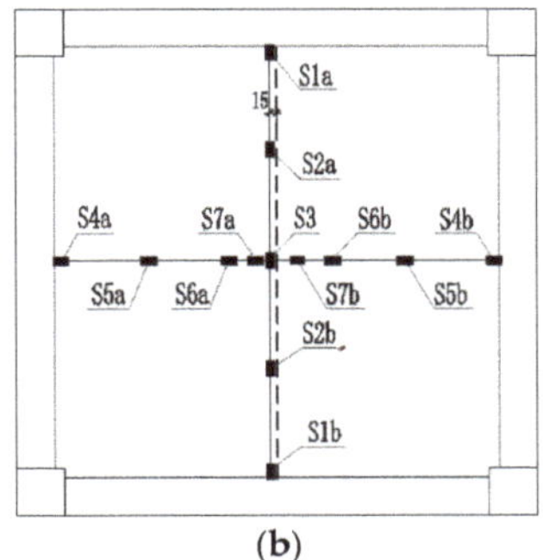

(b)

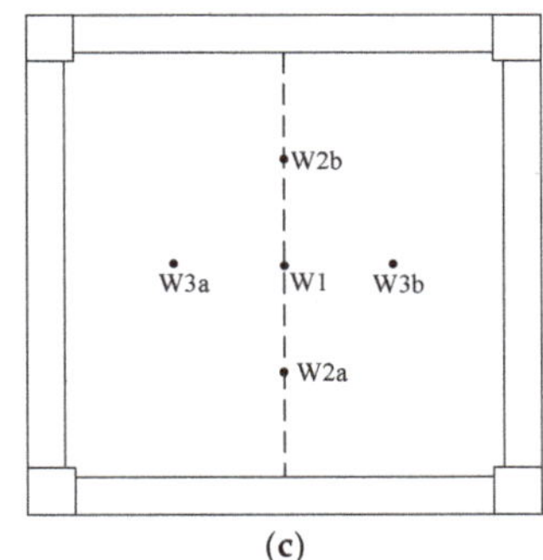

(c)

Figure 7. Arrangement of testing measuring points: (**a**) Strain measuring points of concrete; (**b**) Strain measuring points of steel; (**c**) Deflection measuring point.

The loading scheme is carried out with reference to the "Chinese Standard for Testing Methods of Concrete Structures (GB/T 50152-2012) [18]". A uniformly distributed load is applied by pile loading of sandbags in Figure 8. The serviceability limit state is calculated as 8.4 kN/m^2 and the bearing capacity limit state is calculated as 16.4 kN/m^2 [17]. In order to investigate the ultimate bearing capacity of the composite slab in this study, the maximum value of the uniformly distributed load is chosen as 40 kN/m^2, which is corresponding to 2.5 times of the value of ultimate bearing capacity of the slab. Before the maximum distributed load (40 kN/m^2), the loading procedure (including the weight of the slab and pile loading) is applied progressively by steps, and for each step the load is 2.0 kN/m^2. When the sandbags are piled up, they are distributed evenly to avoid arch effect. It should be noted that for each loading step, sandbags are evenly piled to the slab by forklift. After each loading step, the load should be kept for 15 min until the measured stress, strain and deflection are stable.

Figure 8. Pile loading of the slab.

2.3. Test Results

During the entire loading process of the slab, the crack development of the slab is carefully observed and monitored. The crack distribution at different loading stages is shown in Figure 9. No visible crack is observed under the serviceability limit state (8.4 kN/m^2), which satisfy the crack width limit 0.2 mm of the design code [17]. When the slab is loaded to 10 kN/m^2, the first crack appears in the midspan of the precast bottom plank at 45°, which width is recorded as 0.06 mm. After loaded to 10 kN/m^2, the existing cracks in the midspan of the precast bottom plank gradually extend and pass through the monolithic joint. With further increase of load, the cracks in the precast bottom plank continue to increase, and gradually develop towards the direction of 45°. It is also observed that some paralleled cracks developed near the main crack. Finally, when the load reaches 40 kN/m², the maximum width of the crack at the precast bottom plank is 0.76 mm, located at the joint, and the maximum width of the crack at the joint is 0.10 mm. Under the maximum loading value of this test is 40 kN/m², which is about 2.5 times of the design value of bearing capacity, the crack width of the LGCS does not reach the crack width limit of 1.50 mm corresponding to the bearing capacity limit, and no shear failure is observed near the joint.

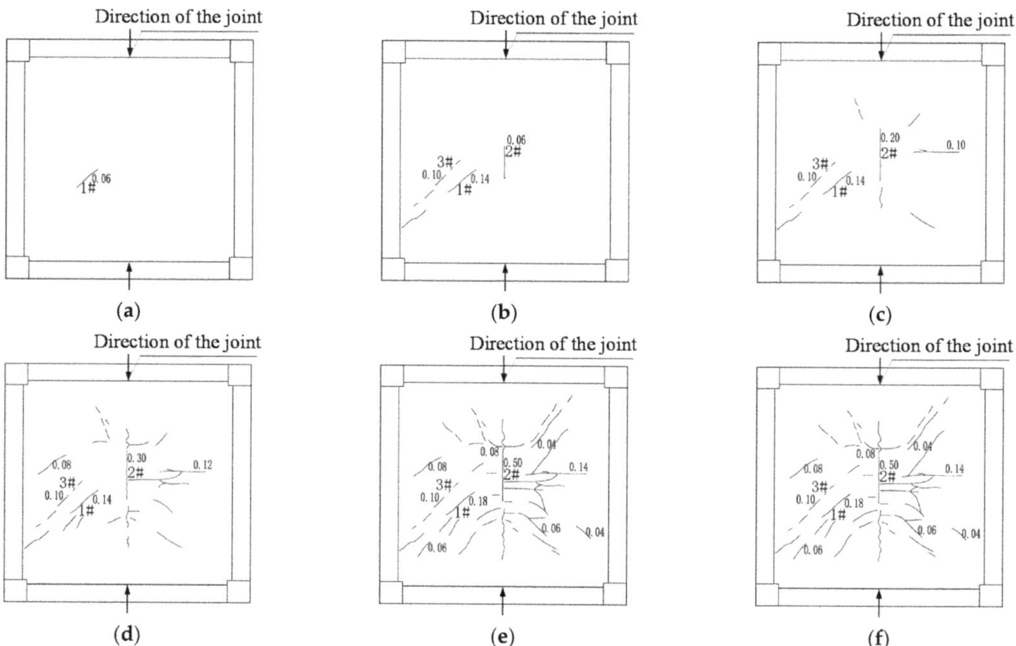

Figure 9. Cracking distribution of bottom plank at different loading stages: (**a**) Crack distribution at load q = 10 kN/m²; (**b**) Crack distribution at load q = 16 kN/m²; (**c**) Crack distribution at load q = 22 kN/m²; (**d**) Crack distribution at load q = 28 kN/m²; (**e**) Crack distribution at load q = 34 kN/m²; (**f**) Crack distribution at load q = 40 kN/m².

Before the distributed load of composite plate reaches 10 kN/m², the deflection increases linearly with the load. The lattice girder slab is a hybrid system and, therefore, its initial stiffness is due to the composite action between the lattice girders, concrete and reinforcement. When cracking in the concrete plank progressively occurs, there is a load redistribution and stiffness degradation. At this stage, the strength of concrete decreases and the stiffness of the slab is primarily provided by the lattice girder and the reinforcement. When loaded to the ultimate state at 40 kN/m², the midspan deflection of the bottom slab reaches 24.07 mm, which is about 1/208 of the span of the slab. However, the deflection is far below the limit of the maximum deflection (1/50 of the span of the slab) according to the Code for Design of Concrete Structures (GB50010-2010) [17]. The load–deflection curve of the latticed girder composite slab is depicted in Figure 10. In this figure, W1 curve represents the load–deflection curve of the midspan from measuring point W1 in Figure 7. W2 and W3 curves represent the mean load–deflection curves of W2a and W2b, W3a and W3b, respectively. In this test, the deflections of x direction and y direction are basically the same in the whole process, indicating that the LGCS has compatible deformation capability.

Under the uniformly distributed load, the vertical deformation of the slab is similar to that of the traditional concrete two-way slab.

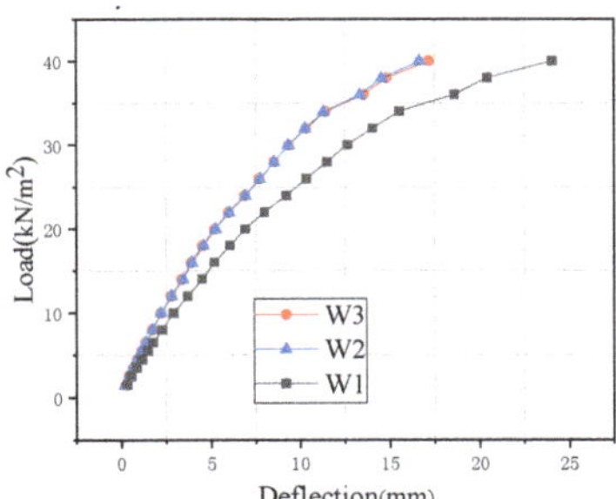

Figure 10. Load–deflection curve.

It can be concluded from both the crack distribution and the load–deflection curve that the design of LGCS meets the requirement of existing design provisions. Admittedly, the safety and serviceability of LGCS under uniformly distributed load can be achieved. However, the crack width and deflection of LGCS is relatively small, even under 2.5 times design value of bearing capacity, which reflect conservative design of LGCS.

According to the ER strain gauges at the bottom plank of concrete, the strain distribution of the concrete can be recorded. It can be seen from Figure 11 that the tensile strain ε_x of concrete which parallels to the joint is greater than that ε_y in the vertical direction after reaches 7.5 kN/m^2, which indicates that the slab presents a certain mechanical characteristic of orthogonal anisotropic due to the existence of monolithic joint.

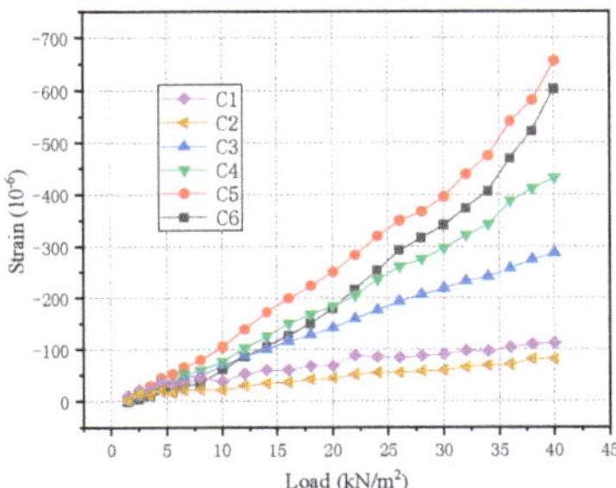

Figure 11. Load–strain curves of concrete at bottom plank.

Figure 12a,b show the load–strain curves of bottom longitudinal reinforcement which are parallel and vertical to monolithic joint, respectively. As for the reinforcement parallel to the joint, it can be seen from that the reinforcement strain is small before loading to 16 kN/m^2. After being loaded to 16 kN/m^2, the strains from all measuring points paralleled to joint increases progressively, and their value exceed that of vertical to joint. The maximum strain value appears in the midspan. As for the reinforcement vertical to the joint, similar strain distribution can also be observed. The load–strain curves of the vertical longitudinal reinforcement are more irregular when comparing with that of paralleled longitudinal reinforcement, probably due to the existence of the joint which cuts off the steel bars and brings the discontinuities.

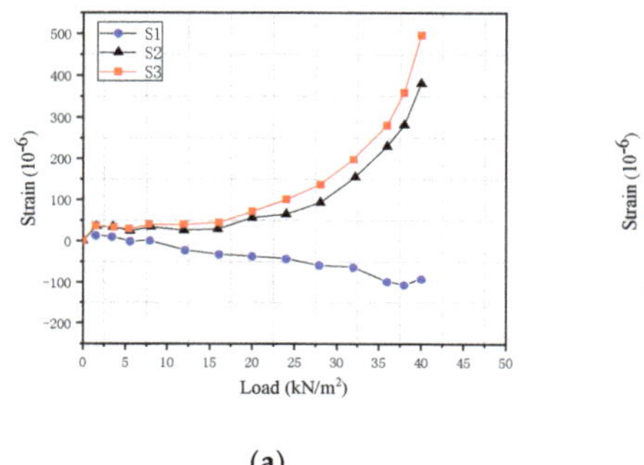 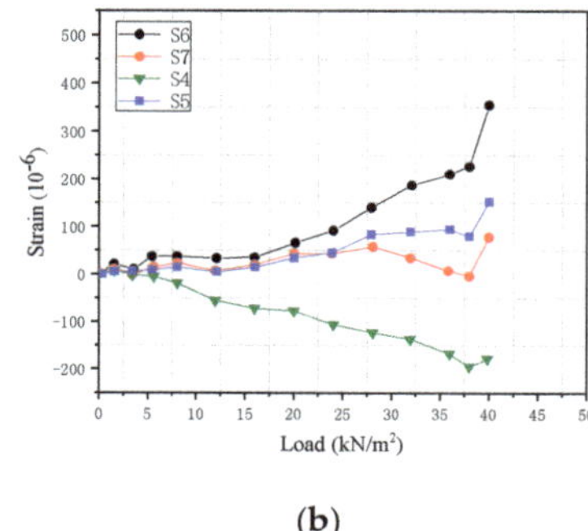

(**a**)   (**b**)

Figure 12. Load–strain curves of reinforcement: (**a**) Load–strain curves paralleled to the joint; (**b**) Load–strain curves vertical to the joint.

3. Numerical Simulation

3.1. Numerical Model

In this study, a finite element (FE) model based on ABAQUS [19] is established to investigate the behavior of the latticed girder composite slab. The main components affecting the behavior of latticed girder composite slab are the thickness of the precast and cast-in-place slab, the set-up of latticed girder, and the material properties. In addition, to obtain accurate results from the FE analysis, the contact between precast slab and cast-in-place slab must be properly modelled. Therefore, the nonlinearities, such as contacts between slabs and material properties are considered in the FE models.

As for the steel material modeling, the von Mises criterion is adopted, where the option (*PLASTIC) in ABAQUS used in association with the plastic flow rule. In the plastic model of ABAQUS, the steel material behavior is initially elastic with Young's modulus E_s followed by strain hardening and then by the yielding criterion. In the modeling, the Young's modulus of steel is chosen as $E_s = 2 \times 10^5$MPa and the Poisson's ratio is $\nu = 0.3$. The yield strength of steel is chosen as $f_{y,r} = 400$Mpa and the ultimate strength is $f_{st,r} = 540$MPa. In the FE analysis of ABAQUS [19], the uniaxial behavior of the steel in can be automatically transformed into a multiaxial stress state.

When comparing with the steel, concrete exhibits complex nonlinear properties, such as stiffness degradation, lateral effect and strain softening. In this study, concrete damage plastic (CDP) [20] model is used to represent the mechanical property of concrete. In CDP model, the stiffness degradation which is represented by damage variables and constitutive relations, can be decoupled from the plastic evolution equations. Two damage variables, namely tensile damage and compressive damage, account for the different stiffness degradation states. The constitutive equations for elastoplastic responses are established from the degradation damage response. To control the evolution of the yield surface, the effective stress function is applied, so that the material parameters can be conveniently calibrated. In the application of CDP model, the stress–strain law and damage law are thoroughly studied and discussed [21,22].

Under the framework of the plastic damage model, the total strain can be divided as

$$\varepsilon = \varepsilon^e + \varepsilon^p \tag{1}$$

$$\varepsilon^e = E^{-1} : \sigma \tag{2}$$

where E is the elastic stiffness tensor; ε^e is the elastic strain and ε^p is the inelastic strain.

The constitutive relationship can be attributed as

$$\sigma = E_0 : (\varepsilon - \varepsilon^p) \tag{3}$$

$$\varepsilon^{\cdot p} = \lambda^{\cdot} \nabla_{\sigma} \Phi(\sigma) \tag{4}$$

$$\kappa^{\cdot} = \lambda^{\cdot} \mathbf{H}(\sigma, \kappa) \tag{5}$$

where σ is the effective stress; E_0 is the initial elastic stiffness tensor with $E = (1 - D)E_0$; $\lambda^{\cdot}$ is a nonnegative function referred to as the plastic consistency parameter; Φ is a scalar plastic potential function; κ is the damage variable tensor which can be the compressive damage variable d_c and tensile damage variable d_t in the scalar damage model; $\mathbf{H}$ is the damage evolution function.

For uniaxial tension and compression, the following stress-strain law can be defined as

$$\sigma_t = (1 - d_t)E_0\varepsilon \tag{6}$$

$$\sigma_c = (1 - d_c)E_0\varepsilon \tag{7}$$

In this study, we adopt the suggested damage law of concrete by Code for Design of Concrete Structures (GB50010-2010/2015) [17]. Under uniaxial compression, the damage law is given as

$$d_c = \begin{cases} 1 - \dfrac{\rho_c n}{n - 1 + x_c^n} & x_c \leq 1 \\[2mm] 1 - \dfrac{\rho_c}{\alpha_c(x_c - 1)^2 + x_c} & x_c > 1 \end{cases} \tag{8}$$

$$\rho_c = \frac{f_{cr}}{E_0\varepsilon_{cr}} \tag{9}$$

$$n = \frac{E_0\varepsilon_{cr}}{E_0\varepsilon_{cr} - f_{cr}} \tag{10}$$

$$x_c = \frac{\varepsilon}{\varepsilon_{cr}} \tag{11}$$

where E_0 is the Young's modulus of concrete; α_c is the concrete compressive damage curve shape parameter; f_{cr} is the ultimate compressive strength of concrete; ε_{cr} is the strain corresponding to the ultimate compressive strength f_{cr}.

Under uniaxial tension, the tensile damage law is given as

$$d_t = \begin{cases} 1 - \rho_t\left(1.2 - 0.2x_t^5\right) & x_t \leq 1 \\[2mm] 1 - \dfrac{\rho_t}{\alpha_t(x_t - 1)^{1.7} + x_t} & x_t > 1 \end{cases} \tag{12}$$

$$x_t = \frac{\varepsilon}{\varepsilon_{tr}} \tag{13}$$

$$\rho_t = \frac{f_{tr}}{E_0\varepsilon_{tr}} \tag{14}$$

where α_t is the concrete tensile damage curve shape parameter; f_{tr} is the ultimate tensile strength of concrete; ε_{tr} is the strain corresponding to the ultimate compressive strength f_{tr}.

In this modeling, we choose the concrete material value as: $E_0 = 2.55 \times 10^4$MPa; $f_{cr} = 26.8$MPa; $\varepsilon_{cr} = 1.64 \times 10^{-3}$; $\alpha_c = 1.36$; $f_{tr} = 2.15$MPa; $\varepsilon_{tr} = 1.02 \times 10^{-4}$; $\alpha_t = 1.51$. In the application of ABAQUS [19], the following parameters should be determined for the CDP model given Table 1.

Table 1. Parameter value of CDP.

Dilation Angle	Eccentricity	f_{b_0}/f_{c_0}	k	Viscosity Parameter
30	0.1	1.16	0.667	0.0005

The concrete is the modelled by eight-node solid FE (C3D8R) in ABAQUS. Additionally, the reinforcement is modelled by Truss Element in ABAQUS. To avoid numerical inaccuracies, the shape of the C3D8R satisfies the limits and aspect ratio as recommended by ABAQUS [19]. The mesh size of the composite plate is selected as

100 mm $\times$ 100 mm $\times$ 20 mm (length $\times$ width $\times$ depth). To model the loading process of the slab, the distributed load is applied on the top surface of the slab. The boundary condition of the composite plate is set as four side fixed support.

To obtain accurate results from the FE analyses, all components in contact with the concrete must be properly modelled. There are two surfaces of interaction: (a) the precast concrete plank and cast-in-place concrete interface; (b) the contacts between concrete and reinforcement, including the latticed girder and longitudinal reinforcement. As for interaction (a), a surface-to-surface contact is chosen, where normal behavior and tangential behavior were considered. Therefore, the default contact option in ABAQUS [19] is utilized. This default contact option consists of a hard contact pressure-over closure relationship. Regarding the tangential direction, the penalty frictional formulation with a friction coefficient equal to 0.3 is employed. When it comes to the interaction (b), the fully coupled contact between reinforcement and concrete is adopted and the embedded region in ABAQUS is used in the simulation.

3.2. Simulation Results

Figure 13a,b depict the maximum principal stress of the concrete and reinforcement stress at the bottom of the slab, respectively. It can be seen that the overall stress distribution of slab shows the strong characteristics of two-way slab, and the reinforcement stress near the support and mid-span of the slab is relatively large. It is found that a region of additional reinforcement near the joint has smaller stress value when comparing with the neighboring region. This is probably due to the arrangement of additional reinforcement near the joint.

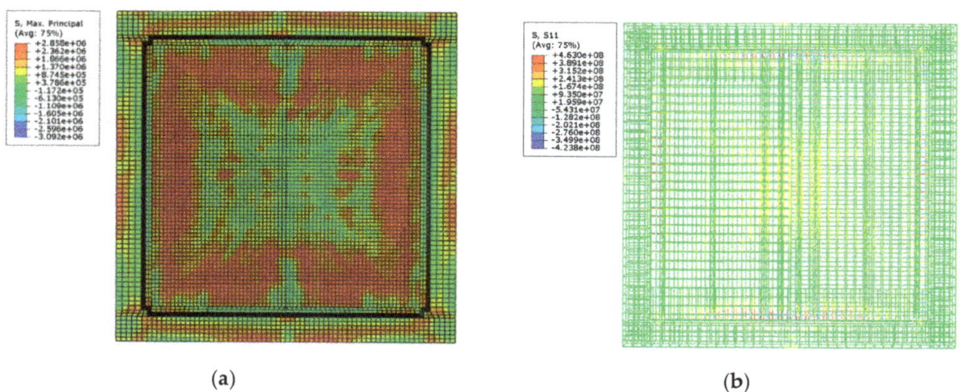

(a)   (b)

Figure 13. (**a**) Maximum principal stress of slab bottom concrete; (**b**) Reinforcement Stress of slab.

Figure 14 demonstrates the simulation results of load–strain curves concrete at bottom plank. As shown, the numerical curves are compared to the experimental results, where numerical results show satisfactory accuracy within the band corresponding to the experimental results, verifying the reliability of the proposed numerical model. The detailed comparison between numerical and experimental results are given in Table 2. As proposed in Table 2, the maximum error is 7.6%, which further indicates accurate simulation results. The midspan load–deflection curve of the composite slab shows the deformation development under the load. Figure 15 demonstrates the comparison between the simulation results and the test results of the load–deflection curve of the mid-span of the composite slab. The simulated maximum deflection of the midspan is 25.13 mm, and the experimental maximum deflection is 24.07 mm. The result is in agreement with the experimental result, which testify correctness and effectiveness of the proposed numerical model.

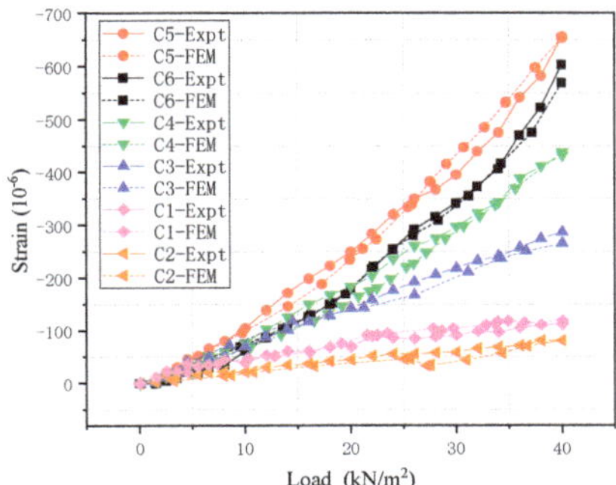

Figure 14. Simulation of concrete strain.

Table 2. Comparison of simulation and experimental value of concrete strain

Strain/(10^{-6})	C1	C2	C3	C4	C5	C6
Experiment	112	81	287	432	656	603
Simulation	118	80	265	438	653	569
Error	5.4%	−1.2%	−7.6%	1.4%	−0.4%	−5.6%

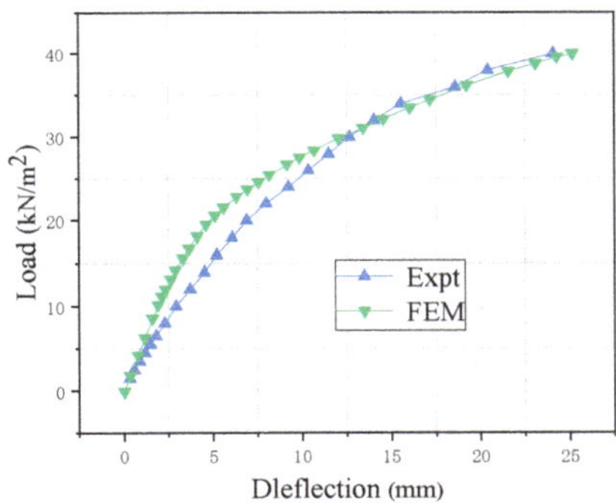

Figure 15. Load–deflection of the test and simulation.

Apart from the bearing capacity and deformation of the concrete structure, the investigations of concrete cracking, including crack spacing and crack width, is also important in the numerical simulation of concrete structure. To estimate the crack spacing and the crack width of simple concrete components, empirical equations had been proposed in some design codes such as Code for Design of Concrete Structures (GB50010-2010) [17]. When it comes to the RC structure with complex reinforcement and loading status, the estimations of crack spacing and crack should rely on numerical methods. In the past decades, several advanced numerical models had been put forward to simulate the cracking process of concrete, ranging from cohesive elements [23,24], element-free methods [25–27] to extended finite element method (XFEM) [28,29]. Admittedly, these numerical methods provided relatively accurate ways to simulate and evaluate the cracks of concrete. However, for the finite element modelling of real structure, which usually contains huge amount elements, the aforementioned numerical methods might bring undesirable computational costs. In the continuum damage models of concrete, the damage variables, the tensile and compressive damage variables, are adopted to represent the tensile and compressive damage mechanisms which brought by the development of cracks. Therefore, the damage variables could partially represent the cracking status without the introduction of additional numerical costs.

The interface between precast and cast-in-place concrete is the weak zone of the composite slab. The bonding in this interface is mainly provided by the chemical adhesive force, interfacial friction, mechanical interaction and the resistance brought by lattice girder, which contains complex mechanisms. Observing the concrete damage at this interface is conducive to determining whether full composite action up to the ultimate loading capacity without interface failure can be achieved. The bottom surface of the precast plank is also intriguing in the simulation that the cracks at bottom surface might induce the leaks and the reinforcement corrosion. Since the cracks in the concrete is usually brought by the tension, the tensile damage of the interface and bottom surface are shown in Figures 16 and 17b. As demonstrated in Figures 16 and 17b, the tensile damage distribution reveals a typical crack pattern of the two-way slab. However, it can be seen from Figure 16 that the damage value is relatively small in the region where additional reinforcement is settled near the joint. This numerical result in Figure 17b is correspondent to the experimental crack pattern depicted in Figure 9 (Figure 17a).

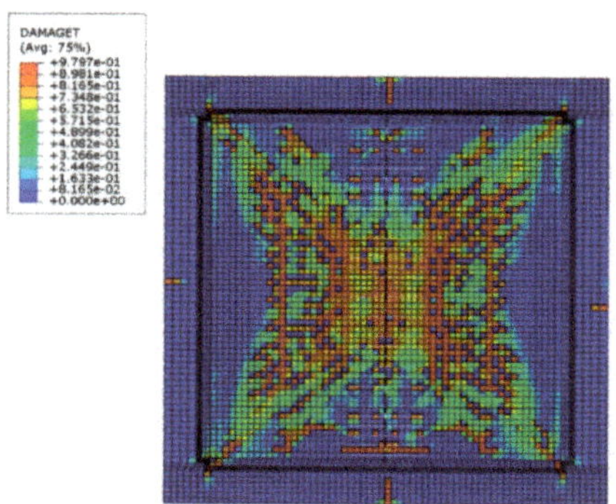

Figure 16. Tensile damage contour at concrete interface.

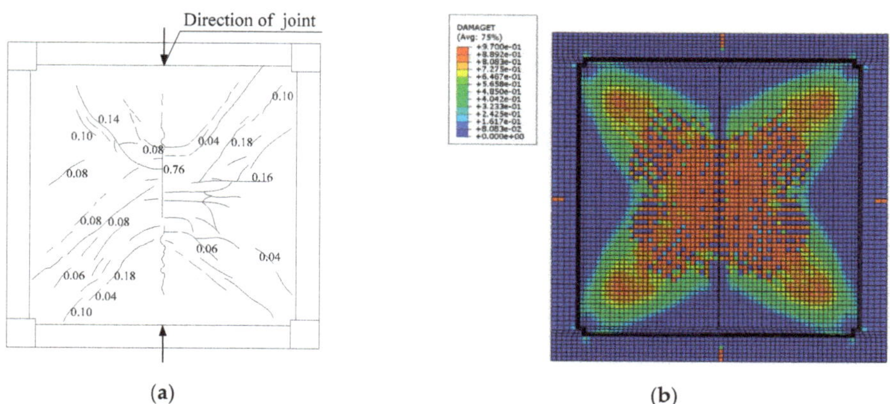

(**a**)   (**b**)

Figure 17. Comparison of experimental and numerical failure modes at concrete bottom plank: (**a**) Experimental cracking distribution; (**b**) Tensile damage contour of concrete bottom plank from FE simulation.

4. Parametric Study

Once the FE model is testified by the experimental tests, a parametric study can be performed by using the aforementioned numerical model of the slab. The parametric simulation analyzes (1) the influence of the lattice girder to the mechanical behavior of slab; (2) the influence of the additional reinforcement near the joint (Figure 5) to the stiffness and load–deflection curve of slab. Three FE models are generated in this parametric study that M1 is the original FE model, M2 is the FE model without additional longitudinal reinforcement and M3 is the FE model without lattice girder. It is emphasized that the material properties, sizes, boundary conditions and the surface contacts are the same in these FE models.

The comparison of load–deflection curves is given in Figure 18. It can be seen in this figure that M1 and M2 curves are similar. As for M2, the maximum deflection is 26.41 mm, which increases 5.1% compared to the deflection of M1. It is implied that the existence of the additional reinforcement has only slight influence on the resistance and stiffness of the slab. However, the additional reinforcement plays a part in preventing the cracks near joint. Tensile damage distribution of M2 is shown in Figure 19 that the maximum value is bigger than M1 given in Figure 17b. Furthermore, the damage of M2 near the joint is severer than M1 (Figure 17b) with the absence of additional reinforcement.

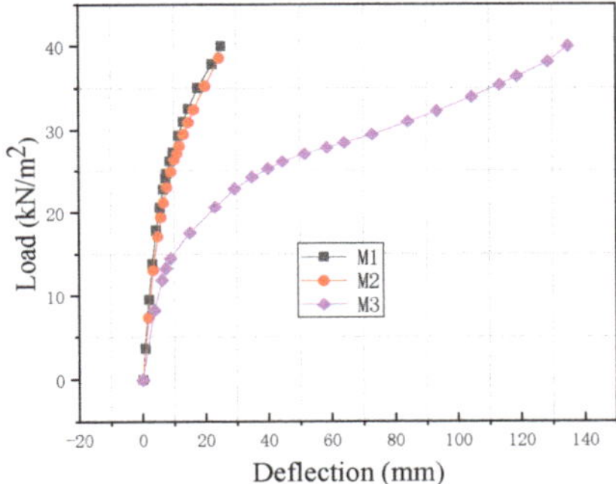

Figure 18. Comparison of load–deflection curves from FE modeling.

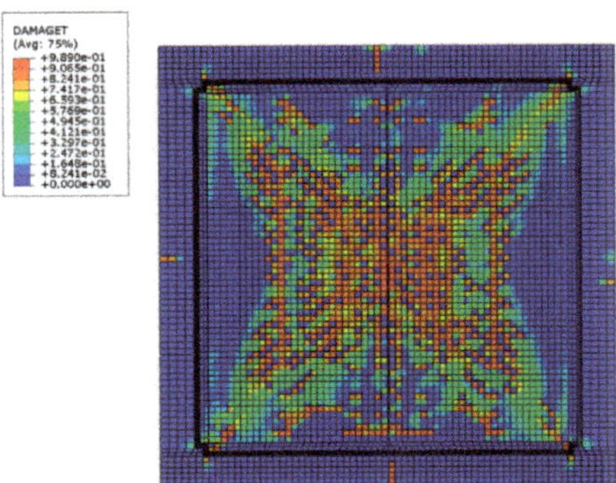

Figure 19. Tensile damage contour of concrete bottom plank from M2 simulation result.

As for M3, the shape of load–deflection curve and maximum deflection in Figure 18 are significantly different from M1 and M2, which indicate a much lower stiffness of the slab without the lattice girder. The maximum deflection of M3 reaches 134.90 mm (1/37 of the span of the slab), almost 5 times that of M1, which exceeds the limit of the maximum deflection (1/50 of the span of the slab) according to the Code for Design of Concrete Structures (GB50010-2010) [17]. Therefore, the existence of the lattice girder plays an essential role in the mechanical behavior of the slab.

5. Conclusions

A comprehensive experimental and numerical modelling has been conducted to investigate the structural performance of LGCS. A full-scale experiment is performed to investigate the bending resistance, deflection and the final crack distribution of LGCS under uniformly distributed load. The experimental results are used for the validation of FE models, which were then employed to conduct a series of parametric studies to extend the current test data over a broader range of the influences of lattice girder and additional reinforcement. The main conclusions are summarized below:

(1) Before the distributed load reaches 10 kN/m^2, the deflection increases linearly with the load. When cracking in the concrete plank progressively occurs, there is a load redistribution and stiffness degradation. The final midspan deflection LGCS of the bottom slab is 24.07 mm (about 1/208 of the span) which is far below the limit of the maximum deflection (1/50 of the span of the slab) according to the Code for Design of Concrete Structures (GB50010-2010), under 2.5 times of the bearing capacity limits (40 kN/m^2). It also can be observed by the load–deflection curve that the LGCS has sufficient stiffness and bending resistance. Therefore, the LGCS meets the requirement of existing design code [16,17]. Experimental test data in the presenting paper can be used to determine the load–deflection of the LGCS that can result in significant efficiencies for propping arrangements on site.

(2) No visible crack is observed under serviceability limit state, which is below the limit of maximum crack width (0.2 mm) under the design code. The GLCS satisfies the criterion of the serviceability. The final cracks reach to the 4 corners of the slabs along the direction of 45°, which illustrate typical cracking pattern of the two-way slab. Under 2.5 times of the bearing capacity limits (40 kN/m^2), the final maximum crack appears at the monolithic joint with the value of 0.76 mm which is much bigger than the other areas.

(3) An elaborate FE model to simulate the LGCS is presented in this study. From the numerical analyses, it has been demonstrated that the numerical model successfully predict the composite slab's resistance capacity, the load–deflection behavior and the final cracking pattern of the on-site test. The proposed FE model can be applied in the numerical modelling of LGCS in the further precast structural nonlinear analysis.

(4) A parametric study is performed to observe the following conclusions: (a) the existence of lattice girder provides significant stiffness to the slab and helps the precast plank and cast-in-place layer working together; (b) the additional reinforcement to the joint has slight effect on increasing the stiffness of slab, but it prevents the cracks near joint.

Author Contributions: Conceptualization, S.L.; methodology, X.Z. and H.L.; software, X.Z. and H.L.; validation, S.L and H.Z.; writing—original draft preparation, S.L.; writing—review and editing, S.L.; All authors have read and agreed to the published version of the manuscript.

Funding: This research was funded by National Science Foundation of China, grant number 51808499; Science Foundation of Zhejiang Province of China, grant number LGF20E080019.

Data Availability Statement: Data is contained within the article.

Acknowledgments: This study is supported by Engineering Research Centre of Precast Concrete of Zhejiang Province. The help of all members of the Engineering Research Centre is sincerely appreciated. We would also like to express our sincere appreciations to the anonymous referee for valuable suggestions and corrections.

Conflicts of Interest: The authors declare no conflict of interest.

References

1. Girhammar, U.A.; Pajari, M. Tests and analysis on shear strength of composite slabs of hollow core units and concrete topping. *Constr. Build. Mater.* **2008**, *22*, 1708–1722. [CrossRef]
2. Baran, E. Effects of cast-in-place concrete topping on flexural response of precast concrete hollow-core slabs. *Eng. Struct.* **2015**, *98*, 109–117. [CrossRef]
3. Rahimi Mansour, F.; Abu Bakar, S.; Ibrahim, I.S.; Marsono, A.K.; Marabi, B. Flexural performance of a precast concrete slab with steel fiber concrete topping. *Constr. Build. Mater.* **2015**, *75*, 112–120. [CrossRef]
4. Ibrahim, I.S.; Elliott, K.S.; Abdullah, R.; Kueh, A.B.H.; Sarbini, N.N. Experimental study on the shear behaviour of precast concrete hollow core slabs with concrete topping. *Eng. Struct.* **2016**, *125*, 80–90. [CrossRef]
5. Lam, S.S.E.; Wong, V.; Lee, R.S.M. Bonding assessment of semi-precast slabs subjected to flexural load and differential shrinkage. *Eng. Struct.* **2019**, *187*, 25–33. [CrossRef]
6. Du, H.; Hu, X.; Meng, Y.; Han, G.; Guo, K. Study on composite beams with prefabricated steel bar truss concrete slabs and demountable shear connectors. *Eng. Struct.* **2020**, *210*, 110419. [CrossRef]
7. Newell, S.; Goggins, J. Experimental study of hybrid precast concrete lattice girder floor at construction stage. *Structures* **2019**, *20*, 866–885. [CrossRef]
8. Nahar, M.; Islam, K.; Billah, A.M. Seismic collapse safety assessment of concrete beam-column joints reinforced with different types of shape memory alloy rebars. *J. Build. Eng.* **2020**, *29*, 101106. [CrossRef]
9. Yu, J.; Tan, K.H. Numerical analysis with joint model on RC assemblages subjected to progressive collapse. *Mag. Concr. Res.* **2014**, *66*, 1201–1218. [CrossRef]
10. Celik, O.C.; Ellingwood, B.R. Modeling Beam-Column Joints in Fragility Assessment of Gravity Load Designed Reinforced Concrete Frames. *J. Earthq. Eng.* **2008**, *12*, 357–381. [CrossRef]
11. Abdullah, R.; Samuel Easterling, W. New evaluation and modeling procedure for horizontal shear bond in composite slabs. *J. Constr. Steel Res.* **2009**, *65*, 891–899. [CrossRef]
12. Tzaros, K.A.; Mistakidis, E.S.; Perdikaris, P.C. A numerical model based on nonconvex–nonsmooth optimization for the simulation of bending tests on composite slabs with profiled steel sheeting. *Eng. Struct.* **2010**, *32*, 843–853. [CrossRef]
13. Ríos, J.D.; Cifuentes, H.; Martínez-De La Concha, A.; Medina-Reguera, F. Numerical modelling of the shear-bond behaviour of composite slabs in four and six-point bending tests. *Eng. Struct.* **2017**, *133*, 91–104. [CrossRef]
14. Feng, D.; Wu, G.; Lu, Y. Finite element modelling approach for precast reinforced concrete beam-to-column connections under cyclic loading. *Eng. Struct.* **2018**, *174*, 49–66. [CrossRef]
15. Karam, M.S.; Yamamoto, Y.; Nakamura, H.; Miura, T. Numerical Evaluation of the Perfobond (PBL) Shear Connector Subjected to Lateral Pressure Using Coupled Rigid Body Spring Model (RBSM) and Nonlinear Solid Finite Element Method (FEM). *Crystals* **2020**, *10*, 743. [CrossRef]
16. *Technical Specification for Application of Lattice Girder Slab*; T/CECS 715-2020; China Construction Industry Press: Beijing, China, 2020.
17. *Code for Design of Concrete Structures*; GB50010-2010/2015; China Construction Industry Press: Beijing, China, 2015.
18. *Chinese Standard for Testing Methods of Concrete Structures*; China Construction Industry Press: Beijing, China, 2012.
19. *Abaqus, 6.14; User's Manual*; SIMULIA: Providence, RI, USA, 2014.
20. Lee, J.; Fenves, G.L. Plastic-damage model for cyclic loading of concrete structures. *J. Eng. Mech.* **1998**, *124*, 892–900. [CrossRef]
21. Kytinou, V.K.; Chalioris, C.E.; Karayannis, C.G. Analysis of Residual Flexural Stiffness of Steel Fiber-Reinforced Concrete Beams with Steel Reinforcement. *Materials* **2020**, *13*, 2698. [CrossRef] [PubMed]
22. Sun, Y.; Liu, Y.; Wu, T.; Liu, X.; Lu, H. Numerical Analysis on Flexural Behavior of Steel Fiber-Reinforced LWAC Beams Reinforced with GFRP Bars. *Appl. Sci.* **2019**, *9*, 5128. [CrossRef]
23. Xu, X.; Needleman, A. Numerical simulations of fast crack growth in brittle solids. *J. Mech. Phys. Solids* **1994**, *42*, 1397–1434. [CrossRef]
24. Camacho, G.T.; Ortiz, M. Computational modelling of impact damage in brittle materials. *Int. J. Solids. Struct.* **1996**, *33*, 2899–2938. [CrossRef]
25. Belytschko, T.; Lu, Y.Y.; Gu, L. Element-free Galerkin methods. *Int. J. Numer. Meth. Eng.* **1994**, *37*, 229–256. [CrossRef]
26. Liu, W.K.; Jun, S.; Zhang, Y.F. Reproducing kernel particle methods. *Int. J. Numer. Meth. Fluids* **1995**, *20*, 1081–1106. [CrossRef]
27. Chen, J.; Pan, C.; Wu, C.; Liu, W.K. Reproducing kernel particle methods for large deformation analysis of non-linear structures. *Comp. Meth. Appl. Mech. Eng.* **1996**, *139*, 195–227. [CrossRef]
28. Sukumar, N.; Moes, N.; Moran, B.; Belytschko, T. Extended finite element method for three-dimensional crack modelling. *Int. J. Numer. Meth. Eng.* **2000**, *48*, 1549–1570. [CrossRef]
29. Moes, N.; Belytschko, T. Extended finite element method for cohesive crack growth. *Eng. Fract. Mech.* **2002**, *69*, 813–833. [CrossRef]

Article

Acoustic Emission Study on the Damage Evolution of a Corroded Reinforced Concrete Column under Axial Loads

Ye Chen [1], Shuang Zhu [1], Shenghua Ye [1], Yifeng Ling [2,*], Dan Wu [3], Geqiang Zhang [1], Nianfu Du [1], Xianyu Jin [3] and Chuanqing Fu [4]

1   POWERCHINA Huadong Engineering Corporation Limited, Hangzhou 310034, China; chen_y8@ecidi.com (Y.C.); zhu_s@ecidi.com (S.Z.); ye_sh@ecidi.com (S.Y.); zhang_gq@ecidi.com (G.Z.); du_nf@ecidi.com (N.D.)
2   School of Qilu Transportation, Shandong University, Jinan 250002, China
3   College of Civil Engineering and Architecture, Zhejiang University, Hangzhou 310027, China; wudan@126.com (D.W.); xianyu@zju.edu.cn (X.J.)
4   College of Civil Engineering and Architecture, Zhejiang University of Technology, Hangzhou 310034, China; chqfu@zjut.edu.cn
*   Correspondence: jemeryling@gmail.com

Abstract: In this paper, the damage of a reinforced concrete (RC) column with various levels of reinforcement corrosion under axial loads is characterized using the acoustic emission (AE) technique. Based on the AE rate process theory, a modified damage evolution equation of RC associated with the axial load and different corrosion rates is proposed. The experimental results show that the measured AE signal parameters during the loading process are closely related to the damage evolution of the RC column as well as the reinforcement corrosion level. The proposed modified damage evolution equation enables dynamic analysis for the damage of corrosion on a RC column under axial loading for a further real-time quantitative evaluation of corrosion damage on reinforced concrete.

Keywords: acoustic emission; AE rate process theory; corrosion rate; damage evolution; axial load

Citation: Chen, Y.; Zhu, S.; Ye, S.; Ling, Y.; Wu, D.; Zhang, G.; Du, N.; Jin, X.; Fu, C. Acoustic Emission Study on the Damage Evolution of a Corroded Reinforced Concrete Column under Axial Loads. *Crystals* **2021**, *11*, 67. https://doi.org/10.3390/cryst11010067

Received: 30 December 2020
Accepted: 12 January 2021
Published: 15 January 2021

Publisher's Note: MDPI stays neutral with regard to jurisdictional claims in published maps and institutional affiliations.

1. Introduction

The safety and durability of reinforced concrete (RC) structures during their service lifetime are mainly dependent on reinforcement corrosion, which can lead to concrete cracking [1–3]. Such corrosion-induced cracking will harm the internal concrete structure, resulting in a decline in the service life of the concrete [4,5]. Hence, developing a convenient and precise evaluation method for the corrosion of RC structures has become an urgent problem to be solved [6,7].

Acoustic emission (AE) refers to the phenomenon of elastic waves released during material fracture [8–10]. The AE method can capture the generation process of microcracks inside the concrete structure dynamically and in real time [11,12]. It can also detect the internal damage of a concrete structure and provide early warning. Thus, the AE rate process theory has been widely used to quantitatively investigate the internal damage evolution of concrete materials.

Thus far, many efforts have been made to investigate the characteristics of AE parameters during the loading process of concrete [13–15]. Ohtsu et al. [16] proposed the AE rate theory to evaluate the compressive strength of actual concrete structures and the process of steel corrosion with the moment tensor theory [17]. Suzuki et al. [18] took samples from an existing bridge structure and performed AE monitoring in the uniaxial compression process. They noted that the variation of damage was consistent with that of the compressive strength. Suzuki et al. [19] also studied the uniaxial compression process of concrete samples, which were taken from a canal wall, by applying the AE monitoring method. Based on the AE rate process theory and the damage mechanism, they put forward the implementation steps to determine the relative damage and further evaluate the damage to the samples. Zhou et al. [20] nondestructively obtained the elastic modulus and initial material damage by combining the AE rate process theory and the basic damage mechanics theory. In order to achieve a real-time quantitative assessment of steel strand damage, Deng et al. [21] assumed the number of AE events and stress levels of steel strands in a mathematical model and established the AE probability density function and damage evolution model for steel strands, enabling the real-time monitoring of the tensile damage of steel strands by AE technology [22].

This research conducted axial load experiments of RC columns with 0%, 5%, 10% and 15% corrosion rates. The corrosion rate and AE parameters were introduced to the AE rate process theory as the damage characteristic parameters of reinforced concrete. Then, a modified damage evolution model of a corroded RC column was proposed. This model provides real-time monitoring of the damage of corroded RC and quantitatively evaluates the damage degree from corrosion.

2. Damage Model

The AE phenomenon can reflect the damage and deterioration of a concrete structure under loading [23]. In the literature [24], the AE rate process theory was introduced into the mathematical model for concrete under uniaxial compression. When the load level increases from V to $V + dV$, the probability density function $f(V)$ of the AE event can be expressed as follows:

$$f(V)dV = dN/N \tag{1}$$

where N is the total number of AE events with the load level V increasing from the initial state.

In Ohtsu's model [16], $f(V)$ can be used to represent the AE rate and is approximately expressed by hyperbolic functions as follows:

$$f(V) = a/V + b \tag{2}$$

where a and b are experimental parameters. Substituting Equation (2) into Equation (1) will give the relationship between the load level V and the total number of AE events N as follows:

$$N = cV^a \exp(bV) \tag{3}$$

where a, b and c are AE test parameters, and a reflects the number of microcracks in the material. When $a > 0$, a higher AE rate can appear at lower stress levels, which indicates multiple cracks. When $a < 0$, a low AE rate appears at lower stress levels, which corresponds to no cracks or a small number of cracks.

In Dai-Labuz's model [25], the relationship between V and N can be expressed as follows:

$$V = aN + c\ln(1 + qN) \tag{4}$$

where a, c and q are related parameters for the AE.

After substituting Equation (4) into Equation (2), the probability density function $f(V)$ can be expressed as follows:

$$f(V) = \frac{1}{N_0}\left(\frac{1 + qN}{a + cq + aqN}\right) \tag{5}$$

After integrating the probability density function over a range of load level V, the damage degree of the specimen under this load level can be obtained:

$$D = \int_0^V f(V)dV \tag{6}$$

During the loading process of reinforced concrete, a large number of cracks will appear at the internal structure due to steel corrosion. Meanwhile, the corrosion degree of reinforcement will have an influence on the number of cracks on the structure during the entire loading process. However, limited research has been conducted on the damage evolution of corroded reinforced concrete [26], and there is not enough convincing evidence to show the diversity in the structural damage evolution of RC under different corrosion degrees [27]. Thus, it is necessary to fill these gaps. In this research, on the basis of Dai-Labuz's model [25], the load level and accumulative AE hit number in the loading process can be expressed as follows:

$$N = s(V) \tag{7}$$

where $s(V)$ is determined by experiments.

After taking the derivative of Equation (7) and substituting it into Equation (2), the following equation can be obtained:

$$f(V) = \frac{1}{N_0} \frac{ds(V)}{dV} \tag{8}$$

where N_0 is the accumulative AE hit number when the load level reaches the ultimate load.

Finally, according to Equation (6), the damage degree of RC at load level V can be obtained as follows:

$$D = \frac{1}{N_0} \int_0^V ds(V) \tag{9}$$

3. Experimental Program

3.1. Specimen Preparation

Type I (ASTM C150) ordinary Portland cement (OPC) was used [28]. The fine aggregate was river sand with a fineness modulus of 2.6, and the coarse aggregate was crushed limestone with a maximum aggregate size of 20 mm. A commercial high-range water reducer (HRWR) was added to improve the workability and flowability of the concrete. The measured 28-day compressive strength of the concrete was 32.0 MPa on 150 mm cubes. Table 1 presents the mix proportion of the concrete. The longitudinal reinforcements in the concrete beam was a plain bar made of HPB235, with a nominal diameter of 10 mm, and the stirrups comprised a plain bar made of HPB235, with a nominal diameter of 6 mm. Then, eight RC column specimens of 100 mm × 100 mm × 400 mm were cast. A detailed configuration of the specimens is shown in Figure 1.

Table 1. Mix proportions of concrete.

OPC (kg/m³)	Fine Aggregate (kg/m³)	Coarse Aggregate (kg/m³)	HRWR (kg/m³)	Water (kg/m³)
372	698	1116	3.71	175

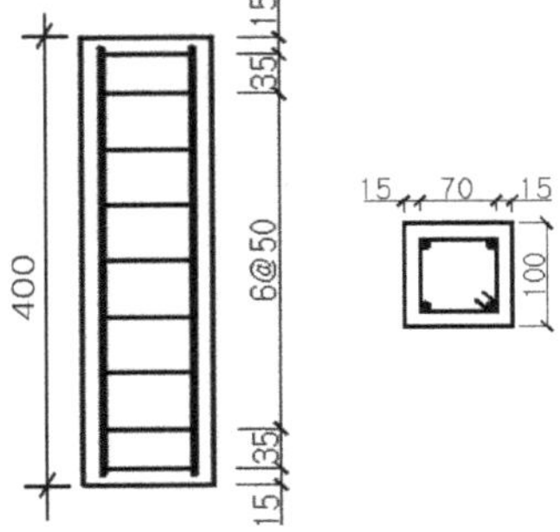

Figure 1. Reinforcement configuration (units mm).

3.2. Corrosion Acceleration

The impressed current method was adopted to induce reinforcement corrosion in the RC specimens to reach the target level of corrosion. The layout of the corrosion acceleration apparatus is shown in Figure 2. The surfaces of the specimens were wrapped with sponge and further contained in a stainless-steel cage. In order to keep the RC specimens fully wet prior to inducing the impressed current, the specimens were soaked in a 3.5% NaCl solution for 72 h. During accelerated corrosion, the steel embedded in the specimen was connected to an anode of the direct current (DC)-regulated power supply, and the stainless-steel cage was connected to the cathode. A 3.5% NaCl solution was sprayed using a sponge every day.

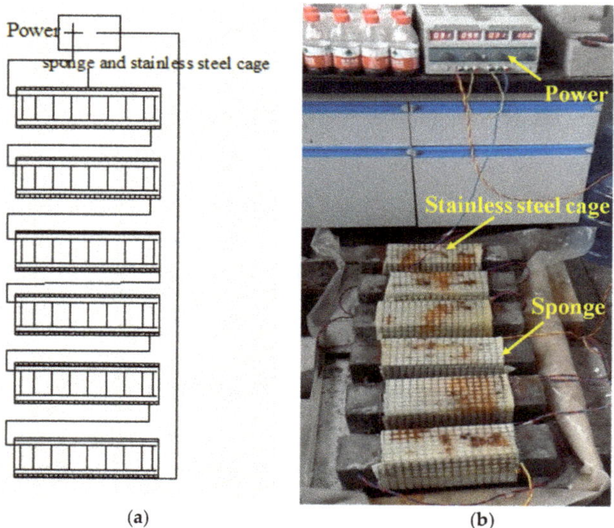

Figure 2. Accelerated corrosion test.

A total of eight RC column specimens were prepared and divided into four groups with different levels of reinforcement corrosion, namely, 0% (Z-0), noncorroded as a reference; 5% (Z-5); 10% (Z-10); and 15% (Z-15). According to Faraday's law, electrification duration of 30 days, 60 days and 90 days, respectively, represents the corrosion level of 5%, 10% and 15%.

3.3. Test System

The test system included the AE signal data acquisition system, the loading system and the load and displacement recording system. The DS2-AE acquisition system produced by Beijing Science and Technology Company was adopted in the test. Eight RS-35C sensors with a frequency of 150 kHz and a 40 dB preamplification were used in the AE system. The sampling frequency was set at 3 MHz. As shown in Figure 3, the eight sensors were arranged on the two sides of the columns. The characterization of the AE source was therefore measured and determined by analyzing the features of the waveform [29].

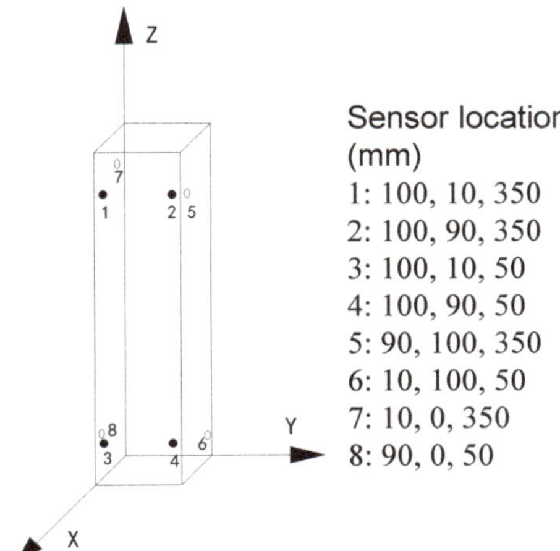

Figure 3. Sensor location in the reinforced concrete.

The load and displacement acquisition system consisted of a load sensor, resistance strain gauge, displacement sensor, displacement transmitter and recorder. The details are shown in Figure 4. The loading device was a 100 T high-performance testing machine, and the static loading was executed in the test. According to the standard of the concrete structure test method (GB/T50152-2012) [30], loading in the displacement control was conducted at a rate of 0.5 mm/min.

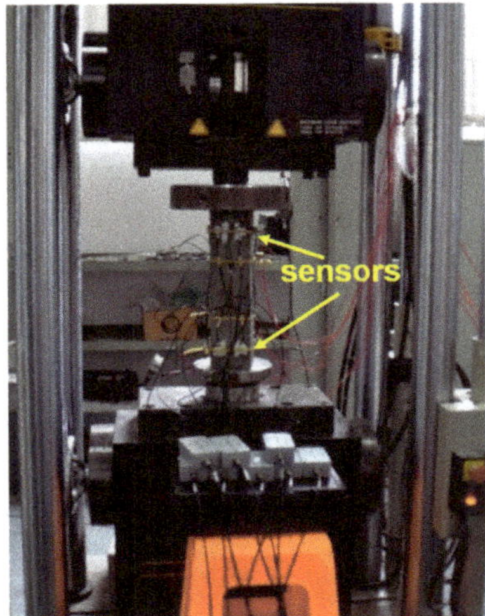

Figure 4. Experimental system.

4. Results and Discussion

4.1. Accumulative AE Hit Number Analysis

The relationship between the accumulative AE hit number and load levels of the RC columns is shown in Figure 5. The effect of corrosion rate on this relationship is depicted in Figure 5a–d.

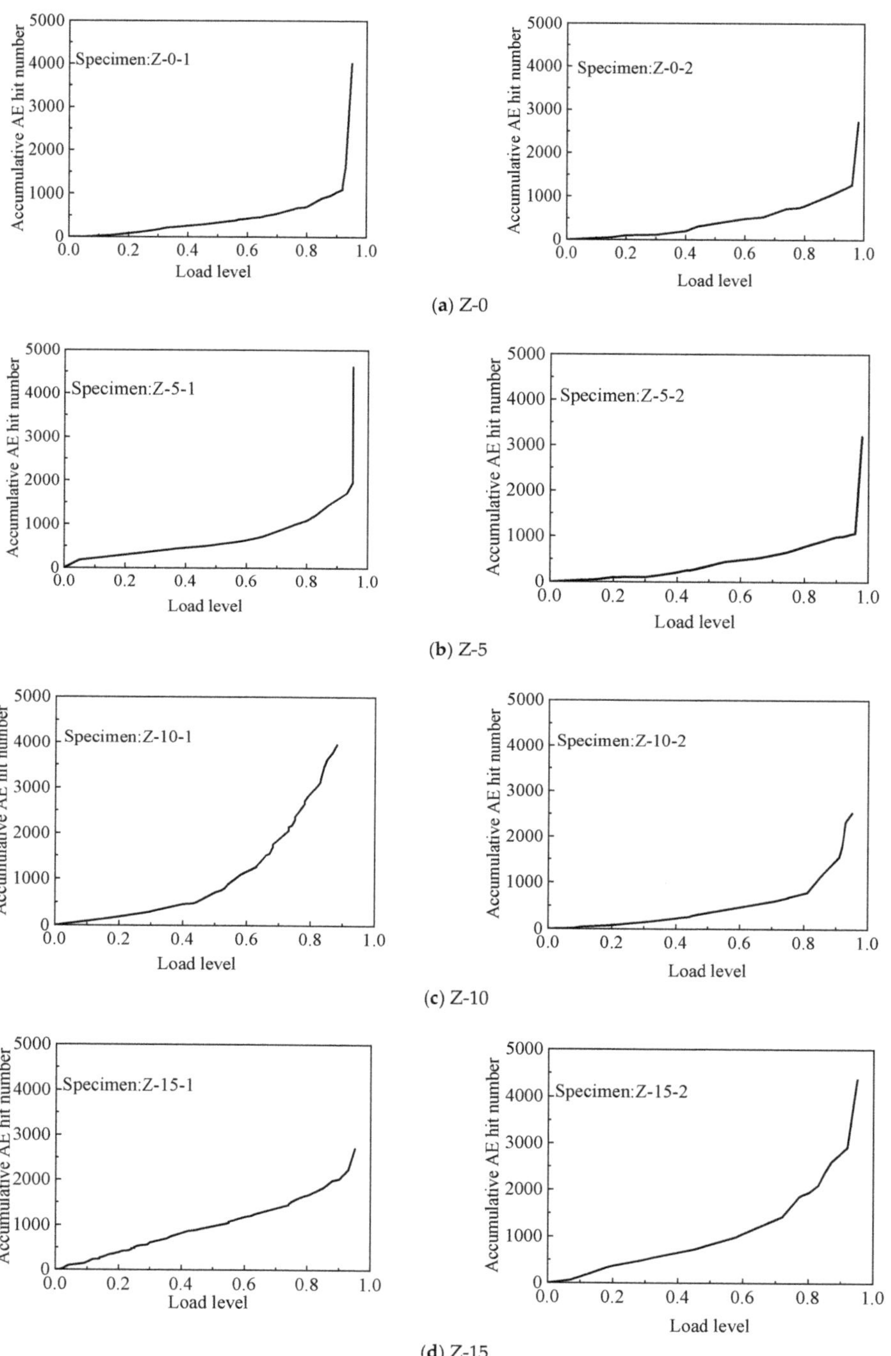

Figure 5. Correlation curve of the acoustic emission (AE) accumulative hit number and load for different corrosion rates

It can be seen from Figure 5 that the development of experimental curves in the four groups is similar. The AE accumulative hit number increases with the increase of the load level. However, there are also significant differences in the curves at different corrosion rates. The curve of Z-0 is similar to that of Z-5. Both curves in these groups show slow growth in the beginning, and a sharp increase occurs when the load level reaches 0.9. This indicates that, when the corrosion rate is below 5%, the internal damage accumulation is slower and lasts for a longer time during the loading process. Therefore, these two groups of specimens show a brittle characteristic, which means internal microcracks are rare before loading. It should be noted that although there is little variation between the two specimens for each group, the trends among different corrosion rates are significant.

Compared to the previous two groups' tests (Z-0, Z-5), the curves of Z-10 and Z-15 are significantly different, showing no obvious turning points. This reveals that the AE signals of these two groups increase in the whole loading process. The initial growth rate of the AE signals is stable and increases until specimen failure. It can be found that with the increase of the corrosion rate, the curve shape changes from steep growth to steady growth. This is because the corrosion of the steel causes a large number of corrosion cracks in the columns [31]. The greater the corrosion rate, the more the cracks occur. During the loading process, with the increase of the load level, the propagation speed of the cracks increases, which leads to a significant increase in the acoustic emission signal.

4.2. Analysis of AE Energy Distribution

From the data in Figure 5, the AE accumulative energy distribution of the RC column was calculated on the area under three ranges of the load level (0–0.3, 0.3–0.6 and 0.6–0.9), as shown in Figure 6. In order to reduce the influence of machine error and environmental noise on the critical failure, this paper focuses on the analysis of energy release within the load level of 0–0.9 [32]. The loading process is divided into three phases: preloading (0–0.3), middle loading (0.3–0.6) and late loading (0.6–0.9). The proportion of the AE accumulative energy in three phases is calculated accordingly.

As shown in Figure 6, regardless of the degree of corrosion, the proportion of energy released in the late loading period is the largest but the smallest in the preloading period. The result shows that the internal damage degree of the RC column increases with the increase of the load level. Comparing among the four groups, the energy proportion of Z-0 is 5% and 6%. When the corrosion rate reaches 15% (Z-15), the preloading energy proportion only increases by 6%, which indicates that the corrosion of the reinforcement has little impact on internal damage in the preloading period. Although the internal crack increases with the increment of the corrosion degree on the reinforcement, the crack propagation is not obvious due to the low load level, which may explain the small increase in energy [33].

In the middle loading period, the energy proportion of Z-0 is 6% and 8%. As the corrosion rate increases, the energy proportion in the middle loading period can increase to 26% and 23% at a growth rate of 20%. This indicates that the load in the middle loading period can enlarge the original microcracks, resulting in a large AE energy release. The corrosion in Z-10 and Z-15 is much more than Z-0 and Z-5, which strengthens the release of AE energy from Z-5 to Z-10. In this case, the AE signals before the middle loading period can be affected by the corrosion degree of specimens. Hence, the AE signals in the middle loading period can reflect the initial health condition of the corroded RC column.

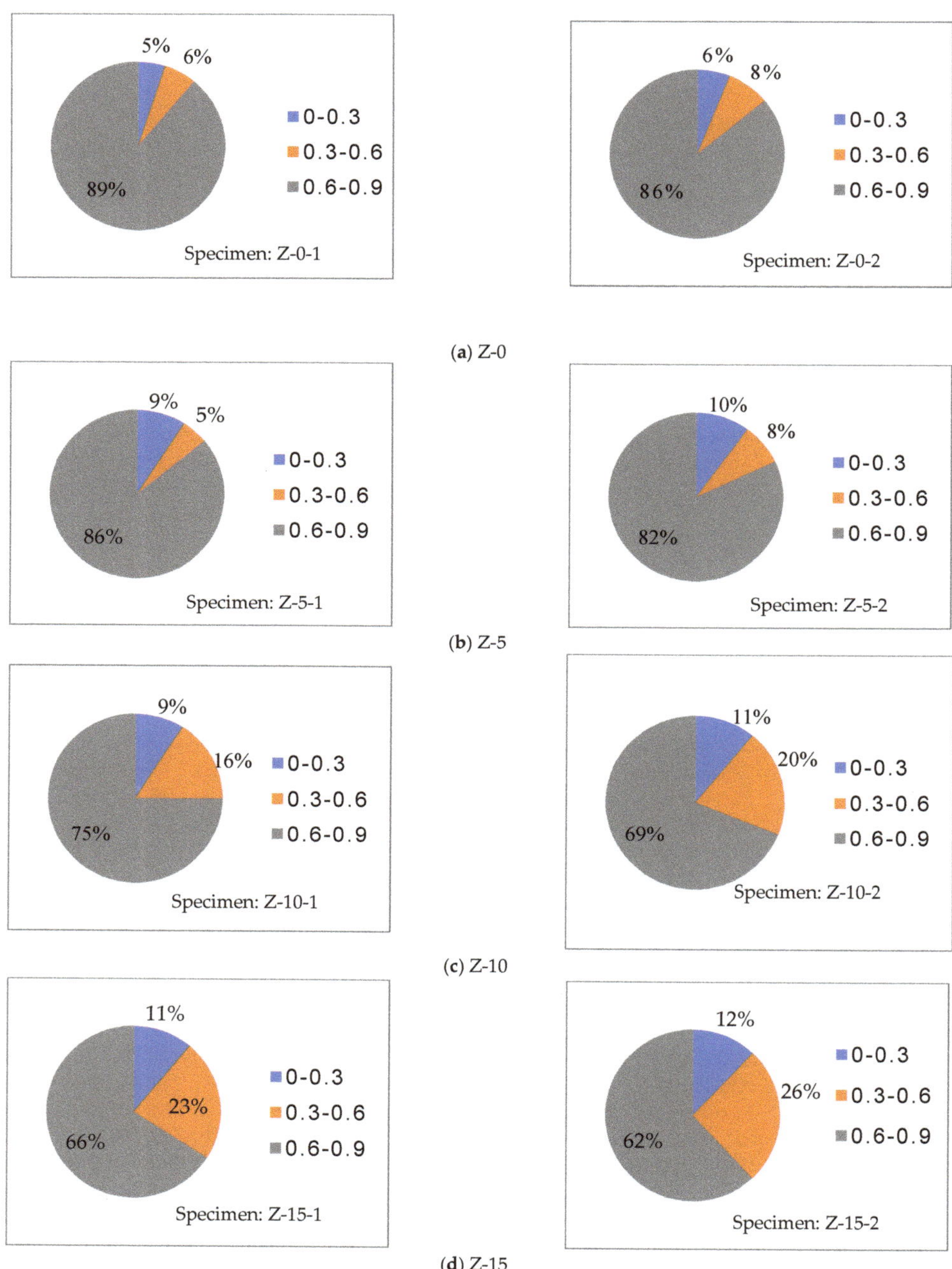

Figure 6. AE accumulative energy distribution of the reinforced concrete column for different corrosion rates (two specimens per corrosion rate).

4.3. Damage Evolution Model of Corroded RC

After taking the derivative of the curve shown in Figure 7a, the curve of slope change could be obtained, shown in Figure 7b. It can be seen that as the corrosion rate increases, at the load level of 0.9, the curve form changes from exponential to linear. As shown in Figure 7b, the slope of the load over the accumulative hit curve increases linearly with the increase of the load level. According to the results of the experiment, the corrosion rate also has an effect on the initial slope. The initial slopes of Z-0, Z-5, Z-10 and Z-15 are 125, 500, 800 and 1700, respectively. It can be found that within a certain corrosion rate range, the higher the corrosion rate, the greater the initial slope of the AE accumulation signal can be achieved. Meanwhile, since the initial slope represents AE activity at the initial low load level, a higher corrosion rate refers to stronger AE activity of the RC columns during the initial loading period. Although there is little variation between the two specimens for each corrosion rate, trends among different corrosion rates are apparent.

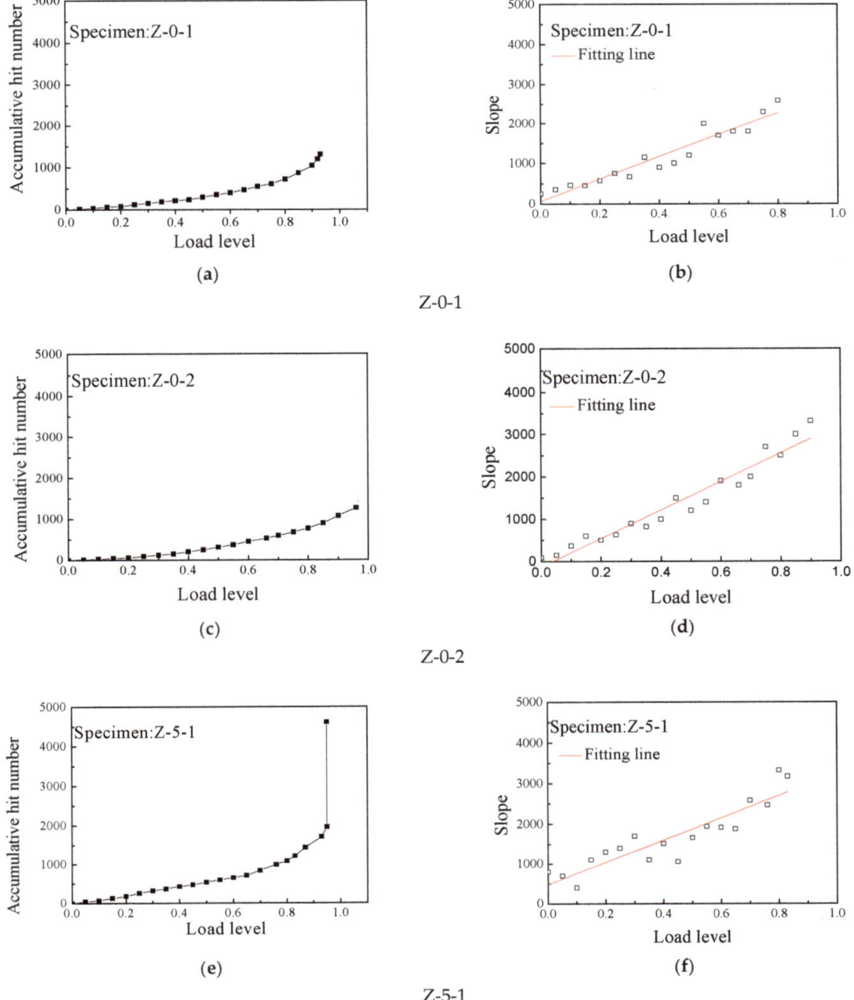

Figure 7. *Cont.*

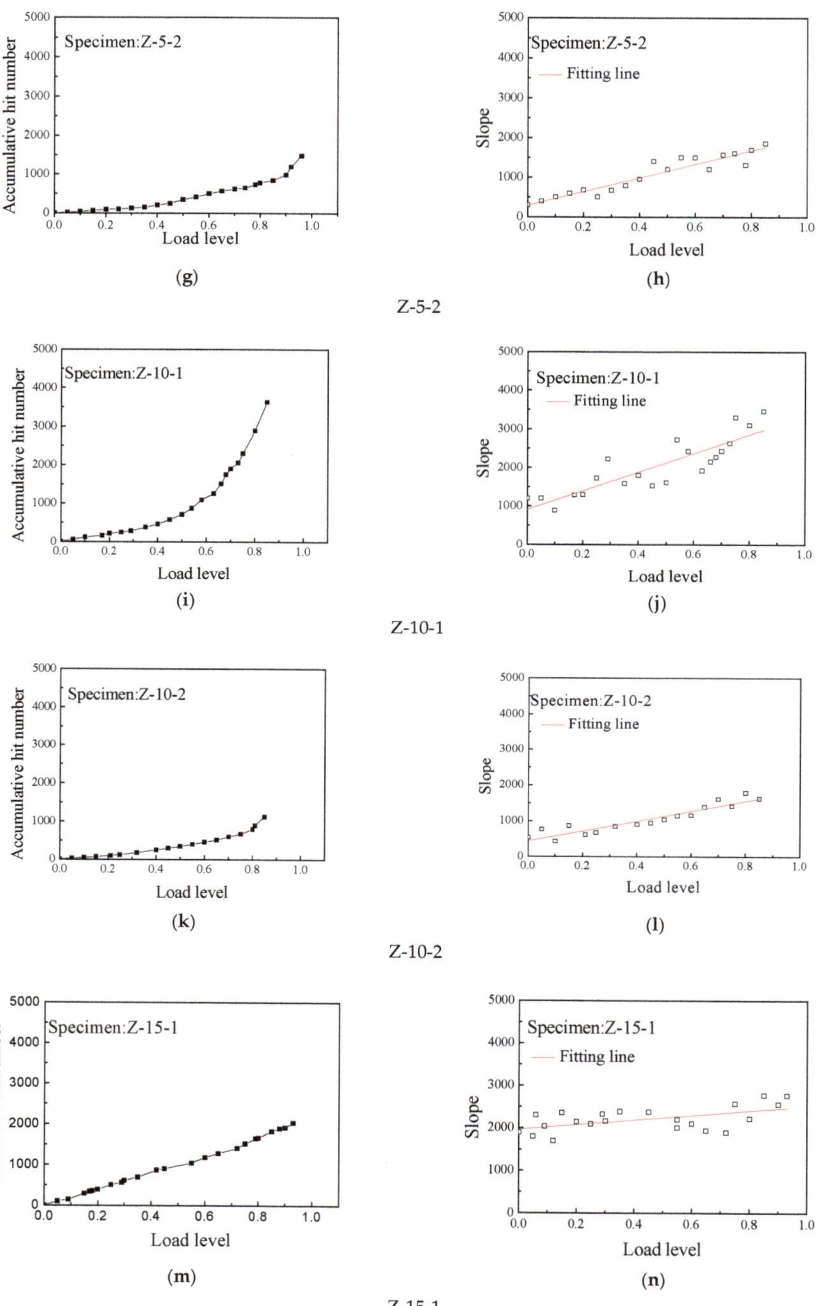

(g)

(h)

Z-5-2

(i)

(j)

Z-10-1

(k)

(l)

Z-10-2

(m)

(n)

Z-15-1

Figure 7. *Cont.*

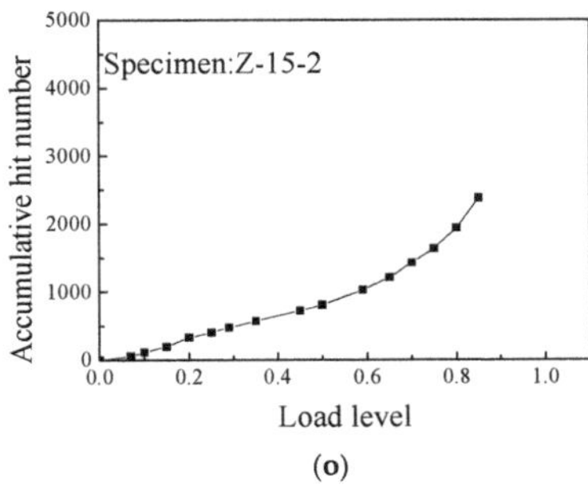
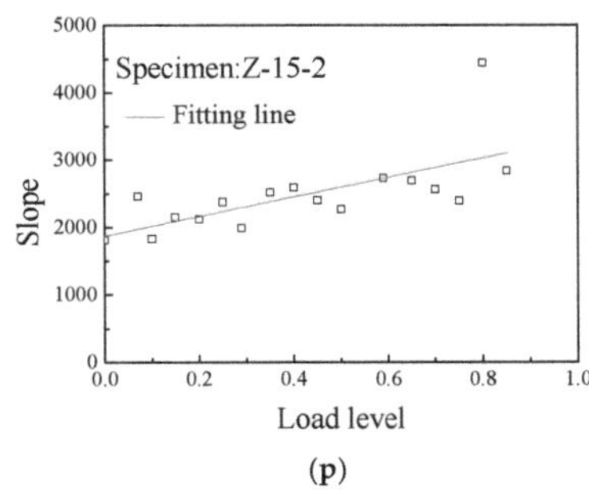

Figure 7. Load accumulative hit curve and slope curve for different corrosion rates (two specimens per corrosion rate).

Comparing the slope changes of each group, it can be seen that the slope of Z-0 increases from 125 to 2750. The slope of the Z-15 increases little from 1750 to 2800, but the growth rate remains at a high level. Therefore, it can be inferred that corrosion of the reinforcement can affect the whole service life of RC columns. When the corrosion rate is low or zero, the deterioration of internal damage is a dynamic process, from slow to fast. As a response, the AE signal changes from low to high. With the increase of the corrosion rate, the internal damage evolution of the RC column is accelerated, and the AE signal maintains a high level.

After considering the slope change trend, two slope relationship models are selected to fit the data [31]. The fitting results are shown in Table 2. Comparing these two models, the R^2 value of $N' = pe^{-V} + q$ is generally lower than that of $N' = mV + n$. Therefore, based on the characteristics of the above relationships, the binomial is adopted to establish the slope relation curve.

$$N' = mV + n \tag{10}$$

where N' is the slope of accumulative AE hit number, and m and n are the AE accumulative parameters, which can be obtained by fitting, as shown in Table 3.

Table 2. Fitting results of the slope.

Group Number	Fitting Equation	R^2
Z-0-1	$N' = mV + n$	0.91057
	$N' = pe^{-V} + q$	0.80362
Z-0-2	$N' = mV + n$	0.92083
	$N' = pe^{-V} + q$	0.84135
Z-5-1	$N' = mV + n$	0.88712
	$N' = pe^{-V} + q$	0.80169
Z-5-2	$N' = mV + n$	0.84131
	$N' = pe^{-V} + q$	0.75148
Z-10-1	$N' = mV + n$	0.77161
	$N' = pe^{-V} + q$	0.71694
Z-10-2	$N' = mV + n$	0.63221
	$N' = pe^{-V} + q$	0.55148
Z-15-1	$N' = mV + n$	0.60132
	$N' = pe^{-V} + q$	0.45985
Z-15-2	$N' = mV + n$	0.58443
	$N' = pe^{-V} + q$	0.40871

Table 3. AE accumulative parameters.

Parameter	Z-0	Z-0	Z-5	Z-5	Z-10	Z-10	Z-15	Z-15
m	2985	3324	1684	1935	1885	1382	683	1256
n	−120	24	256	504	832	450	1833	1836

The variation of the parameters with the corrosion rate is shown in Figure 8. As the corrosion increases, m decreases from 3500 to 540, and at the same time, n increases from −120 to 2000. This indicates that parameters m and n are related to the change of the corrosion rate, and their approximate linear function can be expressed as [34]:

$$m = 2931 - 134\rho \tag{11}$$

$$n = -234 + 118\rho \tag{12}$$

where ρ is the corrosion rate. By substituting m and n into Equation (10), N′ can be expressed as $N' = mV + n = 2931V - 234 - \rho(134V - 118)$.

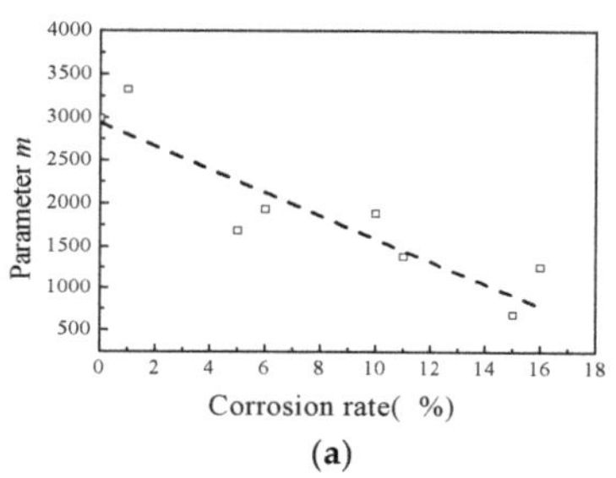

(a)

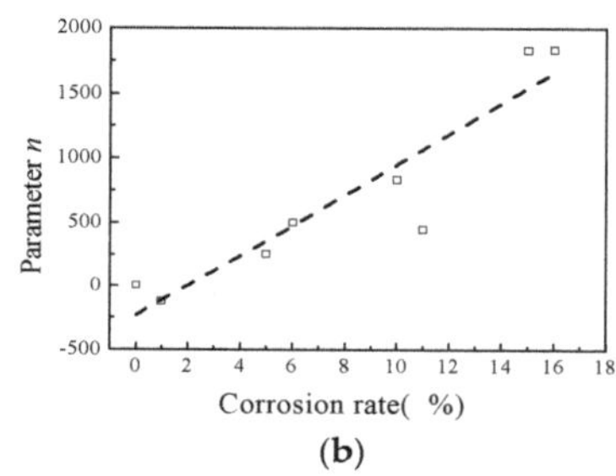

(b)

Figure 8. Parameter variation with the corrosion rate. (**a**) *m* variation with corrosion rate, (**b**) *n* variation with corrosion rate

By integrating Equation (13), the relationship between load and accumulative hit can be expressed as follows:

$$N = \frac{1}{2}mV^2 + nV = 1465V^2 - 234V - \rho(67V^2 - 118V) \tag{13}$$

where N is the accumulative AE hit number, V is the load level, ρ is the corrosion rate and m and n are the AE accumulative parameters. Combining Equations (14) and (2), the probability density function of the AE event can be obtained [35]:

$$f(V) = \frac{1}{N_0}\frac{ds(V)}{dV} = \frac{2931V - 234 - \rho(134V - 118)}{N_0} \tag{14}$$

The integral of Equation (15) is the damage evolution model of corroded RC:

$$D = \frac{1}{N_0}\int_0^V ds(V) = \frac{N}{N_0} \tag{15}$$

The relationship of the load and damage degree under different corrosion rates for the RC column is shown in Figure 9. It can be seen that corrosion of reinforcement has a great influence on the damage evolution of the RC column. At the beginning, at the lower load level, the damage evolution is very slow. As the load level increases, the internal damage increases until the late loading period. Compared with the uncorroded specimens at the lower load level, with the increase in the corrosion rate, the damage factor of specimens is significantly increased, which leads to great damage accumulation before initial loading.

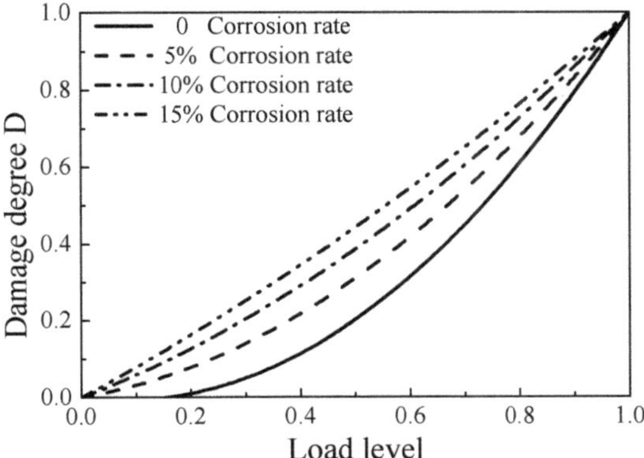

Figure 9. Correlation curve of the load and damage degree.

4.4. Initial Damage of Corroded RC

The number of AE events in different corrosion rates is shown in Figure 10. In order to reduce the effect of system noise and eliminate the instability of the failed concrete, the AE signal is collected under the load level of 0.1–0.9. Based on Ohtsu's model, the data at different corrosion rates are fitted. The fitting parameters a, b and c are shown in Figure 10.

It can be seen that the values of the uncorroded column are −0.15 and −0.86, which indicates that the AE rate of the specimens is very low at the lower load level. The initial microcracks of the two uncorroded columns are very few. The values of specimens at a 5% corrosion rate are −0.06 and 0.88. The AE rate increased with the increase of the corrosion rate, indicating that the number of microcracks also increased. The values of a for specimens at a 10% and 15% corrosion rate are both greater than 0, and the microcracks propagated. Therefore, comparing the results of the AE rate process theory theoretically proves that the different reinforcement corrosion degrees cause different initial damage to RC columns before axial compression. As a result, the diversity in the internal damage evolution of the corroded column obtained by Equation (10) is verified.

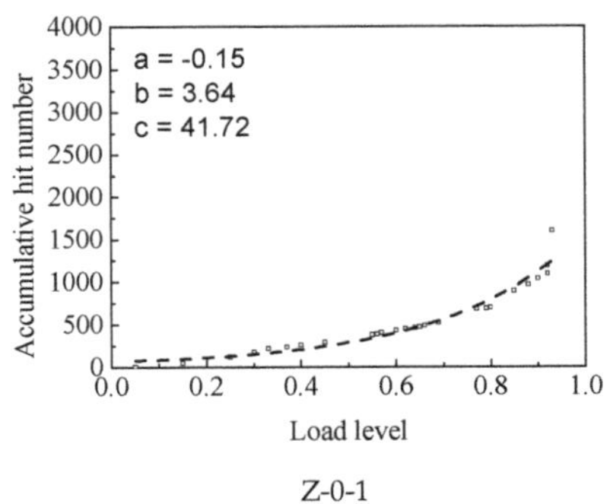

Z-0-1

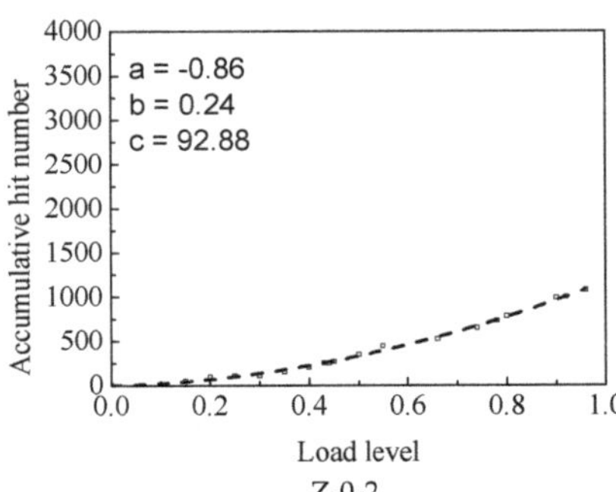

Z-0-2

Figure 10. *Cont.*

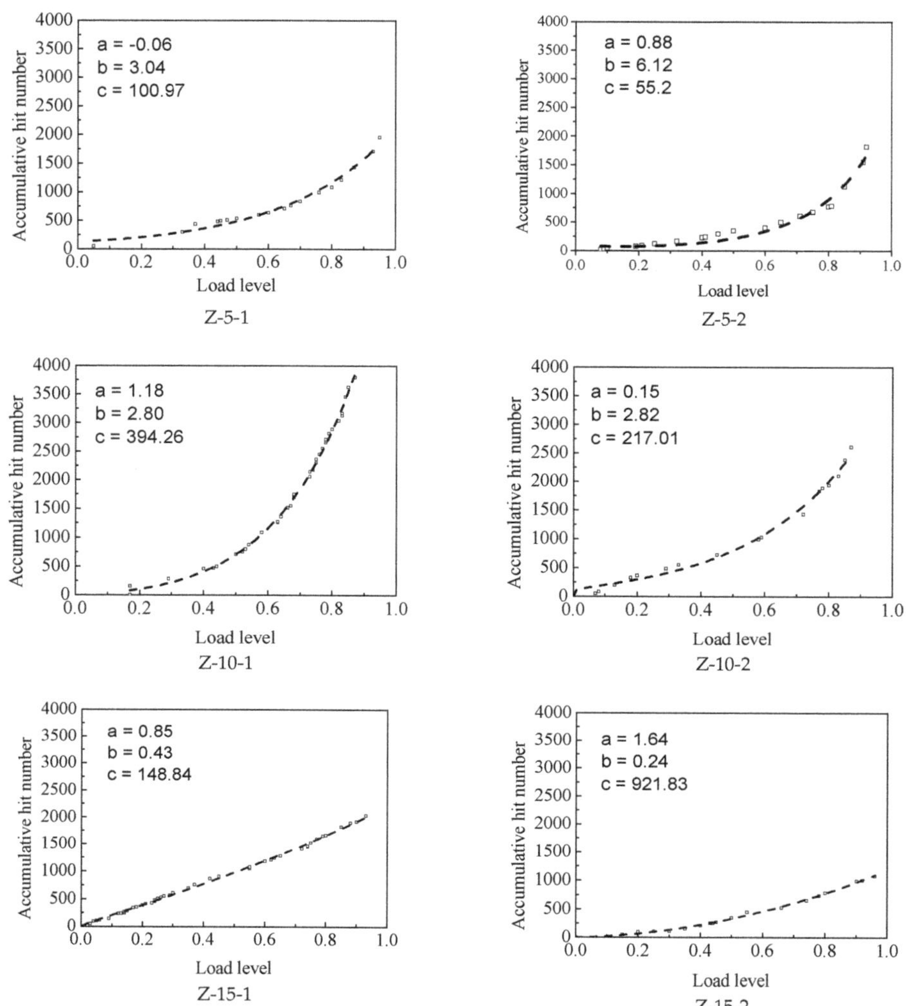

Figure 10. AE events in different corrosion rates and N–V fitting curves.

5. Conclusions

Based on the experimental investigation, the following findings can be drawn from this study:

1. The corrosion degree of reinforcement in RC can be monitored by the AE signal in the whole loading process. The greater corrosion degree causes a stronger AE signal in the middle loading period, which means the initial damage of the RC columns is accelerated in this period.
2. The corrosion rate can be used as a parameter in the V–N model. After regression comparison, the developed model has a high agreement with the actual data.
3. Based on the AE rate process theory and the influence of the corrosion rate, the damage evolution model of the corroded RC column with different corrosion rates is established. It can be used to monitor the internal damage of RC columns at different corrosion rates and quantitatively evaluate the damage degree of the loading process.

Author Contributions: Conceptualization, Y.C. and S.Z.; investigation, Y.C.; data curation, C.F., Y.L. and S.Y.; methodology, X.J.; visualization, Y.L.; formal analysis, D.W.; software, C.F.; writing—original draft, Y.C. and S.Z.; writing—review and editing, Y.L., G.Z., N.D., X.J. and S.Z.; project administration, Y.C.; funding acquisition, C.F.; validation, X.J., Y.L. and G.Z.; resources, Y.C. All authors have read and agreed to the published version of the manuscript.

Funding: The financial support from the National Key R&D Program of China (No. SQ2019YFB160077), and the Natural Science Foundation of Zhejiang Province (Grant No. LR21E080002, LZ20E080003), and the National Natural Science Foundation (Grant Nos. 51678529 and 51978620) are gratefully acknowledged.

Conflicts of Interest: The authors declare no conflict of interest.

References

1. Jiang, G.; Keller, J.; Bond, P.L.; Yuan, Z. Predicting concrete corrosion of sewers using artificial neural network. *Water Res.* **2016**, *92*, 52–60. [CrossRef] [PubMed]
2. Fu, C.; Jin, X.; Ye, H.; Jin, N. Theoretical and experimental investigation of loading effects on chloride diffusion in saturated concrete. *J. Adv. Concr. Technol.* **2015**. [CrossRef]
3. Li, B.; Cai, L.; Zhu, W. Predicting Service Life of Concrete Structure Exposed to Sulfuric Acid Environment by Grey System Theory. *Int. J. Civ. Eng.* **2018**, *16*, 1017–1027. [CrossRef]
4. Fu, C.; Ling, Y.; Wang, K. An innovation study on chloride and oxygen diffusions in simulated interfacial transition zone of cementitious material. *Cem. Concr. Compos.* **2020**. [CrossRef]
5. Huo, L.; Li, X.; Chen, D.; Li, H. Structural health monitoring using piezoceramic transducers as strain gauges and acoustic emission sensors simultaneously. *Comput. Concr.* **2017**, *20*, 595–603. [CrossRef]
6. Mather, B. Concrete durability. *Cem. Concr. Compos.* **2004**, *26*, 3–4. [CrossRef]
7. Zohari, M.H.; Epaarachchi, J.A.; Lau, K.T. Modal Acoustic Emission investigation for progressive failure monitoring in thin composite plates under tensile test. In *Key Engineering Materials*; Trans Tech Publications Ltd.: Baech, Switzerland, 2013; Volume 558, pp. 65–75. [CrossRef]
8. Aggelis, D.G.; Shiotani, T.; Momoki, S.; Hirama, A. Acoustic emission and ultrasound for damage characterization of concrete elements. *ACI Mater. J.* **2009**, *106*, 509–514. [CrossRef]
9. Iturrioz, I.; Lacidogna, G.; Carpinteri, A. Acoustic emission detection in concrete specimens: Experimental analysis and lattice model simulations. *Int. J. Damage Mech.* **2014**, 327–358. [CrossRef]
10. Nair, A.; Cai, C.S. Acoustic emission monitoring of bridges: Review and case studies. *Eng. Struct.* **2010**, *32*, 1704–1714. [CrossRef]
11. Abouhussien, A.A.; Hassan, A.A.A. Evaluation of damage progression in concrete structures due to reinforcing steel corrosion using acoustic emission monitoring. *J. Civ. Struct. Heal. Monit.* **2015**, *5*, 751–765. [CrossRef]
12. Vidya Sagar, R.; Raghu Prasad, B.K. A review of recent developments in parametric based acoustic emission techniques applied to concrete structures. *Nondestruct. Test. Eval.* **2012**, *27*, 47–68. [CrossRef]
13. Desa, M.S.M.; Ibrahim, M.H.W.; Shahidan, S.; Ghadzali, N.S.; Misri, Z. Fundamental and assessment of concrete structure monitoring by using acoustic emission technique testing: A review. In Proceedings of the 4th International Conference on Civil and Environmental Engineering for Sustainability (IConCEES 2017), Langkawi, Malaysia, 4–5 December 2017; Volume 140. [CrossRef]
14. Asamene, K.; Hudson, L.; Sundaresan, M. Influence of attenuation on acoustic emission signals in carbon fiber reinforced polymer panels. *Ultrasonics* **2015**, *59*, 86–93. [CrossRef] [PubMed]
15. Mainali, G.; Dineva, S.; Nordlund, E. Experimental study on debonding of shotcrete with acoustic emission during freezing and thawing cycle. *Cold Reg. Sci. Technol.* **2015**, *111*, 1–12. [CrossRef]
16. Masayasu, O.; Watanabe, H. Construction and Building Materials -Quantitative damage estimation of concrete by acoustic emission. *Constr. Build. Mater.* **2001**, *15*, 217–224. [CrossRef]
17. Ji, H.; Zhang, T.; Cai, M.; Zhang, Z. Experimental study on concrete damage by dynamic asurement of acoustic emission. *Chin. J. Rock Mech. Eng.* **2000**, *19*, 165–168.
18. Suzuki, T.; Ohtsu, M.; Shigeishi, M. Relative damage evaluation of concrete in a road bridge by AE rate-process analysis. *Mater. Struct. Constr.* **2007**, *40*, 221–227. [CrossRef]
19. Suzuki, T.; Ogata, H.; Takada, R.; Aoki, M.; Ohtsu, M. Use of acoustic emission and X-ray computed tomography for damage evaluation of freeze-thawed concrete. *Constr. Build. Mater.* **2010**, *24*, 2347–2352. [CrossRef]
20. Zhou, X.; Yang, Y.; Li, X.; Zhao, G. Acoustic emission characterization of the fracture process in steel fiber reinforced concrete. *Comput. Concr.* **2016**, *18*, 923–936. [CrossRef]
21. Deng, Y.; Ding, Y.; Li, A. Experimental study on damage evolution of steel strands based on acoustic emission signals and rate process theory. *J. Southeast Univ. (Nat. Sci. Ed.)* **2010**, *40*, 1238–1242. [CrossRef]
22. Li, X.; Huo, L.S.; Li, H.N. AE monitoring for the loading test of reinforced concrete column. *J. Vib. Shock* **2014**, *33*, 12–15. [CrossRef]
23. Li, X.W. Application of working face rock burst prediction of grey modeling cusp catastrophe analysis based on the acoustic emission. *Appl. Mech. Mater.* **2013**, *373–375*, 689–693. [CrossRef]

24. Qiang, L.I.; Zhonggou, C.H.E.N.; Dan, W.U. Temporal and Spatial Evolution of Acoustic Emission in the Damage Process of Reinforced Concrete Column. *Appl. Mech. Mater.* **2014**, *638–640*, 275–278. [CrossRef]

25. Dai, S.T.; Labuz, J.F. Damage and failure analysis of brittle materials by acoustic emission. *J. Mater. Civ. Eng.* **1997**, *9*, 200–205. [CrossRef]

26. Barrios, F.; Ziehl, P.H. Cyclic load testing for integrity evaluation of prestressed concrete girders. *ACI Struct. J.* **2012**, *109*, 615–624. [CrossRef]

27. Wiedmann, A.; Weise, F.; Kotan, E.; Müller, H.S.; Meng, B. Effects of fatigue loading and alkali–silica reaction on the mechanical behavior of pavement concrete. *Struct. Concr.* **2017**, *18*, 539–549. [CrossRef]

28. *ASTM C150/C150M: Standard Specification for Portland Cement*; ASTM International: West Conshohocken, PA, USA, 2020.

29. Ohno, K.; Ohtsu, M. Crack classification in concrete based on acoustic emission. *Constr. Build. Mater.* **2010**, *24*, 2339–2346. [CrossRef]

30. *GB/T50152-2012: Standard for Test Method of Concrete Structures*; TransForyou Co., Ltd.: Beijing, China, 2012.

31. Rodríguez, P.; Celestino, T.B. Application of acoustic emission monitoring and signal analysis to the qualitative and quantitative characterization of the fracturing process in rocks. *Eng. Fract. Mech.* **2019**, *210*, 54–69. [CrossRef]

32. Mostafapour, A.; Davoodi, S.; Ghareaghaji, M. Acoustic emission source location in plates using wavelet analysis and cross time frequency spectrum. *Ultrasonics* **2014**, *54*, 2055–2062. [CrossRef]

33. Fan, X.; Hu, S.; Lu, J. Damage and fracture processes of concrete using acoustic emission parameters. *Comput. Concr.* **2016**, *18*, 267–278. [CrossRef]

34. Abdelrahman, M.A.; ElBatanouny, M.K.; Rose, J.R.; Ziehl, P.H. Signal processing techniques for filtering acoustic emission data in prestressed concrete. *Res. Nondestruct. Eval.* **2019**, *30*, 127–148. [CrossRef]

35. Liu, X.; Zhang, W.; Gu, X.; Ye, Z. Probability distribution model of stress impact factor for corrosion pits of high-strength prestressing wires. *Eng. Struct.* **2021**, *230*, 111686. [CrossRef]

 *crystals*

Article

Energy Evolution Analysis and Brittleness Evaluation of High-Strength Concrete Considering the Whole Failure Process

Ruihe Zhou [1],***, Hua Cheng** [1,2]**, Mingjing Li** [1]**, Liangliang Zhang** [1] **and Rongbao Hong** [1]

1   School of Civil Engineering and Architecture, Anhui University of Science and Technology, Huainan 232001, China; hcheng@aust.edu.cn (H.C.); lmj201007@126.com (M.L.); zllaust@163.com (L.Z.); cherishrb2020@163.com (R.H.)
2   School of Resources and Environmental Engineering, Anhui University, Hefei 230022, China
*   Correspondence: zrhaust@163.com; Tel.: +86-187-5602-3671

Received: 23 October 2020; Accepted: 27 November 2020; Published: 30 November 2020

Abstract: In this work, we aimed to solve the problems that exist in the brittleness evaluation method of high-strength concrete through a triaxial compression test of C60 and C70 high-strength concrete. Then, the relationship between the energy evolution of its elastic energy, dissipative energy, pre-peak total energy and additional energy and its axial strain, confining pressure, and concrete strength grade was analyzed. Taking the accumulation rate of pre-peak elastic strain energy and the dissipation rate of dissipative energy, and the release rate of post-peak elastic energy, as the evaluation indicators to characterize the brittleness of high-strength concrete. A brittleness evaluation method that reflects the whole failure process of high-strength concrete is proposed and verified by experiments. The results show that with the increase of the confining pressure, the proportion of elastic energy in the whole process of high-strength concrete failure gradually decreases. The storage rate of pre-peak elastic energy and the release rate of post-peak elastic energy are gradually reducing, the brittleness index gradually decreases, and the confining pressure inhibits the brittleness of high-strength concrete. Under the same confining pressure, the brittleness index of C70 concrete is greater than that of C60 concrete, which indicates that, with the increase of the strength grade, the brittleness level of concrete gradually increases and the ductility decreases. These findings have a certain theoretical significance for the scientific design of high-strength concrete structures and the improvement of their safety in the future.

Keywords: high-strength concrete; energy evolution; elastic strain energy; brittleness evaluation index

1. Introduction

In recent years, high-strength concrete has been widely used in civil engineering, transportation, water conservancy, and municipal engineering. Concrete is a quasi-brittle material, and its failure mode is not only affected by the strength level of concrete, but is also closely related to its stress state [1–5]. Studies have shown that, compared with ordinary concrete, high-strength concrete presents obvious brittle characteristics [6–8].

To date, domestic and foreign scholars have carried out a great deal of research on the characterization methods of concrete brittleness. Jenq and Shah [9] proposed a critical material constant Q that was positively related to the brittleness index based on the fracture toughness and elastic modulus, showing that the larger the value of Q, the greater the brittleness of the material. Tasdemir et al. [10] used the energy method to evaluate the brittleness of concrete and proposed the ratio of the recoverable elastic deformation energy to unrecoverable plastic deformation energy as the brittleness evaluation index B_1 of concrete. Zhang et al. [11] proposed to use the ratio of the pre-peak recoverable elastic energy to the total energy as the brittleness evaluation index B_2 of concrete. Yan et al. [12] used mechanical parameters and area methods to evaluate the brittleness of concrete and proposed that the ratio of the fracture energy to nominal stress σ_N should be taken as the brittleness index B_3; furthermore, they established the brittleness evaluation index B_4 based on the roughness of the fracture surface [13]. Their results showed that these two brittleness indicators were linearly correlated, and it was considered that the greater the roughness of the fracture surface, the greater the unrecoverable deformation energy consumed by fracture failure and the lower the brittleness of concrete. Yao et al. [14] used the strength method to evaluate the brittleness of concrete and proposed that the tension–compression ratio of concrete specimens should be taken as their brittleness index B_5; furthermore, they considered that the smaller the tension–compression ratio, the greater the brittleness of concrete. Guo and Huang et al. [15–17] considered the size effect of strength and established a brittleness index η that was proportional to the size of the plastic zone; furthermore, they considered that the smaller the value of η, the greater the brittleness of the material. Han et al. [18] proposed the evaluation of the brittleness of concrete by a morphology method based on the analysis of the insufficient brittleness of the concrete cross-sectional area ratio and established a new brittleness evaluation method of concrete based on the area fraction of section stones. In summary, most of the existing concrete brittleness evaluation indicators are suitable for low-strength concrete, and there are few studies on the brittleness evaluation of high-strength concrete. The physical meaning of some concrete brittleness evaluation indicators is not clear, and the evaluation indicators cannot be obtained through conventional mechanical tests of concrete; thus, their practical application and promotion have certain limitations. Some concrete brittleness evaluation indicators only consider the pre-peak or post-peak stage and cannot fully reflect the brittleness characteristics of the entire failure process of concrete.

Based on the experimental results of C60 and C70 high-strength concrete under different confining pressures, the evolution of elastic strain energy, dissipative energy, and input energy with axial strain during the deformation process is analyzed in this paper. Combined with these three kinds of energy, a brittleness evaluation index which can comprehensively reflect the whole deformation and failure process of high-strength concrete is established. Furthermore, a brittleness evaluation method for high-strength concrete is proposed. This has a certain theoretical significance for the scientific design of high-strength concrete structures and the improvement of their safety in the future.

2. Triaxial Compression Test of C60 and C70 High-Strength Concrete

2.1. Experimental Process

According to the "Specification for Concrete Mix Design" (JGJ55-2011) [19], high-strength concrete with strength grades of C60 and C70 was prepared. Cement used was ordinary Portland cement P II52.5 R, and the coarse aggregate used was basalt gravel. The size of the crushed stone was 5–20 mm and the bulk density was 1455 kg/m^3. The fine aggregate adopts medium-coarse sand with a fineness modulus of 2.73 and a mud content of 1.45%. The admixture was an NF–F compound high-efficiency admixture, and the proportions of slag and silica fume in C60 and C70 were the same, 73% and 20%, respectively. The mix proportion of C60 and C70 high-strength concrete was determined by an orthogonal experimental design (see Table 1).

Table 1. Mixture proportions of C60 and C70 high-strength concrete.

Strength Grade	Cement (kg/m^3)	Admixture (kg/m^3)	Cementitious Materials (kg/m^3)	Sand (kg/m^3)	Gravel (kg/m^3)	Water (kg/m^3)	Sand Rate	Dosage of Admixture
C60	415	135	550	620	1105	175	0.36	32.5%
C70	425	145	570	615	1095	170	0.36	32.5%

The conventional triaxial loading equipment for high-strength concrete specimens was the ZTCR-2000 low-temperature rock triaxial system (Figure 1). During the test, the concrete specimens were first preloaded to 0.5 MPa, and then confining pressure was applied to the predetermined value at the loading speed of 50 N/s according to the "load" control mode. The test confining pressure was set at 0, 5, 10, 15, and 20 MPa. After the confining pressure reached the predetermined value, the pressure was stabilized for 30 s, and finally, the axial pressure was applied at the loading speed of 0.05 mm/min according to the "displacement" control mode until the concrete specimens was destroyed.

Figure 1. Rock mechanics test system.

2.2. Analysis of Test Results

Figure 2 shows the stress–strain curves of C60 and C70 high-strength concrete specimens under different confining pressures. It can be seen from Figure 2 that, when the confining pressure is 0 MPa, there is no obvious yield point in the pre-peak curve of high-strength concrete; after reaching the peak strength, the stress drops rapidly, and the post-peak curve is steep, indicating that the brittleness of the concrete is relatively strong at this time and close to an ideal brittle material. With the increase of confining pressure, obvious yield steps appear in the pre-peak curve, and the post-peak curve and the stress drop trends gradually tend to be gentle. The post-peak phase gradually transforms from the strain softening state to the ideal plastic state, and its brittleness level gradually reduces, showing that, with the increase of confining pressure, the brittleness of high-strength concrete develops from high to low. This shows that there are obvious differences in the brittleness characteristics of concrete under different confining pressures.

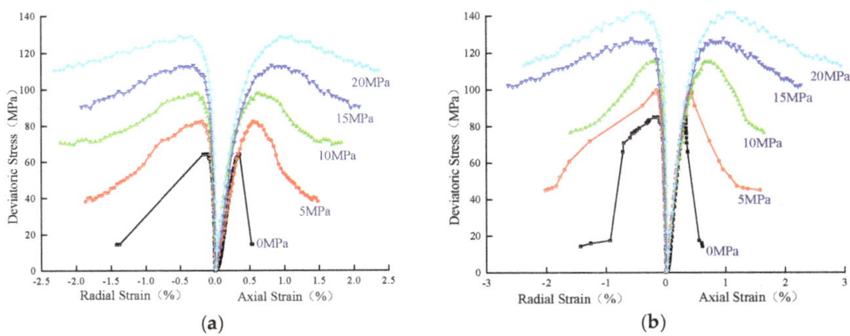

Figure 2. Full stress–strain curves of high-strength concrete under different confining pressures. (**a**) C60; (**b**) C70.

According to the full stress–strain curve of high-strength concrete, the basic mechanical parameters of C60 and C70 high-strength concrete are shown in Table 2.

Table 2. Mechanical parameters of C60 and C70 high-strength concrete.

σ_3 (MPa)	σ_{1p} (MPa)		σ_{1r} (MPa)		ε_{1p} (10^{-3})		ε_{1r} (10^{-3})		E (GPa)		μ	
	C60	C70	C60	C70	C60	C70	C60	C70	C60	C70	C60	C70
0	65.38	84.41	14.66	14.35	0.343	0.337	0.523	0.611	28	31	0.31	0.29
5	81.96	100.22	38.93	45.68	0.540	0.419	1.388	1.373	29	32	0.28	0.26
10	98.26	115.98	71.16	76.70	0.631	0.721	1.515	1.640	31	33	0.29	0.26
15	114.01	128.22	89.97	102.27	0.812	0.969	1.953	2.167	33	35	0.28	0.28
20	129.12	141.67	111.13	114.47	1.10	1.081	2.240	2.864	34	36	0.26	0.26

Figure 3 shows the variation of the concrete peak strain with confining pressure. It can be seen from Figure 3 that with the increase of confining pressure, the corresponding strain when the concrete stress reaches the peak strength gradually increases, indicating that it is more difficult for the concrete to enter the brittle failure state, and more energy is needed from the outside to assist the cracks inside the concrete to connect and penetrate to form a macroscopic fracture surface. Therefore, the peak strain reflects the difficulty of achieving brittle failure for high-strength concrete. The greater the peak strain, the higher the threshold of concrete brittleness, which indicates that the confining pressure can limit the brittleness of concrete.

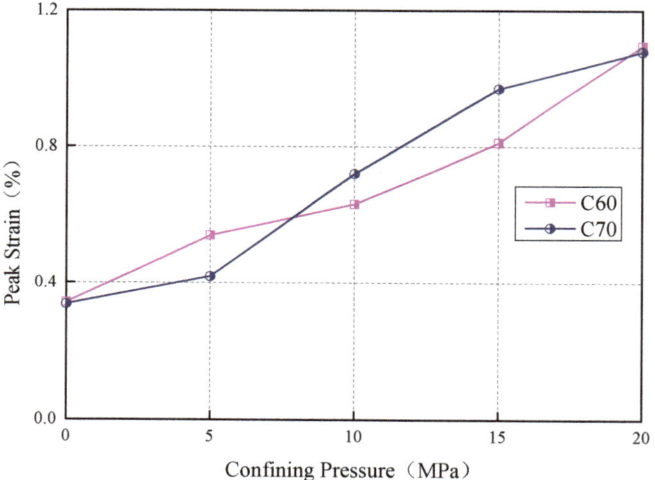

Figure 3. Variation of peak strain with confining pressure.

3. Energy Evolution Law of High-Strength Concrete during Compression

3.1. Theoretical Analysis

Assuming that there is no heat exchange between the concrete system and the external environment, according to the first law of thermodynamics, the energy W_F input to the system by the external force during the concrete deformation and failure process is equal to the elastic strain energy W_E plus the energy W_D dissipated by the concrete specimen during the test [20–22]. The energy input by the external force to the system mainly includes the work done by the axial force when the concrete specimen undergoes axial deformation and the work done by the confining pressure when radial deformation occurs. The dissipative energy is mainly used for the development of internal damage or plastic deformation in the concrete, including the surface energy consumed during the generation, the expansion and penetration of the cracks, the plastic strain energy of the irreversible plastic deformation of concrete specimens, the heat energy generated by friction and slipping between cracks, and various radiation energies, etc. [23]; the magnitude of the dissipative energy is mainly related to the structural properties of the concrete itself and the stress environment. The above-mentioned energy relationship is shown in Equation (1):

$$W_F = W_E + W_D \tag{1}$$

The work done by the axial force and confining pressure during the test is [24]:

$$W_F = \frac{\pi}{4}D^2H\left(\int_0^{\varepsilon_1} \sigma_1 d\varepsilon_1 + 2\int_0^{\varepsilon_3} \sigma_3 d\varepsilon_3\right) = VU_F \tag{2}$$

where σ_1, σ_3 are the axial pressure and confining pressure, respectively; ε_1, ε_3 are the axial and radial strain, respectively; D, H are the diameter and height of the concrete specimen, respectively; V is the volume of the concrete sample; and U_F is the input energy density.

In the same way, the elastic strain energy and dissipative energy are as follows:

$$\begin{cases} W_E = \frac{\pi}{4}D^2HU_E = VU_E \\ W_D = \frac{\pi}{4}D^2HU_D = VU_D \end{cases} \tag{3}$$

where U_E is the elastic strain energy density and U_D is the dissipative energy density.

According to the elastic theory [24], the elastic strain energy density is:

$$U_E = \frac{1}{2}\left(\sigma_1\varepsilon_1^e + \sigma_2\varepsilon_2^e + \sigma_3\varepsilon_3^e\right) \tag{4}$$

The three-dimensional constitutive relationship of concrete is:

$$\varepsilon_{ij}^e = \frac{1+\mu}{E_{ij}}\sigma_{ij} - \frac{\mu}{E_{ij}}\sigma_{kk}\delta_{ij} \tag{5}$$

where $\varepsilon_{ij}^e (i, j = 1, 2, 3)$ is the elastic strain in the direction of the main stress, $\sigma_{ij}(i, j = 1, 2, 3)$ is the main stress, $\sigma_{kk} = \sigma_1 + \sigma_2 + \sigma_3$; δ_{ij} is the Kronecker tensor and $E_{ij}(i, j = 1, 2, 3)$ is the unloading modulus of elasticity—for convenience of calculation, the initial elastic modulus E can be used instead [24]—and μ is the Poisson's ratio.

Equations (4) and (5) can be substituted into Equation (3) to obtain the elastic strain energy W_E:

$$W_E = \frac{1}{2E}\left(\sigma_1^2 + \sigma_2^2 + \sigma_3^2 - 2\mu\sigma_1\sigma_2 - 2\mu\sigma_1\sigma_3 - 2\mu\sigma_2\sigma_3\right)V \tag{6}$$

In the conventional triaxial compression test, where $\sigma_2 = \sigma_3$, Equation (6) can be simplified to:

$$W_E = \frac{1}{2E}\left[\sigma_1^2 + 2(1-\mu)\sigma_3^2 - 4\mu\sigma_1\sigma_3\right]V \tag{7}$$

Substituting Equation (7) into Equation (1) and combining this with Equation (2) to obtain the dissipative energy results in the following:

$$W_D = \left\{\int_0^{\varepsilon_1} \sigma_1 d\varepsilon_1 + 2\int_0^{\varepsilon_3} \sigma_3 d\varepsilon_3 - \frac{1}{2E}\left[\sigma_1^2 + 2(1-\mu)\sigma_3^2 - 4\mu\sigma_1\sigma_3\right]\right\}V \tag{8}$$

The energy conversion diagram of the concrete failure process is shown in Figure 4. When the load reaches the yield stress σ_{1c}, the elastic strain energy accumulated inside the concrete is $W_{E(A)}$, showing that the concrete is in the linear elastic stage, without damage and energy dissipation; as the load increases, when the peak strength σ_{1p} is reached, the total elastic strain energy accumulated in the pre-peak stage is $W_{E(B)}$. One part of the work $W_{F(pre)}$ done by the external testing machine is converted into dissipative energy W_D, which is irreversible energy that is consumed in the plastic yield stage of concrete, and the other part is transformed into stored elastic strain energy, which is reversible. The energy relationship in the pre-peak stage is as follows:

$$W_{F(pre)} = W_{E(B)} + W_D \tag{9}$$

$$W_{F(pre)} = \left(\int_0^{\varepsilon_{1p}} \sigma_1 d\varepsilon_1 + 2 \int_0^{\varepsilon_{3p}} \sigma_3 d\varepsilon_3 \right) V = \left(\int_0^{\varepsilon_{1p}} \sigma_1 d\varepsilon_1 + 2\sigma_3 \varepsilon_{3p} \right) V \tag{10}$$

$$W_{E(B)} = \frac{1}{2E} \left[\sigma_{1p}^2 + 2(1-\mu)\sigma_3^2 - 4\mu\sigma_{1p}\sigma_3 \right] V \tag{11}$$

$$W_D = \left(\int_0^{\varepsilon_{1p}} \sigma_1 d\varepsilon_1 + 2\sigma_3 \varepsilon_{3p} \right) V - \frac{1}{2E} \left[\sigma_{1p}^2 + 2(1-\mu)\sigma_3^2 - 4\mu\sigma_{1p}\sigma_3 \right] V \tag{12}$$

where $W_{F(pre)}$ is the area $S_1 + S_2$ enclosed by the stress–strain curve and the axial strain axis from the initial loading point to the peak strength, $W_{E(B)}$ is the area S_2 enclosed by the peak point unloading curve and the axial strain axis, and W_D is the area S_1 between the loading curve and the unloading curve.

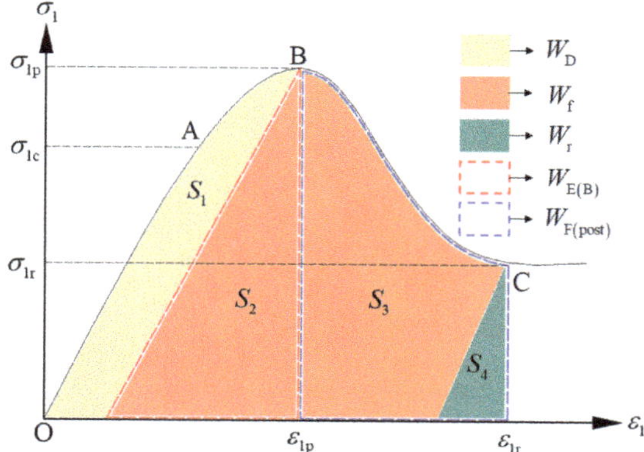

Figure 4. Schematic diagram of energy conversion during concrete failure.

When the concrete stress reaches the peak strength, the specimen enters the failure state, and its internal storage of elastic strain energy $W_{E(B)}$ is not sufficient to support its complete failure: therefore, external work is required to provide additional energy $W_{F(post)}$. When the concrete stress reaches the residual strength, the internal storage of elastic strain energy $W_{E(B)}$ and the external additional energy $W_{F(post)}$ are converted into the energy W_f required for the failure of concrete, and at this time, a part of the elastic strain energy W_r will remain inside the concrete specimen. The energy relationship in the post-peak stage is as follows:

$$W_f = W_{F(post)} + W_{E(B)} - W_r \tag{13}$$

$$W_{F(post)} = \left(\int_{\varepsilon_{1p}}^{\varepsilon_{1r}} \sigma_1 d\varepsilon_1 + 2 \int_{\varepsilon_{3p}}^{\varepsilon_{3r}} \sigma_3 d\varepsilon_3 \right) V = \left[\int_{\varepsilon_{1p}}^{\varepsilon_{1r}} \sigma_1 d\varepsilon_1 + 2\sigma_3 \left(\varepsilon_{3r} - \varepsilon_{3p} \right) \right] V \tag{14}$$

$$W_r = \frac{1}{2E} \left[\sigma_{1r}^2 + 2(1-\mu)\sigma_3^2 - 4\mu\sigma_{1r}\sigma_3 \right] V \tag{15}$$

$$W_f = \left[\int_{\varepsilon_{1p}}^{\varepsilon_{1r}} \sigma_1 d\varepsilon_1 + 2\sigma_3 \left(\varepsilon_{3r} - \varepsilon_{3p} \right) \right] V + \frac{1}{2E} \left[\sigma_{1p}^2 - \sigma_{1r}^2 - 4\mu\left(\sigma_{1p} - \sigma_{1r} \right)\sigma_3 \right] V \tag{16}$$

where $W_{\text{F(post)}}$ is the area S_3 enclosed by the stress–strain curve between the peak strength and the residual strength and the axial strain axis, W_r is the area S_4 enclosed by the unloading curve of the residual strength point and the axial strain axis, and W_f is the area shown by the orange area in the figure.

3.2. Relationship between Energy Evolution and Axial Strain

By substituting the triaxial compression test results into Equations (2), (7), and (8), the evolution curves of the input energy, elastic energy and dissipative energy with the axial strain of high-strength concrete under different confining pressures are obtained, as shown in Figures 5 and 6.

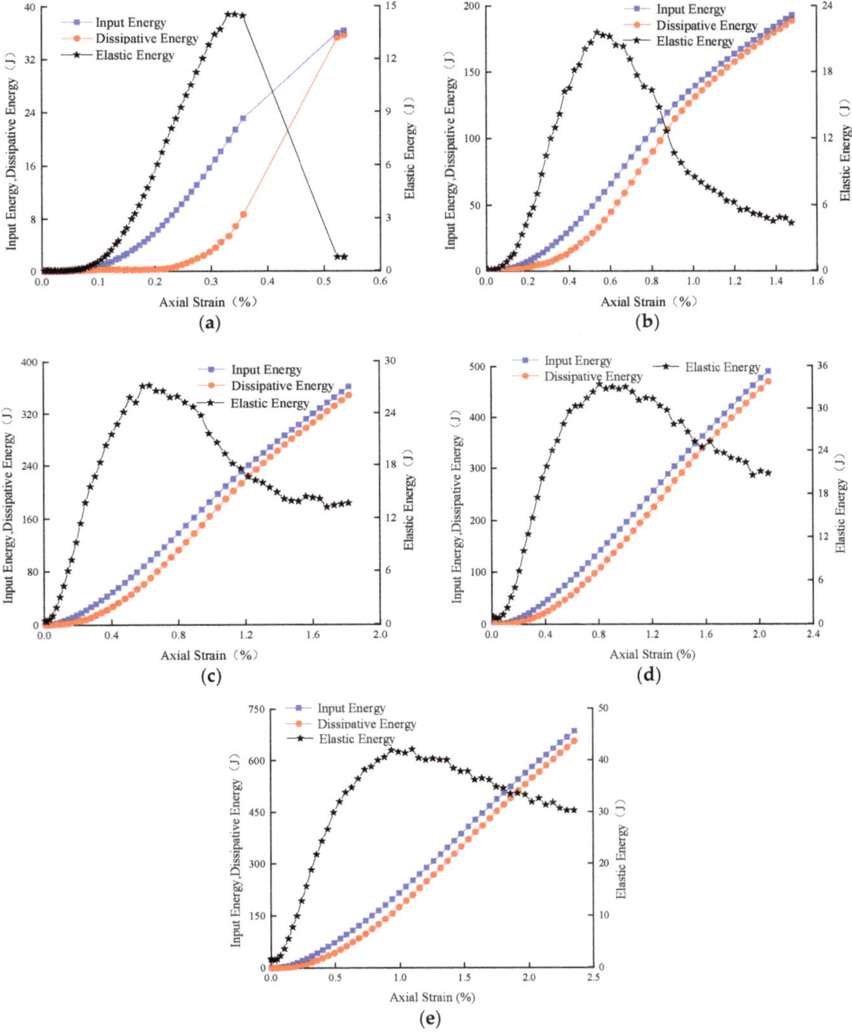

Figure 5. Energy evolution curves of C60 high-strength concrete under different confining pressures. (**a**) 0 MPa; (**b**) 5 MPa; (**c**) 10 MPa; (**d**) 15 MPa; (**e**) 20 MPa.

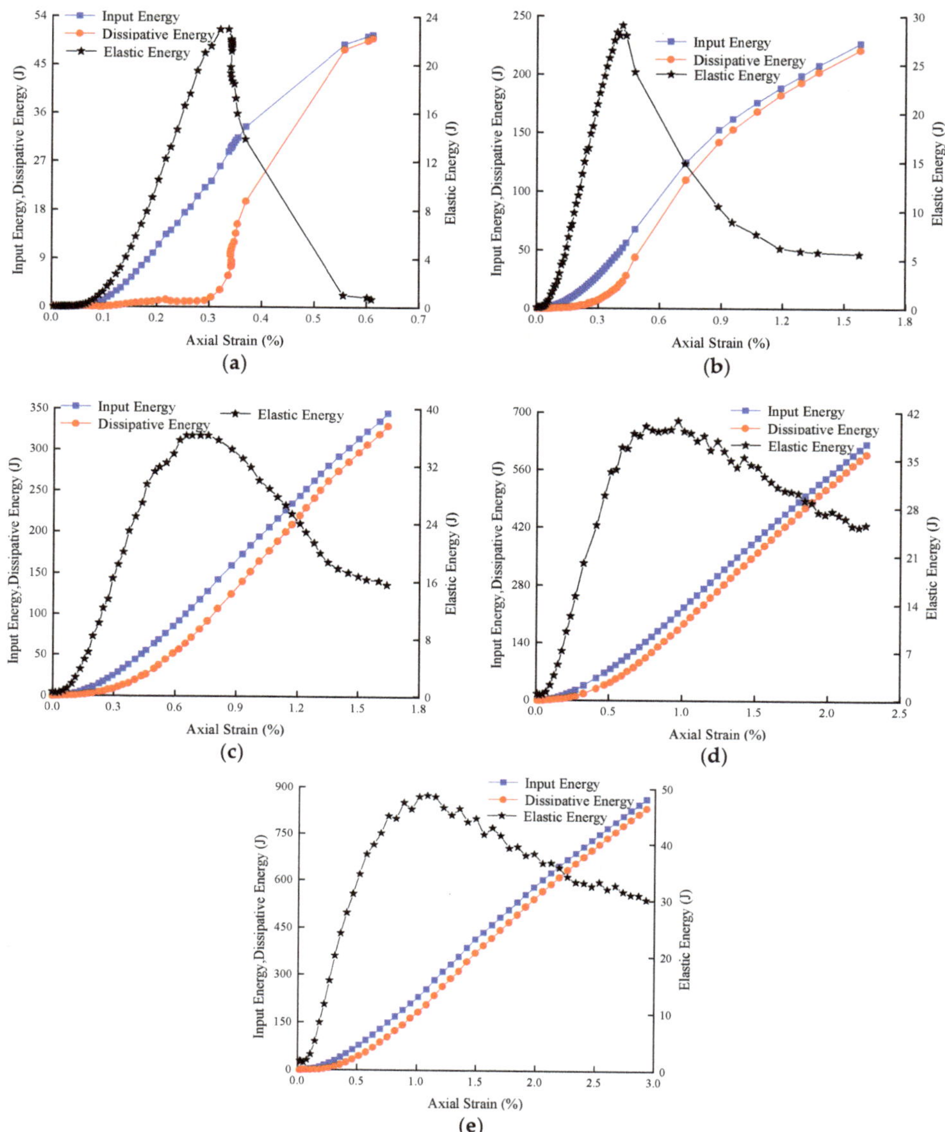

Figure 6. Energy evolution curves of C70 high-strength concrete under different confining pressures. (**a**) 0 MPa; (**b**) 5 MPa; (**c**) 10 MPa; (**d**) 15 MPa; (**e**) 20 MPa.

It can be seen from Figures 5 and 6 that under different confining pressures, both the input energy and the dissipative energy increase with the increase of the axial strain, and the elastic strain energy first increases and then decreases. In the initial stage of loading, the initial pores inside the concrete gradually close under the action of external load, and most of the work done by external load is converted into elastic energy, which is stored inside the specimen. The elastic strain energy gradually increases with the axial strain, the dissipative energy is very low at this stage and the evolution curves of input energy, elastic energy and dissipative energy all show an upward concave trend. As the external load increases, the concrete specimen enters the linear elastic deformation stage, and the external work is basically transformed into elastic strain energy; furthermore, the slopes of the three energy evolution curves reach the maximum value. This stage is the main stage of energy storage in the whole process of concrete failure. When the external load reaches the yield limit of concrete, new cracks will appear inside the specimen, and the initiation and diffusion of cracks need to dissipate part of energy; thus, the slope of the elastic strain energy curve gradually decreases at this stage. When the external load reaches the peak strength, the elastic strain energy reaches the maximum value, and then the concrete specimen is destroyed and the stored elastic strain energy in the pre-peak stage is released rapidly; thus, the post-peak elastic strain energy gradually decreases with the axial strain, and most of the external input energy is dissipated by the cracks intersecting each other to form a macroscopic fracture surface. When the specimen reaches the peak strength, the elastic strain energy and dissipative energy gradually decrease and increase, respectively, until the specimen failure reaches the maximum value and minimum value.

3.3. Relationship between Energy Evolution and Confining Pressure

Based on the triaxial compression test data of high-strength concrete under different confining pressures, the characteristic energy of high-strength concrete is calculated by substituting Equations (9)–(16), and the results are shown in Table 3. It can be seen from Table 3 that when the confining pressure is 0 MPa, the additional energy $W_{F(post)}$ provided from the outside for the post-peak fracture of C60 and C70 concrete specimens is 14.65 J and 21.83 J, respectively, accounting for 52% and 49% of the fracture energy, respectively, which indicates that the elastic energy stored before the peak is the source power of the specimen failure. When the confining pressure is 20 MPa, the additional energy $W_{F(post)}$ required for the post-peak fracture of C60 and C70 concrete specimens is 400.11 J and 593.85 J, respectively, accounting for 97% of the fracture energy, indicating that the energy required for the failure of C60 and C70 concrete specimens at this time is mainly provided by external work, that the self-sustaining fracture ability of the specimens is poor and that the brittleness level is low.

Table 3. Calculation results of the characteristic energy of C60 and C70 high-strength concrete.

σ_3 (MPa)	$W_{E(B)}$ (J)		$W_{F(pre)}$ (J)		W_D (J)		$W_{F(post)}$ (J)		W_f (J)		W_r (J)	
	C60	C70	C60	C70	C60	C70	C60	C70	C60	C70	C60	C70
0	14.52	22.94	21.38	28.80	6.86	5.86	14.65	21.83	28.42	44.14	0.75	0.64
5	21.58	29.05	54.81	52.22	33.23	23.17	129.47	155.86	146.53	179.13	4.51	5.78
10	27.22	36.21	97.30	118.63	70.08	82.42	205.88	227.01	219.23	247.66	13.86	15.56
15	33.43	40.73	142.69	214.17	109.26	173.44	320.11	382.98	333.00	398.29	20.54	25.41
20	42.08	48.57	251.92	255.41	209.84	206.83	400.11	593.85	411.51	611.59	30.68	30.84

The evolution laws of elastic energy, dissipative energy, pre-peak total energy, additional energy, fracture energy, and residual elastic strain energy with confining pressure are shown in Figure 7. It can be seen from Figure 7 that, with the increase of confining pressure, the six kinds of energy increase at different rates, and the storage rate of elastic energy gradually increases, which indicates that the confining pressure has a more obvious limiting effect on crack propagation, thus keeping the concrete from reaching the failure state and finally reducing the brittleness of concrete. When the confining pressure is 0 MPa, the elastic strain energy stored inside the specimen is released suddenly at the peak value, which can make the specimen fully fracture, and the additional energy consumed is very small; at this time, the self-sustaining fracture ability of concrete is stronger and the brittleness is higher. When the confining pressure is 20 MPa, the additional energy consumed by the specimen failure is much greater than the releasable elastic energy stored before the peak, and the further failure after the peak of the specimen will require more energy than the elastic deformation energy accumulated inside; this means that the concrete specimen will not only have no sudden release of energy, but will also require more external work to further destroy it, and the specimen will show obvious ductility characteristics.

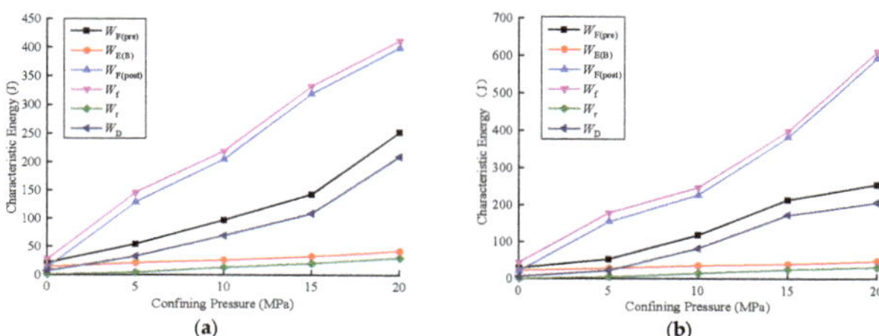

Figure 7. The characteristic energy of high-strength concrete under different confining pressures. (**a**) C60; (**b**) C70.

3.4. Relationship between Energy Evolution and Concrete Strength Grade

Compared with Figures 5 and 6, it can be seen that under the same confining pressure, the slope of the elastic strain energy evolution curve of C70 concrete is greater than that of C60 concrete, indicating that the energy storage rate of C70 concrete in the pre-peak stage and the energy release rate in the post-peak stage are both higher than that of C60 concrete. Figure 8 shows the characteristic energy evolution trend of high-strength concrete with different strength grades. It can be seen from Figure 8 that, with the increase of concrete strength grade, the elastic strain energy, the pre-peak total energy and the fracture energy all increase, while the residual elastic strain energy shows little difference. When the confining pressure is 10 MPa, the elastic strain energy $W_{E(B)}$ stored before the peak of C60 and C70 concrete is 27.22 J and 36.21 J, respectively, and the proportion of energy in the total energy before the peak is 28% and 31%, respectively. This shows that with the increase of the strength grade, the storage capacity of the pre-peak elastic energy of the concrete specimen has increased. After reaching the peak strength, the residual elastic strain energy of the C60 and C70 concrete specimens is 13.86 J and 15.56 J, respectively, and the elastic strain energy release rate is 49% and 57%, respectively, indicating that the elastic strain energy accumulated in C70 concrete is released more completely. The releasable elastic energy of C60 and C70 concrete is 13.36 J and 20.65 J, respectively, which accounts for 6% and 8%, respectively, of the fracture energy. This shows that the C70 concrete specimen has a greater ability to maintain self-fracturing and its brittleness is higher. Since the energy evolution characteristics under the five confining pressures are the same, this section only takes a confining pressure of 10 MPa as an example, and other confining pressure levels will not be repeated.

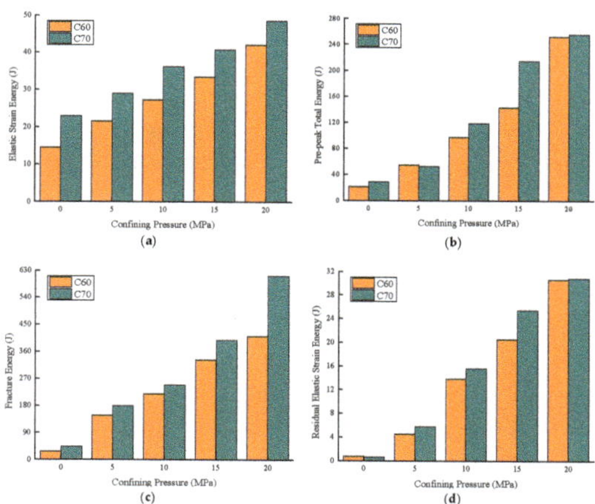

Figure 8. The characteristic energy of high-strength concrete with different strength grades. (**a**) Elastic strain energy; (**b**) pre-peak total energy; (**c**) fracture energy; (**d**) residual elastic strain energy.

4. Evaluation Method of Concrete Brittleness

4.1. Brittleness Evaluation Index

The performance of a material's brittleness is affected by its own properties as well as the specimen shape, size effect, and loading conditions. In the evaluation of material brittleness, attention should be paid to its ability to resist plastic deformation before the peak, the slope of the post-peak stress–strain curve and residual strength, and other characteristic factors, and the two stages before and after the peak need to be considered comprehensively when characterizing the brittleness of materials. Therefore, an approach based on the full stress–strain characteristics of high-strength concrete is an effective brittleness evaluation method to characterize the difficulty of brittle failure and the degree of brittleness of high-strength concrete in the form of energy parameters.

In the energy evolution process of concrete, the more strain energy is absorbed before the peak, the faster the stress–strain curve will drop after the peak, and the less additional energy will be provided by the external work required for concrete fracture after the peak; the energy required to cause concrete fracture mainly comes from the elastic strain energy absorbed before the peak, while the dissipative energy before the peak and the magnitude of the external work after the peak directly affect the fracture degree of the concrete after the peak. Based on this, it is proposed that the pre-peak elastic strain energy accumulation rate and the pre-peak dissipative energy dissipation rate should be used to define the pre-peak brittleness of concrete based on the whole process of concrete failure. The pre-peak brittleness index can be expressed as:

$$B_{pre} = 1 - \frac{W_D}{W_{E(B)} + W_D} = \frac{W_{E(B)}}{W_{E(B)} + W_D} = \frac{\sigma_{1p}^2 + 2(1 - \mu)\sigma_3^2 - 4\mu\sigma_{1p}\sigma_3}{2E\left(\int_0^{\varepsilon_{1p}} \sigma_1 d\varepsilon_1 + 2\sigma_3\varepsilon_{3p}\right)} \tag{17}$$

where $W_{E(B)}$ is the elastic strain energy at the peak point and W_D is the dissipative energy before the peak.

B_{pre} represents the ability to store elastic strain energy before the peak, and the value range is $(0,1)$. When the concrete is in the ideal elastic state, $B_{pre} = 1$, whereas when the concrete is in the fully plastic state, $B_{pre} = 0$. Therefore, the larger the value of B_{pre}, the higher the brittleness level of concrete.

Regarding the energy required for concrete failure, the less additional energy is provided from the outside after the peak, the higher the proportion of elastic strain energy that is stored before the peak and the more energy is provided for the self-fracture of concrete after the peak, which indicates that the brittleness of concrete is increased. Therefore, the release rate of post-peak elastic energy is used to define the post-peak brittleness of concrete, and the post-peak brittleness evaluation index can be expressed as

$$B_{post} = 1 - \frac{W_{F(post)}}{W_{E(B)} + W_{F(post)} - W_r} = \frac{W_{E(B)} - W_r}{W_{E(B)} + W_{F(post)} - W_r} = \frac{W_{E(B)} - W_r}{W_f}$$
$$= \frac{\sigma_{1p}^2 - \sigma_{1r}^2 - 4\mu(\sigma_{1p} - \sigma_{1r})\sigma_3}{2E\left[\int_{\varepsilon_{1p}}^{\varepsilon_{1r}} \sigma_1 d\varepsilon_1 + 2\sigma_3(\varepsilon_{3r} - \varepsilon_{3p})\right] + \left[\sigma_{1p}^2 - \sigma_{1r}^2 - 4\mu(\sigma_{1p} - \sigma_{1r})\sigma_3\right]} \tag{18}$$

where $W_{F(post)}$ is the additional energy provided by the external testing machine, W_r is the residual elastic strain energy, and W_f is the fracture energy.

B_{post} characterizes the ability of concrete to maintain self-fracture and crack propagation at the post-peak stage, and the value range is $(0, 1)$. When concrete is in an ideal plastic state, $B_{post} = 0$; when concrete is in an ideal brittle state, $B_{post} = 1$. Therefore, the larger the value of B_{post}, the more obvious the brittleness of concrete.

4.2. Brittleness Evaluation Method

Since both B_{pre} and B_{post} are positively correlated with the brittleness of concrete, a multi-index multiplicative synthesis method is used to establish parameters that can comprehensively characterize the pre-peak and post-peak brittleness index. The evaluation results of this method are continuous and monotonous, i.e., the product of B_{pre} and B_{post} is used to characterize the brittleness index of the whole stress–strain process as follows:

$$B = B_{\mathrm{pre}}B_{\mathrm{post}} \tag{19}$$

The value range of the brittleness index B is (0, 1). For the ideal brittle state of concrete, $B = B_{\mathrm{pre}} = B_{\mathrm{post}} = 1$; for the ideal plastic state of concrete, $B = B_{\mathrm{pre}} = B_{\mathrm{post}} = 0$, so when the brittleness index increases from 0 to 1, the failure behavior of concrete changes from plastic to brittle. The method of comprehensively characterizing the brittleness evaluation index combines the pre-peak and post-peak brittleness evaluation indexes to establish a brittleness evaluation method that reflects the whole process of concrete failure.

4.3. Verification and Analysis

4.3.1. Experimental Verification

According to the calculation method of the brittleness evaluation index B, the triaxial compression test data of C60 and C70 concrete under different confining pressures are analyzed. The calculation results are shown in Table 4. When the confining pressure is 0 MPa, the brittleness evaluation indexes (B) of C60 and C70 concrete are 0.329 and 0.402, respectively; since the brittleness evaluation index is positively correlated with the brittleness level, the brittleness of the concrete is also increased at this time. When the confining pressure is 20 MPa, the brittleness evaluation indexes (B) of C60 and C70 concrete are 0.005 and 0.006, respectively, indicating that the brittleness of concrete is low at this time. Under the same confining pressure, the brittleness evaluation index B of C70 concrete is greater than that of C60, indicating that the strength of concrete is positively correlated with its brittleness index. The higher the concrete strength, the higher the brittleness level. Figure 9 shows the variation of brittleness index with confining pressure; the brittleness indexes B_{pre}, B_{post} and B of C60 and C70 concrete decrease with the increase of confining pressure. When the confining pressure increases from 0 MPa to 20 MPa, the brittleness index of C60 and C70 concrete decreases by 98% and 99%, respectively, indicating that the confining pressure inhibits the development of the brittleness level of concrete, and the brittleness evaluation index of C70 concrete is more sensitive to the inhibition of the confining pressure. By fitting the calculated value of the brittleness evaluation index B, the relationship between the brittleness index and the confining pressure of C60 and C70 high-strength concrete is obtained, respectively, as shown below. Equations (20) and (21) reflect the brittleness characteristics of high-strength concrete under different confining pressures to a certain extent.

$$\mathrm{C60}: \ B = 0.008 + 0.32 \cdot e^{-\frac{\sigma_3}{2.36}}, \mathrm{R}^2 = 0.999 \tag{20}$$

$$\mathrm{C70}: \ B = 0.008 + 0.39 \cdot e^{-\frac{\sigma_3}{2.81}} \tag{21}$$

Table 4. Calculation results of different brittleness evaluation indexes.

σ_3/MPa	B_{pre}		B_{post}		B		B_1		B_2		B_3	
	C60	C70	C60	C70	C60	C70	C60	C70	C60	C70	C60	C70
0	0.679	0.796	0.484	0.505	0.329	0.402	2.116	3.911	0.403	0.453	0.442	0.518
5	0.394	0.556	0.116	0.130	0.046	0.072	0.649	1.254	0.117	0.140	1.777	1.795
10	0.280	0.305	0.061	0.083	0.017	0.025	0.388	0.439	0.090	0.105	2.238	2.154
15	0.234	0.190	0.039	0.038	0.009	0.007	0.306	0.235	0.072	0.068	2.941	3.115
20	0.167	0.190	0.028	0.029	0.005	0.006	0.201	0.235	0.065	0.057	3.187	4.305

Notes: the standard deviation of brittleness evaluation indexes for C60 and C70 is 0.0029 and 0.0044, respectively.

Comparing the brittleness evaluation indexes calculated by different evaluation methods in Figure 9, we can see that the brittleness index B in this paper has a high consistency with the trend of B_1 and B_2, and the values of B, B_1, and B_2 gradually decrease as the confining pressure increases, indicating that the brittleness of the concrete specimen gradually decreases, which is basically consistent with the experimental trend, indicating that the three brittleness indicators are negatively correlated with the confining pressure. However, B_1 and B_2 only consider the impact of recoverable elastic energy on the brittleness of concrete at the pre-peak stage, ignoring the effect of the additional energy provided by external work at the post-peak stage on the brittleness level of concrete; thus, they cannot well reflect the brittle behavior of the whole process of concrete failure, which indicates that these two brittleness characterization methods B_1 and B_2 have certain limitations. The brittleness level reflected by B_3 is contrary to reality; the value of B_3 gradually increases with the increase of confining pressure, indicating that B_3 is positively correlated with the confining pressure, and the value range of B_3 is larger, which is not conducive to measuring the brittleness level of the material, as the numerator (fracture energy) and denominator (nominal stress) in the expression do not belong to the same order of magnitude, so the physical meaning of B_3 is not sufficiently clear. The brittleness evaluation index B proposed in this paper is established based on the whole process of concrete energy evolution with a clear physical meaning: the value range is (0, 1), and the value of B is continuous and monotonous, so it has good adaptability. In summary, the brittleness level represented by the brittleness index B is more consistent with the physical test results, and it has a better evaluation effect on the brittleness of high-strength concrete materials under different confining pressures.

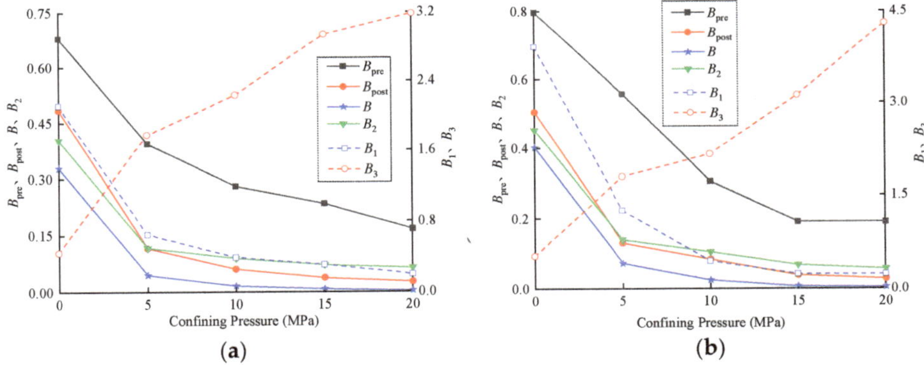

Figure 9. Relationship between brittleness evaluation index and confining pressure of high-strength concrete. (a) C60; (b) C70.

4.3.2. Analysis of Influencing Factors

The relationship between the characteristic strength of concrete under triaxial compression and the brittleness evaluation index B is shown in Figure 10. The figure shows that the brittleness evaluation index B is negatively correlated with the peak strength and residual strength, showing an exponential function relationship. With the increase of the peak strength and residual strength, the brittleness evaluation index B gradually decreased, and the brittleness level of concrete gradually decreased, indicating that the confining pressure can inhibit the initiation and propagation of micro-cracks in concrete and increase the threshold stress for micro-crack initiation and propagation in concrete, thereby improving the load-bearing capacity of concrete and reducing its brittleness level.

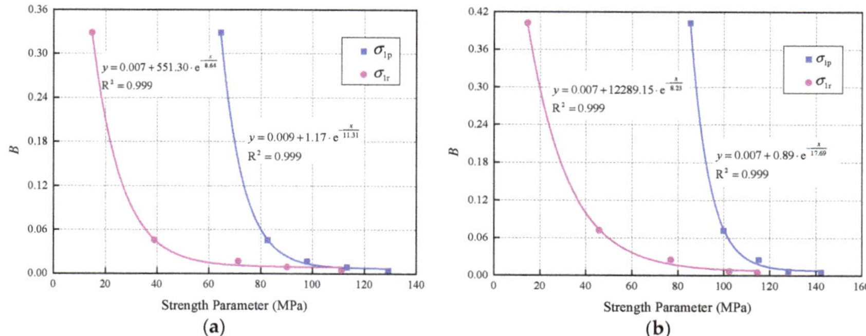

Figure 10. Relationship between brittleness evaluation index and strength parameters. (**a**) C60; (**b**) C70.

The relationship between the brittleness index B and the elastic energy stored before the peak, the total energy before the peak, the additional energy and the fracture energy is shown in Figure 11. It can be seen from Figure 11 that the brittleness evaluation B has an exponential function and negative correlation with the elastic energy stored before the peak, the total energy before the peak, the additional energy, and the fracture energy. In the pre-peak stage, with the increase of confining pressure, the load-bearing capacity of the concrete increases. At this time, the propagation and penetration of the micro-cracks inside the specimen needs to dissipate more energy, so W_D gradually increases with the increase of confining pressure. The ratio of W_D to $W_{F(pre)}$ increases gradually, while that of $W_{E(B)}$ to $W_{F(pre)}$ decreases gradually and the storage capacity of elastic energy decreases gradually, resulting in the gradual weakening of the brittleness level of concrete. In the post-peak stage, as the confining pressure increases, W_r and $W_{F(post)}$ gradually increases and the ratio of $W_{F(post)}$ to W_f gradually increases. This is because the confining pressure increases the threshold stress for crack initiation and improves the residual strength of the concrete specimen at the post-peak stage; furthermore, the self-sustaining fracture ability of the specimen weakens at the post-peak stage, thus weakening its brittleness.

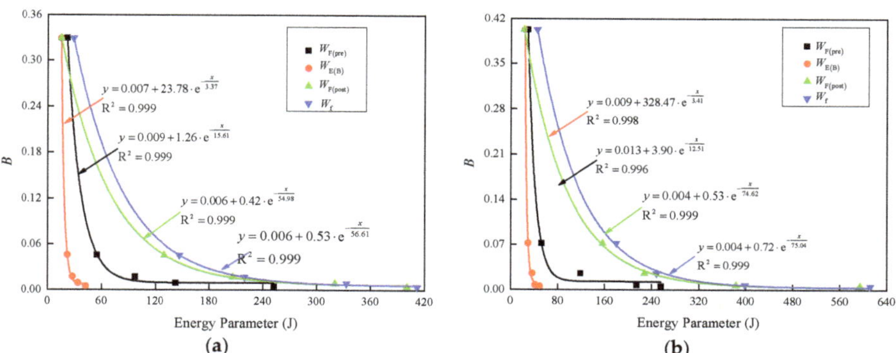

Figure 11. Relationship between brittleness evaluation index and energy parameters. (**a**) C60; (**b**) C70.

5. Conclusions

1. Under different confining pressures, the input energy and dissipative energy of C60 and C70 high-strength concrete specimens increase with the increase of axial strain, and the elastic strain energy shows a trend of first increasing and then decreasing. After the specimen reaches the peak strength, the elastic strain energy decreases gradually, and the dissipative energy increases gradually, reaching maximum and minimum values until the failure of the specimen. When the

high-strength concrete specimen is damaged, the ratio of the additional energy $W_{F(post)}$ provided by the outside world to the fracture energy is proportional to the confining pressure.

2. Based on the energy evolution law in the full stress–strain process of high-strength concrete, the pre-peak and post-peak brittleness indexes B_{pre} and B_{post} are defined; then, according to the positive correlation between B_{pre}, B_{post} and the brittleness of the concrete, a method to characterize the brittleness index B of the whole stress–strain process by the product of B_{pre}, B_{post} is proposed.

3. The brittleness evaluation index B is negatively correlated with the peak strength and residual strength, showing an exponential function relationship, and it has an exponential function and negative correlation with the elastic energy stored before the peak, the total energy before the peak, additional energy, and fracture energy. The comparative analysis of different brittleness evaluation methods shows that the brittleness evaluation index method proposed in this paper presents a continuous and monotonous brittleness index and can better characterize the brittleness level of high-strength concrete under different stress conditions.

Author Contributions: The author contributions of the paper are as follows: R.Z., H.C., and M.L. proposed the conceptualization and methodology; R.Z. and R.H. conducted the tests; L.Z. and M.L. analyzed the data; R.Z. wrote the paper; H.C. reviewed and edited the paper. All authors have read and agreed to the published version of the manuscript.

Funding: This research was supported by the National Natural Science Foundation of China (NO.51474004; NO.51874005).

Conflicts of Interest: The authors declare no conflict of interest.

References

1. Morley, A. *Strength of Materials*; Longmans Green and Company: London, UK, 1944; pp. 71–72.
2. Hetenyi, M. *Handbook of Experimental Stress Analysis*; John Wiley: New York, NY, USA, 1966; pp. 23–25.
3. Ramsay, J.G. *Folding and Fracturing of Rocks*; McGraw-Hill: London, UK, 1967; pp. 44–47.
4. Lawn, B.R.; Marshall, D.B. Hardness, toughness, and brittleness: An indentation analysis. *J. Am. Ceram. Soc.* **1979**, *62*, 347–350. [CrossRef]
5. Quinn, J.B.; Quinn, G.D. Indentation brittleness of ceramics: A fresh approach. *J. Mater. Sci.* **1997**, *32*, 4331–4346. [CrossRef]
6. Li, X.; Xue, W.; Fu, C. Mechanical properties of high-performance steel-fibre-reinforced concrete and its application in underground mine engineering. *Materials* **2019**, *12*, 2470. [CrossRef] [PubMed]
7. Chen, Z.; Tang, J.; Zhou, X.; Zhou, J.; Chen, J. Interfacial bond behavior of high strength concrete filled steel tube after exposure to elevated temperatures and cooled by fire hydrant. *Materials* **2019**, *13*, 150. [CrossRef] [PubMed]
8. Eid, R.; Muravin, B.; Kovler, K. Acoustic emission monitoring of high-strength concrete columns subjected to compressive axial loading. *Materials* **2020**, *13*, 3114. [CrossRef] [PubMed]
9. Jenq, Y.; Shah, S.P. Two parameter fracture model for concrete. *J. Eng. Mech.* **1985**, *111*, 1227–1241. [CrossRef]
10. Tasdemir, C.; Tasdemir, M.A. Effects of silica fume and aggregate size on the brittleness of concrete. *Cem. Concr. Res.* **1996**, *26*, 63–68. [CrossRef]
11. Zhang, B.; Bicanic, N.; Pearce, C.J. Relationship between brittleness and moisture loss of concrete exposed to high temperatures. *Cem. Concr. Res.* **2002**, *32*, 363–371. [CrossRef]
12. Yan, A.; Wu, K.; Zhang, D.; Yao, W. The relationship between the brittleness and fracture surface characteristics of high-strength concrete. *J. Tongji Univ.* **2002**, *30*, 66–70. (In Chinese)
13. Lange, D.A.; Hamlin, M.; Surendra, P.S. Relationship between fracture surface roughness and fracture behavior of cement paste and mortar. *J. Am. Ceram. Soc.* **1993**, *76*, 589–597. [CrossRef]
14. Yao, M.; Zhan, B.; Huang, X. Experimental study on the effect of curing on the brittleness of high performance concrete. *J. Hefei Univ. Technol.* **2004**, *11*, 1438–1442. (In Chinese)
15. Guo, X. Study on the Brittleness Evaluation Method and Toughening Measures of High-Strength Concrete. Ph.D. Thesis, Wuhan University, Wuhan, China, 2005. (In Chinese).
16. Huang, Y.; Qian, J. Brittleness and strength size effect of high-strength and ultra-high-strength concrete. *Ind. Constr.* **2005**, *35*, 15–17. (In Chinese)

17. Huang, Y.; Qian, J.; Zhou, X. Study on the brittleness index of quasi-brittle materials based on the strength size effect. *Eng. Mech.* **2006**, *23*, 38–42. (In Chinese)
18. Han, J.; Yan, P. A new method of concrete brittleness evaluation-fractured stone area fraction method. *Concrete* **2011**, *2*, 1–3. (In Chinese)
19. Ministry of Housing and Urban-Rural Development of China. *Specification for Design of Concrete Mixture Ratio*; JGJ55-2011; China Construction Industry Press: Beijing, China, 2011. (In Chinese)
20. Xie, H.P.; Ju, Y.; Li, L.Y.; Peng, R.D. Energy mechanism of deformation and failure of rock masses. *Chin. J. Rock Mech. Eng.* **2008**, *27*, 1729–1740. (In Chinese)
21. Zhang, L.; Gao, S.; Wang, Z.; Cong, Y. Analysis of marble failure energy evolution under loading and unloading conditions. *Chin. J. Rock Mech. Eng.* **2013**, *32*, 1572–1578. (In Chinese)
22. Tian, Y.; Yu, R. Energy analysis of limestone during triaxial compression under different confining pressures. *Rock Soil Mech.* **2014**, *35*, 118–122. (In Chinese)
23. Xie, H.P.; Li, L.Y.; Ju, Y.; Peng, R.D.; Yang, Y.M. Energy analysis for damage and catastrophic failure of rocks. *Sci. China Technol. Sci.* **2011**, *54*, 199–209. [CrossRef]
24. Wu, J. *Elasticity*; Higher Education Press: Beijing, China, 2011. (In Chinese)

Publisher's Note: MDPI stays neutral with regard to jurisdictional claims in published maps and institutional affiliations.

Article

A Numerical Study on Chloride Diffusion in Cracked Concrete

Qiannan Wang [1], Guoshuai Zhang [1], Yunyun Tong [1,*] and Chunping Gu [2,3]

1   School of Civil Engineering and Architecture, Zhejiang University of Science and Technology, Hangzhou 310023, China; wangqiannan@zust.edu.cn (Q.W.); 212002814023@zust.edu.cn (G.Z.)
2   College of Civil Engineering, Zhejiang University of Technology, Hangzhou 310023, China; guchunping@zjut.edu.cn
3   Zhejiang Construction Investment Group, Hangzhou 310013, China
*   Correspondence: 112013@zust.edu.cn

Abstract: The cracks in concrete are a fast transport path for chlorides and influence the service life of concrete structures in chloride environments. This study aimed to reveal the effect of crack geometry on chloride diffusion in cracked concrete. The chloride diffusion process in cracked concrete was simulated with the finite difference method by solving Fick's law. The results showed that the apparent chloride diffusivity was lower in more tortuous cracks, and the cracks with more narrow points also showed lower apparent chloride diffusivity. For tortuous cracks, a higher crack width meant relatively more straight cracks, and consequently, higher apparent chloride diffusivity, while a lower crack width resulted in more tortuous cracks and lower apparent chloride diffusivity. The crack depth showed a more significant influence on the chloride penetration depth in cracked concrete than crack geometry did. Compared with rectangular and V-shaped cracks, the chloride diffusion process in cracked concrete with a tortuous crack was slower at the early immersion age. At the same crack depth, the crack geometry showed a marginal influence on the chloride penetration depth in cracked concrete during long-term immersion.

Keywords: cracked concrete; crack width; crack depth; tortuosity; numerical simulation

Citation: Wang, Q.; Zhang, G.; Tong, Y.; Gu, C. A Numerical Study on Chloride Diffusion in Cracked Concrete. *Crystals* **2021**, *11*, 742. https://doi.org/10.3390/cryst11070742

Academic Editors: Yifeng Ling, Chuanqing Fu, Peng Zhang, Peter Taylor and Tomasz Sadowski

Received: 11 May 2021
Accepted: 23 June 2021
Published: 25 June 2021

1. Introduction

In chloride environments such as marine areas, the chloride diffusivity of concrete is considered to be the key point that determines the service life of reinforced concrete structures [1–3]. The chloride concentration threshold value that initiates the corrosion process is designated as the critical chloride concentration with respect to the binder content [4]. This value is usually taken in the range of 0.5–0.9% for tidal and splash zones, and 1.6–2.3% for submerged structures [5]. Normally, cracks always exist in concrete due to shrinkage, external load or other reasons [6,7]. The crack in concrete could act as a path for the fast transport of chlorides, thus accelerating the chloride transport in concrete [8].

Numerous experimental studies have shown that the chloride diffusivity in cracked concrete is significantly influenced by the crack geometry [9,10]. Crack width is regarded as the most important factor that influences the chloride diffusivity in cracked concrete [8]. General findings on this topic have been obtained [11–13]. The chloride diffusivity in a concrete crack is not influenced by the crack width when the crack width is very small (smaller than a low threshold). Under this circumstance, the cracks do not act as a fast transport path for chlorides, and the chloride diffusivity in concrete cracks is close to that in sound concrete. When the crack width is very large (bigger than a high threshold), the chloride diffusivity in a concrete crack is also independent of the crack width. The chlorides could be transported very quickly in the cracks, and the chloride diffusivity in concrete cracks is considered to be equal to the chloride diffusivity in bulk crack solution. When the crack width is between the low and high thresholds, the chloride diffusivity is obviously influenced by the crack width. However, the values for the low and high thresholds, and the relationships between the chloride diffusivity in a concrete crack and crack width, varied in previous studies [8].

Djerbi et al. [14] adopted a steady-state migration test to study the influence of crack width on chloride diffusivity in cracked concrete. The chloride diffusivity in concrete cracks was also calculated in their study, and the relationship between the chloride diffusivity in concrete cracks and crack width was established. The low and high thresholds in their study were 30 µm and 80 µm, respectively, and the calculated chloride diffusivity in concrete cracks increased linearly with the crack width when the crack width was between 30 µm and 80 µm. Non-steady-state migration and diffusion methods were adopted by several researchers to study the effect of crack width on the chloride diffusivity in concrete crack [15,16]. The low threshold ranged from 30 µm to 120 µm, and the high threshold was between 80 µm and 680 µm. There is still no consensus on the exact values of the low and high thresholds. The different concrete compositions, crack generation methods and chloride diffusion test methods may be the reasons for the different results.

In addition to the crack width, the crack depth, tortuosity, connectivity and surface roughness could also influence the chloride transport process [17–21]. Marsavina et al. [22] found that the chloride penetration depth increased with an increasing (artificial) crack depth. This effect was more pronounced for longer test durations. Similar results were also found by Audenaert [13]. The chloride diffusivity in real concrete cracks was lower than that in artificial rectangular cracks and V-shaped cracks, and the chloride diffusivity in C30 cracks was higher than that in C80 cracks [11]. This was mainly because the cracks in concrete with a low water-to-binder ratio may be blocked due to the self-healing phenomenon, thus hampering the chloride transport process [23].

Due to the uncertainties during experimental studies, especially the geometry of the induced cracks in concrete, the influence of crack geometry on chloride diffusivity in concrete cracks has not been clearly revealed yet. In experimental studies, the crack geometry cannot be controlled when introducing real cracks in concrete, and human influences always exist when experiments are performed. For example, when studying the effect of crack width, cracks with the same width at the concrete surface could be created, but the crack geometry will vary in different concrete specimens. Hence, it is difficult to systemically study the effect of geometry on chloride diffusion in cracked concrete with experiments. Under this circumstance, numerical methods could provide a more promising solution to this question [8]. With simulation studies, experimental errors can be eliminated, and cracks with desired geometries can be implemented in the simulations. In this study, the chloride diffusion process in cracked concrete was simulated. The apparent chloride diffusivity in concrete cracks with different geometries was calculated, and the effect of crack geometry on the chloride diffusion in cracked concrete was discussed.

2. Simulation Methods

2.1. Numerical Method

The chloride diffusion process in cracked concrete follows the mass conservation equation and Fick's diffusion law [1]. In general, the chloride diffusion process in cracked concrete can be described as follows:

$$\frac{\partial C}{\partial t} = div(D \cdot gradC) \tag{1}$$

where C is the chloride concentration, t is time and D is the chloride diffusivity in sound concrete or crack solution.

To overcome the computational limitations, the chloride transport process in concrete can be simulated with simplified 2D elements [24]. For the case of 2D, Equation (1) becomes:

$$\frac{\partial C}{\partial t} = \frac{\partial C}{\partial x}\left(D\frac{\partial C}{\partial x}\right) + \frac{\partial C}{\partial y}\left(D\frac{\partial C}{\partial y}\right) \tag{2}$$

The Crank–Nicholson finite difference method was used to solve Equation (2). The finite difference approximation of Equation (2) can be written as follows:

$$\frac{C_{i,j}^{n+1} - C_{i,j}^{n}}{\Delta t} = \frac{D_{(i+1)/2,j}C_{i+1,j}^{n} - \left(D_{(i+1)/2,j} + D_{(i-1)/2,j}\right)C_{i,j}^{n} - D_{(i-1)/2,j}C_{i-1,j}^{n}}{2 \times (\Delta x)^2}$$

$$+ \frac{D_{(i+1)/2,j}C_{i+1,j}^{n+1} - \left(D_{(i+1)/2,j} + D_{(i-1)/2,j}\right)C_{i,j}^{n+1} - D_{(i-1)/2,j}C_{i-1,j}^{n+1}}{2 \times (\Delta x)^2}$$

$$+ \frac{D_{i,(j+1)/2}C_{i,j+1}^{n} - \left(D_{i,(j+1)/2} + D_{i,(j-1)/2}\right)C_{i,j}^{n} - D_{i,(j-1)/2}C_{i,j-1}^{n}}{2 \times (\Delta y)^2}$$

$$+ \frac{D_{i,(j+1)/2}C_{i,j+1}^{n+1} - \left(D_{i,(j+1)/2} + D_{i,(j-1)/2}\right)C_{i,j}^{n+1} - D_{i,(j-1)/2}C_{i,j-1}^{n+1}}{2 \times (\Delta y)^2} \tag{3}$$

where $C_{i,j}^{n}$ is the chloride concentration at nod (i,j) at time step n; $D_{i,j}$ is the chloride diffusivity at nod (i, j); $D_{(i+1)/2,j}$, $D_{(i-1)/2,j}$, $D_{i,(j+1)/2}$ and $D_{i,(j-1)/2}$ are the harmonic means of $D_{i+1,j}$ and $D_{i,j}$, $D_{i-1,j}$ and $D_{i,j}$, $D_{i,j+1}$ and $D_{i,j}$, $D_{i,j-1}$ and $D_{i,j}$, respectively [25]. By solving the implicit difference equations, the chloride concentration distribution in cracked concrete at different times, i.e., the chloride diffusion process, can be obtained. A self-written MATLAB (MathWorks, Natick, MA, United States) program was used to solve the diffusion equation. The Crank-Nicholson difference scheme is stable unconditionally, therefore, the numerical solutions are always convergent [26].

2.2. Simulation of the Steady-State Chloride Diffusion Process

In order to reveal the influence of crack geometry on the chloride transport in cracked concrete, steady-state diffusion in cracked concrete with different crack geometries was simulated. The flux through the outlet surface at a steady state can be determined and can then be used to calculate the chloride diffusivity in the specimen.

For a cracked concrete specimen, this method can be used to determine the chloride diffusivity in the cracks. This calculated chloride diffusivity is regarded as the apparent chloride diffusivity in crack $D_{cr}{}'$, which is influenced by the crack geometry [11]. On the other hand, the chloride diffusivity in crack solution D_{cr} is independent of crack geometry but is determined, instead, by the characteristic of the crack solution [27].

Figure 1 shows a schematic illustration of the simulated steady-state diffusion in concrete with a crack in 2D. A 1 cm × 1 cm square is used to represent the cracked concrete specimen. The whole domain was digitized into a 1000 × 1000 mesh when performing the finite difference analysis. The initial chloride concentration at every nod was 0 mol/L. Dirichlet boundary conditions [25] were applied in the simulations. The chloride concentrations at the inlet surface (x = 0 mm) and outlet surface (x = 10 mm) were set to be 1 mol/L and 0 mol/L, respectively. The upper and lower surfaces were considered to be impermeable, which meant that chloride flux through these two surfaces was zero. A crack existed at the center of the specimen, and the chloride diffusivity in the crack solution D_{cr} was assumed to be 1.61×10^{-9} m^2/s, which is the chloride diffusivity in dilute NaCl solution at 25 °C [28]. In order to study the effect of crack geometry on the chloride diffusion in the crack, the chloride diffusivity in the sound concrete (D_0) was set to be zero for simplicity. The time-step was set to be 10^{-4} years.

The simulation in MATLAB program stopped when the chloride concentration in concrete stopped changing, which meant that steady-state diffusion had been achieved. Consequently, the chloride concentration distribution in this cracked concrete and the flux J through the outlet surface at a steady state could be determined through simulation.

The chloride diffusivity of the whole specimen D can be calculated according to Fick's law:

$$D = \frac{J}{A}\frac{L}{\Delta C} \tag{4}$$

where A is the area of the outlet surface, L is the length of the specimen and ΔC is the chloride concentration difference between the inlet and outlet surfaces. Assuming that the sound concrete is impermeable, the chloride only diffuses through the crack. Therefore,

$$J_{cr} = J \tag{5}$$

where J_{cr} is the flux through the crack at a steady state and it can be calculated as follows:

$$J_{cr} = -D_{cr}\frac{\partial C_{cr}}{\partial x}\Big|_{x=outlet} A_{cr} \tag{6}$$

where A_{cr} is the area of the crack at the outlet surface. The flux J_{cr} and chloride concentration distribution can be obtained with the simulation. Based on Fick's law, the apparent chloride diffusivity in the crack $D_{cr}{}'$ can be determined as follows:

$$D'_{cr} = \frac{J_{cr}}{A_{cr}}\frac{L}{\Delta C} \tag{7}$$

By substituting Equation (4) into Equation (5), the relationship between $D_{cr}{}'$ and D_{cr} can be described as follows:

$$D'_{cr} = -D_{cr}\frac{\partial C_{cr}}{\partial x}\Big|_{x=outlet}\frac{L}{\Delta C} \tag{8}$$

It should be pointed out that, $D_{cr}{}'$ may be equal to D_{cr} when the crack is a straight rectangle with parallel crack surfaces, which is almost impossible for cracks in real concrete structures. In other cases, $D_{cr}{}'$ is lower than D_{cr}, and the difference between them is dependent on the crack geometry.

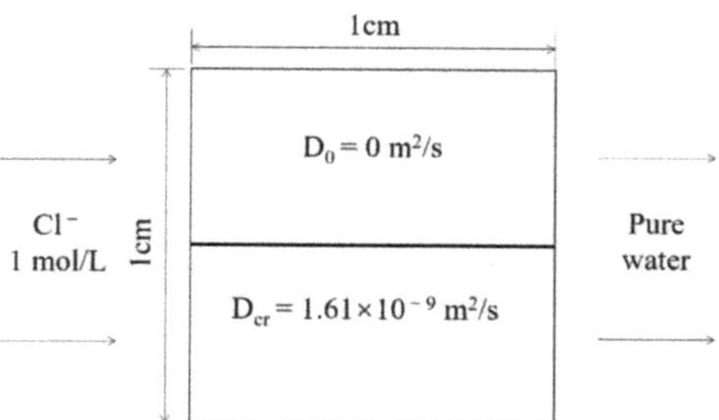

Figure 1. Illustration of the simulated steady-state diffusion in cracked concrete.

2.3. Simulation of Non-Steady-State Chloride Diffusion in Cracked Concrete

The chloride diffusion process in cracked concrete actually determines the service life of concrete structures in chloride environments. Assuming that a cracked concrete specimen is immersed in NaCl solution, an illustration of the simulation settings is shown in Figure 2. The whole domain was digitized into a 1000×500 mesh when performing the finite difference analysis. The initial and boundary conditions included the following: the chloride concentrations at the left surface and right surface were 0.555 mol/L and 0 mol/L, respectively, while other surfaces were sealed; the initial chloride concentration in the crack was considered to be 0.555 mol/L since the solution would enter into the crack due to capillary suction, and the initial chloride concentration in the concrete was set as 0 mol/L. The chloride diffusion process was also simulated by solving Equation (1) with the finite difference method. The chloride diffusivity in the sound concrete was set as 1.0×10^{-11} m^2/s, which is a typical value for widely used concrete with a water-to-binder ratio of over 0.45 [29]. The chloride diffusivity in crack solution D_{cr} was set to be 1.61×10^{-9} m^2/s [28]. The time-step was 10^{-4} years. The crack width at the concrete surface was 0.6 mm. Rectangular, V-shaped and real cracks were investigated in the simulation.

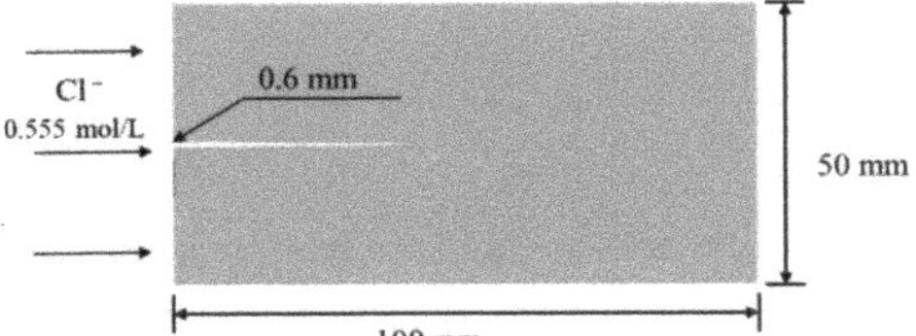

Figure 2. Illustration of the simulation settings for the non-steady-state chloride diffusion simulation in cracked concrete.

3. Results and Discussion

3.1. Effect of Crack Geometry on the Apparent Chloride Diffusivity in the Crack

In order to reveal the effect of crack geometry on the chloride diffusivity in concrete cracks, the chloride diffusion process in cracks with different widths and geometries was simulated. The input crack geometries are shown in Figure 3. Crack 1 was a straight crack with parallel crack surfaces. Crack 2 was a V-shaped crack with different widths at the chloride inlet and outlet surfaces. Cracks 3, 4 and 5 were tortuous cracks with folding lines as the crack surfaces. The tortuosity of these cracks followed the order of crack 5 > crack 4 > crack 3 > crack 1. Crack 6 was a real crack in C30 concrete from [30]. The crack width d ranged from 30 μm to 5 mm, and the width of crack 2 at the outlet end (d_1) was set to be 100 μm.

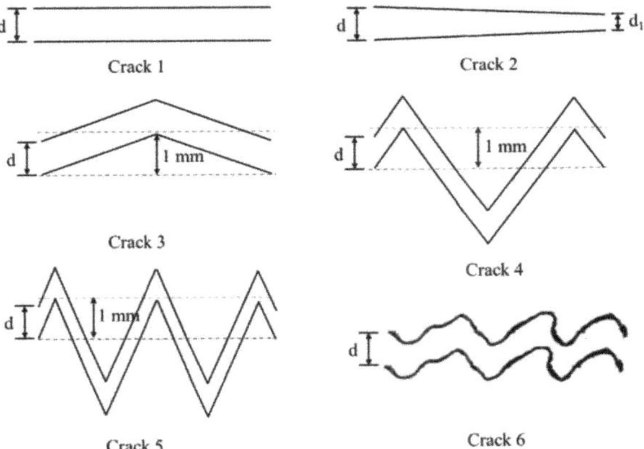

Figure 3. Crack geometries adopted in the simulation.

The apparent chloride diffusivity of the different cracks is shown in Figure 4. For the straight crack, the apparent chloride diffusivity in the crack was constant and equal to the chloride diffusivity in the crack solution. This means that the crack width did not influence the apparent chloride diffusivity in the straight crack. For the V-shaped crack, the apparent chloride diffusivity in the crack increased with the crack width at the inlet surface. It became a straight crack when the crack width at the inlet surface became 100 μm, which was equal to the width at the outlet surface. The apparent chloride diffusivity in the V-shaped crack kept increasing and exceeded the chloride diffusivity in crack solution when the crack width at the inlet surface was higher than 100 μm. This was because the higher crack width at the inlet surface led to higher chloride flux through the inlet and outlet surfaces, resulting in higher apparent chloride diffusivity in the crack.

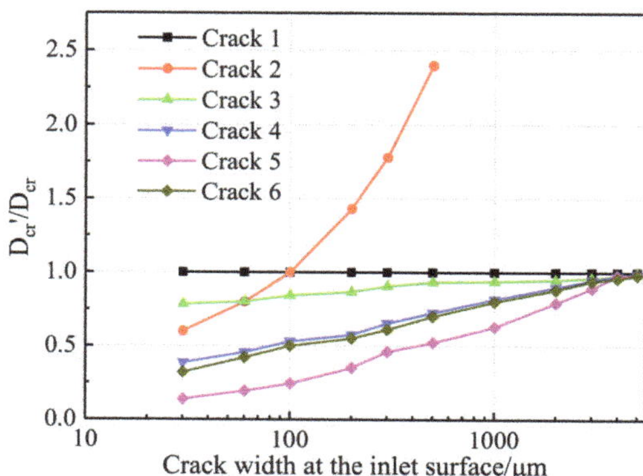

Figure 4. The effect of crack width on the apparent chloride diffusivity of concrete crack.

For tortuous cracks, the apparent chloride diffusivity in the cracks also increased with crack width, but it did not exceed the chloride diffusivity in crack solution. When the width of a tortuous crack was large enough (i.e., 5 mm), the apparent chloride diffusivity of the crack became close to the chloride diffusivity in crack solution. The large crack width made the tortuous crack similar to the straight crack with parallel crack surfaces. The apparent chloride diffusivity in the crack decreased with the increase in crack tortuosity. No matter how large the crack width was, the apparent chloride diffusivity in the cracks followed the order of crack 1 > crack 3 > crack 4 > crack 5. The results are in good agreement with findings obtained from experimental studies [11]. When the crack width was very small, the crack became more tortuous, and the apparent chloride diffusivity in the crack decreased. When the crack width was 30 μm, the apparent chloride diffusivities in cracks 3, 4 and 5 were 78%, 39% and 14% of the chloride diffusivity in crack solution, respectively. Hence, when the crack width is very small (e.g., <30 μm) and the crack geometry is complex, the chloride cannot diffuse quickly in the crack. The apparent chloride diffusivity in the crack would be close to that in sound concrete.

The real crack was also tortuous, and the apparent chloride diffusivity in the real crack was close to that of crack 4, which implies that real cracks could be simplified as fold-line cracks when simulating the chloride diffusion process. The detailed geometry of this fold-line necessitates further study to assure better agreement with reality. Moreover, the apparent chloride diffusivity in rectangular cracks was obviously higher than that in real cracks, especially when the crack width was under 1 mm. The apparent chloride diffusivity D_{cr}' in rectangular cracks is width-independent, while D_{cr}' in real cracks is governed by crack width.

When simulating the chloride diffusion process in cracked concrete, determination of the crack geometry and the chloride diffusivity in the crack is the most important aspect. Normally, in simulations, the crack geometry could be set as rectangular, V-shaped or real-shaped. If a real tortuous crack is simplified to be a rectangular crack, the chloride diffusivity in the crack should be width-dependent to reflect the effect of crack geometry on the chloride diffusivity [31].

In addition to the cracks shown in Figure 3, the chloride diffusion process in cracks with other geometries (as shown in Figure 5) was simulated. Cracks 7, 8 and 9 were tortuous cracks with narrow points. The width at the narrow points (d_2) was smaller than the crack width at the surface (d). In the simulation of chloride diffusion, d was set to be 100 μm, while the d_2 values were set as 20 μm, 40 μm, 60 μm and 80 μm, respectively, to investigate the effect of narrow points' width on the chloride diffusion in the cracks. The calculated apparent chloride diffusivities in the cracks with narrow points are shown in Figure 6.

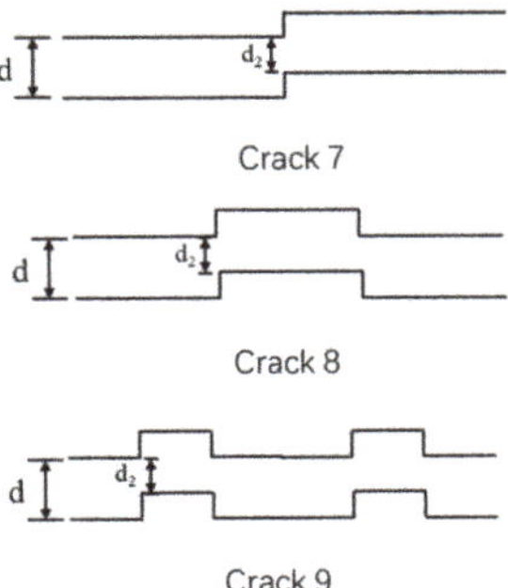

Figure 5. Geometries of tortuous cracks with narrow points adopted in the simulation.

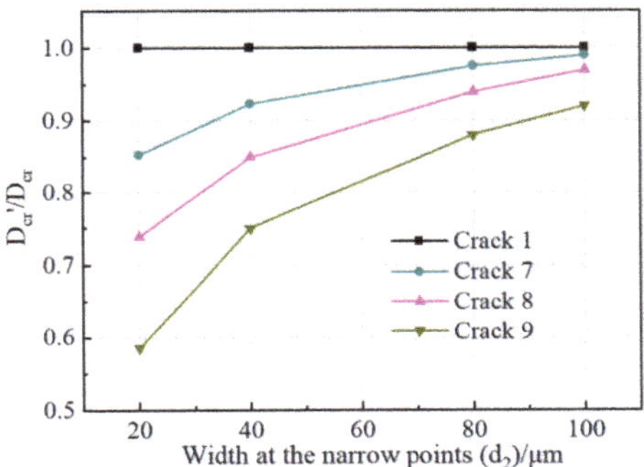

Figure 6. The apparent chloride diffusivity in cracks with narrow points.

For cracks with narrow points, the apparent chloride diffusivity decreased with the reduced width at the narrow points. For example, in crack 7, when the narrow point was 80 μm wide, the apparent chloride diffusivity in the crack was slightly lower than the chloride diffusivity in crack solution, whereas when the width of the narrow point was 20 μm, the apparent chloride diffusivity in the crack reduced to about 85% of the chloride diffusivity in crack solution. In addition, with more narrow points existing in the crack, the reduction in apparent chloride diffusivity with the reduced width at the narrow points was more significant. As for crack 8, the apparent chloride diffusivity in the crack was lower than that in crack 7; it decreased to 74% of chloride diffusivity in crack solution when the narrow points were 20 μm wide. The apparent chloride diffusivity in crack 9 was 59% of chloride diffusivity in crack solution when the width of the narrow points was 20 μm. It was, therefore, obvious that the narrow points in the cracks also showed a significant influence on the apparent chloride diffusivity in cracks.

Based on the simulations of the steady-state diffusion process, it can be concluded that the crack geometry, including the crack width, tortuosity and narrow points, showed a great impact on the apparent chloride diffusivity in cracks. When the chlorides enter the narrow points, the interaction between the crack surface and the chloride will be more pronounced and will hinder the fast diffusion of chlorides [32].

When the width of a real crack is small, the crack is generally more tortuous and has more narrow points. Therefore, the apparent chloride diffusivity in the crack will be very small, and may be close to the chloride diffusivity in sound concrete. When the crack width is large enough, the crack is more like a straight crack, whose apparent chloride diffusivity is close to the chloride diffusivity in crack solution.

3.2. Effect of Crack Geometry on the Chloride Diffusion Process in Cracked Concrete

The non-steady-state chloride diffusion process in cracked concrete was simulated using the finite difference method. Rectangular, V-shaped and real cracks were adopted in the simulations. The detailed geometries of these cracks are shown in Figure 7. The rectangular cracks had different crack depths (i.e., 25 mm, 35 mm and 45 mm). The depth of the V-shaped and real cracks was set to be 45 mm. The geometry of real crack was adopted from [33]. All of the cracks had a crack width of 0.6 mm at the left surface of the specimen.

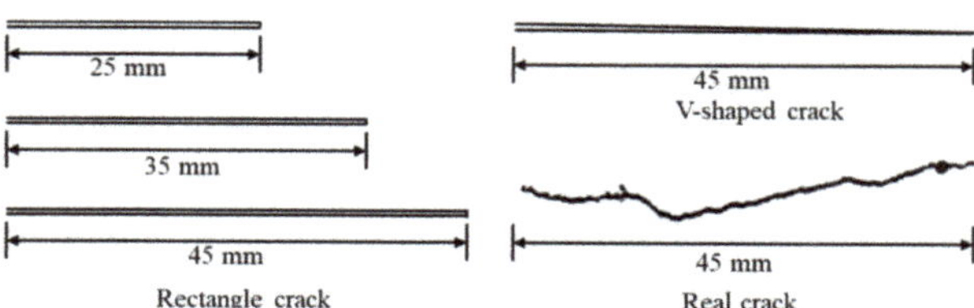

Figure 7. The crack geometries used in non-steady-state diffusion simulation (crack width at the left surface: 0.6 mm).

The simulated chloride concentration distributions in concrete with rectangular cracks after one year of immersion in NaCl solution are shown in Figure 8. The chloride penetrated the concrete specimen quickly through the crack, and the chloride penetration depth was largely dependent on the crack depth. A higher crack depth led to a deeper chloride penetration depth. Hence, the crack depth is the key factor that influences the service life of concrete structures in chloride environments. The chlorides also penetrated concrete through the crack surfaces; hence, the chloride concentration in concrete near the crack surfaces was also increased due to the presence of cracks.

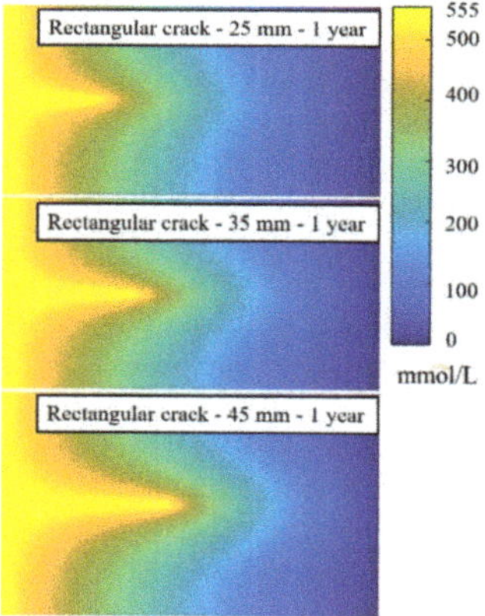

Figure 8. The simulated chloride concentration in cracked concrete with rectangular cracks after one year of immersion in NaCl solution.

The chloride concentration distributions in cracked concrete with different crack geometries are shown in Figure 9. At the immersion age of 0.01 years, it was quite obvious that the chloride concentration in the real tortuous crack was lower than that in the rectangular and V-shaped cracks. The tortuous crack inhibited the fast chlorides' diffusion in concrete cracks. When the immersion age increased, the chloride concentration distributions in different cracked concretes were close to each other. The chloride concentrations at the height of 25 mm in cracked concrete (i.e., at the center of the cracked concrete in the vertical direction) are shown in Figure 10. At the immersion age of 0.01 years, the chloride concentration at the real crack tip was much lower than that in the rectangular and V-shaped cracks, and for the real crack, the chloride concentration beyond the crack tip was also lower than that in the other two. At the immersion age of one year, the chloride concentration distributions in cracked concrete with rectangular and V-shaped cracks were almost the same. The chloride concentration in the real crack was only slightly lower than that in the rectangular and V-shaped cracks. This suggested that the crack geometry influenced the chloride diffusion process in cracked concrete soon after the chloride penetrated the crack. However, for a long-term immersion with a given crack depth, the crack geometry did not show much influence on the chloride penetration depth in cracked concrete. In comparison, the influence of crack depth was more significant than that of crack geometry on the chloride penetration depth in cracked concrete in the long term.

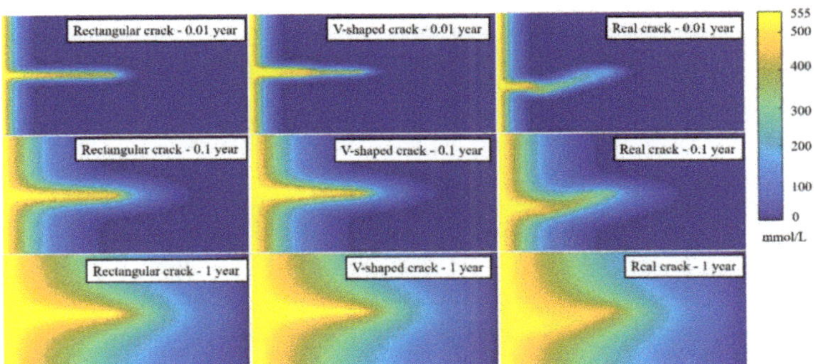

Figure 9. The chloride concentration distribution in concrete with different crack geometries.

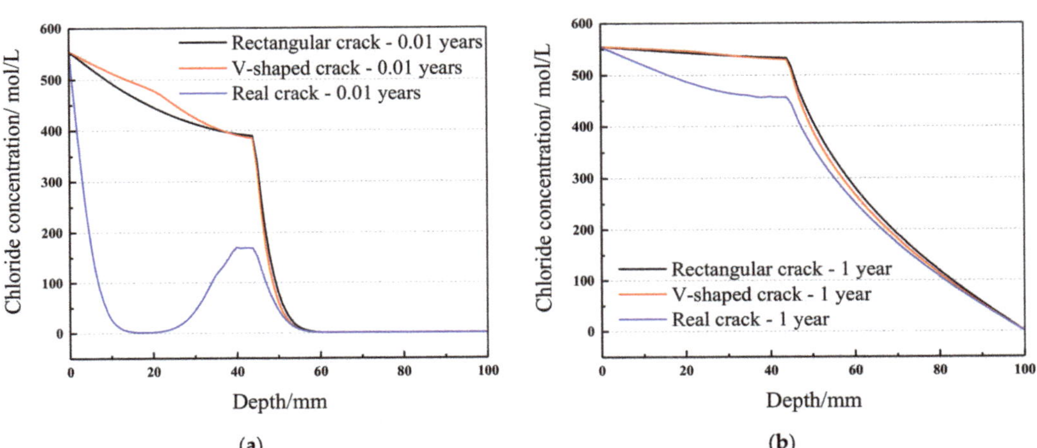

Figure 10. The chloride concentrations at the height of 25 mm in cracked concrete: (**a**) at the immersion age of 0.01 years; (**b**) at the immersion age of one year.

Generally, in cracked concrete, the crack width influences the short-term chloride diffusion process, but for long-term chloride diffusion, the influence of crack depth is more significant. That is, the crack depth is the main factor that influences the residual service life of a cracked concrete structure in chloride environments. Moreover, when predicting the residual service life of cracked concrete structures with numerical methods, the real crack could be simplified as rectangular (or other shaped) crack in the simulations.

4. Conclusions

Based on the simulations of steady and non-steady states of chloride diffusion in cracked concrete, the effects of crack geometry on the chloride diffusion behavior in cracked concrete were discussed. The following conclusions can be drawn.

(1) The apparent chloride diffusivity was lower in more tortuous concrete cracks. In addition, the narrow points in the cracks also showed a significant influence on the apparent chloride diffusivity in the cracks.

(2) When the crack width was very large, the apparent chloride diffusivity in tortuous cracks became almost equal to that in crack solution. When the crack width was small and the crack was more tortuous, the apparent chloride diffusivity in the crack reduced remarkably.

(3) The crack depth showed a more significant influence on the chloride penetration depth in cracked concrete than crack geometry did.

(4) Compared with rectangular and V-shaped cracks, the chloride diffusion process in cracked concrete with a tortuous crack was slower at the early immersion age, whereas with the same crack depth for a long term, the crack geometry showed a marginal influence on the chloride penetration depth in cracked concrete.

Author Contributions: Conceptualization, Q.W. and C.G.; methodology, Q.W. and C.G.; software, C.G.; validation, G.Z. and Y.T.; writing—original draft, Q.W. and C.G.; writing—review and editing, G.Z. and Y.T. All authors have read and agreed to the published version of the manuscript.

Funding: This research was funded by the National Natural Science Foundation of China, grant numbers 52008372 and 51708502.

Data Availability Statement: Data are contained within the article.

Conflicts of Interest: The authors declare no conflict of interest.

References

1. Yang, C.Y.; Li, L.; Li, J.P. Service life of reinforced concrete seawalls suffering from chloride attack: Theoretical modelling and analysis. *Constr. Build. Mater.* **2020**, *263*, 120172. [CrossRef]
2. Yang, D.H.; Li, G.P.; Yi, T.H.; Li, H.N. A performance-based service life design method for reinforced concrete structures under chloride environment. *Constr. Build. Mater.* **2016**, *124*, 453–461. [CrossRef]
3. Zhang, J.Z.; Huang, J.; Fu, C.Q.; Huang, L.; Ye, H.L. Characterization of steel reinforcement corrosion in concrete using 3D laser scanning techniques. *Constr. Build. Mater.* **2020**, *270*, 121402. [CrossRef]
4. Reiterman, P.; Keppert, M. Effect of various de-icers containing chloride ions on scaling resistance and chloride penetration depth of highway concrete. *Roads Bridges-Drogi i Mosty* **2020**, *19*, 51–64. [CrossRef]
5. Hájková, K.; Šmilauer, V.; Jendele, L.; Cervenka, J. Prediction of reinforcement corrosion due to chloride ingress and its effects on serviceability. *Eng. Struct.* **2018**, *174*, 768–777. [CrossRef]
6. Yu, H.F.; Lu, J.L.; Qiao, P.Z. Localization and size quantification of surface crack of concrete based on Rayleigh wave attenuation model. *Constr. Build. Mater.* **2021**, *280*, 120172. [CrossRef]
7. Fu, C.Q.; Fang, D.M.; Ye, H.L.; Huang, L.; Wang, J.D. Bond degradation of non-uniformly corroded steel rebars in concrete. *Eng. Struct.* **2021**, *226*, 111392. [CrossRef]
8. Gu, C.P.; Ye, G.; Sun, W. A review of the chloride transport properties of cracked concrete: Experiments and simulations. *J. Zhejiang Univ-Sci. A* **2015**, *16*, 81–92. [CrossRef]
9. Li, K.F.; Li, L. Crack-altered durability properties and performance of structural concretes. *Cement Concrete Res.* **2019**, *124*, 105811. [CrossRef]
10. Zhu, H.G.; Fan, J.C.; Pang, S.; Chen, H.Y.; Yi, C. The Depth-Width Correlation for Shrinkage-Induced Cracks and Its Influence on Chloride Diffusion into Concrete. *Materials* **2020**, *13*, 2751. [CrossRef]
11. Ismail, M.; Toumi, A.; François, R. Effect of crack opening on the local diffusion of chloride in cracked mortar samples. *Cement Concrete Res.* **2008**, *38*, 1106–1111. [CrossRef]
12. Yoon, I.S.; Schlangen, E. Experimental examination on chloride penetration through micro-crack in concrete. *KSCE J. Civ. Eng.* **2014**, *18*, 188–198. [CrossRef]
13. Audenaert, K.; Schutter, G.D.; Marsavina, L. Influence of cracks and crack width on penetration depth of chlorides in concrete. *Eur. J. Environ. Civ. Eng.* **2009**, *13*, 561–572. [CrossRef]
14. Djerbi, A.; Bonnet, S.; Khelidj, A.; Baroghel-bouny, V. Influence of traversing crack on chloride diffusion into concrete. *Cement Concrete Res.* **2008**, *38*, 877–883. [CrossRef]

15. Rodriguez, O.G.; Hooton, R.D. Influence of cracks on chloride ingress into concrete. *ACI Mater. J.* **2003**, *100*, 120–126.

16. Weiss, J.; Couch, J.; Pease, B.; Laugesen, P. Influence of Mechanically Induced Cracking on Chloride Ingress in Concrete. *J. Mater. Civ. Eng.* **2017**, *29*, 04017128. [CrossRef]

17. Benkemoun, N.; Hammood, M.N.; Amiri, O. A meso-macro numerical approach for crack-induced diffusivity evolution in concrete. *Constr. Build. Mater.* **2017**, *14*, 72–85. [CrossRef]

18. Akhavan, A.; Rajabipour, F. Quantifying Permeability, Electrical Conductivity, and Diffusion Coefficient of Rough Parallel Plates Simulating Cracks in Concrete. *J. Mater. Civ. Eng.* **2017**, *29*, 04017119. [CrossRef]

19. Yang, K.H.; Singh, J.K.; Lee, B.Y.; Kwon, S.J. Simple Technique for Tracking Chloride Penetration in Concrete Based on the Crack Shape and Width under Steady-State Conditions. *Sustainability* **2017**, *9*, 282. [CrossRef]

20. Yang, K.H.; Cheon, J.H.; Kwon, S.J. Modeling of chloride diffusion in concrete considering wedge-shaped single crack and steady-state condition. *Comput. Concrete* **2017**, *19*, 211–216. [CrossRef]

21. Ye, H.L.; Jin, N.G.; Jin, X.Y.; Fu, C.Q. Model of chloride penetration into cracked concrete subject to drying-wetting cycles. *Constr. Build. Mater.* **2012**, *36*, 259–269. [CrossRef]

22. Marsavinaa, L.; Audenaertb, K.; Schutterb, G.D.; Faura, N.; Marsavinac, D. Experimental and numerical determination of the chloride penetration in cracked concrete. *Constr. Build. Mater.* **2009**, *23*, 264–274. [CrossRef]

23. Xu, J.; Peng, C.; Wan, L.J.; Wu, Q. Effect of Crack Self-Healing on Concrete Diffusivity: Mesoscale Dynamics Simulation Study. *J. Mater. Civil. Eng.* **2020**, *32*, 04020149. [CrossRef]

24. Šavija, B.; Luković, M.; Schlangen, M. Lattice modeling of rapid chloride migration in concrete. *Cem. Concr. Res.* **2014**, *61–62*, 49–63. [CrossRef]

25. Ukrainczyk, N.; Koenders, E.A.B.; Vanbreugel, K. Morphological nature of diffusion in cement paste. In *Concrete Repair, Rehabilitation and Retrofitting*; Alexander, M.G., Beushausen, H.D., Dehn, F., Eds.; Taylor & Francis Group: London, UK, 2012; pp. 321–327.

26. Song, H.W.; Shim, H.B.; Petcherdchoo, A.; Park, S.K. Service life prediction of repaired concrete structures under chloride environment using finite difference method. *Cem. Concr. Comp.* **2009**, *31*, 120–127. [CrossRef]

27. Singh, M.B.; Dalvi, V.H.; Gaikar, V.G. Investigations of clustering of ions and diffusivity in concentrated aqueous solutions of lithium chloride by molecular dynamic simulations. *RSC Adv.* **2015**, *5*, 15328–15337. [CrossRef]

28. Gu, C. Chloride Transport Property and Service Life Prediction of UHPFRCC under Flexural Load. Ph.D. Thesis, Southeast University, Nanjing, China, 2016. (In Chinese).

29. Shafikhani, M.; Chidiac, S.E. Quantification of concrete chloride diffusion coefficient—A critical review. *Cem. Concr. Comp.* **2019**, *99*, 225–250. [CrossRef]

30. Chen, S. Study on Transport Behavior of Chlorides in Concrete Cracks. Master's Thesis, Zhejiang University of Technology, Hangzhou, China, 2020. (In Chinese).

31. Jin, W.L.; Yan, Y.D.; Wang, H.L. Chloride diffusion in the cracked concrete. In *Fracture Mechanics of Concrete and Concrete Structures—Assessment, Durability, Monitoring and Retrofitting of Concrete Structures*; Oh, B.H., Choi, O.C., Chung, L., Eds.; Korea Concrete Institute: Seoul, Korea, 2010; pp. 880–886.

32. Maekawa, K.; Ishida, T.; Kishi, T. *Multi-Scale Modeling of Structural Concrete*; Taylor & Francis: Oxon, UK, 2009; pp. 291–352. [CrossRef]

33. Ni, T.; Zhou, R.; Gu, C.; Yang, Y. Measurement of concrete crack feature with android smartphone APP based on digital image processing techniques. *Measurement* **2020**, *150*, 107093. [CrossRef]

Article

Influence of Temperature on the Moisture Transport in Concrete

Qingzhang Zhang [1,2], Zihan Kang [1], Yifeng Ling [3,*], Hui Chen [4] and Kangzong Li [1]

[1] College of Civil Engineering, Henan University of Technology, Zhengzhou 450001, China; zqz313@haut.edu.cn (Q.Z.); kzhsgdsb@163.com (Z.K.); lkz857@163.com (K.L.)
[2] Key Laboratory of Performance Evolution and Control for Engineering Structures of Ministry of Education, Tongji University, Shanghai 200092, China
[3] School of Qilu Transportation, Shandong University, Jinan 250002, China
[4] Department of Architecture and Civil Engineering, Oujiang College, Wenzhou University, Wenzhou 325035, China; chenhui0306@wzu.edu.cn
* Correspondence: jemeryling@gmail.com

Abstract: Moisture with harmful ions penetrates into the interior of concrete, which causes deterioration of the concrete structure. In this study, a moisture saturation equilibrium relationship of concrete was tested under different temperatures and relative humidity conditions to develop moisture absorption and desorption curves. Based on experimental data and numerical simulation, a model of moisture transport in concrete was established. The results from the model indicate that the moisture absorption rate was lower at higher temperatures and largely dependent on the saturation gradient, while the desorption was increased at higher temperatures and mostly affected by the saturation gradient. The proposed model was highly in agreement with the experimental data.

Keywords: concrete; humidity; moisture absorption; moisture desorption; numerical simulation

Citation: Zhang, Q.; Kang, Z.; Ling, Y.; Chen, H.; Li, K. Influence of Temperature on the Moisture Transport in Concrete. *Crystals* **2021**, *11*, 8. https://dx.doi.org/10.3390/cryst11010008

Received: 9 December 2020
Accepted: 21 December 2020
Published: 23 December 2020

Publisher's Note: MDPI stays neutral with regard to jurisdictional claims in published maps and institutional affiliations.

1. Introduction

The inevitable ingress of moisture with harmful ions into concrete could reduce the pH value of pore solution. Temperature is a key factor in the rate of moisture transport in concrete because it can change pore pressures to cause concrete spalling at a critical degree [1–3]. The rate of moisture transport in concrete directly affects the time and degree of concrete deterioration. Therefore, understanding the moisture transport process in concrete is essential to design durable concrete structures [4–6]. The main factors determining the rate of moisture transport in concrete include temperature, relative humidity (RH), microstructure and porosity of concrete.

The isothermal absorption and desorption of moisture vapor in a cement-based material reflect the ability of its pore structure to absorb and desorb moisture [7]. Therefore, the moisture absorption and desorption processes are typically characterized using isothermal absorption–desorption curves [8]. Concrete under an environment with different RH levels will finally attain an equilibrium state at a constant temperature when the pore structure reaches a particular moisture saturation level [9]. The Young-Laplace equation describes the relationship between the capillary pressure and aperture, while the Kelvin equation expresses gas–liquid equilibrium relationship between the curvature of liquid surface and vapor pressure. These two equations can be used to transform the isothermal absorption–desorption curve so as to represent the relationship between the capillary pressure and saturation [10,11]. Trabelsi et al. developed an isothermal absorption–desorption curve to describe the moisture desorption using statistical and finite element methods [12]. Neithalath et al. [13] calculated the intrinsic permeability of concrete based on the porosity, specific surface area and tortuosity. Baroghel-Bouny et al. [14] obtained isothermal desorption curves for concrete and determined a theoretical relationship between the relative permeability coefficient and saturation based on findings in the literature [15]. The diffusion coefficient of moisture in concrete was determined, and then the moisture transfer curves under the drying process were obtained.

Zhou and Li [16] studied the concrete permeability using a three-phase composite random aggregate concrete simulation model based on the finite element method. Wang and Ueda [17] discretely divided concrete at the mesoscale to characterize moisture transport and investigated the influence of the interfacial transition zone on the capillary absorption of the concrete. Li et al. [18] established a three-dimensional mesoscale model to evaluate concrete permeability.

In previous numerical studies on moisture transport, the parameters that determine the influences of temperature and RH in the driving force of capillary pressure, have not been derived. In this paper, isothermal adsorption and desorption experiments were carried out at three temperatures (20 °C, 35 °C, 50 °C) to obtain the adsorption–desorption curves. The influences of different temperatures and RHs on the moisture transport process of concrete were analyzed. Further, a model in function of RH and saturations of adsorption and desorption process was regressed based on experimental data. Using the Kelvin equation, the relationship between capillary pressure and saturation was evaluated. A moisture transport model in respect of capillary pressure and saturation of concrete was established and verified with the experiment results. In addition, the moisture transport behavior of concrete at different temperatures and RHs was simulated.

2. Materials and Methods

2.1. Test Materials

Concrete was made of cement, mineral powder, fine aggregate, and coarse aggregate. The binder material comprised P.O42.5 ordinary Portland cement and mineral powder with densities of 3060 and 2890 kg/m^3 respectively, in equal mass fraction. The chemical composition of cement is shown in Table 1. The fine aggregate was well-graded medium sand with a measured apparent density of 2623 kg/m^3 (Table 2). The coarse aggregate had a 5–20 mm continuous particle gradation, a maximum particle size of 20 mm (Table 3), and a measured apparent density of 2710 kg/m^3. The fresh concrete was cast into 120 molds in dimensions of 100 mm × 100 mm × 100 mm. After curing at 20 ± 2 °C and 95% RH for 28 days, each specimen was cut horizontally into three pieces in a dimension of 100 mm × 100 mm × 30 mm. To avoid separation from vibration, only the middle piece was selected for testing. In total, 60 absorption and 60 desorption specimens were prepared. Table 4 provides the mix proportion of concrete in accordance with previous studies [19–21].

Table 1. Chemical compositions of cement in mass.

Composition	SiO_2	Fe_2O_3	Al_2O_3	CaO	MgO	Na_2O	SO_3
%	22.4	3.7	4.7	60.3	2.7	0.13	2.1

Table 2. Particle size distribution of fine aggregate.

Size of Screen Mesh (mm)	4.75	2.36	1.18	0.60	0.30	0.15
Passing percent (%)	99.5	87.7	72.0	45.9	19.6	2.1

Table 3. Particle size distribution of coarse aggregate.

Size of Screen Mesh (mm)	26.5	16	9.5	4.75	2.36
Passing percent (%)	100.0	76.0	14.4	0.3	0.0

Table 4. Concrete mix ratio for the absorption and desorption specimens.

Water kg/m³	Binder Material kg/m³	Fine Aggregate (Sand) kg/m³	Coarse Aggregate kg/m³
261.0	562.0	514.0	960.0

2.2. Test Method

The tests were performed as per ISO12571-2013 [22]. The specimens for the absorption test were first oven-dried at 100 °C for 24 h and then placed in a dryer to cool down to 20 °C (room temperature) and weighed. Subsequently, all specimens were oven-dried for another 12 h, cooled down to 20 °C in the dryer and weighed. These steps were repeated until the change in mass was less than 0.5% to obtain a dry state. For the desorption test, all dry specimens were placed in water for seven days to be saturated, surface-dried with a towel and weighed. Then, the specimens were returned to the water for 24 h and weighed. This was repeated until the change in the specimen mass was less than 0.5% to obtain a saturated state.

Afterward, the specimens were moved to glass containers in a constant-temperature/humidity chamber with a temperature accuracy of ±1 °C and 50 ± 2% RH as shown in Figure 1. During the test, saturated salt solutions of LiCl, $MgCl_2$, KBr and KNO_3 were added at the bottom of glass containers to secure 10%, 35%, 80% and 95% RH respectively. It should be noted that for the specimens at 50% RH, there was no saturated salt solution. The testing temperatures were 20 °C, 35 °C and 50 °C.

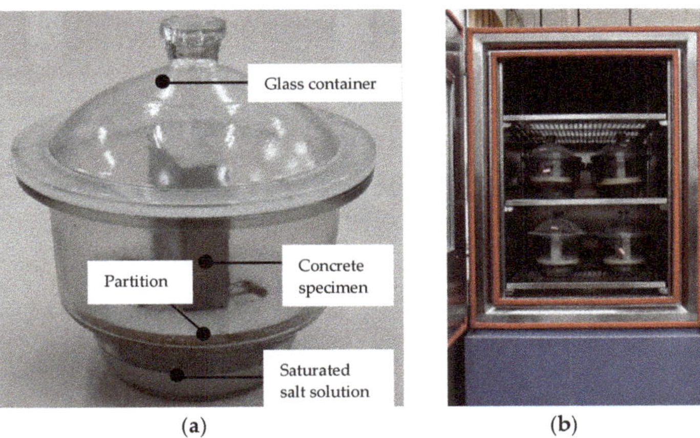

Figure 1. Isothermal adsorption devices (**a**) glass container, (**b**) constant-temperature/humidity chamber.

Each specimen was weighed at designated time interval until the change in mass was less than 0.1%. It should be noted that since the mass change was decreasing over time, the time interval of weight measurement in the early stage was shorter than in the later stage. In total, there were 12 weight measurements for the absorption test and 13 weight measurements for the desorption test.

The specimens were designated as X-% and P-% for the absorption and desorption tests, respectively. The percentage represented the RH conditions. For example, specimens X-10% is the absorption specimens for 10% RH.

2.3. Determination of Saturation

The moisture content of a specimen can be defined as follows [8]:

$$S = \frac{V_S}{V_\phi} \tag{1}$$

where S is the moisture saturation, V_S is the volume occupied by water in the pores of the specimen, and V_ϕ is the volume of pores in the specimen.

In order to determine adsorption and desorption isotherms, the moisture saturation S can be expressed based on the water content [23]:

$$S = \frac{m_w - m_d}{m_{ws} - m_d} \tag{2}$$

where m_w is the mass of specimen underwater (kg), m_d is the mass of the specimen in a dry state (kg) and m_{ws} is the mass of the specimen under saturation state (kg).

3. Results and Discussion

3.1. Absorption–Desorption Curves

Absorption mainly refers to physical absorption, which is the phenomenon that moisture enters concrete through capillary pressure, while desorption is the phenomenon that moisture in concrete is transferred from liquid to gas and released from concrete.

Absorption–desorption curves representing the changes in the moisture contents of the specimens over time at 20, 35, and 50 °C are shown in Figure 2.

Figure 2a shows that, at the same temperature, the absorption curves tended to be consistent between the different RH conditions, as the specimen saturation increased significantly during the initial stage of the absorption process. As absorption progressed, the saturation between the interior and exterior of concrete decreased due to decreased moisture absorption capacity and eventually stabilized. The moisture absorption behaviors of the specimens were similar at different temperatures. For a given temperature, a higher RH resulted in faster moisture absorption. When the RH was lower than 50%, the moisture-absorption process of the specimen was gradual, which reveals that temperature and RH had little effect on the time required to reach the equilibrium. The absorption rate and capacity increased sharply when the RH was greater than 50%. At 80% RH, the moisture absorption capacity and equilibrium time both increased as temperature decreased. Under low-RH conditions (i.e., less than 50% RH), the concrete exhibited less moisture absorption capacity with saturation less than 0.1, and temperature's effect on the moisture absorption rate was marginal. When RH was greater than 50%, temperature's effect on the moisture absorption became significant.

During the desorption test, a lower RH came with a greater saturation between the interior and exterior of concrete, and a longer time to reach equilibrium, as shown in Figure 2b. When the RH was lower than 80%, the increment of temperature increased desorption capacity and the time to reach equilibrium. When RH was higher than 80%, the effect of temperature on the moisture transport rate decreased. Regardless of temperature, the saturation at equilibrium was approximately 0.9.

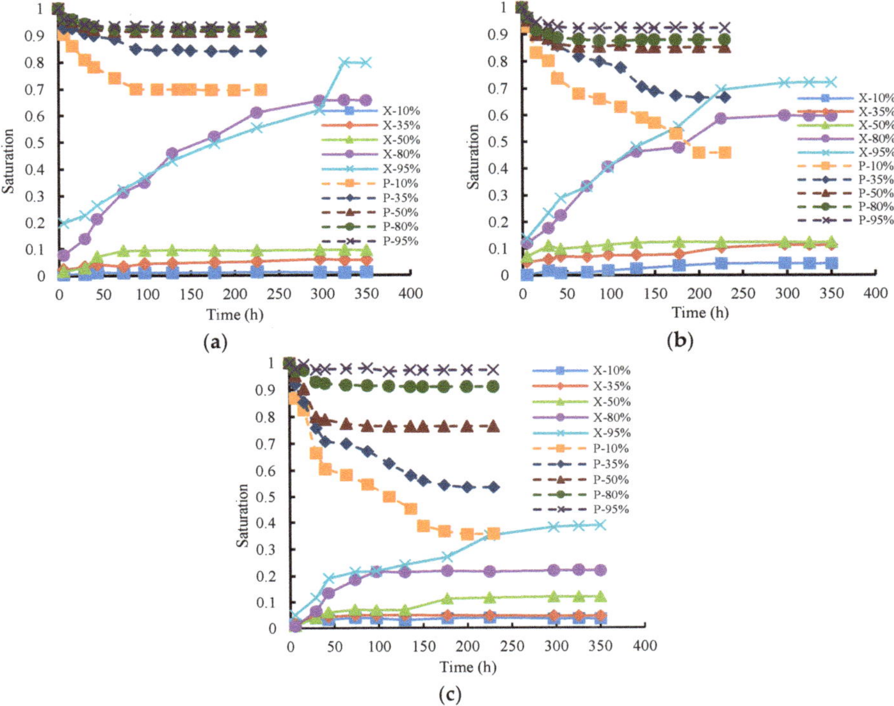

Figure 2. Absorption and desorption over time: (**a**) 20 °C, (**b**) 35 °C, (**c**) 50 °C.

3.2. RH–Saturation Equilibrium

The RH–saturation curves during absorption and desorption for different RH conditions are shown in Figure 3. The curves were substantially similar for different temperatures. As RH increased, the saturation gradually increased. When RH was below 50%, the absorption process was gradual. The absorption capacity significantly increased when RH was greater than 50%. Irrespective of temperature, 50% RH was an inflection point in the moisture absorption process. Based on the Kelvin equation, the maximum pore size that could be saturated was approximately 4 nm at 50% RH, and the boundary between the gel pores and capillary pores was generally 10 nm (i.e., gel pores < 10 nm < capillary pores). Therefore, when RH was below 50%, only part of the gel pores was saturated, while for RH above 50%, the capillary pores saturated gradually, thereby rapidly increasing the saturation [24]. Further, the saturation decreased with decreasing RH. As temperature decreased, the saturation considerably decreased. At 80% RH, the equilibrium saturation decreased substantially, which indicates that an RH greater than 80% could greatly affect the equilibrium of the moisture desorption process in concrete.

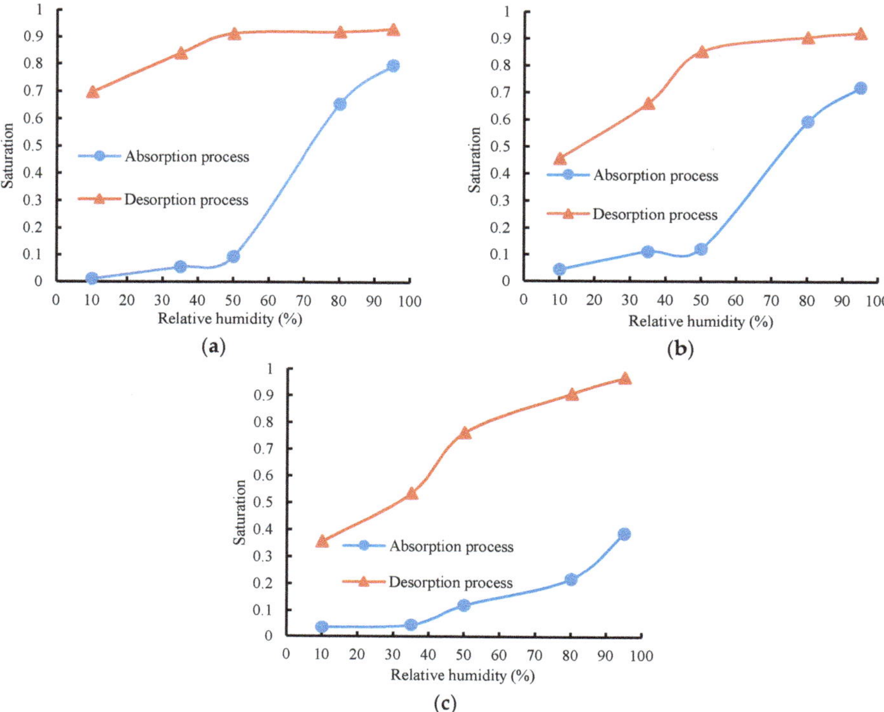

Figure 3. Absorption and desorption equilibrium curves: (**a**) 20 °C, (**b**) 35 °C, (**c**) 50 °C.

In Figure 3, since the desorption curve is above the absorption curve, it indicates a higher saturation for a given RH. The desorption process displays a significantly hysteretic nature compared to the absorption process, which implies that the gas–liquid equilibrium in pores is nonhomogeneous. The hysteresis represents the "ink bottle effect," which refers to the effect that a small pore (bottleneck) exerts stress upon the water in a connected large pore (bottle) [25]. Larger concrete capillary pores require a higher RH to reach equilibrium saturation than small pores. For initially saturated pores, the liquid water in large pores transports through the liquid water in small pores without forming a gas–liquid interface. Therefore, the water in large pores can be discharged only when the ambient RH is below the saturated RH of small pores. For a given RH, the saturation equilibrium of the moisture absorption process is lower than that of the moisture desorption process [26].

For a given RH, the capillary condensation phenomenon is more likely to occur during the absorption process, and a higher saturation can be attained at a lower temperature. During desorption, at a higher temperature, the saturation is lower because the pore water evaporates more easily [27]. According to a study by Zeng [8], the relationship between the capillary pressure and RH is:

$$p_c = \frac{\rho_l RT}{M} \ln h \tag{3}$$

where M is the molar mass of water (kg/mol), R is the gas constant (given as 8.314 J/mol/K), p_c is the capillary pressure (Pa), h is the RH, T is the absolute temperature (K) and ρ_l is the density of liquid water (kg/m^3).

Using Equation (3), the isothermal equilibrium curve of RH and concrete saturation can be converted into a moisture characteristic curve, as shown in Figure 4. When the concrete was at the same saturation through absorption and desorption, the capillary pressure of the desorbed concrete was greater than that of the absorbed concrete, reflecting the hysteretic nature of the desorption process. When the temperature was lower during both the absorption and desorption processes, the capillary pressure was greater, as well as the equilibrium saturation corresponding to the same capillary pressure.

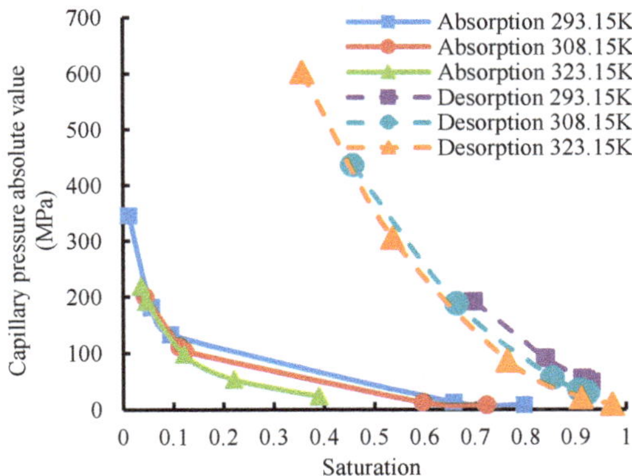

Figure 4. Equilibrium relationship between the capillary pressure and saturation.

3.3. RH and Saturation Equilibrium Model

In [28], a model representing the absorption of water by the hygroscopic material and RH is shown as follows:

$$W = \left[\frac{-\ln(1-h)}{A(T+B)} \right]^{\frac{1}{C}} \tag{4}$$

where W is the equilibrium moisture content and A, B and C are coefficients.

By replacing the equilibrium moisture content with the saturation, Equation (4) can be applied to find the inverse function. The RH and saturation relationship is:

$$h = 1 - \exp(-A(T+B)S^C) \tag{5}$$

where S is the saturation.

The model of the relationship between RH and equilibrium saturation can be obtained by substituting the data from the experimental results as seen in Figure 3 into Equation (5). Nonlinear regression was used to yield Equations (6) and (7) for the absorption and desorption processes, respectively.

$$h_{as} = 1 - \exp(-0.03(T - 180.7)S^{0.837}) \qquad R^2 = 0.91 \tag{6}$$

$$h_{ds} = 1 - \exp(-0.05(T - 258.76)S^{5.04}) \qquad R^2 = 0.91 \tag{7}$$

4. Moisture Transport Model

4.1. Moisture Movement in Unsaturated Concrete

The transport of liquid water in unsaturated concrete can be expressed by Darcy's law. For saturated concrete, the permeability coefficient is constant and a function of saturation for unsaturated concrete.

According to the Darcy–Buckingham equation, the relationship between transport rate of liquid water and capillary pressure is expressed as [11]:

$$v_l = -\frac{k_l}{\mu}\text{grad}(p_c) \tag{8}$$

where v_1 is the rate (m/s); k_1 is the effective permeability for moisture transport (m^2), which is related to the moisture content in concrete and micro-geometric parameters of the pore occupied by moisture; μ is the dynamic viscosity coefficient of water (Pa·s).

If the conversion of liquid water to water vapor in the pores of concrete is neglected, the moisture transport in the concrete conforms to the law of mass conservation. The moisture transport equation in concrete can be written as below [29,30]:

$$\frac{\partial \theta}{\partial t} = \text{div}\left[\frac{k_1}{\mu}\text{grad}(p_c)\right]$$ (9)

Converting the left side of Equation (9) to pore moisture content and the right side into the pore moisture saturation gradient, it becomes:

$$\frac{\partial \theta}{\partial t} = \text{div}\left[\frac{k_1}{\mu}\frac{\partial p_c}{\partial S}\text{grad}(S)\right]$$ (10)

where θ is the moisture content of the concrete.

The effective permeability to moisture transport (k_1) can be expressed as

$$k_1 = kk_{rl}$$ (11)

Therefore, Equation (10) can be rewritten as follows:

$$\frac{\partial \theta}{\partial t} = \text{div}\left[\frac{kk_{rl}}{\mu}\frac{\partial p_c}{\partial S}\text{grad}(S)\right]$$ (12)

where k is the intrinsic permeability of concrete (m^2), and k_{rl} is the relative permeability, which is a parameter associated with the saturation.

Equation (12) can be expressed as:

$$\frac{\partial \theta}{\partial t} = \text{div}[D(S)\text{grad}(S)]$$ (13)

$$D(S) = \frac{k}{\mu}k_{rl}\frac{\partial p_c}{\partial S}$$ (14)

where $D(S)$ is the transport coefficient of liquid water (m^2/s).

The relationship between the dynamic viscosity coefficient of water (μ), and temperature can be expressed as follows: [31]

$$\mu = \frac{0.001775}{[1 + 0.0837(T - 273.15) + 0.000221(T - 273.15)^2]}$$ (15)

The intrinsic permeability (k) of concrete can be determined based on the porosity, specific surface area and tortuosity of concrete using the Kozeny–Carman model [13]:

$$k = \frac{\phi^3}{F_s \tau_c^2 \Omega^2 (1 - \phi)^2}$$ (16)

where F_s is the factor representing the influence of pore shape (2 for a circular and tubular pore), τ_c is the tortuosity and Ω is the specific surface.

4.2. Relationship between the Moisture Transport Parameters

By substituting Equations (6) and (7), which represent the equilibrium relationship between RH and saturation, into Equation (3), the relationship between the capillary pressure and saturation during the moisture absorption and desorption processes can be expressed as Equations (17) and (18), respectively.

$$p_{as} = \frac{\rho_l RT}{M} \ln(1 - \exp(-0.03(T - 180.7)S^{0.837}))$$ (17)

$$p_{ds} = \frac{\rho_l RT}{M} \ln(1 - \exp(-0.05(T - 258.76)S^{5.04}))$$ (18)

The relationship between the relative permeability ($k_{rl}(S)$), and relative saturation can be established based on previous studies [14,15,32]:

$$p_c = A_a(S^{-1/m} - 1)^{1-m}$$ (19)

$$k_{rl}(S) = \sqrt{S}[1 - (1 - S^{1/m})^m]^2$$ (20)

where m and A_a are undetermined coefficients.

The value of m can be determined through regression of the capillary pressure versus saturation curve. The absolute value of capillary pressure m for absorption and desorption processes at different temperatures can be derived based on Equation (19) through nonlinear fitting.

Using the data in Table 5 to fit the relationship between m and T for the absorption and desorption processes gives Equations (21) and (22), respectively.

$$m_{as} = 1/(2.89\ln(T) - 14.89)$$ (21)

$$m_{ds} = 1/(21.06 - 3.16\ln(T))$$ (22)

Table 5. m values corresponding to the absorption and desorption process.

	Moisture Absorption			Moisture Desorption		
T(K)	293.15	308.15	323.15	293.15	308.15	323.15
m	0.672	0.579	0.573	0.328	0.324	0.362
R^2	0.985	0.988	0.991	0.999	0.997	0.992

Equations (21) and (22) are then substituted into Equation (20) to express the relationship between the relative permeability and saturation during the moisture absorption process.

$$k_{rl-as}(S) = \sqrt{S}[1 - (1 - S^{2.89\ln(T)-14.89})^{1/[2.89\ln(T)-14.89]}]^2$$ (23)

Thus, the relationship between the relative permeability and saturation during the moisture desorption process is:

$$k_{rl-ds}(S) = \sqrt{S}[1 - (1 - S^{21.06-3.16\ln(T)})^{1/[21.06-3.16\ln(T)]}]^2$$ (24)

4.3. Moisture Transport Model

The relationships between capillary pressure and saturation during the process of moisture absorption and desorption are shown in Equations (17) and (18). These equations are then substituted into Equation (13) to develop the moisture transport model for absorption and desorption processes.

$$\frac{\partial \phi S}{\partial t} = \text{div}[D(S)\text{grad}(S)]$$ (25)

where ϕ is the porosity of concrete.

Transport coefficient $D(S)$ of Equation (25) in absorption and desorption processes are then represented as Equations (26) and (27), respectively.

$$D_{as}(S) = \frac{0.02511kk_{rl-as}\rho_l RT(T-180.7)S^{-0.163}\exp(-0.03(T-180.7)S^{0.837})}{\mu M(1-\exp(-0.03(T-180.7)S^{0.837}))} \tag{26}$$

$$D_{ds}(S) = \frac{0.252kk_{rl-ds}\rho_l RT(T-258.76)S^{4.04}\exp(-0.05(T-258.76)S^{5.04})}{\mu M(1-\exp(-0.05(T-258.76)S^{5.04}))} \tag{27}$$

4.4. Heat Balance Equation

The heat balance equation correlated to the saturation can be established based on Fourier's law [11,33]:

$$c\rho\frac{\partial T}{\partial t} + \nabla \cdot [-\lambda_S \nabla T] = Q \tag{28}$$

where c is the specific heat of material (J/kg·K), λ_S is the thermal conductivity at different saturations (W/m·K), ρ is the material density (kg/m^3) and Q is the heat flux (J/m^2·s). The negative sign indicates that the direction of heat flow is opposite to the direction of the temperature gradient.

The transient plane source (TPS) technique was used to determine the thermal conductivity for the concrete in the same mix proportion as in the present study and at the same saturation values of 0, 0.3, 0.5, 0.7, 0.9 and 1.0 as in the moisture absorption and desorption tests described in Section 2.1. The relationship between the thermal conductivity and saturation is:

$$\lambda_S = (1 + 0.3432S^2)\lambda_a \tag{29}$$

where λ_a is the thermal conductivity under dry state (W/m·K).

5. Numerical Simulation

5.1. Model Parameters and Mesh Generation

The concrete was considered as a uniform and isotropic continuous medium; thus the concrete skeleton was assumed as an impermeable material that did not react with gas or liquid phases. COMSOL®® software (COMSOL Inc., Stockholm, Sweden) was used to establish a three-dimensional numerical model based on Equations (25)–(28), and one-dimensional transmission was adopted. The physical parameters of heat and moisture transport are shown in Tables 6 and 7. The size of the simulated specimen was the same as that of the experimental specimen used in the test, namely 30 mm × 100 mm × 100 mm. A total of 9520 tetrahedral elements were used in COMSOL®® software to develop the finite element model with DOFs of 28,186 and convergence criteria of 10^{-5}, as shown in Figure 5.

Table 6. Heat transfer physical parameters.

Parameter	Dry Thermal Conductivity, λ_a (W/m·K)	Specific Heat, c (J/kg·K)	Concrete Density, ρ	Porosity, ϕ
Value	1.578	900	2297	0.1796

Table 7. Moisture transport physical parameters.

Parameter	Intrinsic Permeability, k (m^2)	Water Density, ρ_1 (kg/m^3)	Gas Constant R (J/mol·K)	Molar Mass of Water, M (kg/mol)
Value	1.413×10^{-21}	1000	8.3144	0.018

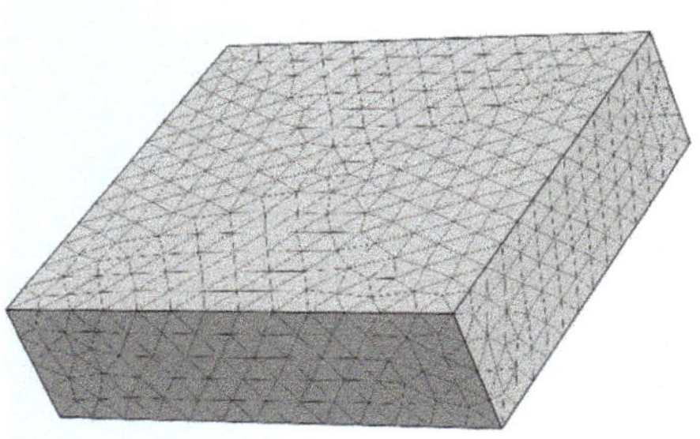

Figure 5. Mesh elements division of concrete.

5.2. Moisture Transport Model Validation

The experimental results shown in Figure 2 were simulated for 80% and 95% RH at 20, 35 and 50 °C to validate the moisture transport model. The initial internal saturation of the concrete was set as 0.08, and the boundary saturation was the corresponding equilibrium saturation. The initial temperature of inside concrete was 283.15 K, and the simulation time was the hygroscopic equilibrium time for the three temperatures, which was approximately 350 h. The simulated results were compared with the experimental results, as shown in Figure 6.

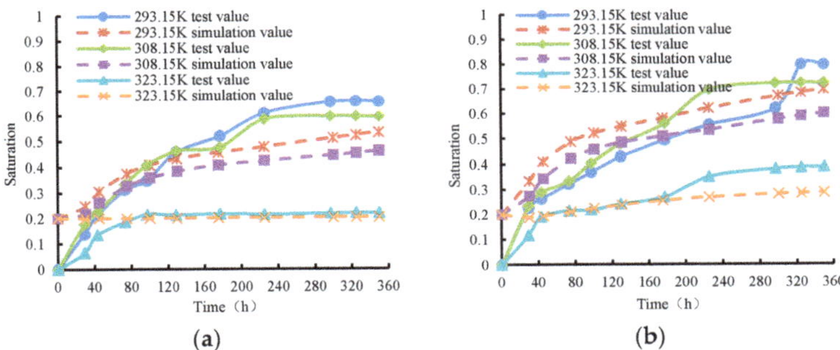

Figure 6. Test and simulation values of moisture absorption process: (**a**) is 80% RH, (**b**) is 95% RH.

The simulated and experimental results show a similar trend, and the values also agreed well. The simulated results were slightly lower than the experimental results. Under the simulated conditions, the moisture absorption rate decreased, and the moisture absorption capacity was slightly lower than that in the experiment. This could be because the moisture transport model only considers the transport of liquid water, not the moisture vapor transport that occurred under low-saturation conditions.

The experimental results provided in Figure 2 were applied to validate the moisture desorption model. The simulation was performed under the conditions of 10% and 35% RH at 20, 35 and 50 °C. The initial temperature inside the concrete was 283.15 K. The concrete interior was initially saturated and the boundary saturation was the corresponding equilibrium saturation. The simulation time was 250 h.

Figure 7 provides a comparison of the simulated and experimental results, which shows that the simulated values were in good agreement with the experimental values under 35% RH and were slightly higher than the experimental values at 10% RH. The overall results demonstrated that the simulated model of moisture desorption was valid. This also indicates that the simulated results remained accurate when the concrete was at a higher saturation level. When the saturation decreased, the simulated moisture desorption rate was slightly lower than the experimental rate, implying that the water vapor transport in the concrete exerted limited influence under the low-saturation condition. In the marine environment, it is difficult for concrete to reach a low saturation state due to high RH. Therefore, the simulated model of moisture transport was accurate for the applications in the high RH environment.

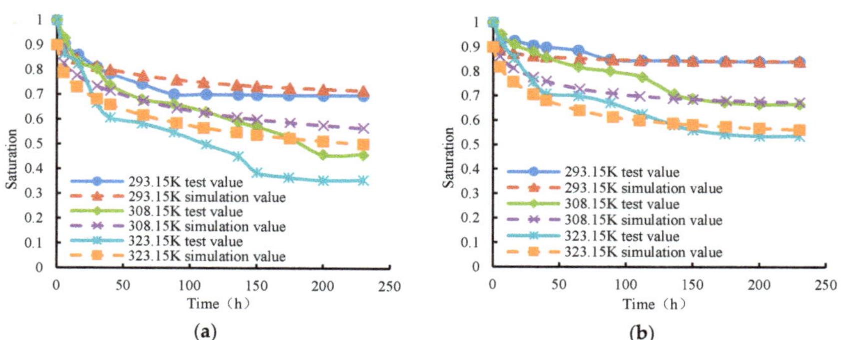

Figure 7. Test and simulation values of moisture desorption process: (**a**) is 10% RH, (**b**) is 35% RH.

5.3. Absorption

The upper, lower and right ends of the two-dimensional geometric model of concrete were defined as heat and moisture insulation. Therefore, moisture absorption was from left to inside. The effect of temperature on the moisture absorption process at 95% RH was evaluated under the temperatures of 293.15, 308.15 and 323.15 K and the boundary saturation values of 0.80, 0.72, and 0.39, respectively. The simulation had an initial temperature of 273.15 K, saturation of 0.2 and time of 500 days.

The simulated results are shown in Figure 8. It can be seen that the saturation was the lowest at 323.15 K (i.e., the highest temperature) but the highest at 293.15 K (i.e., the lowest temperature). For a given RH, an increased temperature decreased boundary saturation, moisture transport rate, moisture transport capacity, moisture transport distance and moisture absorption. The saturation gradient is the main driving potential of moisture transfer. A higher temperature reduced moisture absorption rate. Meanwhile, with temperature increasing, the equilibrium saturation of the concrete boundary further decreased. Such a combined effect decreased moisture absorption of concrete.

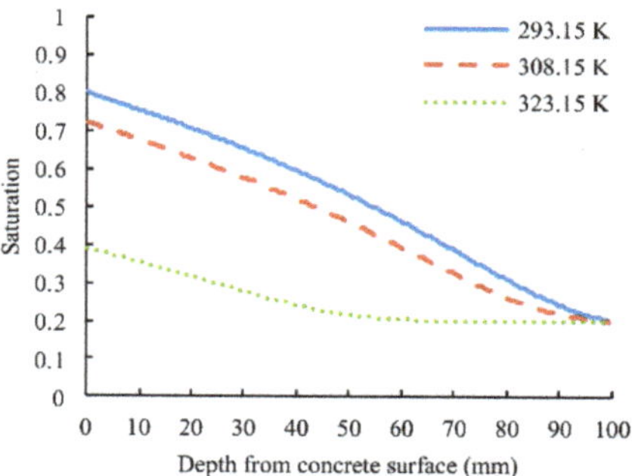

Figure 8. Simulated distribution at different temperatures.

The combined influences of the temperature and saturation in the moisture transport process were investigated at three conditions: high temperature and RH (boundary temperature T = 323.15 K, boundary saturation S = 0.9), medium temperature and RH (boundary temperature T = 308.15 K, boundary saturation S = 0.75) and low temperature and RH (boundary temperature T = 293.15 K, boundary saturation S = 0.6). The simulation had an initial temperature of 273.15 K, saturation of 0.2 and time of 500 days. The simulated results are provided in Figure 9, which shows the same trend among the three conditions. The transport rate is directly correlated with temperature and RH. It also shows that the saturation gradient mostly affected moisture transport rate. Although the saturation was large at high temperature, the moisture transport rate was notably high. Therefore, the effect of RH on the moisture transport rate in the concrete was greater than that of temperature. The transport process of moisture in concrete at the temperature of 293.15 K and saturation 0.6 for simulation at 500d is shown in Figure 10. It can be seen that the moisture is gradually transferred from the left end of the concrete to the interior, and the saturation is reduced with depth increasing, which is coherent with the trend in Figure 9.

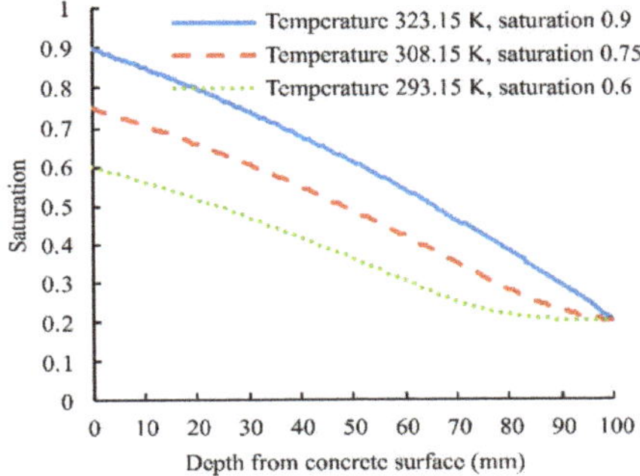

Figure 9. Saturation distribution during the absorption simulation.

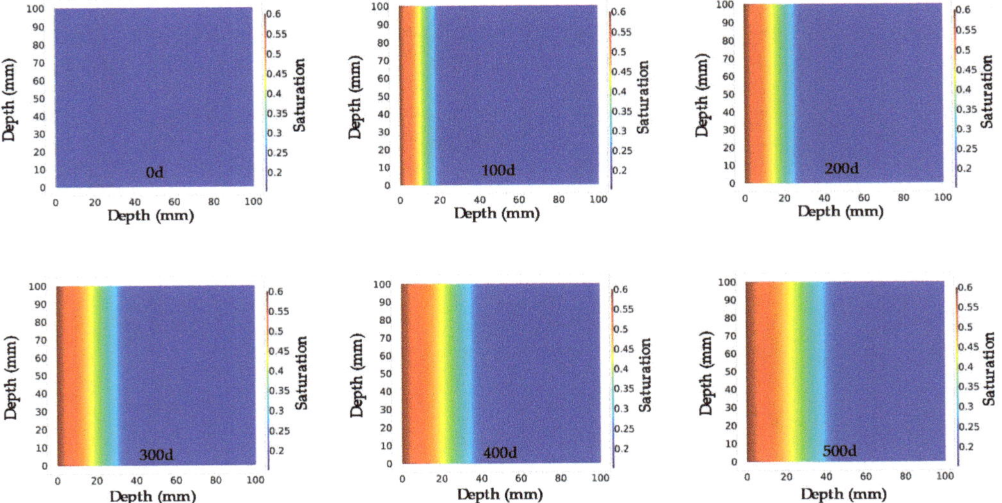

Figure 10. Cloud atlas of moisture in absorption process at temperature of 293.15 K and saturation of 0.6.

5.4. Desorption

The upper and lower ends of the two-dimensional geometric model of concrete were treated as heat and moisture insulation. The direction of moisture desorption was from inside to both sides. The effect of temperature on the moisture desorption process at 10%

RH was analyzed at temperatures of 293.15, 308.15 and 323.15 K and boundary saturation values of 0.70, 0.46 and 0.36, respectively. The simulation time was 200 days with an initial temperature of 273.15 K and saturation of 0.9.

Figure 11 presents the simulated results. It illustrates that, for a given RH, higher temperatures lowered the boundary saturation. When the internal saturation of concrete was higher, the difference in saturation was greater between the outside and inside of concrete, resulting in a significant loss of moisture in the concrete. The boundary saturation was lower at higher temperatures, which increased the moisture desorption rate as well as the moisture desorption capacity.

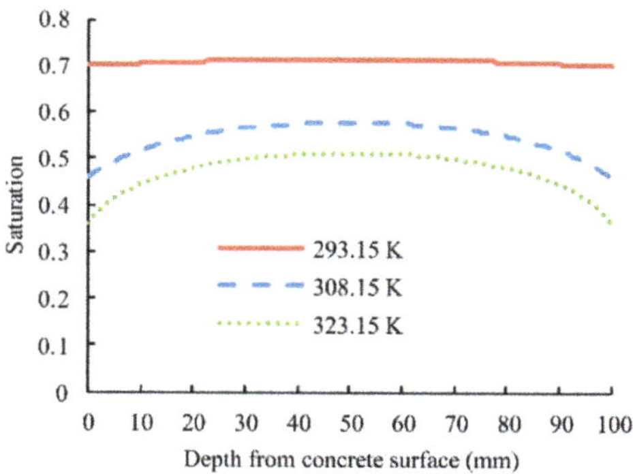

Figure 11. Saturation distribution during the desorption simulation.

Additionally, the combined effects of temperature and saturation on the moisture transport process during desorption were investigated for three conditions: low temperature and high RH (boundary temperature T = 293.15 K, internal saturation S = 0.9), medium temperature and RH (boundary temperature T = 308.15 K, internal saturation S = 0.75) and high temperature and low RH (boundary temperature T = 323.15 K, internal saturation S = 0.6). The simulation time was 500 days with an initial temperature of 273.15 K and saturation of 0.2.

As shown in Figure 12, the three curves displayed a similar trend. The saturation values for the three conditions at the internal depth of 50 mm were 0.552, 0.481 and 0.417, respectively. Even at low temperature, the greatest moisture transport rate was observed at the highest RH. The temperature had a greater effect on the moisture desorption process than saturation. The change of saturation of moisture in the desorption process at a temperature of 308.15 K and saturation of 0.75 is shown for the simulation at 500d in Figure 13. This indicates that the moisture was uniformly desorbed to both sides. The rate of saturation reduction was high close to both sides, while it was slow in the middle. This is due to the fact that the boundary temperature is higher at the concrete's surface, resulting in faster moisture transport.

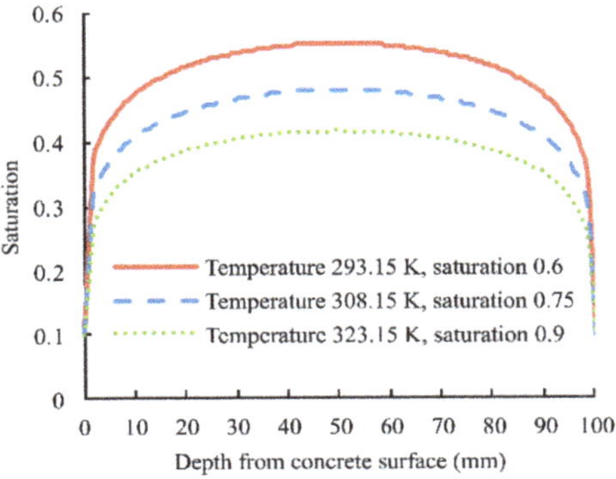

Figure 12. Internal average saturation distribution during the desorption process.

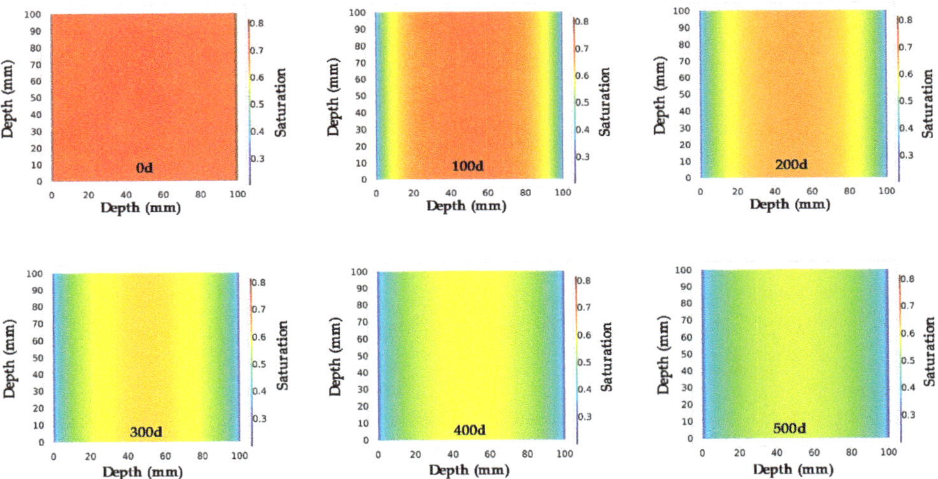

Figure 13. Cloud atlas of moisture in the desorption process at a temperature of 308.15 K and saturation of 0.75.

6. Conclusions

This study sought to elucidate the mechanisms of moisture transport in concrete. A numerical model was developed. Based on the experimental and simulated results, the following conclusions can be drawn.

(1) Under isothermal conditions, a higher RH increased the moisture absorption rate, and the moisture absorption capacity was greater at lower temperatures. When RH exceeded 50%, the moisture absorption rate increased significantly. Similarly, a lower RH resulted in a faster desorption, and higher temperatures increased the moisture desorption capacity.

(2) A model of the relationship between RH and equilibrium saturation in absorption and desorption processes was proposed. A formula in the function of capillary pressure and moisture saturation of concrete was established, and a model of moisture transport in the concrete was developed. The experimental results verified the accuracy of the moisture transport model.

(3) Based on the simulated results, an increased temperature reduced the rate of moisture absorption. The saturation had more effects on the absorption rate than tempera-

ture. Although an increment of temperature increased the moisture desorption rate, the saturation had more effects on the moisture desorption than temperature.

(4) Under low saturation, the transport of water vapor affected the moisture desorption rate in concrete. Hence, the transport of water vapor should be further considered in the moisture transport model.

Author Contributions: Conceptualization, Q.Z. and H.C.; investigation, Z.K.; data curation, Q.Z., Z.K. and K.L.; methodology, Q.Z.; visualization, Y.L.; formal analysis, Z.K.; software, K.L.; writing—original draft, Z.K.; writing—review and editing, Y.L., Q.Z., Z.K., H.C. and K.L.; project administration, Q.Z.; funding acquisition, Q.Z.; validation, Q.Z., Y.L. and H.C.; resources, H.C. All authors have read and agreed to the published version of the manuscript

Funding: This work was financially supported by the National Natural Science Foundation of China [grant number 51509084]; the Foundation of Key Laboratory of Performance Evolution and Control for Engineering Structures of Ministry of Education, Tongji University [grant number 2018KF-2]; the Cultivation Plan for Youth Backbone Teachers of Institution of Higher Education by Henan Province [grant number 2019GGJS086]; the Fundamental Research Funds for the Henan Provincial Colleges and Universities in Henan University of Technology [grant number 2017RCJH03]; the Cultivation Plan for Youth Backbone Teachers by Henan University of Technology; and the Innovative Funds Plan of Henan University of Technology (Grant No. 2020ZKCJ05).

Conflicts of Interest: The authors declare no conflict of interest.

References

1. Kodur, V.; Naser, M. *Structural Fire Engineering*; MGH Press: New York, NY, USA, 2020. Available online: https://www.researchgate.net/publication/340183424_Structural_Fire_Engineering (accessed on 22 December 2020).
2. Khoury, G. Passive fire protection in tunnels. *Concrete* **2003**, *37*, 31–36.
3. Kodur, V.; Phan, L. Critical factors governing the fire performance of high strength concrete systems. *Fire Saf. J.* **2007**, *42*, 482–488. [CrossRef]
4. Zhang, Y.; Ye, G. A model for predicting the relative chloride diffusion coefficient in unsaturated cementitious materials. *Cem. Concr. Res.* **2019**, *115*, 133–144. [CrossRef]
5. Homan, L.; Ababneh, A.N.; Xi, Y. The effect of moisture transport on chloride penetration in concrete. *Constr. Build. Mater.* **2016**, *125*, 1189–1195. [CrossRef]
6. Li, K.; Li, C.; Chen, Z. Influential depth of moisture transport in concrete subject to drying–wetting cycles. *Cem. Concr. Compos.* **2009**, *31*, 693–698. [CrossRef]
7. Baroghel-Bouny, V. Water vapour sorption experiments on hardened cementitious materials: Part I: Essential tool for analysis of hygral behaviour and its relation to pore structure. *Cem. Concr. Res.* **2007**, *37*, 414–437. [CrossRef]
8. Zeng, Q.; Wang, Y.; Li, K. Uniform model for moisture transport in porous materials and its application to concrete at selected Chinese regions. *J. Mater. Civ. Eng.* **2014**, *26*, 05014001. [CrossRef]
9. Li, C. Study on Water and Ionic Transport Processes in Cover Concrete under Drying-Wetting Cycles. Ph.D. Thesis, Tsinghua University, Beijing, China, 2009.
10. Ishida, T.; Maekawa, K.; Kishi, T. Enhanced modeling of moisture equilibrium and transport in cementitious materials under arbitrary temperature and relative humidity history. *Cem. Concr. Res.* **2007**, *37*, 565–578. [CrossRef]
11. Černý, R.; Rovnanikova, P. *Transport Processes in Concrete*, 1st ed.; CRC Press: London, UK, 2002. [CrossRef]
12. Trabelsi, A.; Belarbi, R.; Turcry, P.; Aït-Mokhtar, A. Water vapour desorption variability of in situ concrete and effects on drying simulations. *Mag. Concr. Res.* **2011**, *63*, 333–342. [CrossRef]
13. Neithalath, N.; Sumanasooriya, M.S.; Deo, O. Characteriz ing pore volume, sizes, and connectivity in pervious concretes for permeability prediction. *Mater. Charact.* **2010**, *61*, 802–813. [CrossRef]
14. Baroghel-Bouny, V.; Mainguy, M.; Lassabatere, T.; Coussy, O. Characterization and identification of equilibrium and transfer moisture properties for ordinary and high-performance cementitious materials. *Cem. Concr. Res.* **1999**, *29*, 1225–1238. [CrossRef]
15. Van Genuchten, M.T. A closed-form equation for predicting the hydraulic conductivity of unsaturated soils. *Soil Sci. Soc. Am. J.* **1980**, *44*, 892–898. [CrossRef]
16. Zhou, C.; Li, K. Numerical and statistical analysis of permeability of concrete as a random heterogeneous composite. *Comput. Concr.* **2010**, *7*, 469–482. [CrossRef]
17. Wang, L.; Ueda, T. Mesoscale modeling of water penetration into concrete by capillary absorption. *Ocean Eng.* **2011**, *38*, 519–528. [CrossRef]
18. Li, X.; Xu, Y.; Chen, S. Computational homogenization of effective permeability in three-phase mesoscale concrete. *Constr. Build. Mater.* **2016**, *121*, 100–111. [CrossRef]

19. Zhang, Q.; Tian, S.; Li, K.; Li, Z. Experiment and analysis of thermal conductivity of marine concrete with different composition. *IOP Conf. Ser. Earth Environ. Sci.* **2019**, *242*, 1–6. [CrossRef]
20. Wu, Q.; Yu, H. Rebar corrosion rate estimation of reinforced concrete components exposed to marine environment. *Strength Mater.* **2019**, *51*, 653–659. [CrossRef]
21. Cheewaket, T.; Jaturapitakkul, C.; Chalee, W. Initial corrosion presented by chloride threshold penetration of concrete up to 10 year-results under marine site. *Constr. Build. Mater.* **2012**, *37*, 693–698. [CrossRef]
22. BSI Standards Publication. ISO 12571:2013 Hygrothermal Performance of Building Materials and Products—Determination of Hygroscopic Sorption Properties, CEN-CENELEC Management Centre. 2013. Available online: https://www.iso.org/cms/render/live/en/sites/isoorg/contents/data/standard/06/13/61388.html (accessed on 22 December 2020).
23. Wei, X.; Jin, L.; Zhang, Y.; Ding, Z. Water transport in cracked unsaturated concrete. *J. Build. Mater.* **2018**, *21*, 725–731. [CrossRef]
24. Schiller, P.; Wahab, M.; Bier, T.; Mögel, H.-J. A model for sorption hysteresis in hardened cement paste. *Cem. Concr. Res.* **2019**, *123*, 105760. [CrossRef]
25. Jiang, Z.; Xi, Y.; Gu, X.; Huang, Q.; Zhang, W. Modelling of water vapour sorption hysteresis of cement-based materials based on pore size distribution. *Cem. Concr. Res.* **2019**, *115*, 8–19. [CrossRef]
26. Ranaivomanana, H.; Verdier, J.; Sellier, A.; Bourbon, X. Toward a better comprehension and modeling of hysteresis cycles in the water sorption–desorption process for cement based materials. *Cem. Concr. Res.* **2011**, *41*, 817–827. [CrossRef]
27. Poyet, S. Experimental investigation of the effect of temperature on the first desorption isotherm of concrete. *Cem. Concr. Res.* **2009**, *39*, 1052–1059. [CrossRef]
28. Yu, C. *Numerical Analysis of Heat and Mass Transfer for Porous Materials*, 1st ed.; THU Press: Beijing China, 2011.
29. Baroghel-Bouny, V. Water vapour sorption experiments on hardened cementitious materials. Part II: Essential tool for assessment of transport properties and for durability prediction. *Cem. Concr. Res.* **2007**, *37*, 438–454. [CrossRef]
30. Daian, J.-F. Condensation and isothermal water transfer in cement mortar Part I—Pore size distribution, equilibrium water condensation and imbibition. *Transp. Porous Media* **1988**, *3*, 563–589. [CrossRef]
31. Ji, Z.; Zhu, R.; Li, D. *Transmission Principle*, 1st ed.; HIT Press: Harbin, China, 2017.
32. Savage, B.M.; Janssen, D.J. Soil physics principles validated for use in predicting unsaturated moisture movement in Portland cement concrete. *Mater. J.* **1997**, *94*, 63–70. [CrossRef]
33. Jerman, M.; Černý, R. Effect of moisture content on heat and moisture transport and storage properties of thermal insulation materials. *Energy Build.* **2012**, *53*, 39–46. [CrossRef]

Article

Effect of Synthetic Quadripolymer on Rheological and Filtration Properties of Bentonite-Free Drilling Fluid at High Temperature

Jiangfeng Wang [1], Mengting Chen [1], Xiaohui Li [2], Xuexuan Yin [1] and Wenlong Zheng [1,*]

[1] College of Geosciences and Engineering, North China University of Water Resources and Electric Power, Zhengzhou 450018, China; wangjiangfeng@ncwu.edu.cn (J.W.); zjytgzy@126.com (M.C.); yinxueyan2022@163.com (X.Y.)

[2] The Fifth Institute of Resources and Environment Investigation of Henan Province, No. 99, Nanyang Road, Jinshui District, Zhengzhou 450053, China; xiaohuili2022@163.com

* Correspondence: 15138480305@163.com; Tel.: +86-151-3848-0305

Abstract: High temperature would dramatically worsen rheological behaviors and increase filtration loss volumes of drilling fluids. Synthetic polymers with high temperature stability have attracted more and more attention. In this paper, a novel quadripolymer was synthesized using 2-acrylamido-2-methylpropanesulfonic acid (AMPS), acrylamide (AM), sodium styrene sulfonate (SSS), and dimethyl diallyl ammonium chloride (DMDAAC). Firstly, the molecular structure was studied by Fourier transform–infrared spectroscope (FT-IR) and nuclear magnetic resonance (^{1}H-NMR) analysis. It was shown that the synthetic polymer contained all the designed functional groups. Moreover, the effect of temperature and the quadripolymer concentration on the rheological behavior and filtration loss of the bentonite-free drilling fluid were investigated. It was experimentally established that when the adding amount of the quadripolymer was 0.9 wt%, the prepared drilling fluid systems exhibited relatively stable viscosities, and the filtration losses could be controlled effectively after hot rolling aged within 180 °C. Further, it was confirmed that the bentonite-free drilling fluid containing the synthesized quadripolymer had good reservoir protection performance. In conclusion, the synthetic quadripolymer is a promising rheology modifier and a filtrate reducer for the development of the bentonite-free drilling fluid at high temperature.

Keywords: synthetic polymer; high temperature; bentonite-free drilling fluid; rheology; filtration

Citation: Wang, J.; Chen, M.; Li, X.; Yin, X.; Zheng, W. Effect of Synthetic Quadripolymer on Rheological and Filtration Properties of Bentonite-Free Drilling Fluid at High Temperature. *Crystals* **2022**, *12*, 257. https://doi.org/10.3390/cryst12020257

Academic Editor: Mikhail Osipov

Received: 29 December 2021
Accepted: 12 February 2022
Published: 14 February 2022

Publisher's Note: MDPI stays neutral with regard to jurisdictional claims in published maps and institutional affiliations.

1. Introduction

Drilling fluid is indispensable in oil and gas drilling engineering to maintain wellbore stability, carry and transport drilled cuttings, and reduce water loss. It is divided into water-based drilling fluid system (WBDF), oil-based drilling fluid system (OBDF), and gas-based drilling fluid system (GBDF) [1]. Among them, a water-based working fluid is the most commonly used one because of its environmental friendliness and low cost, and it is mainly composed of bentonite clay and different functional polymer treating agents [2–4]. Bentonite as the most essential component in drilling fluid exhibits incomparable advantages in viscosity and filtration loss control [5,6]. Furthermore, clay particles can plug pores and fractures in the near wellbore area and form a thin and low-permeability filtration cake. However, as the number of deep and ultra-deep wells increases, one of the inevitable problems is that drilling fluid must endure high temperatures [7]. It is worth noting that high bentonite content of water-based drilling fluids at high temperatures would give rise to serious detrimental effects [8–10], for example, deterioration of rheological behavior, increase in filtration volume seeping into formation, particularly formation damages caused by dispersive clay particles. Thus, less or no bentonite in drilling fluids capable of providing necessary performance in harsh operating conditions should be contained to maintain the desired properties [11]. Xiao et al. prepared a bentonite-free drilling fluid with amphoteric polymer (FA367) as the main treating agent, and the filed application results in the shallow formations of Anpeng oilfield showed that the drilling speed greatly increased, and the complex conditions such as leakage and sticking were effectively alleviated [12].

In order that a drilling fluid with low or no bentonite could be appropriate for deep well drilling project, one or more polymers are required to take place of bentonite and provide satisfactory performance, such as proper rheological parameters, low filtration loss volume, and good salt tolerance at high temperature conditions [13–15]. However, one disadvantage of polymers is related to the thermal degradation at elevated temperatures (above 150 °C) [16,17], which gives rise to the rheology and filtration loss being difficult to meet the demands of field application. Thus, it is necessary to develop novel polymers with high temperature resistance to contribute to rheological stability and low filtration loss for bentonite-free drilling fluids.

Synthetic polymers have attracted researchers' attentions for a long time. Water soluble polymers are universally used in water based drilling fluids. The rheological behavior and filtrates of drilling fluid can be adjusted primarily by adding them However, with the increase of well depth, one disadvantage of natural polymer is related to the thermal degradation and oxidation under high temperature (150 °C) [18,19] which could not control the rheology and filtration loss to meet the demands of the field application. Thus, regarding these disadvantages of natural polymer, some researchers were directed to synthetic polymer that could contribute to rheological stability and low fluid loss in water based fluids at HTHP conditions. Perricone et al. used acrylamido-methylpropanesulfonic acid (AMPS), acrylamide (AM), and alkyl acrylamide monomers to synthesize a copolymer (named COP) through inverse micro-emulsion polymerizations [20]. In the field application, the drilling fluid with this synthetic product showed good high temperature resistance. Furthermore, Tao et al. (2011) synthesized a terpolymer of AM/acrylic acid/sodium styrene sulfonate (SSS) and evaluated its high temperature and high pressure rheological property with various salt concentrations [21]. Wu et al. synthesized AM/AMPS/itaconic acid (IA)/N-vinyl caprolactam (NVCap) by solution polymerization. It is reported that drilling fluid containing this terpolymer could control fluid loss and rheological properties after aging at 220 °C [22]. Some other synthetic polymers were developed and were used as rheology modifiers and filtrate reducers, which indicated that the sulfonate structure can be resistant to high temperature and salt [23–25]. The excellent performance of the drilling fluid decides the formation protection and drilling efficiency in deep wells [26].

In this paper, a new quadripolymer was synthesized with monomers of AMPS, AM, DMDAAC, and SSS by aqueous solution polymerization. Among these, amide group provided by AM monomer is mainly responsible for forming the main chain structure. The sulfonic acid group of AMPS plays a role of hydration, and the side group -$(CH_3)_2CH_2SO_3Na$ can enhance the rigidity of the molecular chain, thus improving the thermal stability of the product. The rigid group of benzene ring contained in SSS monomer has a strong hindering effect [27,28]. Then, two double bonds in the molecule of DMDAAC contribute to form mesh structure. Thus, such a quadripolymer was characterized with Fourier transform infrared spectroscopy and nuclear magnetic resonance. Firstly, the ability of the quadripolymer in adjusting rheological and filtration performance of bentonite-free fluids at elevated temperatures were studied by thermal aging test. Meanwhile, the salt tolerance of the bentonite-free drilling fluid was evaluated through an investigation of the viscosity change of the fluid system in different salinity of sodium chloride solution. Additionally, the reservoir protection performance of the bentonite-free drilling fluid containing the quadripolymer was evaluated experimentally.

2. Materials and Methods

2.1. Materials and Experiment Instruments

Acrylamide(AM), 2-acrylamido-2-methylpropanesulfonic acid (AMPS), dimethyl diallyl ammonium chloride(DMDAAC), sodium styrene sulfonate (SSS), chemical grade, were purchased from Aladdin Reagent Co., Ltd. (Shanghai, China); Ammonium persulfate ($(NH_4)_2S_2O_8$) and sodium hydrogen sulfite ($NaHSO_3$), sodium hydroxide ($NaOH$), sodium carbonate (Na_2CO_3), analytical grade, were employed without further purification.

Drispac and HE150 were purchased from Chevron Phillips Chemical Company, Texas, America; 80A51, PAM, LOCKSEAL, PAC-LV and sized $CaCO_3$ were collected from Jiahua Technology Co., Ltd., Jingzhou City, China. The experiment instruments used for this study are listed in Table 1.

Table 1. Experiment instruments and the providers.

Instrument	Provider
Nicolet 6700 Fourier Transform Infrared Spectrometer	Thermo Fisher Scientific Co., Ltd., Waltham, MA, USA
Angilent 400 MHz NMR Spectrum	Angilent Technologies Co., Ltd., California, USA
GJSS-B12K multi–spindle mixer ZNN-D6 viscometer SD–4 API filtration apparatus GCS71-A HTHP filtration apparatus XGRL-4A Hot roller oven	Qingdao Haitongda Special Purpose Instrument Co., Ltd., Qingdao City, Shandong Province, China
ZDY50-180 Core flow tester	Nantong Yichuang Experimental Instrument Co., Ltd., Nantong City, Jiangsu Province, China

2.2. Methods

2.2.1. Synthesis of Quadripolymer

The quadripolymer of AM, AMPS, SSS, DMDAAC was synthesized by solution free radical polymerization. The effect of molar ratio of four monomers, dosage of initiator, pH and reaction temperature on optimizing the best synthesis condition was determined (AM: AMPS: SSS: DMDAAC is 10:3:3:2, redox initiator dosage($(NH_4)_2S_2O_8$) is 0.2 wt%, temperature is 80 °C and reactants system pH is 7) by dosage experiment.

Firstly, a desired amount of mixture monomers was mixed well in a reaction flask and deoxygenated with nitrogen. Secondly, sodium hydroxide was used to adjust the pH value (pH = 7) of the reaction system. The whole reaction process took place in a constant temperature oil bath. Next, a redox initiator ($(NH_4)_2S_2O_8$) dosage of 0.2 wt% was added into the above solution with a constant stirring speed of 200 r/min. After 5 h of reaction time, the white product was filtered off and extracted with acetone and methanol for three times, and finally dried powders were obtained.

2.2.2. Characterization of Molecular Structure

FT-IR spectrum of the quadripolymer was recorded on a Nicolet 6700 Fourier Transform Infrared Spectrometer. A pellet sample made from a mixture of 1 mg quadripolymer and about 100 mg of potassium bromide (KBr) was prepared under a pressure of 100 psi and tested in the optical range of 400–4000 cm^{-1}.

^1H NMR spectrum of the quadripolymer was measured by Angilent 400 MHz Nuclear magnetic resonance spectrometer. Mass of 10 mg of sample powder was dissolved into a test tube containing 0.65 mL D_2O, and then transferred to a sample cavity. ^1H NMR spectrum measurement was carried out with controlled heating and cooling steps.

Thermal stability: The thermal stability of the quadripolymer was tested with HCT-1 Differential Thermobalance analyzer (Beijing Henven Instrument Plant, Beijing, China) in nitrogen gas atmosphere. The heating rate was 10 °C/min, and the temperature was in the range of 30–600 °C.

2.2.3. Sample Preparation

Base formula of bentonite-free drilling fluid (marked as formula1#) was prepared as follows: 1000 mL deionized water, 0.1 wt% NaOH, 0.15 wt% Na_2CO_3, desired amount of polymer (the quadripolymer or Drispac), 2 wt% PAC-LV, 2 wt% LOCK-SEAL, 4 wt% super-fine $CaCO_3$, 0.5 wt% Na_2SO_3 were weighted and added sequentially at high speed stirring for 20 min with a GJSS-B12K multi-spindle mixer, and then the solution was stood still for 24 h at room temperature for complete dissolution.

2.2.4. Drilling Fluid Performance Measurements

Rheological and filtration properties of bentonite-free drilling fluid were evaluated according to the American Petroleum Institute test program [29], Formulas (1)–(3) were applied to calculate the rheological parameters, including apparent viscosity (AV), plastic viscosity (PV) and yield point (YP), for the fluid system. The readings at 600 rpm and 300 rpm obtained with a ZNN-D6 six-speed rotational viscometer were marked as $\Phi600$ and $\Phi300$, respectively. The initial and final gel strength were also measured and marked as G_1 and G_2. Before thermal stability evaluation, the bentonite-free drilling fluid was required to be heated at a given temperature for 16 h with a XGRL-4A hot roller oven and then cooled to ambient temperature.

$$AV \text{ (Apparent viscosity)} = \Phi600/2, \text{mPa·s,} \tag{1}$$

$$PV \text{ (Plastic viscosity)} = \Phi600 - \Phi300, \text{mPa·s,} \tag{2}$$

$$YP \text{ (Yield point)} = 0.511(\Phi300 - PV), \text{Pa,} \tag{3}$$

$$K = \frac{Q\mu l}{\Delta pA}. \tag{4}$$

API filtration loss (FL_{API}) was determined by a SD-4 API filtration apparatus with a pressure difference of 100 psi at ambient temperature for 30 min. HTHP filtration loss (FL_{HTHP}) was measured with a GCS71-A HTHP filtration apparatus at 150 °C and 500 psi for 30 min. After the test, the filtrate was collected into the measuring cylinder and the volume was recorded.

HTHP rheology test: After placing the drilling fluid inside the rheometer, the sample was heated up to the desired temperature and meanwhile pre-shear at 100 s^{-1} was performed before rheological measurement. The pressure was adjusted to 500 psi and the shear stress was measured at the shear rate of 511 s^{-1}.

Reservoir protection test: The core original kerogen permeability K_0 was measured as follows, Formula (4), wherein, Q, flow rate; μ, Kerosene viscosity; l, Core length; Δp, Differential pressure before and after flow through the core; A, core, end surface area.

In step 1, the core saturated with simulated groundwater is placed into the core holder and repelled with filtered and dewatered kerosene at 0.4 times the critical flow rate; subsequently, repelled at 0.8 times the critical flow rate until no water flows out and the pressure is stable. According to the above formula, K_0 can be calculated.

In step 2, the core tested for K_0 is quickly loaded into the core holder of JHDS high temperature and high pressure dynamic water loss instrument, and the formulation drilling fluid is injected in the reverse direction at 90 °C and 3.5 MPa differential pressure, and cycled at a shear rate of 300 s^{-1} for 125 min. The core is loaded into the gripper and repelled at the same flow rate as in step 1, and the contaminated permeability K_1. is calculated after the pressure and flow rate are stabilized.

In step 3, the end face of the core contaminated by drilling fluid in step 2 is cut off for about 1 cm, and step 2 is repeated to obtain the core permeability K_2.

3. Results and Discussion

3.1. Characteristic

The quadripolymer used in this study was milky white dispersion. The scanning electron microscopy (SEM) picture is shown in Figure 1. The microstructure of the quadripolymer had the following characteristic: irregular shapes, different sizes (Figure 1a), and smooth surface (Figure 1b) of the particles were easily recognizable. Among them, the particle size is relatively widely distributed. The largest is around 200 microns and the smallest is just under 10 microns.

(a) (b)

Figure 1. Microscopic images of the quaripolymer particles: (**a**) 200 times, (**b**) 1500 times.

The FT-IR spectrum of the quadripolymer is presented in Figure 2. Moreover, 3444 cm^{-1} and 3300 cm^{-1} are assigned to the N-H stretching vibration of AM and AMPS, respectively. The absorbency at 2935 cm^{-1} results from the characteristic peak of the methyl group. The stretching vibration of C=O at 1699 cm^{-1} is attributed to AM and AMPS. The absorption peak observed at 1454 cm^{-1} is due to the C-H bending vibration from DMDAAC unit. Bands recorded at 1185 cm^{-1} and 1044 cm^{-1} assigned to S=O stretching vibration in the sulfonic group correspond to SSS unit. The absorption peaks at 766 cm^{-1} reveal the bending vibration of a benzene ring of the SSS unit.

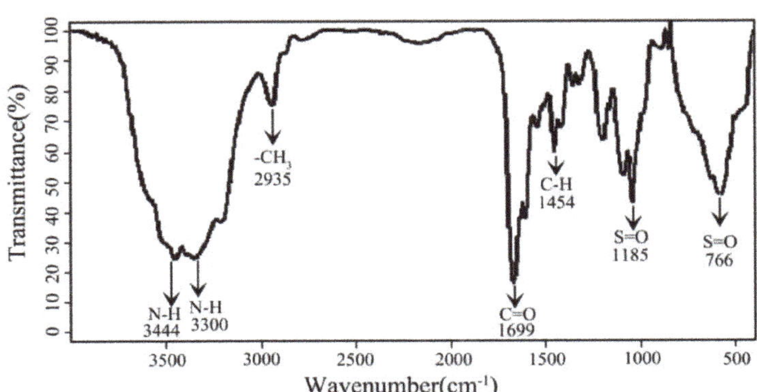

Figure 2. FT−IR spectrum of quadripolymer (AM/AMPS/DMDAAC/SSS).

The ^{1}H-NMR spectrum of the quadripolymer is shown in Figure 3. Bands recorded at 7.48 ppm (a$_1$) and 7.24 ppm (a$_2$) indicate the chemical shift in the benzene ring of SSS units. The N-H proton from 5.87 to 5.92 ppm (b) are related to AM. The peaks between 5.51 and 5.63 ppm (c) are due to the N-H proton vibration of AMPS. The characteristic peak at 4.79 ppm belongs to the chemical shift of D$_2$O protons. Broad peaks between 3.45 and 3.53 ppm (d) are assigned to N-C proton of DMDAAC. Respectively, the peaks at 2.76 ppm and 1.65 ppm correspond to the chemical shift of –CH$_2$ and –CH$_3$ of AMPS (e). The board peak at 2.93 ppm and 1.45 ppm are related to the -CH- linked with the benzene ring and the main chain of SSS, respectively (f). Combined with the FT-IR spectrum, the quadripolymer molecular structure is consistent with the designed one.

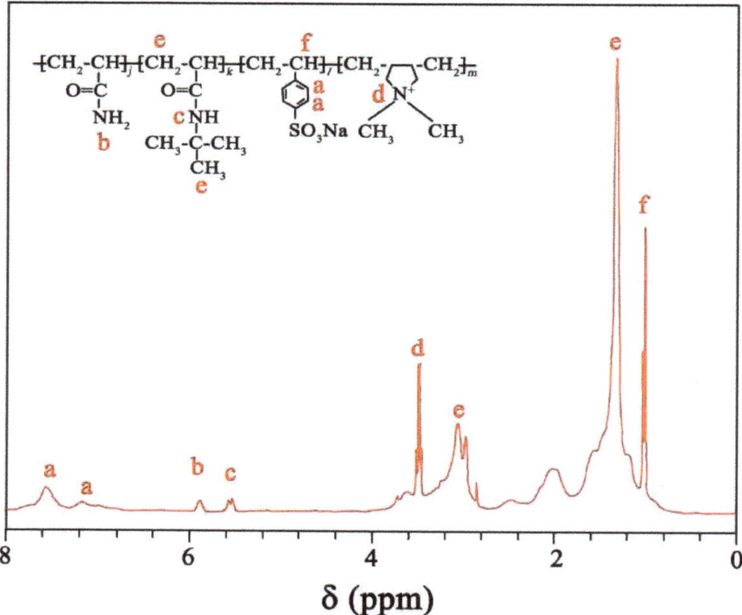

Figure 3. ^{1}H−NMR spectrum of quadripolymer (AM/AMPS/DMDAAC/SSS).

From the TG curve (Figure 4), it can be found that the four stages of the thermal degradation process, first stage, the quadripolymer has 8.5% of mass losses before 260 °C, indicating slightly thermal degradation from chain scissionThe second stage is from 260 °C to 319.5 °C, apparent mass loss of the quadripolymer occurs (27.71%) in the TG curve and the fastest mass loss temperature emerges at 300.5 °C in the DSC curve. The third stage is from 319.5 °C to 528.5 °C, the mass loss declines constantly (35.43%). the release of the non-cyclic anhydrides formation acrylic acid repeat unit or the carboxylic acid pendant group. Moreover, accordingly, there is no significant fluctuation on the DSC curve. The last stage is from 528.5 °C to 600 °C, 28.36% mass is left. This might be due to the thermal degradation of C=C decomposition in the main chain of AM and SSS segments and the breakage of benzene ring in SSS monomers. From the above analysis, the thermal degradation of the quadripolymer is not obvious before 260 °C and only 27.7%, which demonstrates the quadripolymer has strong heat-resistance ability.

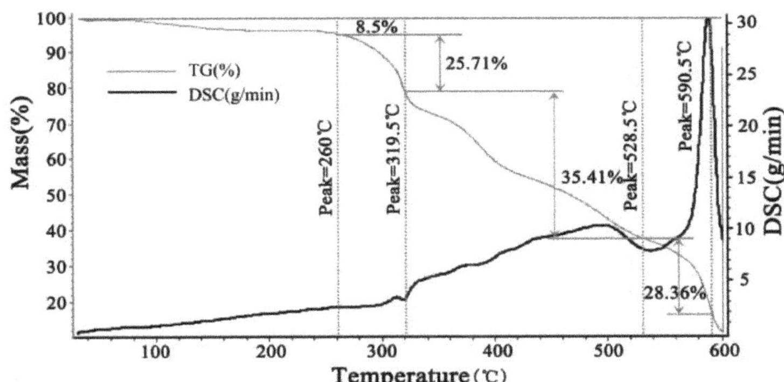

Figure 4. TG- DSC curve.

The data of quadripolymer molecular weight is follow in Table 2. Based on the molecular structure of the quadripolymer. The weight-average molecular weight is 1161,000 g/mol, and the number-average molecular weight is 684,000 g/mol. Besides, the viscosity-average molecular weight is 1090,000 g/mol calculated with Mark–Houwink–Sakurada equation ($M\eta$ = 802*[η]1.25), among which the η value is 3.236 dL/g measured by Ubbelohdo viscometer (the flowing time of 0.1 wt% quadripolymer solution). Moreover, the molecular weight distribution index PDI is 1.6974, indicating that the quadripolymer has narrow molecular weight distribution.

Table 2. Molecular weight indexed of copolymer.

Sample	M_W	M_η	M_N	PDI(M_W/M_N)
quadripolymer	1161,000	1090,000	684,000	1.6974

3.2. Performance Evaluation of Bentonite-Free Drilling Fluid with the Quadripolymer

3.2.1. Rheological Behavior

A drilling fluid having good rheological properties (low plastic viscosity value, suitable gel strength, and high yield point) will provide good cuttings transportation efficiency in drilling operation. To investigate the influence of the quadripolymer on the rheological behavior of the bentonite-free drilling fluid, various amount (from 0.3 to 0.9 wt%) of two polymers were added to the base formula and the rheological parameters, such as the apparent viscosity (AV), plastic viscosity (PV), and yield point (YP), were measured and compared.

As illustrated in Table 3, the two polymer drilling fluids display a remarkable increase in rheological parameters (AV, PV, and YP), as the concentration of polymer increases in the range from 0.3 to 0.9 wt%, which is mainly because that polymer contributes to forming the network structure of fluids and consequently improving the rhelogical properties. Compared with Drispac, PV values of the quadripolymer containing bentonite-free fluid system are relatively lower at the same polymer concentrations on the whole, while YP show relatively higher values. Moreover, the quadripolymer system exhibits a dynamic ratio (YP/PV) of up to 0.80, significantly larger than that of Drispac, at polymer concentration of 0.9 wt%, indicating that the quadripolymer has good shear thinning property. Lastly, gel strength of drilling fluid describes the capacity to suspend the drilling cuttings when the pump is shut down. It can be observed that the increase in concentration of the quadripolymer contributes to a more dramatic increase than Drispac in G_1 and G_2. The enhanced gel strength is attributed to the interactions among polymer molecules. Therefore, the bentonite-free drilling fluid with the quadripolymer has excellent cuttings suspending and carrying capacity.

Table 3. Influence of polymer concentration on rheological behavior of base formula.

Formula	Rheological Parameters				
	AV, mPa·s	*PV,* mPa·s	*YP,* Pa	*YP/PV,* -	G_1/G_2 Pa/Pa
1# + 0.3 wt% quadripolymer	25	16	9.2	0.58	1.5/1.5
1# + 0.6 wt% quadripolymer	35	21	14.3	0.68	2.5/3.5
1# + 0.9 wt% quadripolymer	48	27	21.5	0.80	4.5/5.5
1# + 0.3 wt% Drispac	24	17	7.2	0.42	1.0/1.0
1# + 0.6 wt% Drispac	33	22	11.2	0.51	1.5/2.0
1# + 0.9 wt% Drispac	52	34	18.4	0.54	2.5/3.0

3.2.2. Salt Resistance

It is shown from Table 4 that all bentonite-free polymer drilling fluids with NaCl have lower rheological parameters than that without salt. Furthermore, as the salt content increases, all the rheological parameters display a remarkable decreasing trend. This is mainly because of charge shielding effect. Na^+ has strong ability to compress electric double layer of polymer and neutralize negative charge on ionized polymer, which could reduce an electrostatic repulsion force between particles to lower viscosity values, thus destroying network structure between polymer molecules. Furthermore, with a further increase in salt, large amount of Na^+ will increase the polarity of solution and enhance the hydrophobic associate effect, which can compensate the decrease of viscosity to a certain extent [30]. However, it could be seen clearly that compared with Drispac containing fluids, the quadripolymer shows a less variation in rheological parameters and has higher YP and gel strength values at the same salt concentration of 5 wt% and 10 wt%, respectively, suggesting that the quadripolymer has better salt resistance than Drispac.

Table 4. Influence of salt concentration on rheological behavior of bentonite-free drilling fluid.

Formula	*AV,* mPa·s	*PV,* mPa·s	*YP,* Pa	*YP/PV*	G_1/G_2 Pa/Pa
2#	48	27	21.5	0.80	4.5/5.5
2#+5 wt% NaCl	40	24	16.4	0.68	3.5/4.0
2#+10 wt% NaCl	36	22	14.3	0.65	1.5/1.5
3#	52	34	18.4	0.54	2.5/3.0
3#+5 wt% NaCl	38	28	10.2	0.37	1.0/1.5
3#+10 wt% NaCl	32	24	8.2	0.34	1.0/1.0

Note: 2#: formula 1# + 0.9 wt% quadripolymer; 3#: formula 1# + 0.9 wt% Drispac.

3.2.3. Filtration Property

For filtration property, the API filtration (FL_{API}) experiments were performed under a pressure of 100 psi and room temperature. The effect of salinity and polymers on filtration performance of bentonite-free drilling fluid is presented in Figure 5.

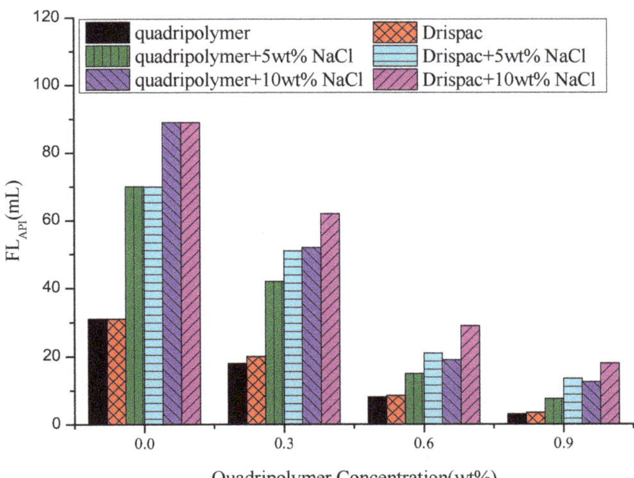

Figure 5. Effect of salinity and polymers on filtration property of bentonite-free drilling fluid.

Dramatic decreases in FL_{API} of both fresh and salt fluid samples are seen as the dosage of polymer increases from 0 to 0.9 wt%. Firstly, in deionized water, FL_{API} of the bentonite-free fluids with 0.9 wt% quadripolymer and Dispac decrease to 3 mL and 3.5 mL, respectively. The two polymers have approximately equal ability to control the filtration loss in fresh water environment. Similarly, in electrolyte solution, the filtration loss variations also present a declining trend with increasing polymer concentration, but the volumes are considerably higher than those in fresh water solution. This is mainly because that the molecular chains of polymer cannot be fully stretched in salt water, and subsequently the repulsive forces between molecular reduce and a high permeable filtration cake is formed. For all that, compared with FL_{API} of Drispac containing salt system (18 mL), at 0.9 wt%, FL_{API} of bentonite-free drilling fluid with the quadripolymer can be controlled within 12.5 mL, which demonstrates that the quadripolymer has better filtration loss control property in brine drilling fluid.

3.2.4. Thermal Stability

In petroleum industry, it is well known that high temperature would dramatically worsen rheological property and increase filtration volume loss of drilling fluids. Therefore, common used polymers PAM, 80A51, HE150 were compared with the quadripolymer to investigate their high temperature resistance. The initial viscosity was recorded as V_1 when polymer solutions were prepared at 0.9 wt%. Then, the polymer solutions were hot rolling aged at different temperatures for 16 h, and the residual viscosity was marked as V_2. The viscosity retention rates R_v at different temperatures are shown in Figure 6.

$$R_v = \left(\frac{V_1 - V_2}{V_1} \right) \times 100\% \tag{5}$$

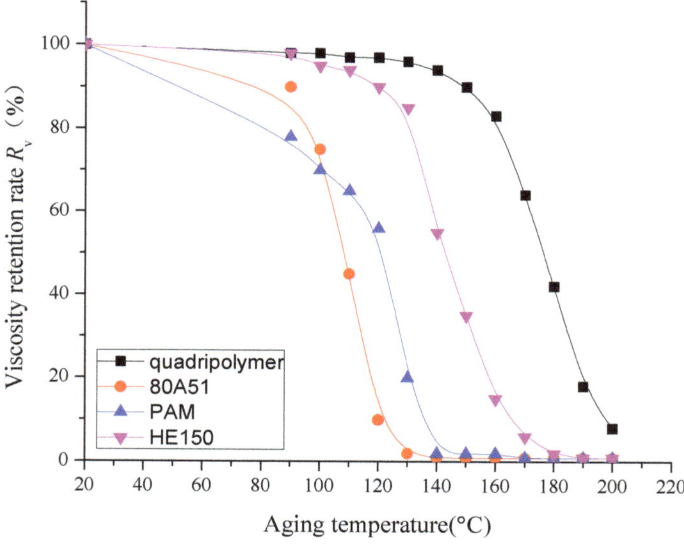

Figure 6. Viscosity retention rate of polymers at different aging temperature.

As illustrated in Figure 6, as the aging temperature increases, the viscosity retention rates of different polymers generally show downward trends, but the change degree varies. Among them, the viscosity of the two polymers, PAM and 80A51, has been basically lost after aging at 140 °C. In contrast, HE150 and the quadripolymer have higher viscosity retention rates at the same temperature, especially for the quadripolymer, the viscosity retention rate is as high as 40% even when the aging temperature reaches to 180 °C, and nearly 10% at 200 °C. Therefore, compared with other three polymers commonly used in drilling fluids, the quadripolymer has better temperature resistance.

In order to further study the effect of the quadripolymer on the thermal stability of the bentonite-free drilling fluid, the rheological and filtration properties were evaluated before and after hot rolling aging at various temperatures (25 °C, 100 °C, 120 °C, 140 °C, 160 °C, 180 °C, 200 °C), and the results are shown in Table 5.

Table 5. Influence of thermal aging temperature on performance of bentonite-free drilling fluid.

T, °C	PV, mPa·s	YP, Pa	G_1/G_2, Pa/Pa	FL_{API}, mL	FL_{HTHP}, mL
25	27	21.5	4.5/5.5	3.0	-
100	25	14.3	4.0/5.0	4.2	14.6
120	24	14.3	3.5/4.5	5.4	17.8
140	23	13.3	3.5/4.0	6.6	19.7
160	22	10.4	2.5/3.5	7.4	21.6
180	20	8.2	1.5/2.5	8.6	24.8
200	12	2.0	0.5/0.5	12.4	46.4

Note: bentnoite-free drilling fluid formulation: deionized water + 0.1 wt% NaOH + 0.15 wt% Na_2CO_3 + 0.9 wt% quadripolymer + 2 wt% PAC–LV + 2 wt% LOCKSEAL + 4 wt% $CaCO_3$ + 0.5 wt% Na_2SO_3.

As shown in Table 5, it can be clearly observed that in the aging temperature range of 100–180 °C, all parameters variation are within the acceptable range, showing a good thermal stability. PV value of the quadripolymer bentonite-free drilling fluid still remains at 20 mPa·s after aging at 180 °C. Meanwhile, FL_{API} and FL_{HTHP} are controlled within 8.6 mL and 24.8 mL, respectively. However, when the aging temperature rises to 200 °C, the rheological parameters of the bentonite-free drilling fluid with the quadripolymer show an obvious decreasing trend, and the amount of filtration loss markedly increases, indicating that the comprehensive properties of drilling fluid are out of control. Based on the above analysis, it can be demonstrated that the bentonite-free drilling fluid prepared by the quadripolymer possesses excellent ability to maintain viscosity and filtration loss at the temperature of 180 °C.

3.2.5. HTHP Rheology

In order to determine the rheological characteristics of the synthetic quadripolymer under high temperature and high pressure environment, the HTHP rheology test of four common high temperature resistant polymers were performed. A constant shear rate ($511\ \text{s}^{-1}$) was applied at different temperatures. Moreover, the curves of viscosity changing with temperature are presented in Figure 7.

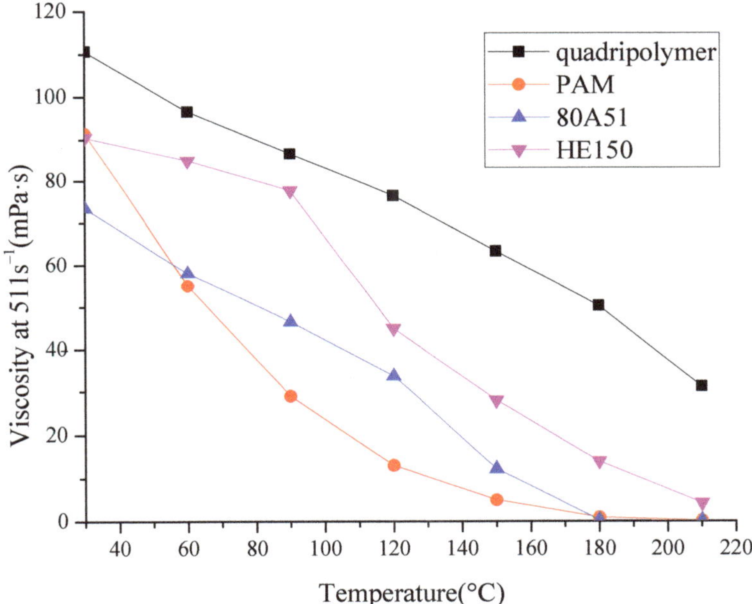

Figure 7. HTHP rheology curves of different polymer solutions.

As shown in Figure 6, heating causes the rapid drop in the viscosity of polymer solutions, especially for PAM, 80A51 and HE150. Among the four polymers, the viscosity reduction rates of the three polymers mentioned above are much higher than that of quadripolymer. When temperature rises to 180 °C, the viscosities of the PAM and 80A51 solution are close to 0. In contrast, the quadripolymer has higher viscosity values at each temperature point than those of the other three polymers in the temperature range of 200 °C, which further suggests that the quadripolymer has good temperature resistance at high temperatures.

3.2.6. Reservoir Protection Performance

ZDY50–180 core flow tester and oil reservoir cores taken from Well TK10 at the depth of 5298.83–5299.50 m, were used for reservoir protection evaluation of the bentonite-free drilling fluid with the new quadripolymer.

The degree of reservoir damage caused by drilling fluid can be characterized by the change of permeability before and after core contamination by drilling fluid, i.e., the lower permeability of core contamination means the higher degree of reservoir damage. It is clear that the permeability of both cores (sample a and sample b) after contamination is restored to more than 85% (Table 6). Moreover, the permeability recovery rate of two cores are obviously improved and reaches up as high as 95.7% and 94.2%, respectively, after cutting off 1 cm of contaminated core section, which indicates that the bentonite-free drilling fluid has a good protective effect on reservoir.

Table 6. Reservoir protection performance test for bentonite-free polymer drilling fluid.

Core Sample	Porosity, %	K_1, mD	K_2, mD	Recovery Rate, %
a	13.51	7.25	6.43	88.69
Cut off 1 cm of contaminated section			6.94	95.72
b	19.20	12.85	10.98	85.45
Cut off 1 cm of contaminated section			12.11	94.24

Note: K_1 is the initial permeability of cores; K_2 is the permeability of contaminated cores.

4. Conclusions

A novel hydrophilic polymer was prepared with AM, AMPS, DMDAAC, and SSS monomers via a method of free radical polymerization in aqueous solution. The new polymer was characterized by ^{1}H NMR and FT-IR analysis. Results indicated the quadripolymer molecular structure was consistent with the designed one. Moreover, the rheological and filtration properties of the bentonite-free drilling fluid containing the quadripolymer and Drispac were examined in deionized and salt water at various temperatures. As a result, the rheological parameters regularly increased with increasing dosage of polymers. Simultaneously, the viscosity of the salt water fluid system was explicit smaller than that of the fresh water system. Additionally, when the adding amount was 0.9 wt%, it was confirmed that FL_{API} of two polymers drilling fluid systems decreased to 3 mL and 3.5 mL, respectively, which are in acceptable range. The thermal stability test of bentonite-free drilling fluid showed that the quadripolymer drilling fluid could maintain a certain viscosity value, and FL_{API} and FL_{HTHP} were also controlled within 8.6 mL and 24.8 mL, respectively. In HTHP rheology test, the synthetic polymer showed superior resistance to high temperature than commonly used polymers. As for its damage to the reservoir, the permeability recovery of cores reached more than 95.72%. In general, the synthesized polymer has good properties of high temperature resistance and salt tolerance in controlling rheology and filtration loss of the bentonite-free drilling fluid system.

Author Contributions: Conceptualization, J.W. and W.Z.; methodology, J.W.; software, M.C.; validation, X.L. and J.W.; formal analysis, J.W.; investigation, X.Y.; resources, J.W.; data curation, X.Y.; writing—original draft preparation, J.W.; writing—review and editing, J.W. and X.L.; supervision, W.Z. All authors have read and agreed to the published version of the manuscript.

Funding: This research received no external funding.

Institutional Review Board Statement: Not application.

Informed Consent Statement: Not application.

Data Availability Statement: Not application.

Conflicts of Interest: The authors declare that they have no known competing financial interests or personal relationships that have appeared to influence the work reported in this paper.

References

1. Caeen, R.; Darley, H.C.H.; Gray, G.R. *Composition and Properties of Drilling and Completion Fluids*, 7th ed.; Gulf Professional Publishing: Waltham, MA, USA, 2017; pp. 36–38.
2. Yan, L.L.; Wang, C.B.; Xu, B.; Sun, J.S.; Yue, W.; Yang, Z.X. Preparation of a Novel Amphiphilic Comb–like Terpolymer as Viscosifying Additive in Low-solid Drilling Fluid. *Mater. Lett.* **2013**, *105*, 232–235. [CrossRef]

3. Liu, F.; Jiang, G.C.; Peng, S.L.; He, Y.B.; Wang, J.X. Amphoteric Polymer as an Anti–calcium Contamination Fluid–loss Additive in Water-based Drilling Fluids. *Energy Fuels* **2016**, *30*, 7221–7228. [CrossRef]

4. Yang, X.J.; Mao, J.C.; Chen, Z.X.; Chen, Y.N.; Zhao, J.Z. Clean fracturing fluids for tight reservoirs: Opportunities with viscoelastic surfactant. *Energy Sources Part A Recovery Util. Environ. Eff.* **2019**, *41*, 1446–1459. [CrossRef]

5. Mahto, V.; Sharma, V.P. Rheological Study of a Water Based Oil Well Drilling Fluid. *J. Petrol. Sci. Eng.* **2004**, *45*, 123–128. [CrossRef]

6. Gautam, S.; Guria, C.; Rajak, D.K.; Pathak, A.K. Functionalization of Fly Ash for the Substitution of Bentonite in Drilling Fluid. *J. Petrol. Sci. Eng.* **2018**, *166*, 63–72. [CrossRef]

7. Sepehri, S.; Soleyman, R.; Varamesh, A.; Valizadeh, M.; Nasiri, A. Effect of Synthetic Water-Soluble Polymers on the Properties of the Heavy Water-based Drilling Fluid at High Pressure-High Temperature (HPHT) Conditions. *J. Petrol. Sci. Eng.* **2018**, *166*, 850–856. [CrossRef]

8. Vipulanandan, C.; Mohammed, A. Effect of Nanoclay on the Electrical Resistivity and Rheological Properties of Smart and Sensing Bentonite Drilling Muds. *J. Petrol. Sci. Eng.* **2015**, *130*, 86–95. [CrossRef]

9. Azouz, K.B.; Bekkour, K.; Dupuis, D. Influence of the Temperature on the Rheological Properties of Bentonite. *Appl. Clay Sci.* **2016**, *123*, 92–98. [CrossRef]

10. Song, K.L.; Wu, Q.L.; Li, M.C.; Wojtanowicz, A.K.; Dong, L.; Zhang, X.; Ren, S.; Lei, T. Performance of Low Solid Bentonite Drilling Fluids Modified by Cellulose Nanoparticles. *J. Nat. Gas Sci. Eng.* **2016**, *34*, 1403–1411. [CrossRef]

11. Zhou, H.; Deville, J.P.; Davis, C.L. Novel Thermally Stable High-density Brine-based Drill-in Fluids for HP/HT Applications. In Proceedings of the SPE Middle East Oil and Gas Show and Conference, Manama, Bahrain, 8–11 March 2015.

12. Xiao, J.F.; Li, Z.J.; Liu, Z.X. Application of Solid Free Drilling Fluid in Anpeng Oilfield. *Drill. Fluid. Complet. Fluid* **2002**, *19*, 82–84.

13. Mustafa, V.K.; Tolga, A. Effect of polymers on the rheological properties of KCl/Polymer Type Drilling fluids. *Energy Sources Part A Recovery Util. Environ. Eff.* **2005**, *27*, 405–415. [CrossRef]

14. Fink, J. *Petroleum Engineer's Guide to Oil Field Chemicals and Fluids*, 2nd ed.; Gulf Professional Publishing/Elsevier: Waltham, MA, USA, 2012; pp. 1–60. ISBN 978-0-12-383844-5.

15. Elkatatny, S. Enhancing the Stability of Invert Emulsion Drilling Fluid for Drilling in High-pressure High-temperature Conditions. *Energies* **2018**, *11*, 2393. [CrossRef]

16. Tehrani, M.A.; Popplestone, A.; Guarneri, A.; Carminati, S. Water Based Drilling Fluid for HT/HP Applications. In Proceedings of the SPE International Symposium on Oilfield Chemistry, Houston, TX, USA, 28 February 2007.

17. Stefano, G.D.; Stamatakis, E.; Young, S. Meeting the Ultrahigh-temperature/Ultrahigh-pressure Fluid Challenge. *SPE Drill. Completion* **2013**, *28*, 86–92. [CrossRef]

18. Li, M.C.; Wu, Q.; Song, K.; Qing, Y.; Wu, Y. Cellulose Nanoparticles as Modifiers for Rheology and Fluid Loss in Bentonite Water-Based Fluids. *ACS Appl. Mater. Interfaces.* **2015**, *7*, 5006–5016. [CrossRef] [PubMed]

19. Taye, S.O.; Johann, P. Preparation and Properties of a Dispersing Fluid Loss Additive Based on Humic Acid Graft Copolymer Suitable for Cementing High Temperature (200°C) Oil Wells. *J. Appl. Polym. Sci.* **2013**, *129*, 2544–2553. [CrossRef]

20. Perricone, A.C.; Enright, D.P.; Lucas, J.M. Vinyl Sulfonate Copolymers for High-temperature Filtration Control of Water-based Muds. *SPE Drill. Eng.* **1986**, *1*, 358–364. [CrossRef]

21. Tao, W.; Jie, Y.; Sun, Z.S.; Wang, L.; Wang, J. Solution and Drilling Fluid Properties of Water Soluble AM-AA-SSS Copolymers by Inverse Microemulsion. *J. Petrol. Sci. Eng.* **2011**, *78*, 334–337. [CrossRef]

22. Wu, Y.M.; Zhang, B.Q.; Wu, T.; Zhang, C.G. Properties of the Forpolymer of N-Vinylpyrrolidone with Itaconic Acid, Acrylamide and 2-Acrylamido-2-Methyl-1-Propane Sulfonic Acid as a Fluid-Loss Reducer for Drilling Fluid at High Temperatures. *Colloid Polym. Sci.* **2001**, *279*, 836–842. [CrossRef]

23. Munawar, K.; Munawar, J.B.K. Viscoplastic Modeling of a Novel Lightweight Biopolymer Drilling Fluid for Underbalanced Drilling. *Ind. Eng. Chem. Res.* **2012**, *51*, 4056–4068. [CrossRef]

24. Bai, X.D.; Yang, Y.; Xiao, D.Y.; Pu, X.L.; Wang, X. Synthesis, Characterization, and Flocculation Performance of Anionic Polyacrylamide P (AM-AA-AMPS). *J. Appl. Polym. Sci.* **2013**, *129*, 1984–1991. [CrossRef]

25. Wu, Y.M.; Sun, D.J.; Zhang, B.Q.; Zhang, C.G. Properties of High-Temperature Drilling Fluids Incorporating Disodium Itaconate-Acrylamide-Sodium 2-Acrylamido-2-Methyl-Propane Sulfonate Terpolymers as Fluid-Loss Reducers. *J. Appl. Poly. Sci.* **2002**, *83*, 3068–3075. [CrossRef]

26. Peng, B.; Peng, S.; Long, B.; Miao, Y.; Guo, W.Y. Properties of High-Temperature Resistant Drilling Fluids Incorporating Acrylamide/(Acrylic Acid)/(2-Acrylamido-2-Methyl-1-Propane Sulfonic Acid) Terpolymer and Aluminum Citrate as Filtration Control Agents. *J. Vinyl. Addit. Technol.* **2010**, *16*, 84–89. [CrossRef]

27. Huang, Y.M.; Zhang, D.Y.; Zheng, W.L. A Synthetic Copolymer (AM/AMPS/DMDAAC/SSS) as Rheology Modifier and Fluid Loss Additive at HTHP for Water-Based Drilling Fluids. *J. Appl. Poly. Sci.* **2019**, *139*, 47813. [CrossRef]

28. Stamatakis, E.; Steve, Y.; Guido, D.S.; Swaco, M.-I. Meeting the Ultra HTHP Fluids Challenge. In Proceedings of the SPE Oil and Gas India Conference and Exhibition, Mumbai, India, 28–30 March 2012.

29. API Recommended Practice. Recommended Practice for Field Testing Water–based Drilling Fluids. In *API Recommended Practice13B-1*, 3rd ed.; API Publishing Services: Washington, DC, USA, 2003; pp. 7–8.

30. Guo, H.; Brûlet, A.; Rajamohanan, P.R.; Marcellan, A.; Sanson, N.; Hourdet, D. Influence of Topology of LCST-Based Graft Copolymers on Responsive Assembling in Aqueous Media. *Polymer* **2015**, *60*, 164–175. [CrossRef]

Article

Effect of Full Temperature Field Environment on Bonding Strength of Aluminum Alloy

Haichao Liu, **Yisa Fan** * and **Han Peng**

School of Mechanical Engineering, North China University of Water Resources and Electric Power, Zhengzhou 450045, China; liuhaichao@ncwu.edu.cn (H.L.); penghan@ncwu.edu.cn (H.P.)
* Correspondence: fanyisa@ncwu.edu.cn; Tel.: +86-173-1977-2020

Abstract: In this paper, the influence of temperature on the bonding strength of aluminum alloy joints under the full temperature field is studied. Based on the service temperature range of vehicle bonding structures, the failure strength of aluminum alloy joints at different temperature points, namely −40 °C, −20 °C, 0 °C, 25 °C (RT), 40 °C, 60 °C and 80 °C, is tested. The results showed that compared with the failure strength of the adhesive at −40 °C, it decreased by 47.69% and 68.15% at RT and 80°C, respectively; the Young's modulus of the adhesive decreased by 57.63% and 75.42% at RT and 80°C, respectively; with the increase of temperature, the young's modulus, tensile strength and failure strain of the adhesive decreased. In addition, the failure strength of aluminum alloy joints varied with temperature. To be specific, the stiffness of joints decreased gradually from 25 °C to 80 °C and increased gradually from −40 °C. Based on the failure strength data of bonded joints at different temperature points, the secondary stress failure criteria of bonded joints at different temperatures were obtained. Then, the surface function of failure criteria under the full temperature field was established to provide reference for failure prediction of bonded structures under different temperatures and stresses.

Keywords: adhesively-bonded joint; temperature aging; residual strength; mechanical behavior; failure criterion

Citation: Liu, H.; Fan, Y.; Peng, H. Effect of Full Temperature Field Environment on Bonding Strength of Aluminum Alloy. *Crystals* **2021**, *11*, 657. https://doi.org/10.3390/cryst11060657

Academic Editors: Yifeng Ling, Chuanqing Fu, Peng Zhang, Peter Taylor and Sergio Brutti

Received: 10 May 2021
Accepted: 5 June 2021
Published: 9 June 2021

1. Introduction

In recent years, with the increasing application of new materials such as aluminum alloy, high strength steel and composite materials in automobiles as well as the continuous development of multi-material hybrid design concept in the automobile industry, the traditional mechanical connection technology (such as welding and riveting) cannot meet the connection requirements between different materials [1,2]. As a new connection method, bonding technology has the advantages of uniform stress distribution, fatigue resistance, light weight, etc. [3–5]. In this case, the connection needs of dissimilar materials can be effectively realized. Therefore, compared with other connection technologies, increasing attention has been paid to bonding technology. As a tough adhesive, polyurethane adhesive is gradually being widely used in automobiles, since it not only has high tear strength, good impact resistance and excellent toughness, but also provides relatively uniform stress distribution due to its low elastic modulus.

However, as a kind of polymer material, adhesive relies on temperature to some extent. The change of temperature will directly affect the mechanical properties of the material, and its failure strength and failure form change with different temperatures [6]. In the process of service, the ambient temperature range of adhesive structure is large. During the process of vehicle operation, the adhesive structure needs to provide enough strength in the service temperature range. The performance of the bonding structure is closely related to the service temperature, and the bonding structure significantly affects the overall strength and fatigue characteristics of the car body. Therefore, the research on the influence of the whole service temperature field on the performance of the body bonding structure is the technical guarantee to realize the lightweight design of the body structure. Scholars at home and abroad have carried out relevant studies on the static performance and strength-checking criteria of temperature bonding structures [7,8].

Temperature is the main factor affecting the performance of the adhesive, and the mechanical properties of the adhesive will change in different temperature ranges. The bonding strength, strain and fracture toughness show temperature sensitivity [9]. The joint strength of the bonding structure is determined by the performance change of the adhesive and the influence of thermal stress [10]. The effect of temperature on the properties of the bonding structure is obvious, especially when the temperature is close to the glass transition temperature (Tg) of the material [11,12]. In addition, when the temperature is higher than Tg, the adhesive is featured with high elasticity, and its failure strength and elastic modulus decrease rapidly, while the elongation increases; however, when the temperature is lower than Tg, its performance is opposite [13]. Adams et al. [14] tested the single lap joint at different temperatures and also compared and analyzed the influence of thermal stress caused by the difference of thermal expansion coefficients and shrinkage stress caused by curing on the joint performance, which lead to the change concerning the stress state of the lap joint; the stress/strain performance of polymer adhesive also changes with the change of temperature. Na et al. [15,16] studied the effect of temperature on the mechanical properties of basalt fiber-reinforced composite/aluminum alloy bonded joints and found that with the increase in temperature, the Young's modulus and tensile strength of the joint decreased, while the tensile strain increased. The closer the temperature to Tg, the more significant the change in mechanical properties. Silva et al. [17] conducted a test on the mechanical properties of the single lap joint at low temperature and high temperature and revealed that the adhesive was brittle at low temperature and ductile at high temperature. They also analyzed the effect of porosity on failure. Banea et al. [12] investigated the stress-strain properties of polyurethane and epoxy adhesives at −40 °C, room temperature and 80 °C. It was found that with an increase in temperature, the failure strength and the Young's modulus of epoxy adhesives decreased, while the failure strain increased, which resulted from the increase of adhesive toughness at high temperature. Zhang et al. [18] conducted a study on the tensile properties of double lap joints in the temperature range of −35~60 °C and found that the load-elongation response was mainly affected by the thermomechanical properties of the adhesive, while it was less affected by the adhesive base material. When the temperature was higher than Tg, the strength and stiffness of the joint decreased, while the elongation increased dramatically, and the failure mechanism changed with the increase in temperature. To be specific, crack growth rate is higher at low temperature. In addition, the critical strain energy release rate for crack initiation and propagation increases continuously with increasing temperature.

Adhesion technology provides technical support for mixed material body design, but it also brings some problems. The service temperature of the bonding structure used in vehicles varies to a great extent in practical application. As a macromolecule material, the performance of the bonding structure is greatly affected by temperature, which causes the mechanical properties of the bonding structure to change with temperature. In order to achieve the safety design of the vehicle, the bonding structure must ensure the reliability of the connection within the full temperature field of the vehicle service. Therefore, temperature is one of the important factors that must be considered in the design of a bonded structure. It is of great significance to study the changing rules of bonded joint performance at different temperatures and propose the failure prediction method of bonded joints under the full temperature field for guiding the design of bonded structures.

2. Material Selection and Specimen Design

2.1. Adhesive and Substrate

The experimental selective adhesive was a modified silane polyurethane adhesive widely used in the window bonding of cars, trucks and trains. ISR-7008 is produced by Bostik China Co., Ltd. The mechanical parameters of the adhesive and adhesive substrate are shown in Table 1. The working temperature range provided in the technical manual is $-40\,°C\sim-90\,°C$. A permanent elastomer is formed by reaction with moisture in the air. 6005A aluminum alloy was selected as the adhesive substrate; it is widely used in automotive body structures. Table 2 shows the main performance parameters of ISR-7008 adhesive (provided by suppliers).

Table 1. Mechanical property parameters of adhesive and adhesive substrate.

Material Attribute	ISR-7008	6005A Aluminum Alloy
Young's modulus (MPa)	4.3	71,000
Poisson's ratio	0.44	0.33
Density (kg/m3)	1400	2730

Table 2. Technical performance parameters of ISR-7008 adhesive.

Performance	ISR-7008
Tensile failure strength (MPa)	2.9
Shear failure strength (MPa)	2.5
Extension at break curing condition	225%
Curing conditions (T/RH)	20 °C/50%RH
Curing rate (mm/24 h)	3
Glass transition temperature (°C)	−59 °C

2.2. Design and Processing of Specimens

To investigate the durability of adhesive joints under different stress states, the single lap joint (SLJ), the scarf joint (SJ) and the butt joint (BJ) were selected. When adherends are isotropic metallic and when the bondline thickness is very thin, the stress of the adhesive is assumed to be uniform and equal to the average values [8,19]. Thus, the SLJ and the BJ represent the shear stress and normal stress, respectively, while the SJ refers to the combined shear and normal stress. Furthermore, the normal and shear stress components with an infinitesimal block of adhesive within the central region of the scarf joints are calculated by assuming the coordinate and stress system as shown in Figure 1. The ratio between the tensile force and shear force of the adhesive layer can be changed by changing the angle α between the adhesive interface and the axis of specimens. It can be seen from the force decomposition that F represents the tensile force on both ends of the specimen. $F\sin\alpha$ is the tensile force component of F on the bonding interface, and $F\cos\alpha$ indicates the shear force component of F on the bonding interface. The normal σ and shear stress τ components are given by Equation (1), where F is the uniaxial failure load, A denotes the bonding area and α refers to the scarf angle.

$$\sigma = \frac{F\sin\alpha}{A},\ \tau = \frac{F\cos\alpha}{A} \tag{1}$$

The docking and lap specimens were designed to study the mechanical properties of the adhesive joints under tensile stress and shear stress, separately. The overall size of the butt joint is $201\times25\times25\ \mathrm{mm}^3$, and the adhesive area is $25\times25\ \mathrm{mm}^2$, while the overall size of the lap joint is $175\times25\times11\ \mathrm{mm}^3$, and the adhesive area is $25\times25\ \mathrm{mm}^2$. The butt joint and the shear joint are shown in Figure 2, where the thickness of the adhesive layer is 1 mm.

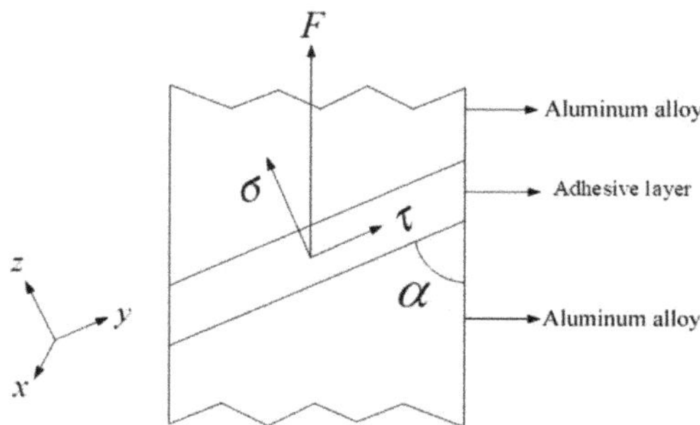

Figure 1. Orientation of stress vectors within adhesive of scarf joints.

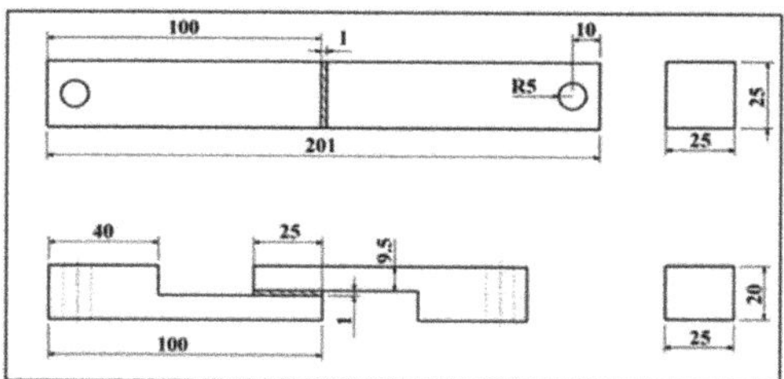

Figure 2. Schematic diagram of docking and lap joints (mm).

In the actual service process, the adhesive structure is often affected by tensile stress and shear stress; thus, it is of great significance to study the failure behavior of adhesive joints under the coupling of tensile stress and shear stress through reasonable joint design, thus being conducive to establishing the failure prediction of adhesive joints under complex stress. In order to study the failure behavior of the adhesive layer under different stress conditions, scarf joints with adhesive angles of 15°, 30°, 45°, 60°and 75° were designed and processed (as shown in Figure 3). The adhesive thickness of all specimens is unified as 1 mm.

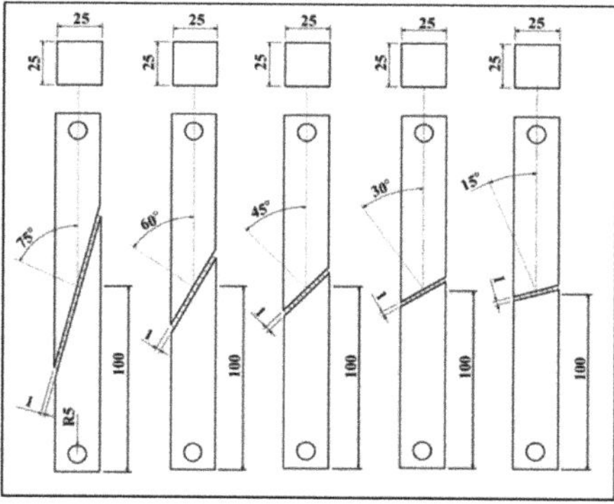

Figure 3. Schematic diagram of 15°, 30°, 45°, 60° and 75° scarf joints (mm).

2.3. Dumbbell Sample

ISR-7008 is a kind of flexible polyurethane adhesive. As the completely cured adhesive sheet is extremely soft, it is not feasible to use machined tensile test samples. Therefore, molding technology is adopted to process the sample for further avoiding the scratch problem in the cutting process and ensuring that the sample is obtained without defects. Apart from that, for achieving the above purpose, the related metal abrasive tool is designed as shown in Figure 4. The grinding tool consists of three parts: the lower part is the base to play a supporting role; the middle is the template to determine the sample size and shape; the upper part is the compression plate and bolts.

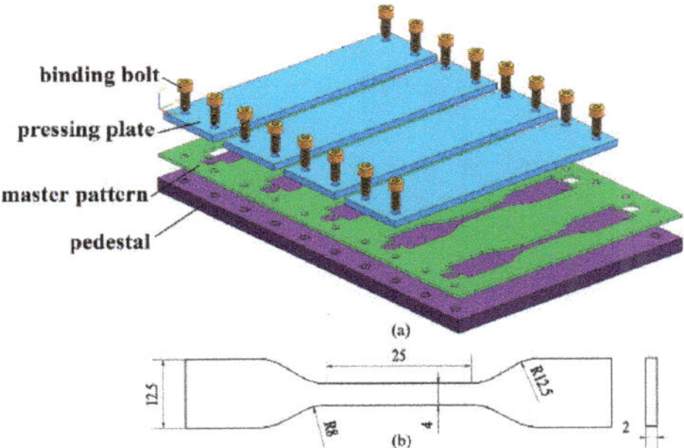

Figure 4. (**a**) Forming mold. (**b**) Dumbbell sample.

During the process of making samples, in order to prevent adhesive from sticking to the mold, a layer of polytetrafluoroethylene (PTFE) material is spread on the upper surface of the base and the lower surface of the pressure plate, and a layer of release agent is applied on all the surfaces of the template. After the die groove is coated with adhesive, the pressure plate on the cover is pressed with bolts. ISR-7008 adhesive is moisture curing. In order to ensure complete curing, it is solidified in 25 °C/50%RH environment for 7 days to remove the pressure plate (the reference manufacturer provides curing conditions), as shown in Figure 4, and then curing continues for 21 days. The geometric dimensions of the dumbbell stretch samples used refer to the NF ISO 527-2 standards.

2.4. Design and Manufacture of Fixture

There are two main difficulties to be overcome in the adhesive process of docking and scarf specimens: the accurate control of neutral and adhesive thickness of the upper and lower square test rods of the specimen. To address these, it is necessary to first design and make the corresponding adhesive fixture. While the fixture is being made, the upper and lower grooves of the fixture are milled with one knife, thus ensuring the neutrality of the upper and lower test rods in the adhesive process of the specimen. At the same time, there is a calibration line next to the groove on one side of the fixture, and it is employed to control the thickness of the adhesive layer. The metal strip in the upper part of the groove is used to fix the adhesive specimen, and the knob on one side is adopted to push the adhesive test rod to the bond, as shown in Figure 5.

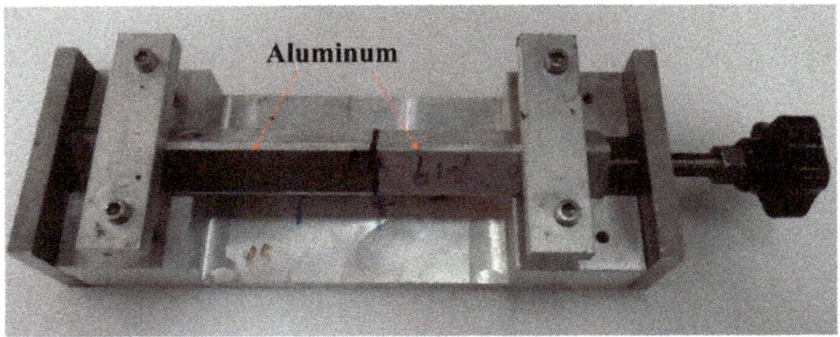

Figure 5. Adhesion fixture for docking and scarf specimens.

There are also some difficulties in the adhesive process of lap specimens; the main ones are the parallelism of the two adhesive test rods and the warping of the upper and lower surfaces. In order to solve these problems, in this paper, the corresponding fixtures, including the lateral fastening fixture and the gantry tightening fixture, are designed as shown in Figure 6. The parallelism of the adhesive specimen is guaranteed by the screw rotation clamping of the lateral fastening fixture, and the metal block between the two rods is used to control the lap width of the shear specimen, while the flatness of the upper and lower planes of the adhesive is ensured by the screw compression of the downward rotating gantry compression fixture.

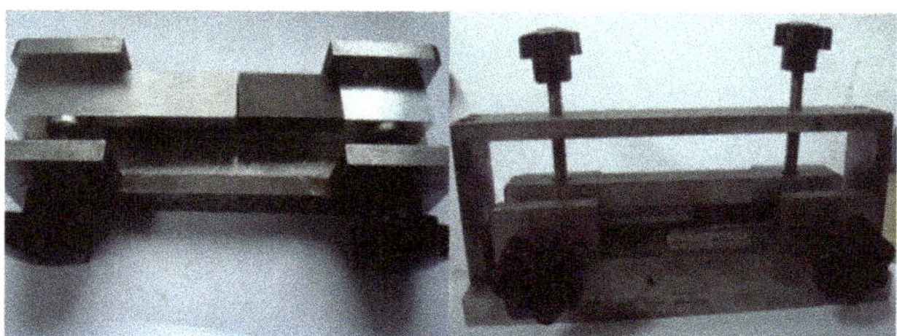

Figure 6. Diagram of lateral fastening jig (**left**) and gantry clamping jig (**right**).

2.5. Bonding Process

All specimens are adhesive in a clean and stable environment (temperature: 25 ± 3 °C; relative humidity: $50 \pm 5\%$). The preparation process is as follows:

1. An 80 mesh sandpaper is used for the cross grind concerning the adhesive surface of aluminum alloy along the 45° direction to increase the surface roughness and facilitate the adhesive.
2. Acetone is used to clean the specimen, thus removing the oil pollution and dust produced in processing. Wiping paper is dipped in acetone, and the adhesive surface is wiped in one direction until the surface of the tissue is clean. The surface is allowed to dry for 10 min.
3. Surface pretreatment coating agent Primer M is used to clean the adhesive surface again. The surface is again allowed to dry for 10 min.
4. As for adhesive ISR-7008, on its surface, the corresponding fixture described above is used to complete the specific adhesive work, and then it is solidified for a period of 4 weeks in the experimental environment.

2.6. Testing of Strength under the Condition of the Full Temperature Field

The cured fully adhesive joints are placed in the hot and humid environment box as shown in Figure 7 and are subject to the test temperature for two hours. To be specific, the temperature range is -40 °C~150 °C, the humidity range is 20% RH~99% RH, the temperature fluctuation is ± 0.1 °C and the humidity fluctuation is $\pm 1\%$. For each test joint, the load and displacement curves are acquired by the tensile test. It should be noticed that all the joints are tested at 25 °C/50%RH to obtain the residual strength of the adhesive joint.

Figure 7. High- and low-temperature hygrothermal cycle experimental chamber.

3. Experimental Data Analysis

3.1. Dumbbell Sample Test

Typical temperature conditions are selected according to the temperature range ($-40\,^\circ$C~$80\,^\circ$C) in body service environment, and typical temperature points of $-40\,^\circ$C, RT and $80\,^\circ$C are chosen. The dumbbell specimens manufactured are placed in the environmental test box according to the temperature conditions for 6 h to achieve full uniformity of the joint temperature. Immediately after taking out the dumbbell samples and loading them into the electronic universal testing machine, the dumbbell-type adhesive specimens are subjected to quasi-static tensile tests at $-40\,^\circ$C, RT and $80\,^\circ$C, and the bonded joints are tested at a constant speed of 5 mm/min until destroyed. The stress and strain curves of dumbbell samples are recorded. The temperature required for the test is provided by a high-low temperature environment chamber; a high temperature environment can be provided by resistance wire heating, while a low temperature environment can be achieved by liquid nitrogen cooling. Apart from this, temperature changes can be accurately controlled by a temperature controller.

To accurately measure the strain of dumbbell specimens during tension, a non-contact full-field strain measurement system (VIC-3D, Correlated Solutions, Inc.) is adopted as shown in Figure 8. The whole-field displacement and strain measurements are carried out by the system, based on three-dimensional digital image correlation technology. The test procedure is as follows: the dumbbell specimen is set to 20 mm in test length and is fixed on the universal testing machine; then, two CCD cameras are installed and calibrated. The strain of the specimen is obtained by analyzing the images collected in the tensile test. Each test is repeated three times to ensure the validity of the data, taking the average as the result.

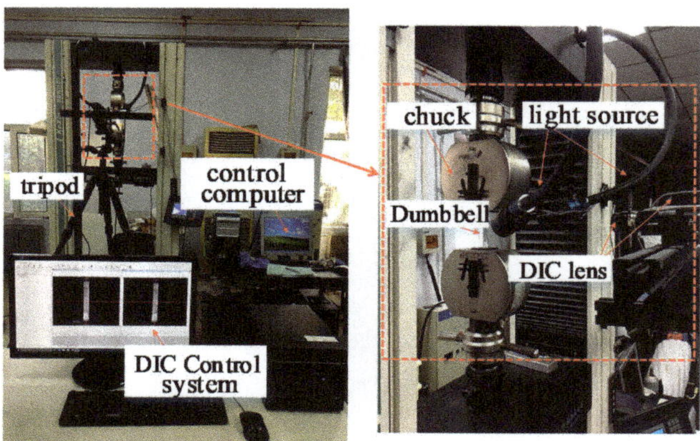

Figure 8. VIC-3D measurement system.

3.2. Mechanics Performance Testing

Table 3 shows the experimental test data concerning seven groups of adhesive specimens from different angles. It can be seen from the data in the table that there is a certain degree of dispersion between the experimental test data of the same group of adhesive specimens. In order to ensure that the failure strength of each adhesive specimen is more accurate and reasonable, the experimental data shown in Table 3 are screened and extracted, and two experimental data extraction methods—the section method and the statistical method—are adopted.

Table 3. Experimental test data of seven groups of adhesive specimens at different angles.

Temperature (°C)	Adhesive Angle (°)	Strength (MPa)					Average Strength (MPa)	ST (%)	CV (%)	Failure Mode
		No. 1	No. 2	No. 3	No. 4	No. 5				
80	90	2.27	2.15	2.19	2.23	2.21	2.21	4.47	2.03	CF
	75	2.11	2.32	2.29	2.03	2.05	2.16	13.60	6.30	CF
	60	1.98	2.11	2.05	2.2	2.11	2.09	8.15	3.90	CF
	45	2.12	1.96	1.97	2.23	2.17	2.09	12.06	5.77	CF
	30	2.11	2.08	1.93	2.05	2.23	2.08	10.82	5.20	CF
	15	2.10	2.25	2.28	1.87	2.00	2.10	17.16	8.17	CF
	0	1.98	2.23	2.05	2.01	2.23	2.10	12.12	5.77	CF
60	90	2.22	2.45	2.54	2.28	2.36	2.37	12.85	5.42	CF
	75	2.26	2.50	2.38	2.49	2.37	2.40	9.87	4.11	CF
	60	1.98	2.11	2.05	2.20	2.11	2.30	14.02	6.09	CF
	45	2.09	2.35	2.24	2.37	2.40	2.29	12.71	5.55	CF
	30	2.11	2.13	2.32	2.40	2.49	2.29	16.66	7.27	CF
	15	2.24	2.31	2.46	2.50	2.24	2.35	12.28	5.23	CF
	0	2.48	2.44	2.31	2.29	2.53	2.41	10.56	4.38	CF
40	90	2.40	2.57	2.64	2.73	2.56	2.58	12.14	4.70	CF
	75	2.63	2.54	2.41	2.70	2.47	2.55	11.73	4.60	CF
	60	2.60	2.61	2.57	2.36	2.46	2.52	10.75	4.26	CF
	45	2.42	2.35	2.56	2.66	2.61	2.52	13.08	5.18	CF
	30	2.51	2.44	2.60	2.38	2.47	2.48	8.21	3.31	CF
	15	2.36	2.48	2.58	2.56	2.77	2.55	15.03	5.89	CF
	0	2.64	2.53	2.66	2.47	2.70	2.60	9.62	3.70	CF
RT	90	2.76	2.61	2.99	2.93	2.61	2.78	17.66	6.35	CF
	75	2.66	2.82	2.57	2.83	2.62	2.70	11.85	4.39	CF
	60	2.56	2.48	2.82	2.77	2.62	2.65	14.25	5.38	CF
	45	2.78	NA	2.64	2.61	2.93	2.74	14.67	5.35	CF
	30	2.56	2.62	2.87	2.91	2.79	2.75	15.38	5.59	CF
	15	3.01	2.92	2.69	2.70	NA	2.83	16.02	5.66	CF
	0	3.01	2.87	3.12	2.78	2.52	2.86	11.42	3.99	CF

Table 3. *Cont.*

Temperature (°C)	Adhesive Angle (°)	Strength (MPa)					Average Strength (MPa)	ST (%)	CV (%)	Failure Mode
		No. 1	No. 2	No. 3	No. 4	No. 5				
0	90	3.01	2.77	2.95	2.82	3.00	2.91	9.65	3.74	CF
	60	3.04	3.13	2.99	2.75	2.69	2.92	14.06	6.52	CF
	45	2.90	2.78	3.12	3.20	3.25	3.05	16.79	6.61	CF
	30	2.88	2.98	2.90	3.17	3.27	3.04	11.45	5.66	CF
	75	3.01	3.08	3.13	2.85	2.83	2.98	10.57	4.53	CF
	15	3.11	3.03	3.00	3.06	3.25	3.09	4.06	3.18	CF
	0	2.81	3.20	3.03	3.34	3.22	3.12	19.78	6.59	CF
−20	90	3.20	3.11	3.05	2.78	3.01	3.03	15.70	5.18	CF
	75	2.96	3.11	3.23	2.92	3.23	3.09	14.61	4.72	CF
	60	3.11	3.21	2.97	2.85	3.16	3.06	14.76	4.82	CF
	45	3.23	3.51	3.37	3.29	3.70	3.42	18.84	5.51	CF
	30	3.01	3.07	3.33	3.42	3.27	3.22	17.46	5.41	CF
	15	3.22	3.44	3.19	3.50	3.20	3.31	14.79	4.47	CF
	0	3.67	3.55	3.29	3.34	3.70	3.51	18.74	5.43	CF
−40	90	3.68	3.92	3.84	4.02	3.74	3.84	13.64	3.55	CF
	75	3.82	3.77	3.90	3.91	3.75	3.83	7.31	1.91	CF
	60	4.04	3.90	3.78	3.82	3.86	3.88	10.00	2.58	CF
	45	4.32	4.18	3.99	3.84	3.72	4.01	24.41	6.08	CF
	30	3.84	3.99	3.72	4.13	4.22	3.98	20.45	5.14	CF
	15	4.12	4.03	4.21	3.87	3.87	4.02	15.09	3.75	CF
	0	3.89	4.16	4.23	3.87	3.90	4.01	17.10	4.26	CF

In the statistical method, the confidence interval in statistics is used to extract the experimental data. The confidence interval refers to the estimated interval of the overall parameters constructed by sample statistics. In this paper, the 95% confidence interval, which is commonly used in engineering data statistics, is selected to screen the experimental data falling into the confidence interval. At the same time, the failure section of the adhesive joint after static tension is analyzed and the test results of cohesion failure are selected as the effective data. Tables 4–10 show the failure strength values concerning seven groups of test specimens at different angles.

Table 4. Failure strength values of seven groups of adhesive specimens at different angles at −40 °C.

Adhesive Angle (°C)	0	15	30	45	60	75	90
Average failure load (N)	2400	8893	4683	3445	2814	2573	2458
Adhesive area (mm^2)	625	2322	1207	859	707	640	613
Normal stress (MPa)	0.00	1.04	1.99	2.83	3.36	3.70	3.84
Shear stress (MPa)	4.01	3.88	3.45	2.83	1.94	0.99	0.00

Table 5. Failure strength values of seven groups of adhesive specimens at different angles at −20 °C.

Adhesive Angle (°C)	0	15	30	45	60	75	90
Average failure load (N)	1894	7175	3693	2938	2277	2118	2152
Adhesive area (mm^2)	625	2322	1207	859	707	640	613
Normal stress (MPa)	0.00	0.86	1.61	2.22	2.65	2.98	3.04
Shear stress (MPa)	3.51	3.20	2.79	2.22	1.53	0.80	0.00

Table 6. Failure strength values of seven groups of adhesive specimens at different angles at 0 °C.

Adhesive Angle (°C)	0	15	30	45	60	75	90
Average failure load (N)	1819	6920	3524	2620	2149	1978	1913
Adhesive area (mm^2)	625	2322	1207	859	707	640	613
Normal stress (MPa)	0.00	0.80	1.52	2.16	2.53	2.88	2.91
Shear stress (MPa)	3.12	2.98	2.63	2.16	1.46	0.77	0.00

Table 7. Failure strength values of seven groups of adhesive specimens at different angles at 25 °C.

Adhesive Angle (°C)	0	15	30	45	60	75	90
Average failure load (N)	1738	6269	3199	2354	1944	1811	1753
Adhesive area (mm^2)	625	2322	1207	859	707	640	613
Normal stress (MPa)	0.00	0.73	1.37	1.94	2.29	2.61	2.78
Shear stress (MPa)	2.86	2.73	2.38	1.94	1.33	0.70	0.00

Table 8. Failure strength values of seven groups of adhesive specimens at different angles at 40 °C.

Adhesive Angle (°C)	0	15	30	45	60	75	90
Average failure load (N)	1613	5921	3042	2165	1753	1632	1594
Adhesive area (mm^2)	625	2322	1207	859	707	640	613
Normal stress (MPa)	0.00	0.66	1.24	1.78	2.18	2.46	2.58
Shear stress (MPa)	2.50	2.43	2.15	1.78	1.26	0.66	0.00

Table 9. Failure strength values of seven groups of adhesive specimens at different angles at 60 °C.

Adhesive Angle (°C)	0	15	30	45	60	75	90
Average failure load (N)	1481	5573	2776	1967	1619	1504	1477
Adhesive area (mm^2)	625	2322	1207	859	707	640	613
Normal stress (MPa)	0.00	0.61	1.15	1.62	1.99	2.32	2.37
Shear stress (MPa)	2.45	2.27	1.98	1.62	1.15	0.62	0.00

Table 10. Failure strength values of seven groups of adhesive specimens at different angles at 80 °C.

Adhesive Angle (°C)	0	15	30	45	60	75	90
Average failure load (N)	1381	5016	2523	1795	1471	1344	1287
Adhesive area (mm^2)	625	2322	1207	859	707	640	613
Normal stress (MPa)	0.00	0.54	1.03	1.48	1.80	2.09	2.21
Shear stress (MPa)	2.09	2.03	1.80	1.48	1.04	0.56	0.00

4. Analysis and Discussion of Experimental Results

4.1. Dumbbell Sample Test

The stress-strain curves at the high temperature of 80 °C, room temperature and the low temperature of −40 °C, as shown in Figure 9, show that the Young's modulus and tensile strength of the adhesive decrease with an increase in temperature, which is contrary to the increase of failure strain of epoxy adhesive with the increase of temperature due to the difference of glass transition temperature between the two kinds of adhesive [20]. The glass transition temperature of epoxy adhesive is higher, and the toughness of the material is enhanced with the increase in temperature. However, for polyurethane adhesive ISR-7008, the glass transition temperature is low (Tg = −59 °C), and the toughness is better at low temperature. The variation of the Young's modulus, tensile strength and failure strain of adhesive ISR-7008 with temperature is displayed in Table 11, and all show a downward trend with an increase in temperature.

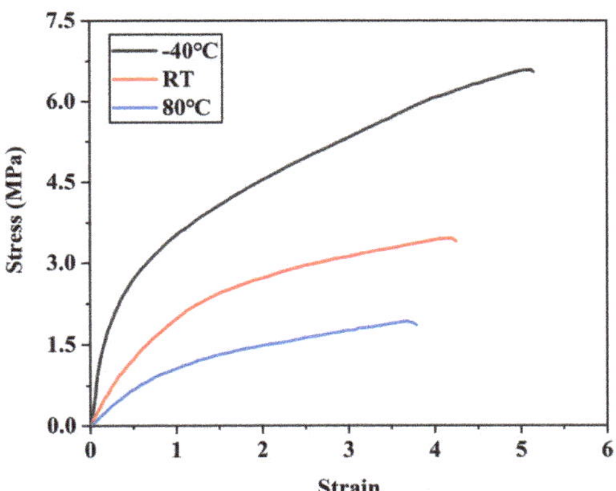

Figure 9. Stress-strain curve of adhesive.

Table 11. Young's modulus, tensile strength and failure strain of adhesives at −40 °C, RT and 80 °C.

Temperature (°C)	Young's Modulus (MPa)	Tensile Strength (MPa)	Failure Strain
−40	11.8 ± 1.61	6.5 ± 0.19	5.2 ± 0.23
RT	5.0 ± 0.27	3.4 ± 0.13	4.2 ± 0.18
80	2.9 ± 0.24	2.07 ± 0.10	3.72 ± 0.11

The failure strength and the Young's modulus of the adhesive vary greatly at different temperatures. The failure strength of the adhesive decreases by 47.69% and 68.15% at RT and 80 °C, respectively, compared with −40 °C, while the Young's modulus of the adhesive decreases by 57.63% and 75.42% at RT and 80 °C, respectively. The increase in temperature results in a declining degree of failure strength and Young's modulus. The closer the glass transition temperature of the adhesive is, the more obvious the change in the properties of the adhesive [14]. For the failure strain of the adhesive, it decreases with increasing temperature, but the decrease is slight. When the temperature rises from a low level, the ductility of the adhesive clearly changes. With the increase in temperature, the change range of ductility decreases, and that of failure strength is smaller. The adhesive fracture occurs before reaching large deformation, which leads to the decrease in failure strain of dumbbell samples with the increase in temperature. Banea et al. [21] studied room temperature silicone sulfide adhesives and similarly found that the failure displacement of the adhesives decreases with increasing temperature.

4.2. BJ and SLJ Tests

The average lap shear strength and tensile strength for joints tested at seven temperature points were obtained. In addition, the variation curves of the average lap shear and tensile strength of joints as a function of temperature are presented in Figure 10.

It was found that the adhesive strength of SLJs and BJs decreased gradually with an increase in temperature: the higher the temperature, the lower the adhesive strength. The highest adhesive strength appeared at the low temperature of −40 °C, and the lowest adhesive strength appeared at the high temperature of 80 °C. For SLJ adhesive strength, the adhesive strengths at −40 °C and −20 °C are 28.68% and 18.52%, respectively; 8.33% higher than that of room temperature. The adhesive strengths at high temperatures of 80 °C, 60 °C and 40 °C are 26.92%, 15.73% and 9.09%, respectively, lower than that at room temperature. In addition, for the adhesive strength of BJs, the adhesive strengths at high temperatures of 80 °C, 60 °C and 40 °C are 20.00%, 14.75% and 7.19% lower than that at room temperature, respectively.

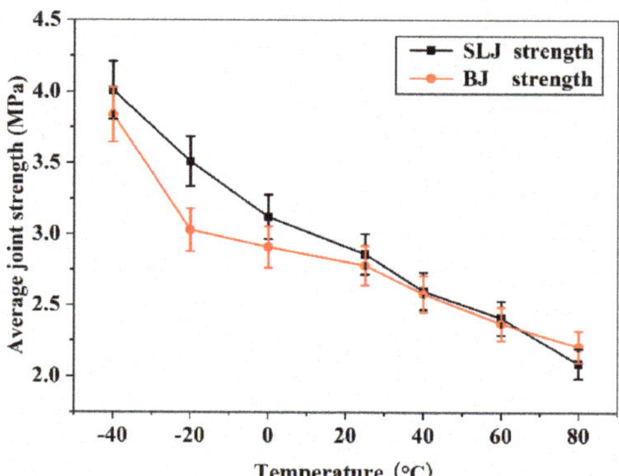

Figure 10. Average SLJ and BJ strength as a function of temperature.

The adhesive strengths at low temperatures of −40 °C, −20 °C and 0 °C are 27.60%, 8.25% and 4.47% higher than that at room temperature, respectively, which is explained by the fact that polyurethane adhesive has low glass transition temperatures (Tg = −60 °C for ISR-7008, provided by the supplier). It remains ductile, and its strength increases at low temperatures, which leads to a higher joint strength. With regard to polyurethane flexible adhesive, the adhesive strength of the joint depends not only on the strength of the adhesive but also on the toughness of the material. By comparing the changing trend concerning tensile strength and shear strength of adhesive joints, it can be found that the shear strength increases more than the tensile strength at low temperature, and it decays more than the tensile strength at high temperature, which indicates that the adhesive strength of the single lap joint is more sensitive to changes in temperature.

4.3. SJ Tests

In practical engineering applications, most stress forms of adhesive structures are basically tensile shear interaction. The adhesive strength of lap joints and butt joints decreases with the increase in temperature, while that of scarf joints under the action of tensile shear need further study. The same quasi-static tests were performed on the 15°, 30°, 45°, 60° and 75° SJs at −40 °C, −20 °C, 0 °C, 25 °C, 40 °C, 60 °C and 80 °C. In order to ensure the credibility of the experimental data, each temperature point has three samples. The average value of failure strength of the three samples is called adhesive strength. By observing the failure forms of all joints, it is found that the failure modes of scarf joints are cohesion failure, and the adhesive strength of 15°, 30°, 45°, 60° and 75° SJs changes with temperature, as shown in Figure 11.

As can be seen from Figure 11, with the decrease of adhesive strength with increased temperature, the change trend of adhesive strength is similar to that of the single tensile joint and the single shear joint. First, the 15° and 30° scarf joints are analyzed. They have the same characteristics: the joint is subjected mainly to shear action, and the joint strength measured at room temperature (RT) is taken as a reference, while the strength of 15° joints decreased by 25.79%, 16.96% and 89.89% with an increase in temperature to 40 °C, 60 °C and 80 °C, respectively. When the temperature drops to 0 °C, −20 °C and −40 °C, the strength of the joint increases by about 29.60%, 14.50% and 8.41%, respectively. When the temperature increases from room temperature (RT) to 40 °C, 60 °C and 80 °C, the adhesive strength of 30° joints decreases by 9.82%, 16.72% and 25.79%, respectively. When the temperature decreases from room temperature (RT) to 0°C, −20 °C and −40°C, the joint strength increases by 9.54%, 14.60% and 30.90%, respectively.

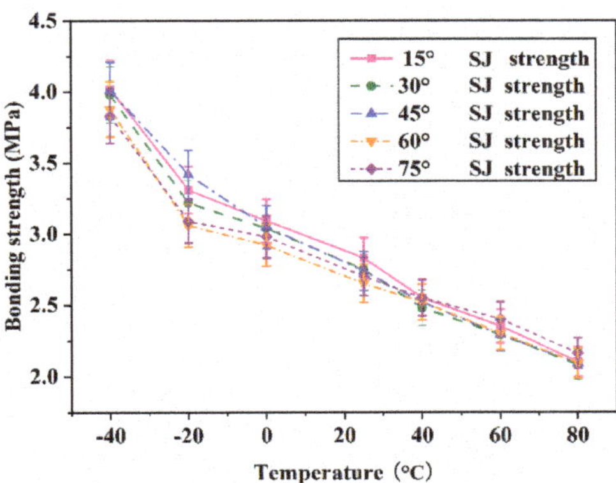

Figure 11. Average 15°, 30°, 45°, 60°, 75° SJ strength as a function of temperature.

The 60° and 75° joints possess the same characteristics: the joint is subjected mainly to tensile action. Taking the joint strength tested at room temperature as a reference, when the temperature reaches 80 °C, the strength of 60° and 75° joints decreases by 21.13% and 20.50%, respectively; when the temperature decreases to −40 °C, the joint strength increases by 31.70% and 29.50%, respectively. The 45° scarf joint is subjected to the same tensile and shear loads. The joint strength tested at room temperature (RT) is taken as a reference. With the temperature rising to 40 °C, 60 °C and 80 °C, the joint strength decreases by 8.03%, 16.42% and 23.72%, respectively. When the temperature decreases to 0°C, −20°C and −40 °C, the adhesive strength increases by 10.16%, 19.88% and 31.70%, respectively. By analyzing the variation trend regarding the adhesive strength of several kinds of scarf joints and combining the strength variation rule of single lap joints and butt joints, it is found that when the temperature rises to the highest level (80 °C) and decreases to the lowest (−40 °C), the attenuation and increase of adhesive strength of the butt joint are at least 20.00% and 27.60%, respectively. It is speculated that the docking joint is the least sensitive to temperature change compared with the lap joint and the butt joint.

4.4. Force-Displacement Curve Comparison

Representative load-displacement curves of SLJs, 15°, 30°, 45°, 60° and 75° scarf joints and BJs as a function of temperature are shown in Figure 12. By comparing the mechanical properties of adhesive and adhesive substrate, it can be found that the elastic modulus of aluminum alloy substrate is 1.6E4 times that of adhesive. In this case, it can be considered that the deformation of adhesive joints is mainly that of the adhesive. The force-displacement curves of seven kinds of joints with different stress forms are nonlinear at all test temperature points. Furthermore, it should be noted that similar findings were obtained by I. Lubowiecka et al. [22–24] for flexible adhesive Terostat MS 9360. From the force-displacement curve, it can be found that the failure displacement of single lap joints, butt joints and scarf joints gradually decreases with the increase in temperature when the temperature changes from the low temperature of −40 °C to the high temperature of 80 °C. The lower the temperature, the larger the failure displacement, and vice versa. Similar findings were obtained by Mariana et al. [25] for Sikaflex-552 and Banea et al. [26] for SikaForce 7888. In addition, the slope of the force-displacement curve is weighed as the stiffness of the joint; it is found that the stiffness of the joint varies slightly with the increase in temperature, which indicates that the change of temperature affects the stiffness of the joint.

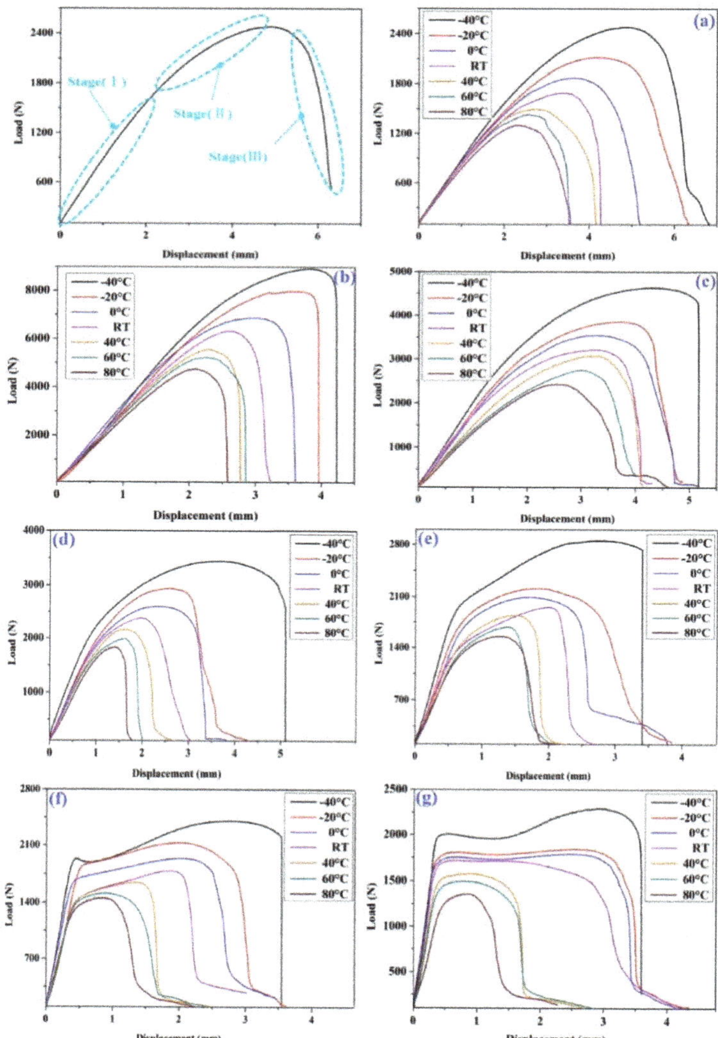

Figure 12. Load-displacement curves: (**a**) SLJ; (**b**) 15° SJ; (**c**) 30° SJ; (**d**) 45° SJ; (**e**) 60° SJ; (**f**) 75° SJ; (**g**) BJ.

The cutoff point is 45° scarf joints with equal tensile and shear loads on the adhesive surface. Based on the SLJs and 15° and 30° scarf joints with shear load on the adhesive surface, the curve of tensile force displacement change was observed. Therefore, it was found that the curve of tensile force displacement basically presented three consecutive stages in both high and low temperature environments: I. near-linear increase; II. nonlinear variation; III. the failure phase (shown in Figure 12a). To be specific, stage I corresponds to the linear elastic behavior, because the elastic modulus of ISR-7008 adhesive is relatively low, and all the curves deviate slightly from a straight line, which is mainly due to the characteristics of the polymer itself. Polyurethane adhesive is featured with nonlinear elastic behavior and low elastic modulus, which is obviously different from the linear elasticity of the epoxy adhesive curve with high elastic modulus [12,27].

In stage II, the joint stiffness is reduced due to the yield of the adhesive, and the slight rotation of SLJs may also affect the joint stiffness during the tensile process: the change from stage I to stage II is obviously dependent on the characteristics of the joint itself, and stage III is mainly the failure and fracture of the joint, which is very important for material characterization. Because of the difference in the characteristics of the joint, the tension-displacement curve of 60° and 75° scarf joints and BJs is different from that of SLJs and 15° and 30° scarf joints, and stage I is closer to a straight line. In addition, the tensile displacement curve concerning the 75° scarf joints and BJs at the low temperature of −40 °C has a yield point that becomes less and less obvious with the increase in temperature, as shown in Figure 12f–g.

4.5. Joint Stiffness

The effect of temperature on the properties of adhesives is evaluated through the change of adhesive joint stiffness. The stiffness of SLJs, 15°, 30°, 45°, 60°, 75° scarf joints and BJs varies with temperature, as shown in Figure 13. It can be clearly seen from the diagram that temperature has a significant effect on the stiffness of the joint. With the increase in temperature, the stiffness of the joint decreases gradually, and vice versa. For SLJs, when the temperature increases from RT to 80 °C, the stiffness of the joint decreases by 8.73%, and when the temperature of the joint decreases from RT to −40°C, the stiffness of the joint increases by 3.83%. For 15°, 30°, 45°, 60° and 75° scarf joints, temperature increases from RT to 80 °C, and the joint stiffness decreases by 10.48%, 30.45%, 19.98%, 9.70% and 0.88%, respectively. When the temperature decreases from RT to −40 °C, the joint stiffness increases by 9.85%, 3.29%, 20.70%, 37.99% and 46.47%, respectively. When the temperature of BJs increases from RT to 80 °C, the stiffness of the joint decreases by 43.97%, while it increases by 8.35% from RT to −40 °C. Through the above analysis, it is found that in addition to the 15° scarf joint, the stiffness of the joint increases gradually with the increase of the specimen angle, or in other words, the greater the proportion of tensile load on the adhesive area of the joint, the greater the stiffness of the joint when the smallest SLJ stiffness appears, which can also explain this law.

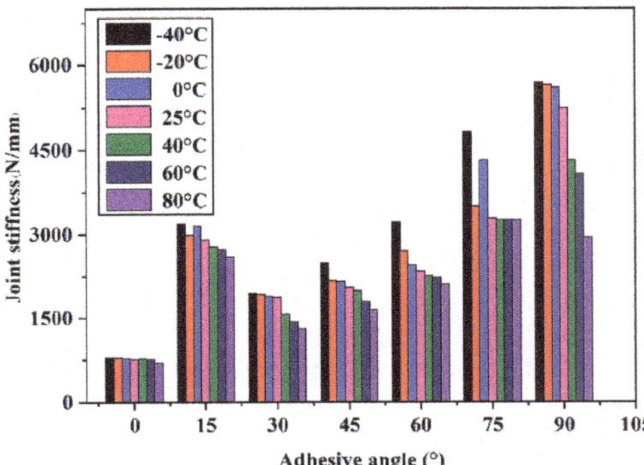

Figure 13. Column diagram of joint stiffness varying with temperature.

4.6. Energy Absorption

The experimental results show that temperature significantly affects the adhesive strength, joint stiffness and joint failure displacement. The area surrounded by the experimental force-displacement curve of the joint is the energy absorbed in the failure process of the joint (called absorption energy). Temperature affects the fracture energy of adhesive joints in a way similar to the maximum load, or in other words, there is an ongoing reduction from low to high temperatures [28]. Temperature strongly affects the energy absorption of adhesive joints. The decrease of adhesive strength is accompanied by the decrease of energy absorption. From the absorption energy histogram, it can be seen that with the reference of room temperature (RT), the fracture energy of all joints decreases with the increase in temperature, and that of all joints increases with the decrease in temperature, as shown in Figure 14. When the temperature of SLJs increases from RT to 40 °C, 60 °C and 80 °C, the fracture energy decreases from 4.811/J to 4.114/J, 3.292/J and 2.882/J, respectively, and when the temperature decreases from RT to 0 °C, −20 °C and −40 °C, the fracture energy increases to 6.46/J, 8.767/J and 10.836/J, with an increase of 34.276%, 82.23% and 125.23%, respectively.

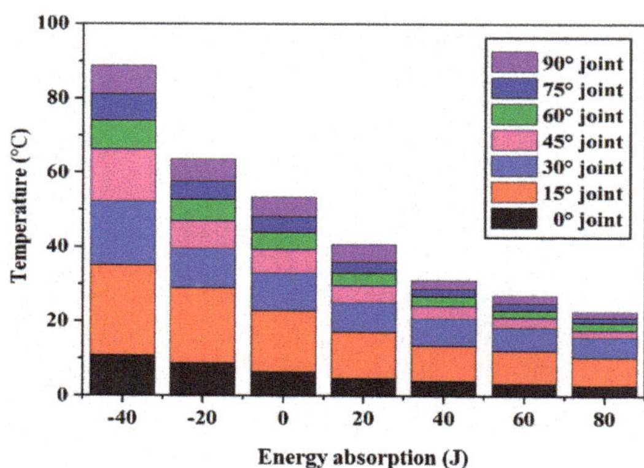

Figure 14. Joint energy absorption histogram as a function of temperature.

When the temperature of 15°, 30°, 45°, 60° and 75° scarf joints rises to 80 °C, the fracture energy of the joint decreases from 12.282/J, 8.063/J, 4.432/J, 3.328/J and 3.078/J to 7.513/J, 5.363/J, 1.899/J, 1.886/J and 1.505/J, respectively, and the attenuation rates are 38.83%, 33.49%, 57.15%, 43.33% and 51.10%, respectively. When the temperature drops to −40 °C, the fracture energy of the joint increases to 24.103/J, 17.368/J, 13.919/J, 7.806/J and 7.213/J, respectively, and the growth rates are 96.25%, 115.40%, 214.06%, 134.56% and 134.34%, respectively. In this case, it is found that the fracture energy of 45° scarf joints at the high temperature of 80 °C decreases the most, while the fracture energy of −40 °C joints increases to the greatest extent. The effect of temperature on the fracture energy of BJs is similar to that of other joints. When the temperature decreases to 0 °C, −20 °C and −40 °C, the fracture energy increases from 4.602/J to 5.245/J, 5.803/J and 7.479/J, respectively, or has an increase of 13.97%, 26.10% and 62.52%. When the temperature rises to 40 °C, 60 °C and 80 °C, the fracture energy decreases to 1.441/J, 2.031/J and 2.301/J, respectively; that is to say, 5.0%, 55.87% and 68.69%. Through the analysis of seven different stress forms of joints, it is found that the fracture energy of BJs is specially affected by temperature; the fracture energy attenuation at the high temperature of 80 °C is the largest, and the fracture energy at the low temperature of −40 °C increases the least.

4.7. Failure Criterion Surface of Adhesive Joint

In order to predict the failure behavior of adhesive joints, realize safety design, and provide further reference and guidance for the practical application of adhesive structures in the automotive industry, it is necessary to establish reasonable failure criteria. The secondary stress criterion is widely used to predict the failure of adhesive joints, and its expressions are shown as Equation (2).

$$\left(\frac{\sigma}{N}\right)^2 + \left(\frac{\tau}{S}\right)^2 = 1 \tag{2}$$

With tangential shear stress τ as the abscissa and normal stress σ as the ordinate, a coordinate system of adhesive positive shear stress was built. According to the experimental data shown in Tables 4–10, the least square method was adopted to fit the secondary stress failure criterion curve of adhesive joints at different temperatures (as shown in Figure 15). It can be seen from the figure that with the increase in temperature, the range between the failure criterion curve and the coordinate axis of the adhesive joint is gradually reduced, which indicates that the adhesive joint is more likely to be destroyed at high temperature. The parameter values (N, S) and corresponding goodness of fit (R2) of the failure criterion formula at seven test temperatures are shown in Table 12.

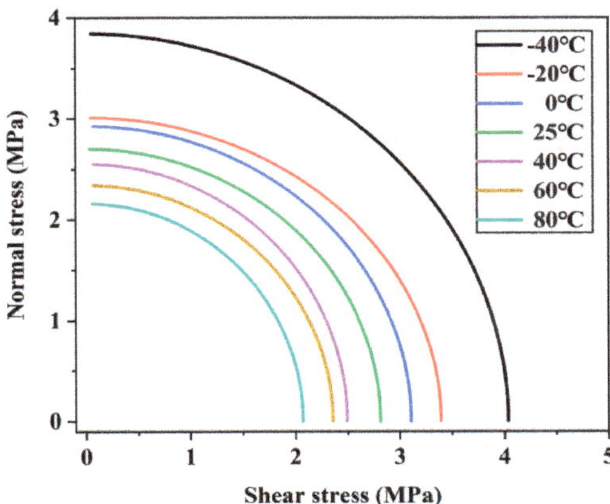

Figure 15. Failure criterion curves of adhesive joints at different temperatures.

Table 12. Parameters of failure criteria and corresponding goodness of fit at seven temperatures.

Temperature (°C)	N	S	R2
−40	3.8443	4.0337	0.9948
−20	3.0142	3.3914	0.9838
0	2.9296	3.1055	0.9903
25	2.7042	2.8139	0.9819
40	2.5549	2.4921	0.9919
60	2.3459	2.3558	0.9709
80	2.1639	2.0695	0.9791

The parameters (N, S) in the failure criterion formula under different temperature conditions were extracted, the change law was fitted by the exponential function (0.9241 and 0.9827, respectively) according to their change characteristics with temperature, and Equations (3) and (4) were obtained as follows. When Equations (3) and (4) are brought in Equation (5), the failure criterion formula of the adhesive joint under the condition of the full temperature field is obtained.

$$N = 2.07 + 0.82 \times 0.98^T \tag{3}$$

$$S = 1.26 + 1.85 \times 0.99^T \tag{4}$$

$$\left(\frac{\sigma}{2.07 + 0.82 \times 0.98^T} \right)^2 + \left(\frac{\tau}{1.26 + 1.85 \times 0.99^T} \right)^2 = 1 \tag{5}$$

where T represents any temperature value between −40 °C and 80 °C. When the ambient temperature is brought into Equations (3) and (4), the failure criterion of adhesive joints at this temperature can be obtained, which provides the basis for the failure prediction of the adhesive structure. In order to reflect the variation of failure criterion with temperature more intuitively, the failure criterion surface of adhesive joints is established by

MATLAB software, as shown in Figure 16. Therefore, it can be seen that with the increase in temperature, the failure criteria of adhesive joints shrink gradually.

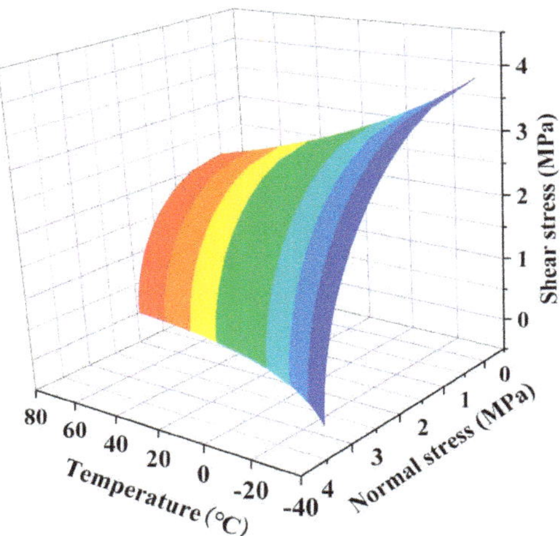

Figure 16. ISR-7008 adhesive joint failure criterion surface.

5. Conclusions

In this paper, the influence of ambient temperature on the mechanical properties and joint strength of the adhesive body was studied, and the failure criterion surface under the full temperature field was also established. The specific research content is summarized as follows:

1. The effect of ambient temperature on the mechanical properties of ISR-7008 adhesive was studied. The results show that the Young's modulus, tensile strength and failure strain of the adhesive decrease with the increase in temperature.
2. Quasi-static tensile tests were carried out on lap joints, scarf joints (15°, 30°, 45°, 60° and 75°) and butt joints at different temperatures. The results show that the strength of the joint decreases with the increase in temperature. Compared with lap joints and scarf joints, the strength of butt joints is the least sensitive to the change in temperature. With the increase of the adhesive angle, the stiffness of the joint increases gradually, while with the increase in temperature, the stiffness of all the tested joints decreases gradually. At the same time, with the increase in temperature, the failure displacement and energy absorbed by the joint also witness a downward trend.
3. Based on the experimental data, the secondary stress failure criteria of adhesive joints at different temperatures are obtained. Considering this, the surface function of failure criteria under the full temperature field environment is proposed, which provides a reference for the failure prediction of adhesive structures under different temperatures and complex stress conditions.

Author Contributions: Conceptualization, Y.F.; methodology, H.L.; writing—original draft preparation, Y.F.; writing—review and editing, Y.F.; project administration, H.P.; funding acquisition, Y.F. All authors have read and agreed to the published version of the manuscript.

Funding: This research was funded by the Science and Technology Project of Henan Province, grant number 202102210044.

Data Availability Statement: The datasets analyzed during the current study are available from the corresponding author on reasonable request.

Conflicts of Interest: The authors declare no conflict of interest.

References

1. Li, Y. Advances in Welding and Joining Processes of Multi-material Lightweight Car Body. *Chin. J. Mech. Eng.* **2016**, *52*, 1–23. [CrossRef]
2. Lai, W.-J.; Pan, J. Failure mode and fatigue behavior of weld-bonded lap-shear specimens of magnesium and steel sheets. *Int. J. Fatigue* **2015**, *75*, 184–197. [CrossRef]
3. Sakundarini, N.S.; Taha, Z.; Abdul-rashid, S.H.; Ghazila, R.A. Optimal multi-material selection for lightweight design of automotive body assembly incorporating recyclability. *Mater. Des.* **2013**, *50*, 846–857. [CrossRef]
4. Banea, M.; Rosioara, M.; Carbas, R.; da Silva, L. Multi-material adhesive joints for automotive industry. *Compos. Part B Eng.* **2018**, *151*, 71–77. [CrossRef]
5. Sousa, J.; Correia, J.; Gonilha, J.; Cabral-Fonseca, S.; Firmo, J.; Keller, T. Durability of adhesively bonded joints between pultruded GFRP adherends under hygrothermal and natural ageing. *Compos. Part B Eng.* **2019**, *158*, 475–488. [CrossRef]
6. Banea, M.D.; De Sousa, F.S.M.; Da Silva, L.; Campilho, R.D.S.G.; De Pereira, A.M.B. Effects of Temperature and Loading Rate on the Mechanical Properties of a High Temperature Epoxy Adhesive. *J. Adhes. Sci. Technol.* **2011**, *25*, 2461–2474. [CrossRef]
7. Kang, S.-G.; Kim, M.-G.; Kim, C.-G. Evaluation of cryogenic performance of adhesives using composite–aluminum double-lap joints. *Compos. Struct.* **2007**, *78*, 440–446. [CrossRef]
8. Fan, Y.; Shangguan, L.; Zhang, M.; Peng, H.; Na, J. Effect of cyclic hygrothermal aging on mechanical behaviours of adhesive bonded aluminum joints. *J. Adhes. Sci. Technol.* **2021**, *35*, 199–219. [CrossRef]
9. Banea, M.D.; Da Silva, L.; Campilho, R. Effect of Temperature on Tensile Strength and Mode I Fracture Toughness of a High Temperature Epoxy Adhesive. *J. Adhes. Sci. Technol.* **2012**, *26*, 939–953. [CrossRef]
10. Viana, G.; Costa, M.; Banea, M.D.; Da Silva, L. A review on the temperature and moisture degradation of adhesive joints. *Proc. Inst. Mech. Eng. Part L J. Mater. Des. Appl.* **2017**, *231*, 488–501. [CrossRef]
11. Grant, L.; Adams, R.; Da Silva, L.F. Effect of the temperature on the strength of adhesively bonded single lap and T joints for the automotive industry. *Int. J. Adhes. Adhes.* **2009**, *29*, 535–542. [CrossRef]
12. Banea, M.D.; Da Silva, L.F.M. The effect of temperature on the mechanical properties of adhesives for the automotive industry. *Proc. Inst. Mech. Eng. Part L J. Mater. Des. Appl.* **2010**, *224*, 51–62. [CrossRef]
13. Marques, E.; Da Silva, L.F.M.; Banea, M.D.; Carbas, R.J.C. Adhesive Joints for Low- and High-Temperature Use: An Overview. *J. Adhes.* **2014**, *91*, 556–585. [CrossRef]
14. Adams, R.; Coppendale, J.; Mallick, V.; Al-Hamdan, H. The effect of temperature on the strength of adhesive joints. *Int. J. Adhes. Adhes.* **1992**, *12*, 185–190. [CrossRef]
15. Na, J.; Mu, W.; Qin, G.; Tan, W.; Pu, L. Effect of temperature on the mechanical properties of adhesively bonded basalt FRP-aluminum alloy joints in the automotive industry. *Int. J. Adhes. Adhes.* **2018**, *85*, 138–148. [CrossRef]
16. Tan, W.; Na, J.X.; Mu, W.L.; Shen, H.; Qin, G. Effect of service temperature on static failure of BFRP/aluminum alloy adhesive joints. *J. Traffic Transp. Eng.* **2020**, *20*, 171–180.
17. Da Silva, L.F.; Adams, R.; Gibbs, M. Manufacture of adhesive joints and bulk specimens with high-temperature adhesives. *Int. J. Adhes. Adhes.* **2004**, *24*, 69–83. [CrossRef]
18. Zhang, Y.; Vassilopoulos, A.; Keller, T. Effects of low and high temperatures on tensile behavior of adhesively-bonded GFRP joints. *Compos. Struct.* **2010**, *92*, 1631–1639. [CrossRef]
19. Pascal, J.; DarqueCeretti, E.; Felder, E.; Pouchelon, A. Rubber-like adhesive in simple shear: Stress analysis and fracture mor-phology of a single lap joint. *J. Adhes. Sci. Technol.* **1994**, *8*, 553–573. [CrossRef]
20. Hu, P.; Han, X.; Li, W.D. Research on the static strength performance of adhesive single lap joints subjected to extreme tem-perature environment for automotive industry. *Int. J. Adhes. Adhes.* **2013**, *41*, 119–126. [CrossRef]
21. Banea, M.D.; Silva, L.F.M.D. Static and fatigue behaviour of room temperature vulcanising silicone adhesives for high tem-perature aerospace applications. *Mater. Werkst.* **2010**, *41*, 325–335. [CrossRef]
22. Lubowiecka, I.; Rodríguez, M.; Rodriguez, E.; Martinez, D. Experimentation, material modelling and simulation of bonded joints with a flexible adhesive. *Int. J. Adhes. Adhes.* **2012**, *37*, 56–64. [CrossRef]
23. Da Silva, L.F.; Carbas, R.; Critchlow, G.; Figueiredo, M.; Brown, K. Effect of material, geometry, surface treatment and environment on the shear strength of single lap joints. *Int. J. Adhes. Adhes.* **2009**, *29*, 621–632. [CrossRef]
24. Karachalios, E.; Adams, R.; da Silva, L.F. Single lap joints loaded in tension with ductile steel adherends. *Int. J. Adhes. Adhes.* **2013**, *43*, 96–108. [CrossRef]
25. Banea, M.D.; Silva, L.F.M.D. Mechanical Characterization of Flexible Adhesives. *J. Adhes.* **2009**, *85*, 261–285. [CrossRef]
26. Banea, M.D.; Silva, L.F.M.D.; Campilho, R.D.S.G. The Effect of Adhesive Thickness on the Mechanical Behaviour of a Structural Polyurethane Adhesive. *J. Adhes.* **2015**, *91*, 331–346. [CrossRef]
27. Loureiro, A.L.; Da Silva, L.F.M.; Sato, C.; Figueiredo, M.A.V. Comparison of the Mechanical Behaviour Between Stiff and Flexible Adhesive Joints for the Automotive Industry. *J. Adhes.* **2010**, *86*, 765–787. [CrossRef]
28. Moroni, F.; Pirondi, A.; Kleiner, F. Experimental analysis and comparison of the strength of simple and hybrid structural joints. *Int. J. Adhes. Adhes.* **2010**, *30*, 367–379. [CrossRef]

Review

Review on the Application of Supplementary Cementitious Materials in Self-Compacting Concrete

Lang Pang [1], Zhenguo Liu [2], Dengquan Wang [3,*] and Mingzhe An [1]

[1]   School of Civil Engineering, Beijing Jiaotong University, Beijing 100044, China;
      20121088@bjtu.edu.cn (L.P.); mzhan@bjtu.edu.cn (M.A.)
[2]   Beijing Urban Construction Group, Beijing 100023, China; liuzhenguothu@163.com
[3]   Department of Civil Engineering, Tsinghua University, Beijing 100084, China
*   Correspondence: wangdq16@mails.tsinghua.edu.cn; Tel.: +86-188-1135-1597

Abstract: For the sustainable development of construction materials, supplementary cementitious materials (SCMs) are commonly added to self-compacting concrete (SCC). This paper reviewed the application techniques and hydration mechanisms of SCMs in SCC. The impacts of SCMs on the microstructure and performance of SCC were also discussed. SCMs are used as a powder material to produce SCC by replacing 10% to 50% of cement. Hydration mechanisms include the pozzolanic reaction, alkaline activation, and adsorption effect. Moreover, the filling effect and dilution effect of some SCMs can refine the pore structure and decrease the temperature rise of concrete, respectively. Specifically, the spherical particles of fly ash can improve the fluidity of SCC, and the aluminum-containing mineral phase can enhance the resistance to chloride ion penetration. Silica fume will increase the water demand of the paste and promote its strength development (a replacement of 10% results in a 20% increase at 28 days). Ground-granulated blast furnace slag may reduce the early strength of SCC. The adsorption of Ca^{2+} by $CaCO_3$ in limestone powder can accelerate the hydration of cement and promote its strength development.

Keywords: self-compacting concrete; supplementary cementitious materials; hydration mechanisms; microstructure; fresh properties

Citation: Pang, L.; Liu, Z.; Wang, D.; An, M. Review on the Application of Supplementary Cementitious Materials in Self-Compacting Concrete. *Crystals* **2022**, *12*, 180. https://doi.org/10.3390/cryst12020180

Academic Editors: Yifeng Ling, Chuanqing Fu, Peng Zhang, Peter Taylor and Helmut Cölfen

Received: 1 December 2021
Accepted: 24 January 2022
Published: 26 January 2022

Publisher's Note: MDPI stays neutral with regard to jurisdictional claims in published maps and institutional affiliations.

1. Introduction

Self-compacting concrete (SCC) is a type of high-performance concrete that can be poured into structural formwork by gravity and compacted without vibration. Okamura et al. [1] pioneered the application of SCC in Japan in 1988. The outstanding features are that SCC eliminates the mechanical vibration process and lowers labor costs as compared to normal vibrating concrete (NVC), and SCC has a high powder material content in the mixture to increase fresh properties (Figure 1). However, using cement solely as a powder material leads to high production cost for SCC, which restricts its wide use. In addition, the high cement content in SCC poses increasing environmental risks as cement production is a high-resource-consuming and waste-discharging process, and its annual production has reached 3000 million tons worldwide [2].

SCC	A	W	Powder	S	G
Air		Water	SCMs can be added	Fine aggregate	Coarse aggregate
NVC	A	W	Cement	S	G

Figure 1. Comparison of component proportions between SCC and NVC [1].

The addition of supplementary cementitious materials (SCMs) as a partial substitute for cement can significantly lower the production cost of SCC as well as relieve the shortages of cement raw materials and solid waste pollution [3]. SCMs can adjust the fresh properties and improve the durability properties [4]. Specifically, SCMs can effectively enhance the microstructure of SCC. Furthermore, SCMs will have superposition effects when two or more of them are used together. Although SCMs have been widely employed in SCC, there is still a lack of a summary in its mix proportions. Moreover, the functions of SCMs in SCC remain unclear, and the systematic analysis of macroscopic properties does not exist yet. Therefore, it is necessary to outline the influences of SCMs in SCC, which can promote the sustainable development of SCC technology and improve the comprehensive utilization of solid wastes. The application techniques and hydration mechanisms of SCMs in SCC were reviewed in this work, and the impacts of fly ash (FA), silica fume (SF), ground granulated blast furnace slag (GBFS), and limestone powder (LP) on the microstructure and performance of SCC were also summarized.

2. Mixture Design of SCC

The fresh paste of SCC should have high fluidity as well as resistance to segregation and bleeding during pouring, especially when the paste flows through the limited space of reinforcing bars. It should be noted that the higher the proportion of coarse aggregate in concrete, the smaller the relative distance between particles, which increases the frequency of collision and friction. As a result, the internal stress caused by coarse aggregate consumes a large amount of energy for flowing, which reduces the fresh properties of the paste and even causes blockage. To avoid this, Okamura et al. [1] initially modified the mix proportion by reducing aggregate content, increasing powder content, and adding superplasticizer. To generate self-compacting concrete, this work first sets the amount of aggregate and next changes the water-to-binder ratio and superplasticizer dosages. Figure 1 depicts the proportion of each component. In China, the mixture proportion design of SCC is often done by the absolute volume technique (Chinese standard: CECS203, 2006). In this technique, the cement paste that meets the performance requirements is prepared first, and the fine and coarse aggregates are added sequentially to produce the appropriate mortar and concrete. In addition, Wu et al. [5] and Nie et al. [6] have proposed a mix design method based on the rheological characteristics of paste.

It can be found that SCC contains a larger proportion of powder material than NVC. Domone [7] summarized a lot of instances from previous research and discovered that roughly 95% of self-compacting concrete had powder masses of more than 400 kg/m^3. This demonstrates that more SCMs can be added to SCC. The application of SCMs reduces the production costs of SCC while improving concrete performance. For example, adding FA can improve the fresh properties of SCC [8], and adding SF can enhance its strength properties [9]. Moreover, the viscosity modifying admixture should be added to stabilize the rheology and setting time when segregation occurs due to the addition of high-content or composite SCMs [10].

3. Material Characteristics

3.1. Characteristics of SCMs

The most common SCMs used in SCC are fly ash, silica fume, ground granulated blast furnace slag, and limestone powder. In addition, some other SCMs with fewer use cases are also used in SCC, such as copper slag [11], zeolite powder [12], etc. From the results of existing studies, the percentages of SiO_2, Al_2O_3, and CaO in FA, SF, GBFS, LP, and cement were compared and shown in Figure 2. FA has the highest Al_2O_3 content; GBFS mainly contains SiO_2 and CaO; the chemical composition of SF and LP is relatively simple, containing only SiO_2 and $CaCO_3$, respectively. It is worth noting that the chemical composition of the same kind of SCM varies greatly due to the considerable variances in the raw ore, manufacturing procedure, and discharge process. Compared with SCMs, Portland cement is more concentrated in the areas shown in Figure 2, and its chemical composition is more stable.

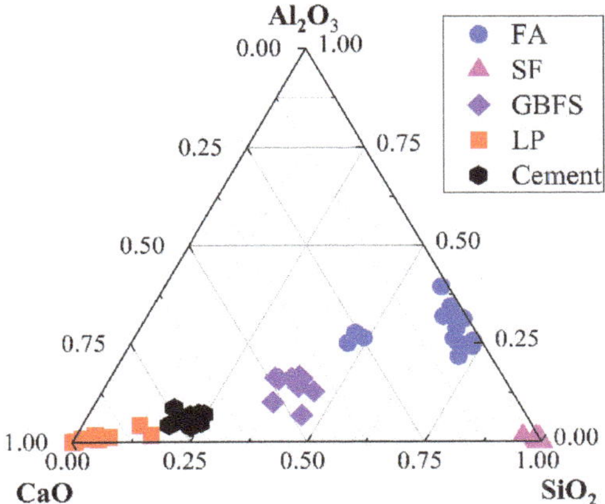

Figure 2. Chemical composition of FA [8,13–25], SF [9,26–41], GBFS [42–54], LP [55–63], and cement [19–28,38–40,64–66].

The microscopic images of FA, SF, GBFS, and LP are shown in Figure 3. It can be found that the FA is mostly smooth and spherical particles, and this spherical shape can play a ball-bearing role in the freshly mixed SCC. The particle sizes of SF are usually small, and it will typically increase the water demand of fresh paste. GBFS and LP show obvious irregular and angular shapes due to mechanical grinding.

Figure 3. Microscopic image of SCMs: (**a**) FA [23], (**b**) SF [17], (**c**) GBFS [50], (**d**) LP [67].

3.2. Hydration Mechanisms

The hydration processes of Portland cement can be divided into three periods: the induction period, the acceleration period, and the deceleration period [68]. When SCMs are incorporated into SCC, the hydration process of the composite system is significantly influenced by SCMs (Figure 4). The hydration process in the composite system usually involves the cement first hydrating to produce primary hydration products; then, the products react with SCMs to produce secondary hydration products. For example, the hydration of GBFS is triggered by the deconstruction of the glass by OH^- generated from the hydration of cement, which serves as an alkaline activator. This process releases the ions in the glass (Ca^{2+}, Al^{3+}, SiO_4^{4-}, etc.) into the solution for subsequent hydration. The main reaction product of GBFS by alkaline activation is a type of aluminum-substituted C-A-S-H gel, which presents a disordered tobermorite-like C-S-H type structure. In addition, the active Al_2O_3 in GBFS will further react with $Ca(OH)_2$ and gypsum ($CaSO_4$) to form ettringite (AFt) [69,70]. These reactions are shown in Equations (1)–(3) as follows:

$$3CaO{\cdot}SiO_2 + nH_2O \rightarrow xCaO{\cdot}SiO_2{\cdot}yH_2O + (3 - x)Ca(OH)_2 \tag{1}$$

$$CaO\text{-}Al_2O_3\text{-}SiO_2 + Ca(OH)_2 + NaOH \rightarrow C\text{-}(N)\text{-}A\text{-}S\text{-}H \tag{2}$$

$$Al_2O_3 + 3Ca(OH)_2 + 3(CaSO_4{\cdot}2H_2O) + 23H_2O \rightarrow 3CaO{\cdot}Al_2O_3{\cdot}3CaSO_4{\cdot}32H_2O. \tag{3}$$

For the C-A-S-H gel, existing studies have shown that when the paste contains Na^+, the chemically bound Ca^{2+} in C-A-S-H will be replaced by Na^+ to form a C-(N)-A-S-H type gel [71]. Myers et al. [72] proposed a structural model to simulate this gel based on the cross-linked and non-cross-linked structure properties of C-(N)-A-S-H, as shown in Figure 5.

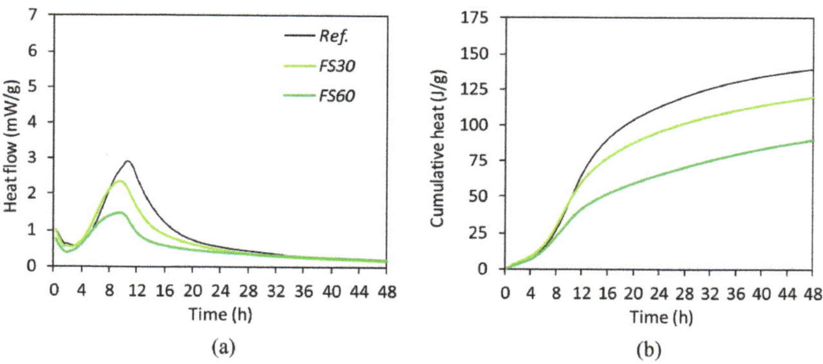

Figure 4. Isothermal calorimetry curves for the paste mixed with GGBS: (**a**) heat flow and (**b**) cumulative heat [73].

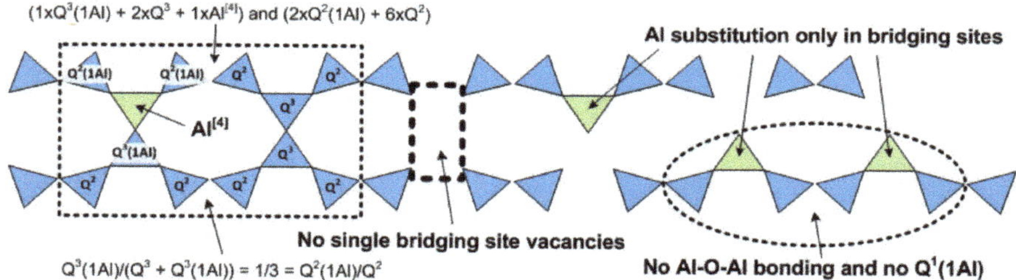

Figure 5. The cross-linked and non-cross-linked structures of the C-(N)-A-S-H type gel [72].

Similar to the hydration process of GBFS, the hydration of FA occurs under the alkali activity of OH⁻ produced by cement hydration. However, FA is less active due to its high amount of SiO_2 and Al_2O_3 (low Ca/(Si + Al) ratio) and stable Si-O and Al-O bonds. Its contribution to hydration is generally apparent at a later stage. Fernandez et al. [74] developed a microstructural model for the hydration process of FA, as illustrated in Figure 6. The attack on the incomplete spherical glassy shell of FA particles by OH⁻ causes the formation of reaction products both inside and outside, and finally, FA particles with various reaction degrees will be embedded in the microstructure. It is worth pointing out that SCMs can offer nucleation sites for cement hydration in addition to the secondary reaction with the primary products. For instance, the chemical adsorption of Ca^{2+} by the surface of limestone particles can effectively enhance the nucleation and growth of C-S-H; silica fume can adsorb Ca^{2+} via electrostatic force and boost the formation of hydration products [75].

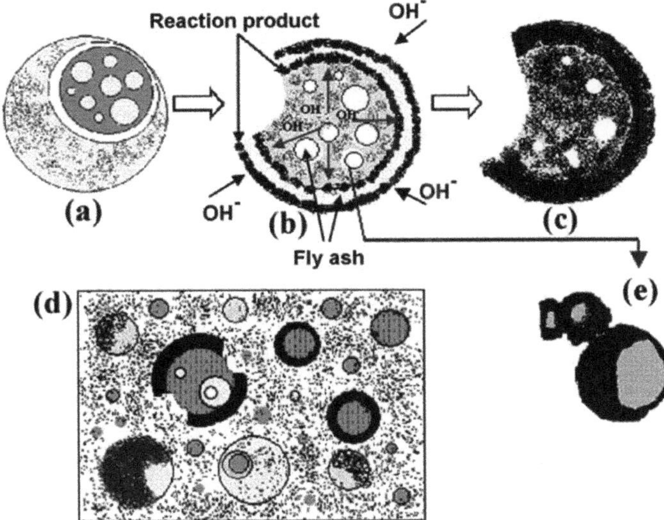

Figure 6. Descriptive model of the reaction process of fly ash: (**a**) initial chemical attack, (**b**) bi-directional alkaline attack, (**c**) reaction product, (**d**) several morphologies and (**e**) products covers certain portions [74].

4. Influence of SCMs in SCC

4.1. Microstructure

Concrete consists of three parts: aggregate, interfacial transition zone (ITZ), and cement paste. ITZ is a thin shell wrapped around the surface of the aggregate and is the lowest-strength part of concrete, whose microstructure determines the performance of concrete [76]. The incorporated SCMs can influence the formation and development of the ITZ in SCC. Figure 7 shows the comparison of the ITZ between SCC without SCMs and mixed with 20% and 30% fly ash [25]. The influences originate from three effects of SCMs: namely, the filling effect, pozzolanic reaction, and dilution effect. The filling effect can increase the packing density and optimize the mixture proportion of concrete. At the same time, the pozzolanic reaction of SCM can consume the CH generated by cement hydration while creating C-S-H, thus improving the interfacial transition zone. In addition, the fact of a general slower hydration reaction of SCM allows for more water to participate in cement hydration, resulting in an adequate reaction of cement. Simultaneously, SCMs can offer nucleation sites for cement hydration, resulting in more evenly dispersed reaction products (Figure 8).

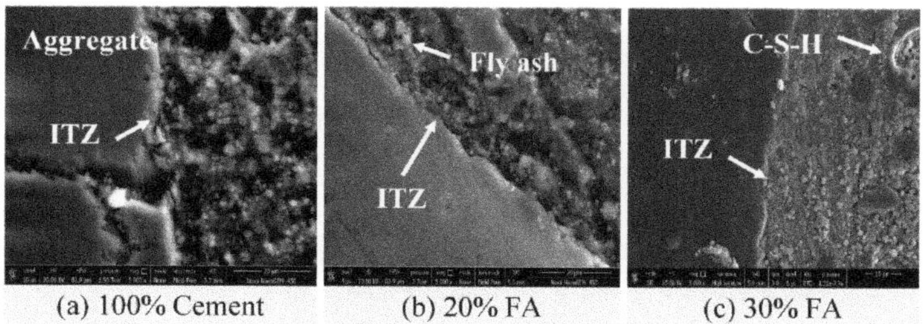

(a) 100% Cement   (b) 20% FA   (c) 30% FA

Figure 7. SEM images of SCC mixes at different levels of fly ash [25].

(a) 100% Cement   (b) 15% SF   (c) 10% LP

Figure 8. SEM images of SCC mixed with SF and LP [33].

4.2. Fresh Properties

Fluidity is the most common performance of fresh concrete, which engineers always primarily consider. On one hand, SCMs in SCC may fill the pores and lower the pore water content, increasing the quantity of free water and wrapping around the surface of the powder particles to produce a uniform water film that lubricates the paste and minimizes friction [77]. On the other hand, the increased specific surface area of some ground SCMs will improve water absorption and decrease free water content. Therefore, the fineness and proportion of SCMs have a significant impact on the fresh properties of SCC. Furthermore, because certain SCMs may reduce fluidity, superplasticizers are often necessary in SCC to improve the performance of the fresh mixture. The superplasticizer mainly disperses the powder particles via steric repulsion. When one end of the superplasticizer is adsorbed to the surface of the powder particles, the long chain at the other end generates physical barriers to prevent the surrounding cement particles from aggregation [78,79]. It is worth noting that the superplasticizers added to the SCC may have negative consequences. Existing studies have shown that the adsorption of these chemical admixtures on cement particles might retard the early hydration and result in a delayed setting [80].

A good deal of research has demonstrated that fly ash can increase the fluidity of the fresh SCC. This effect can be attributed to the smooth and spherical particles of FA [81]. Jain et al. [25] reported that the spherical particles of FA provided a ball-bearing action in newly mixed SCC, reducing the frictional resistance of aggregate particles. Promsawat et al. [23] reported similar findings. In addition, Sonebi [82] proposed that adding FA could enhance the fluidity by thickening the water film. However, FA may have negative effects on the fresh properties of SCC. Duran-Herrera et al. [22] found that FA prolongs the setting time of SCC.

Unlike fly ash, silica fume increases the water demand and decreases the fluidity due to its larger specific area. Choudhary et al. [17] observed the microscopic morphology of FA and SF (Figure 3) and discovered that spherical FA helps to reduce the friction between particles and improve fluidity. However, SF decreases fluidity and increases the superplasticizer demands due to its larger specific surface area and rough surface texture. As shown in Figure 9, Mustapha et al. [8] investigated the effects of FA and SF on the fresh properties of SCC and discovered that FA improved the slump flow diameter (fluidity) but decreased the V-Funnel time (viscosity), whereas SF decreased the fluidity but increased the viscosity of SCC.

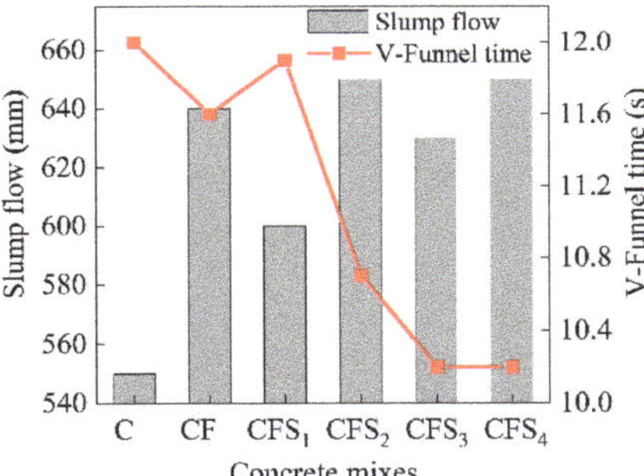

Figure 9. The slump flow and V-funnel time of SCC, adding FA and SF [8].

The impact of GBFS on the fresh properties of SCC is connected to the fineness and admixture proportion. Through the study of Boukendakdji [44], GBFS with a specific surface area of 350 m^2/kg was utilized to substitute cement (Figure 10), with an ideal admixture rate of 15%. Furthermore, it was shown that GBFS has a superior water retention effect at increasing fineness, which minimizes the time-dependent loss of slump flow diameter. In contrast, Selvarani et al. [83] discovered that GBFS inclusion lowers the fluidity of SCC. Moreover, Ofuyatan et al. [84] reported that GBFS had a minimal influence on fresh properties, with an increase in incorporation from 10% to 30%, causing a 5% drop in slump flow diameter. This discrepancy can be attributed to two aspects: the differences in filling effect and specific surface area of varying fineness, and the changes in GBFS adsorption capability by various superplasticizers.

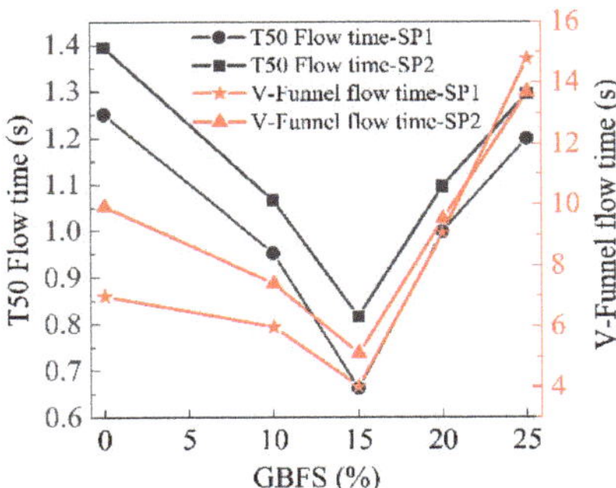

Figure 10. The effect of GBSF content on the T50 flow time and V-Funnel flow time of SCC (SP1 is a polycarboxylate-based superplasticizer and SP2 is a naphthalene sulphonate-based superplasticizer) [44].

The influence of LP on the fresh properties of SCC is mostly determined by its fineness. The filling effect of LP enhances the fluidity of the paste, but a higher content of finer powders increases the water requirement. Sua-iam et al. [56,57] found that mixing LP with a median particle size of 15.63 μm ($D_{50\ cement}$ = 23.30 μm) increased water demand and reduced fluidity. Faheem [61] discovered that mixing LP with a median particle size of 11.60 μm ($D_{50\ cement}$ = 15.32 μm) also reduced fluidity. Celik et al. [58,59] showed that LP with an average particle size of 48.1 μm (D_{cement} = 10.4 μm) enhances fluidity and shortens setting time, and that about 15% of the LP in the experiment enhanced slump flow diameter by 7%. In addition, some research revealed that there is a discrepancy in the compatibility of LP with various superplasticizers [62].

4.3. Strength Properties

The effect of SCMs on the strength properties of SCC is influenced by the filler effect, the pozzolanic reaction, and the dilution effect. The partial substitution of cement by SCMs reduces the cementitious component of the paste, hence slowing the development of early strength. FA is predominantly composed of SiO_2 and Al_2O_3 with minimal pozzolanic reactivity, and it typically facilitates post-strength growth [81]. Mahalingam et al. [21] reported that using 40% FA replacement of cement has a considerable diluting effect and lowers compressive strength. Esquinas et al. [15] also found that using FA delayed the strength development of SCC. According to the research of Altoubat et al. [85], incorporating 50% FA into SCC not only had a negative influence on early strength but also increased shrinkage. The ideal quantity of FA should not exceed 35%; otherwise, silica fume should be added to encourage the development of strength (Figure 11).

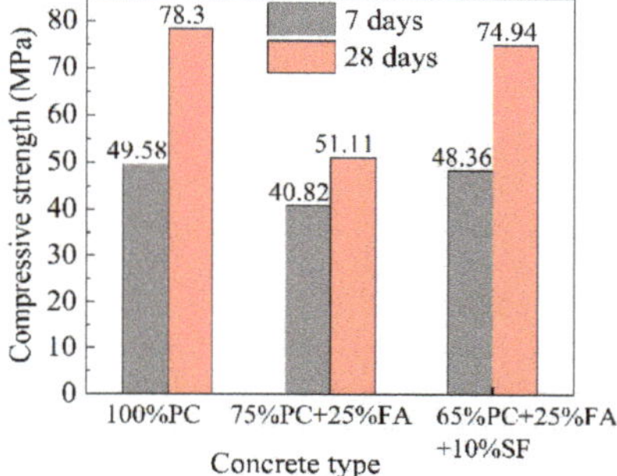

Figure 11. Effect of FA and SF on the compressive strength of SCC [8].

In contrast, SF with high fineness and pozzolanic reactivity frequently enhances the strength properties of SCC [9]. From the results of Fakhri [86,87], the incorporation of 10% SF can improve the compressive strength by more than 30% and increase the flexural strength by around 20%; the addition of 25% SF can optimize the flexural strength by approximately 40%. According to Karthik et al. [88], this is because SF contains a high content of reactive SiO_2, which might enhance the hydration. Furthermore, SF may efficiently enhance the filler effect and improve the interfacial transition zone. Esfandiari et al. [33] indicated that the reactive SiO_2 in SF could effectively react with $Ca(OH)_2$ in the paste to generate C-S-H gels.

The substitution of GBFS for cement in SCC will delay the early strength growth of the paste [8]. Boukendakdji et al. [44] researched the influence of GBFS on SCC and discovered that the early compressive strength of concrete reduced as GBFS admixture increased, but the post-strength (at 56 days and 90 days) did not appreciably decrease (Figure 12). Altoubat et al. [89] also discovered that GBFS lowers the strength of SCC, with the 28-day compressive strength decreasing by around 10% at 70% admixture, while Ofuyatan et al. [84] discovered that GBFS has a negligible influence on the strength development of SCC, with 28-day compressive strength rising by just 1% at 20% substitution and dropping by 4% at 30% substitution. Furthermore, Dadsetan et al. [47] revealed that the effect of GBFS on the strength of SCC is related to the water–cement ratio and the content of GBFS. The experimental results showed that at a water–cement ratio of 0.4, the 7-day and the 28-day compressive strengths of concrete with 10%, 20%, and 30% GBFS admixture were all reduced. However, at a water–cement ratio of 0.45, the 7-day and 28-day compressive strengths of concrete with all three admixtures increased. With the 30% admixture, the 28-day strength increased by about 30%.

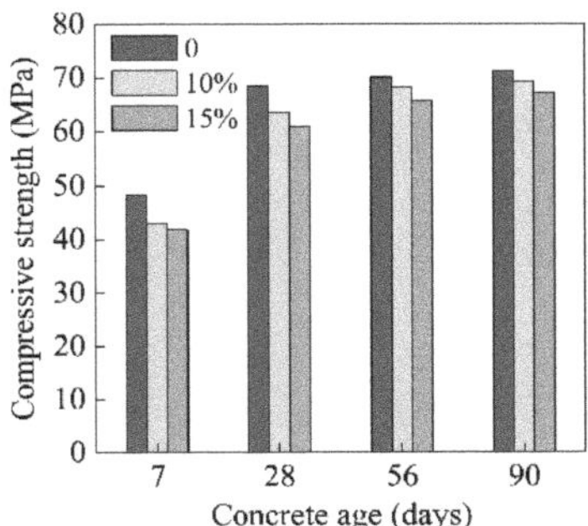

Figure 12. Variations of compressive strength with GBFS content of concrete [44].

LP stimulates the hydration of cement and enhances the strength development of concrete primarily through the adsorption of Ca^{2+} by $CaCO_3$, which promotes the dissolution of C_3S. At the beginning of the cement hydration, the dissolution and hydration of C_3S produce a substantial quantity of Ca^{2+}. Since Ca^{2+} has a stronger migration capacity than SiO_3^{2-}, the chemical adsorption of $CaCO_3$ to Ca^{2+} happens when Ca^{2+} diffuses toward the surface of LP particles. On one hand, this adsorption decreases the concentration of Ca^{2+} around the C_3S particles, allowing C_3S to dissolve and hydrate more quickly. On the other hand, the connection of Ca^{2+} with the surface of LP particles causes them to form ion clusters, which continue to expand to construct the crystal nucleus and gradually grow into C-S-H gels. Therefore, the adsorption of Ca^{2+} by the surface of LP greatly promotes the hydration of cement and leads to the high-density nucleation and directional growth of C-S-H [75].

4.4. Durability Properties

The compactness of the paste greatly influences the durability of concrete [90,91]. The diluting effect of SCMs decreases the quantity of cement, lowering the temperature increase in concrete and minimizing the risks of cracking [82]. Altoubat et al. [89] demonstrated that incorporating GBFS and FA decreased the internal stress of SCC, resulting in better cracking resistance (Figure 13). The filling effect of SCMs may reduce the porosity and permeability by packing the pores of the paste, thus improving the durability of SCC. For example, SF may increase the durability of SCC by filling the pores to reinforce the microstructure. According to the research of Karthik et al. [88], SCC exhibited higher resistance to sulfate attack and chloride ion penetration with the addition of SF admixture. Esfandiari et al. [33] suggested that this is not only due to the filling effect of SF but also the pozzolanic reaction, which is SF reacting with CH to produce C-S-H. Additionally, Sideris et al. [49] reported that the filling effect with the addition of GBFS in SCC decreased the porosity and reduced the carbonation depth. Zhang [92] proposed that GBFS could effectively improve the sulfate attack resistance of concrete. Moreover, Zhu et al. [62] claimed that the smaller the particle size of the LP, the better it filled the pores and the greater the packing density. The adsorption effect of LP on cement hydration is beneficial for enhancing the microstructure.

It should be noted that FA has a significant impact on the resistance to chloride ion penetration of SCC. Gnanaraj et al. [93] discovered that FA could greatly enhance its resistance to chloride ion penetration. The studies of Mahalingam et al. [21] and Esquinas et al. [14,15] also came up with similar outcomes. Dinakar et al. [24] indicated that the resistance to both sulfate attack and chloride ion penetration is noticeably increased with the increasing FA content. This is because the permeability of chloride ions relies on the chloride binding capacity of the substance. The Cl^- penetrates into the interior concrete along with the water, and some of the chlorides can react with the mineral phase in cement (mainly aluminum-containing) to form stable chlorine-containing complexes. However, when the chloride is supersaturated, a certain amount of the free state Cl^- is produced, which will corrode the concrete. It is widely known that FA contains a high proportion of aluminum-containing mineral phases. Figure 2 also illustrates that the concentration of Al_2O_3 in FA is greater than that in cement. Therefore, as the FA proportion rises, the chloride-binding capacity of SCC also increases, limiting the concentration of free Cl^- that might cause corrosion. Similarly, GBFS contains more aluminum phase minerals than cement, and Sideris et al. [49] reported that GBFS can increase the resistance to chloride ion penetration of SCC.

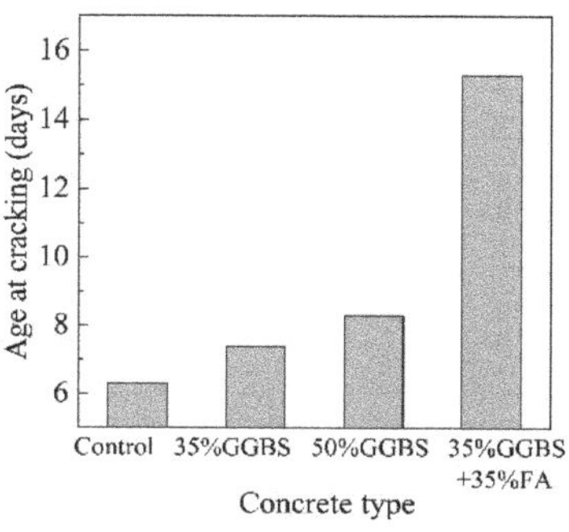

Figure 13. Ages at cracking of SCC, adding GBFS and FA [87].

5. Conclusions

The application techniques and hydration mechanisms of SCMs in SCC were discussed, and the impacts of FA, SF, GBFS, and LP on the microstructure and performance of SCC were reviewed. The major conclusions are as follows:

- SCC has a high content of powder material, and it is feasible to reduce the quantity of cement and thus decrease the production costs by incorporating SCMs. FA, SF, GGBS, and LP are the most frequently utilized SCMs in SCC. FA contains a high proportion of aluminum phase and predominantly spherical particles; SF primarily contains SiO_2 and has a high specific surface area. GBFS mainly contains SiO_2 and CaO; LP chiefly consists of $CaCO_3$, and both of them show obvious irregular and angular shapes due to mechanical grinding. The hydration mechanisms of these SCMs in SCC include pozzolanic reaction, alkaline activation, and adsorption effect. Moreover, the filling effect and dilution effect of some SCMs on the paste will contribute to reducing the porosity and limiting the temperature rise of concrete, respectively.
- The spherical particles of FA improve the fluidity of the freshly mixed paste, whereas SF increases the water demand and reduces fluidity due to its large specific surface area. The effect of GBFS on the fresh properties of SCC is related to the fineness and blending amount. The impact of LP is determined by the fineness, and LP will typically increase water consumption. Furthermore, superplasticizers are often added into SCC to increase the fresh properties, and superplasticizers might retard the early hydration and result in a delayed setting.
- The low pozzolanic reactivity of FA typically decreases the strength properties, particularly the early strength; the active SF usually enhances strength. The effect of GBFS on strength is dependent on the water–cement ratio and admixture amount, and it usually reduces the early strength while having little effect on post-strength. The adsorption effect of $CaCO_3$ on Ca^{2+} in LP will accelerate the hydration of cement and improve the development of the early strength.
- The pozzolanic reaction and filling effect of SCMs reduce the porosity of the hardened paste, resulting in a denser microstructure in the interfacial transition zone, thus increasing the durability of SCC. Furthermore, because of the high aluminum phase composition, FA and GBFS are typically capable of improving the resistance to chloride ion penetration and sulfate attack of SCC.

It is worth mentioning that although the employment of multiple SCMs in a composite system may provide a superposition effect, the chemical composition of different SCMs varies widely, and the issues of compatibility may rise. The rules governing the influence of composite SCMs added to SCC are not consistent and require additional investigation.

Author Contributions: Conceptualization, Z.L. and D.W.; validation, Z.L., M.A. and D.W.; investigation, L.P.; resources, Z.L.; writing—original draft preparation, L.P.; writing—review and editing, D.W.; visualization, L.P.; supervision, D.W.; project administration, D.W. All authors have read and agreed to the published version of the manuscript.

Funding: This work was funded by Natural Science Foundation of China [No. 51478248].

Acknowledgments: The authors would like to acknowledge the National Natural Science Foundation of China (No. 51478248).

Conflicts of Interest: The authors declare that they have no known competing financial interests or personal relationships that could have appeared to influence the work reported in this paper.

References

1. Okamura, H.; Ouchi, M. Self-compacting concrete. *J. Adv. Concr. Technol.* **2003**, *1*, 5–15. [CrossRef]
2. Global Cement Production Top Countries 2020 | Statista. Available online: https://www.statista.com/statistics/267364/world-cement-production-by-country/ (accessed on 1 December 2021).
3. Wang, D.; Wang, Q.; Xue, J. Reuse of hazardous electrolytic manganese residue: Detailed leaching characterization and novel application as a cementitious material. *Resour. Conserv. Recy.* **2020**, *154*, 104645. [CrossRef]
4. Xie, T.; Mohamad Ali, M.S.; Elchalakani, M.; Visintin, P. Modelling fresh and hardened properties of self-compacting concrete containing supplementary cementitious materials using reactive moduli. *Constr. Build. Mater.* **2021**, *272*, 121954. [CrossRef]
5. Wu, Q.; An, X. Development of a mix design method for SCC based on the rheological characteristics of paste. *Constr. Build. Mater.* **2014**, *53*, 642–651. [CrossRef]
6. Nie, D.; An, X. Optimization of SCC mix at paste level by using numerical method based on a paste rheological threshold theory. *Constr. Build. Mater.* **2016**, *102*, 428–434. [CrossRef]
7. Domone, P.L. Self-compacting concrete: An analysis of 11 years of case studies. *Cem. Con. Comp.* **2006**, *28*, 197–208. [CrossRef]
8. Mustapha, F.A.; Sulaiman, A.; Mohamed, R.N.; Umara, S.A. The effect of fly ash and silica fume on self-compacting high-performance concrete. *Mater. Tod. Proceed.* **2021**, *39*, 965–969. [CrossRef]
9. Benaicha, M.; Roguiez, X.; Jalbaud, O.; Burtschell, Y.; Alaoui, A.H. Influence of silica fume and viscosity modifying agent on the mechanical and rheological behavior of self compacting concrete. *Constr. Build. Mater.* **2015**, *84*, 103–110. [CrossRef]
10. Lachemi, M.; Hossain, K.M.A.; Lambros, V.; Nkinamubanzi, P.C.; Bouzoubaâ, N. Self-consolidating concrete incorporating new viscosity modifying admixtures. *Cem. Con. Res.* **2004**, *34*, 917–926. [CrossRef]
11. Gupta, N.; Siddique, R.; Belarbi, R. Sustainable and Greener Self-Compacting Concrete incorporating Industrial By-Products: A Review. *J. Clean. Prod.* **2021**, *284*, 124803. [CrossRef]
12. Sai Teja, G.; Ravella, D.P.; Chandra Sekhara Rao, P.V. Studies on self-curing self-compacting concretes containing zeolite admixture. *Mater. Tod. Proceed.* **2021**, *43*, 2355–2360. [CrossRef]
13. Ponikiewski, T.; Gołaszewski, J. The influence of high-calcium fly ash on the properties of fresh and hardened self-compacting concrete and high performance self-compacting concrete. *J. Clean. Prod.* **2014**, *72*, 212–221. [CrossRef]
14. Esquinas, A.R.; Álvarez, J.I.; Jiménez, J.R.; Fernández, J.M. Durability of self-compacting concrete made from non-conforming fly ash from coal-fired power plants. *Constr. Build. Mater.* **2018**, *189*, 993–1006. [CrossRef]
15. Esquinas, A.R.; Ledesma, E.F.; Otero, R.; Jiménez, J.R.; Fernández, J.M. Mechanical behaviour of self-compacting concrete made with non-conforming fly ash from coal-fired power plants. *Constr. Build. Mater.* **2018**, *182*, 385–398. [CrossRef]
16. Singh, N.; Kumar, P.; Goyal, P. Reviewing the behaviour of high volume fly ash based self compacting concrete. *J. Build. Eng.* **2019**, *26*, 100882. [CrossRef]
17. Choudhary, R.; Gupta, R.; Nagar, R. Impact on fresh, mechanical, and microstructural properties of high strength self-compacting concrete by marble cutting slurry waste, fly ash, and silica fume. *Constr. Build. Mater.* **2020**, *239*, 117888. [CrossRef]
18. Choudhary, R.; Gupta, R.; Alomayri, T.; Jain, A.; Nagar, R. Permeation, corrosion, and drying shrinkage assessment of self-compacting high strength concrete comprising waste marble slurry and fly ash, with silica fume. *Structures* **2021**, *33*, 971–985. [CrossRef]
19. Mohammed, A.M.; Asaad, D.S.; Al-Hadithi, A.I. Experimental and statistical evaluation of rheological properties of self-compacting concrete containing fly ash and ground granulated blast furnace slag. *J. King Saud Univ. Eng. Sci.* **2021**. [CrossRef]
20. Prakash, R.; Raman, S.N.; Divyah, N.; Subramanian, C.; Vijayaprabha, C.; Praveenkumar, S. Fresh and mechanical characteristics of roselle fibre reinforced self-compacting concrete incorporating fly ash and metakaolin. *Constr. Build. Mater.* **2021**, *290*, 123209. [CrossRef]

21. Mahalingam, B.; Nagamani, K.; Kannan, L.S.; Mohammed Haneefa, K.; Bahurudeen, A. Assessment of hardened characteristics of raw fly ash blended self-compacting concrete. *Persp. Sci.* **2016**, *8*, 709–711. [CrossRef]

22. Duran-Herrera, A.; De-León-Esquivel, J.; Bentz, D.P.; Valdez-Tamez, P. Self-compacting concretes using fly ash and fine limestone powder: Shrinkage and surface electrical resistivity of equivalent mortars. *Constr. Build. Mater.* **2019**, *199*, 50–62. [CrossRef]

23. Promsawat, P.; Chatveera, B.; Sua-iam, G.; Makul, N. Properties of self-compacting concrete prepared with ternary Portland cement-high volume fly ash-calcium carbonate blends. *Case Stud. Constr. Mat.* **2020**, *13*, e00426. [CrossRef]

24. Dinakar, P.; Babu, K.G.; Santhanam, M. Durability properties of high volume fly ash self compacting concretes. *Cem. Con. Comp.* **2008**, *30*, 880–886. [CrossRef]

25. Jain, A.; Gupta, R.; Chaudhary, S. Sustainable development of self-compacting concrete by using granite waste and fly ash. *Constr. Build. Mater.* **2020**, *262*, 120516. [CrossRef]

26. Akcay, B.; Tasdemir, M.A. Performance evaluation of silica fume and metakaolin with identical finenesses in self compacting and fiber reinforced concretes. *Constr. Build. Mater.* **2018**, *185*, 436–444. [CrossRef]

27. Bani Ardalan, R.; Joshaghani, A.; Hooton, R.D. Workability retention and compressive strength of self-compacting concrete incorporating pumice powder and silica fume. *Constr. Build. Mater.* **2017**, *134*, 116–122. [CrossRef]

28. Wongkeo, W.; Thongsanitgarn, P.; Ngamjarurojana, A.; Chaipanich, A. Compressive strength and chloride resistance of self-compacting concrete containing high level fly ash and silica fume. *Mater. Design.* **2014**, *64*, 261–269. [CrossRef]

29. Mastali, M.; Dalvand, A. Use of silica fume and recycled steel fibers in self-compacting concrete (SCC). *Constr. Build. Mater.* **2016**, *125*, 196–209. [CrossRef]

30. Ghoddousi, P.; Adelzade Saadabadi, L. Study on hydration products by electrical resistivity for self-compacting concrete with silica fume and metakaolin. *Constr. Build. Mater.* **2017**, *154*, 219–228. [CrossRef]

31. Bernal, J.; Reyes, E.; Massana, J.; León, N.; Sánchez, E. Fresh and mechanical behavior of a self-compacting concrete with additions of nano-silica, silica fume and ternary mixtures. *Constr. Build. Mater.* **2018**, *160*, 196–210. [CrossRef]

32. Zarnaghi, V.N.; Fouroghi-Asl, A.; Nourani, V.; Ma, H. On the pore structures of lightweight self-compacting concrete containing silica fume. *Constr. Build. Mater.* **2018**, *193*, 557–564. [CrossRef]

33. Esfandiari, J.; Loghmani, P. Effect of perlite powder and silica fume on the compressive strength and microstructural characterization of self-compacting concrete with lime-cement binder. *Measurement* **2019**, *147*, 106846. [CrossRef]

34. Salehi, H.; Mazloom, M. Opposite effects of ground granulated blast-furnace slag and silica fume on the fracture behavior of self-compacting lightweight concrete. *Constr. Build. Mater.* **2019**, *222*, 622–632. [CrossRef]

35. Sasanipour, H.; Aslani, F. Effect of specimen shape, silica fume, and curing age on durability properties of self-compacting concrete incorporating coarse recycled concrete aggregates. *Constr. Build. Mater.* **2019**, *228*, 117054. [CrossRef]

36. Sasanipour, H.; Aslani, F.; Taherinezhad, J. Effect of silica fume on durability of self-compacting concrete made with waste recycled concrete aggregates. *Constr. Build. Mater.* **2019**, *227*, 116598. [CrossRef]

37. Guo, Z.; Jiang, T.; Zhang, J.; Kong, X.; Chen, C.; Lehman, D.E. Mechanical and durability properties of sustainable self-compacting concrete with recycled concrete aggregate and fly ash, slag and silica fume. *Constr. Build. Mater.* **2020**, *231*, 117115. [CrossRef]

38. Mahalakshmi, S.H.V.; Khed, V.C. Experimental study on M-sand in self-compacting concrete with and without silica fume. *Mater. Tod. Proceed.* **2020**, *27*, 1061–1065. [CrossRef]

39. Faraj, R.H.; Sherwani, A.F.H.; Jafer, L.H.; Ibrahim, D.F. Rheological behavior and fresh properties of self-compacting high strength concrete containing recycled PP particles with fly ash and silica fume blended. *J. Build. Eng.* **2021**, *34*, 101667. [CrossRef]

40. Saba, A.M.; Khan, A.H.; Akhtar, M.N.; Khan, N.A.; Rahimian Koloor, S.S.; Petrů, M.; Radwan, N. Strength and flexural behavior of steel fiber and silica fume incorporated self-compacting concrete. *J. Mater. Res. Technol.* **2021**, *12*, 1380–1390. [CrossRef]

41. Manikanta, D.; Ravella, D.P. Mechanical and durability characteristics of high performance self-compacting concrete containing flyash, silica fume and graphene oxide. *Mater. Tod. Proceed.* **2021**, *43*, 2361–2367. [CrossRef]

42. Vejmelková, E.; Keppert, M.; Grzeszczyk, S.; Skaliński, B.; Černý, R. Properties of self-compacting concrete mixtures containing metakaolin and blast furnace slag. *Constr. Build. Mater.* **2011**, *25*, 1325–1331. [CrossRef]

43. Ting, L.; Qiang, W.; Shiyu, Z. Effects of ultra-fine ground granulated blast-furnace slag on initial setting time, fluidity and rheological properties of cement pastes. *Powder Technol.* **2019**, *345*, 54–63. [CrossRef]

44. Boukendakdji, O.; Kadri, E.-H.; Kenai, S. Effects of granulated blast furnace slag and superplasticizer type on the fresh properties and compressive strength of self-compacting concrete. *Cem. Con. Comp.* **2012**, *34*, 583–590. [CrossRef]

45. Anastasiou, E.K.; Papayianni, I.; Papachristoforou, M. Behavior of self compacting concrete containing ladle furnace slag and steel fiber reinforcement. *Mater. Design.* **2014**, *59*, 454–460. [CrossRef]

46. Valcuende, M.; Benito, F.; Parra, C.; Miñano, I. Shrinkage of self-compacting concrete made with blast furnace slag as fine aggregate. *Constr. Build. Mater.* **2015**, *76*, 1–9. [CrossRef]

47. Dadsetan, S.; Bai, J. Mechanical and microstructural properties of self-compacting concrete blended with metakaolin, ground granulated blast-furnace slag and fly ash. *Constr. Build. Mater.* **2017**, *146*, 658–667. [CrossRef]

48. Patel, Y.J.; Shah, N. Enhancement of the properties of Ground Granulated Blast Furnace Slag based Self Compacting Geopolymer Concrete by incorporating Rice Husk Ash. *Constr. Build. Mater.* **2018**, *171*, 654–662. [CrossRef]

49. Sideris, K.K.; Tassos, C.; Chatzopoulos, A.; Manita, P. Mechanical characteristics and durability of self compacting concretes produced with ladle furnace slag. *Constr. Build. Mater.* **2018**, *170*, 660–667. [CrossRef]

50. de Matos, P.R.; Oliveira, J.C.P.; Medina, T.M.; Magalhães, D.C.; Gleize, P.J.P.; Schankoski, R.A.; Pilar, R. Use of air-cooled blast furnace slag as supplementary cementitious material for self-compacting concrete production. *Constr. Build. Mater.* **2020**, *262*, 120102. [CrossRef]

51. Santamaría, A.; Ortega-López, V.; Skaf, M.; Chica, J.A.; Manso, J.M. The study of properties and behavior of self compacting concrete containing Electric Arc Furnace Slag (EAFS) as aggregate. *Ain Shams Eng. J.* **2020**, *11*, 231–243. [CrossRef]

52. Boukendakdji, O.; Kenai, S.; Kadri, E.H.; Rouis, F. Effect of slag on the rheology of fresh self-compacted concrete. *Constr. Build. Mater.* **2009**, *23*, 2593–2598. [CrossRef]

53. Zhao, H.; Sun, W.; Wu, X.; Gao, B. The properties of the self-compacting concrete with fly ash and ground granulated blast furnace slag mineral admixtures. *J. Clean. Prod.* **2015**, *95*, 66–74. [CrossRef]

54. Wang, D.; Wang, Q.; Huang, Z. New insights into the early reaction of NaOH-activated slag in the presence of $CaSO_4$. *Compos. Part B-Eng.* **2020**, *198*, 108207. [CrossRef]

55. Rizwan, S.A.; Bier, T.A. Blends of limestone powder and fly-ash enhance the response of self-compacting mortars. *Constr. Build. Mater.* **2012**, *27*, 398–403. [CrossRef]

56. Sua-iam, G.; Makul, N. Utilization of limestone powder to improve the properties of self-compacting concrete incorporating high volumes of untreated rice husk ash as fine aggregate. *Constr. Build. Mater.* **2013**, *38*, 455–464. [CrossRef]

57. Sua-iam, G.; Makul, N. Use of increasing amounts of bagasse ash waste to produce self-compacting concrete by adding limestone powder waste. *J. Clean. Prod.* **2013**, *57*, 308–319. [CrossRef]

58. Celik, K.; Jackson, M.D.; Mancio, M.; Meral, C.; Emwas, A.H.; Mehta, P.K.; Monteiro, P.J.M. High-volume natural volcanic pozzolan and limestone powder as partial replacements for portland cement in self-compacting and sustainable concrete. *Cem. Con. Comp.* **2014**, *45*, 136–147. [CrossRef]

59. Celik, K.; Meral, C.; Petek Gursel, A.; Mehta, P.K.; Horvath, A.; Monteiro, P.J.M. Mechanical properties, durability, and life-cycle assessment of self-consolidating concrete mixtures made with blended portland cements containing fly ash and limestone powder. *Cem. Con. Comp.* **2015**, *56*, 59–72. [CrossRef]

60. Benjeddou, O.; Soussi, C.; Jedidi, M.; Benali, M. Experimental and theoretical study of the effect of the particle size of limestone fillers on the rheology of self-compacting concrete. *J. Build. Eng.* **2017**, *10*, 32–41. [CrossRef]

61. Faheem, A.; Rizwan, S.A.; Bier, T.A. Properties of self-compacting mortars using blends of limestone powder, fly ash, and zeolite powder. *Constr. Build. Mater.* **2021**, *286*, 122788. [CrossRef]

62. Zhu, W.; Bartos, P. Microstructure and Properties of Interfacial Transition Zone in SCC. In Proceedings of the First International Symposium on Design Performance and use of Self Consolidating Concrete, RILEM Publications SARL, Changsha, China, 26–28 May 2005; pp. 319–327.

63. Gesoğlu, M.; Güneyisi, E.; Kocabağ, M.E.; Bayram, V.; Mermerdaş, K. Fresh and hardened characteristics of self compacting concretes made with combined use of marble powder, limestone filler, and fly ash. *Constr. Build. Mater.* **2012**, *37*, 160–170. [CrossRef]

64. Zhuang, S.; Wang, Q. Inhibition mechanisms of steel slag on the early-age hydration of cement. *Cem. Con. Res.* **2021**, *140*, 106283. [CrossRef]

65. Yang, S.; Zhang, J.; An, X.; Qi, B.; Shen, D.; Lv, M. Effects of fly ash and limestone powder on the paste rheological thresholds of self-compacting concrete. *Constr. Build. Mater.* **2021**, *281*, 122560. [CrossRef]

66. Chinthakunta, R.; Ravella, D.P.; Sri Rama Chand, M.; Janardhan Yadav, M. Performance evaluation of self-compacting concrete containing fly ash, silica fume and nano titanium oxide. *Mater. Tod. Proceed.* **2021**, *43*, 2348–2354. [CrossRef]

67. Zhu, W.; Gibbs, J.C. Use of different limestone and chalk powders in self-compacting concrete. *Cem. Con. Res.* **2005**, *35*, 1457–1462. [CrossRef]

68. Scrivener, K.; Ouzia, A.; Juilland, P.; Kunhi Mohamed, A. Advances in understanding cement hydration mechanisms. *Cem. Con. Res.* **2019**, *124*, 105823. [CrossRef]

69. Provis, J.L.; Van Deventer, J.S. *Alkali Activated Materials: State-of-the-Art Report*; RILEM TC 224-AAM; Springer: Dordrecht, The Netherlands, 2014; Volume 13, ISBN 978-94-007-7671-5.

70. Provis, J.L.; Palomo, A.; Shi, C. Advances in understanding alkali-activated materials. *Cem. Con. Res.* **2015**, *78*, 110–125. [CrossRef]

71. Ben Haha, M.; Le Saout, G.; Winnefeld, F.; Lothenbach, B. Influence of activator type on hydration kinetics, hydrate assemblage and microstructural development of alkali activated blast-furnace slags. *Cem. Con. Res.* **2011**, *41*, 301–310. [CrossRef]

72. Myers, R.J.; Bernal, S.A.; San Nicolas, R.; Provis, J.L. Generalized structural description of calcium-sodium aluminosilicate hydrate gels: The cross-linked substituted tobermorite model. *Langmuir* **2013**, *29*, 5294–5306. [CrossRef]

73. Yalçınkaya, Ç.; Çopuroğlu, O. Hydration heat, strength and microstructure characteristics of UHPC containing blast furnace slag. *J. Build. Eng.* **2021**, *34*, 101915. [CrossRef]

74. Fernández-Jiménez, A.; Palomo, A.; Criado, M. Microstructure development of alkali-activated fly ash cement: A descriptive model. *Cem. Con. Res.* **2005**, *35*, 1204–1209. [CrossRef]

75. Ouyang, X.; Xu, S.; Ma, Y.; Ye, G. Effect of Aggregate Chemical Properties on Nucleation and Growth of Calciun Silicate Hydrate C-S-H. Kuei Suan Jen Hsueh Pao. *J. Chin. Ceram. Soc.* **2021**, *49*, 972–979. [CrossRef]

76. Mehta, P.K.; Monteiro, P.J. *Concrete: Microstructure, Properties, and Materials*; McGraw-Hill Education: New York, NY, USA, 2014.

77. Li, L.G.; Kwan, A.K.H. Concrete mix design based on water film thickness and paste film thickness. *Cem. Con. Comp.* **2013**, *39*, 33–42. [CrossRef]

78. Zhang, Y.; Kong, X. Correlations of the dispersing capability of NSF and PCE types of superplasticizer and their impacts on cement hydration with the adsorption in fresh cement pastes. *Cem. Con. Res.* **2015**, *69*, 1–9. [CrossRef]

79. Wang, D.; Wang, Q.; Huang, Z. Investigation on the poor fluidity of electrically conductive cement-graphite paste: Experiment and simulation. *Mater. Design.* **2019**, *169*, 107679. [CrossRef]

80. Sha, S.; Wang, M.; Shi, C.; Xiao, Y. Influence of the structures of polycarboxylate superplasticizer on its performance in cement-based materials-A review. *Constr. Build. Mater.* **2020**, *233*, 117257. [CrossRef]

81. Alexandra, C.; Bogdan, H.; Camelia, N.; Zoltan, K. Mix design of self-compacting concrete with limestone filler versus fly ash addition. *Proced. Manufact.* **2018**, *22*, 301–308. [CrossRef]

82. Sonebi, M. Medium strength self-compacting concrete containing fly ash: Modelling using factorial experimental plans. *Cem. Con. Res.* **2004**, *34*, 1199–1208. [CrossRef]

83. Selvarani, B.; Preethi, V. Investigational study on optimum content of GGBS and fibres in fibre non-breakable self compacting concrete. *Mater. Tod. Proceed.* **2021**, *47*, 6111–6115. [CrossRef]

84. Ofuyatan, O.M.; Adeniyi, A.G.; Ijie, D.; Ighalo, J.O.; Oluwafemi, J. Development of high-performance self compacting concrete using eggshell powder and blast furnace slag as partial cement replacement. *Constr. Build. Mater.* **2020**, *256*, 119403. [CrossRef]

85. Altoubat, S.; Talha Junaid, M.; Leblouba, M.; Badran, D. Effectiveness of fly ash on the restrained shrinkage cracking resistance of self-compacting concrete. *Cem. Con. Comp.* **2017**, *79*, 9–20. [CrossRef]

86. Fakhri, M.; Yousefian, F.; Amoosoltani, E.; Aliha, M.R.M.; Berto, F. Combined Effects of Recycled Crumb Rubber and Silica Fume on Mechanical Properties and Mode I Fracture Toughness of Self-compacting Concrete. *Fatigue Fract. Eng. Mater. Struct.* **2021**, *44*, 2659–2673. [CrossRef]

87. Fakhri, M.; Saberi., K.F. The Effect of Waste Rubber Particles and Silica Fume on the Mechanical Properties of Roller Compacted Concrete Pavement. *J. Clean. Prod.* **2016**, *129*, 521–530. [CrossRef]

88. Karthik, D.; Nirmalkumar, K.; Priyadharshini, R. Characteristic assessment of self-compacting concrete with supplementary cementitious materials. *Constr. Build. Mater.* **2021**, *297*, 123845. [CrossRef]

89. Altoubat, S.; Badran, D.; Junaid, M.T.; Leblouba, M. Restrained shrinkage behavior of Self-Compacting Concrete containing ground-granulated blast-furnace slag. *Constr. Build. Mater.* **2016**, *129*, 98–105. [CrossRef]

90. Bossio, A.; Lignola, G.P.; Fabbrocino, F.; Monetta, T.; Prota, A.; Bellucci, F.; Manfredi, G. Nondestructive Assessment of Corrosion of Reinforcing Bars through Surface Concrete Cracks. *Struct. Concr.* **2017**, *18*, 104–117. [CrossRef]

91. Pittella, E.; Angrisani, L.; Cataldo, A.; Piuzzi, E.; Fabbrocino, F. Embedded Split Ring Resonator Network for Health Monitoring in Concrete Structures. IEEE Instrum. *Meas. Mag.* **2020**, *23*, 14–20. [CrossRef]

92. Zhang, Z.; Wang, Q.; Chen, H.; Zhou, Y. Influence of the initial moist curing time on the sulfate attack resistance of concretes with different binders. *Constr. Build. Mater.* **2017**, *144*, 541–551. [CrossRef]

93. Gnanaraj, S.C.; Chokkalingam, R.B.; Thankam, G.L.; Pothinathan, S.K.M. Durability properties of self-compacting concrete developed with fly ash and ultra fine natural steatite powder. *J. Mater. Res. Technol.* **2021**, *13*, 431–439. [CrossRef]

MDPI
St. Alban-Anlage 66
4052 Basel
Switzerland
Tel. +41 61 683 77 34
Fax +41 61 302 89 18
www.mdpi.com

Crystals Editorial Office
E-mail: crystals@mdpi.com
www.mdpi.com/journal/crystals

www.ingramcontent.com/pod-product-compliance
Lightning Source LLC
LaVergne TN
LVHW071611170726
843515LV00009B/2373